Springer INdAM Series

Volume 57

This series will publish textbooks, multi-authors books, thesis and monographs in English language resulting from workshops, conferences, courses, schools, seminars, doctoral thesis, and research activities carried out at INDAM - Istituto Nazionale di Alta Matematica, http://www.altamatematica.it/en. The books in the series will discuss recent results and analyze new trends in mathematics and its applications.

THE SERIES IS INDEXED IN SCOPUS

Michele Correggi • Marco Falconi
Editors

Quantum Mathematics I

Springer

Editors
Michele Correggi
Dipartimento di Matematica
Politecnico di Milano
Milano, Italy

Marco Falconi
Dipartimento di Matematica
Politecnico di Milano
Milano, Italy

ISSN 2281-518X ISSN 2281-5198 (electronic)
Springer INdAM Series
ISBN 978-981-99-5893-1 ISBN 978-981-99-5894-8 (eBook)
https://doi.org/10.1007/978-981-99-5894-8

This Springer imprint is published by the registered company Springer Nature Singapore Pte Ltd.
The registered company address is: 152 Beach Road, #21-01/04 Gateway East, Singapore 189721, Singapore

Paper in this product is recyclable.

Preface

The two volumes "Quantum Mathematics I" and "Quantum Mathematics II" originate from the INdAM Intensive Period "INdAM Quantum Meetings (IQM22)", that was held in Spring 2022 at the Department of Mathematics of Politecnico di Milano. The trimester was the perfect opportunity to restart the social aspects of research after the Covid restrictions, in the mathematical physics community working on the mathematical features of quantum mechanics. After almost two entire years of break due to the Covid 19 pandemia, the project was very successful since the first steps of the organization: almost all the invited scientists gladly accepted to participate, also showing the enthusiasm of meeting in person to form new collaborations as well as to renew existing ones.

The main activities during IQM22 were:

- A kick-off workshop at the beginning of March 2022 focusing on the topics of many-body quantum mechanics, quantum statistical mechanics and open quantum systems;
- A series of short courses given throughout the period March - May 2022 by Z. Ammari (Université de Rennes 1), C. Brennecke (Bonn Universität), J. Dereziński (University of Warsaw), M. Fraas (University of California Davis), M. Merkli (Memorial University of Newfoundland), F. Nier (Université Pais 13), N. Rougerie (ENS Lyon);
- Thematic lectures and seminars;
- A concluding workshop at the end of May 2022 focusing on the topics of semi-classical analysis, quantum field theory, nonlinear PDEs of quantum mechanics and their derivation.

More than 40 invited guests contributed to the activities of the period and gave rise to many fruitful collaborations among themselves and with the local members of the mathematical physics group. The participation of young researchers, postdocs and PhD students was very significant with more than 20 young people (some of which financially supported) showing interest in the scientific programme of the trimester and participating to the activities. Most of the contributions (talks, lectures, courses, etc.) were recorded and made available online.

All the contributions collected in these volumes are linked to IQM22, either as proceedings of its activities, or as brand new works originating and benefiting from the interactions occurred at IQM22. The main theme is the mathematics of quantum mechanics in a broad sense, but the present volume is focused on the following more specific topics:

- *Semiclassical analysis.* The role of semiclassical techniques in the research of the mathematical aspects of quantum mechanics cannot be overstated, given the broad range of models and phenomena where a classical behavior may emerge. We present here some significant examples ranging from waves in random media and fermionic systems to models of spin-field interactions.
- *Operator and spectral theory.* Several key results in quantum mechanics rely on nontrivial investigations of the spectral theory of certain operators – typically but not exclusively, Schrödinger-type operators. Some examples are discussed here: Dirac operators and Schrödinger operators with (periodic, singular or strong) magnetic fields or with contact interactions.
- *Effective nonlinear models.* Both the derivation of effective nonlinear models and their accurate study are behind the mathematical understanding of many phenomena in quantum mechanics. Here the attention is mostly focused on fermionic systems but more general questions are also addressed.

To conclude, we express our gratitude to INdAM and its scientific board for the support to IQM22, which made possible the organization of a very fruitful intensive period and the involvement of a large number of young participants.

Milano, Italy
January 2023

Michele Correggi
Marco Falconi

Contents

Part I
Semiclassical Analysis

Waves in a Random Medium: Endpoint Strichartz Estimates and Number Estimates

S. Breteaux and F. Nier

1 Introduction

The asymptotic analysis or random homogenization of wave propagation in a random medium, in a kinetic or diffusive regime has motivated several works in the recent decades. It is not our purpose here to give an exhaustive list but we think essentially of two different approaches: the one initiated by G. Papanicolaou and coauthors (see e.g. [11, 24, 26]) with a rather complete review by J. Garnier in [13] and the one proposed by L. Erdös, H.T. Yau and later with M. Salmhofer in [8–10]. Those two approaches formulate their results in terms of a kinetic (or diffusive) evolution equation for some weak limit of scaled Wigner functions. The main difference between the two approaches can be summarized as follows: The first approach presented in [13] modeled on the problem of randomly layered media (see [11]) focusses on space-time wave functions, by solving a space-time PDE (it can be a Schrödinger or a wave equation) with random coefficients but with a smooth and essentially deterministic right-hand side. With very strong assumptions on the right-hand side of the equation, essentially deterministic and smooth, a kinetic equation is written for the distributional weak limit of the Wigner function associated with the space-time wave function. The work of [8–10] is concerned with Cauchy problems, at the quantum level for the Schrödinger equation and semiclassically at a classical level for a linear Boltzmann equation in [8] or a heat equation in [9, 10]. The strategy of this second approach consists after writing a Dyson expansion (the iteration of Duhamel's formula), in making an accurate combinatorial analysis of Feynman

S. Breteaux
Université de Lorraine, CNRS, IECL, Metz, France
e-mail: sebastien.breteaux@univ-lorraine.fr

F. Nier (✉)
LAGA, Université de Paris XIII, Villetaneuse, France
e-mail: nier@math.univ-paris13.fr

M. Correggi, M. Falconi (eds.), *Quantum Mathematics I*,
Springer INdAM Series 57, https://doi.org/10.1007/978-981-99-5894-8_1

diagrams which label all the random interaction terms of the expanded Dyson series. This Dyson expansion technique was actually already used for a similar problem by H. Spohn in [29]. The final step which gives the asymptotic behaviour of the Wigner transform, essentially relies on the accurate control and expression of the remaining terms of the series by using stationary phase asymptotic expressions for the many oscillating integrals. The results of this second approach always require strong assumptions on the initial data at the initial time $t = 0$ and prove weak convergence results at the macroscopic time $t \neq 0$. After the first preprint version of this text we were informed of more recent results by F. Hernandez in [18] with more accurate and general statements for this diagrammatic approach. With a more spectral point of view but with some conjectural statements about resonances, related problems were considered by M. Duerinckx and C. Shirley in the recent work [7].

The main difficulty in this problem is concerned with the control of recollisions and especially the proof that the asymptotic evolution is Markovian, or given by some semigroup associated to a kinetic or heat equation, although the multiple scattering process of waves could destroy this markovian aspect. Depending on the asymptotic regime, the effective asymptotic evolution could be affected by some memory or non local in time effect. In the considered asymptotic problems, it must be checked that those memory effects vanish asymptotically. In the approach reviewed in [13] which is concerned with rather general random fields, this is proved by estimating higher moments. In the approach of [8] the combinatorial accurate analysis of Feynman diagrams, is reminiscent of the accurate control of recollision terms by G. Gallavotti in [12] for the classical Lorentz gas problem (Wind tree model). Both approaches bring accurate information about a difficult problem in slightly different frameworks and with various range of applications.

However those results remain unsatifactory from the mathematical point of view and for the following reason: The dynamics of (quantum) waves is given by a semigroup (actually a unitary group when there is no dissipation) and the asymptotic kinetic or diffusive limits are also given by well defined (semi)-groups. In the Cauchy problem approach, one does not yet understand the dynamically stable class of initial data which makes the derivation of a classical kinetic or heat equation possible. Actually the results of [8–10] are themselves puzzling because with very specific initial data at time $t = 0$, they prove the asymptotic expected behaviour at the macroscopic time $t \neq 0$. But this means that the time evolved quantum state at the macroscopic time $t/2 \neq 0$, enters in the class of admissible initial data for which the asymptotic evolution can be proved for a nonzero time interval (at the macroscopic scale). Such initial data do not enter in the very specfic class considered at time $t = 0$. In the recent work of F. Hernandez [18], the class of initial data has been significantly improved: Actually it works for any deterministic initial data. But they are still deterministic while the initial time $t/2$ a priori allows stochastic initial data. So the general question remains unsolved. In the space time approach reviewed in [13] the strong assumptions on the right-hand side compared with the weak convergence results of the wave function, have been considered in a negative way. Actually what is called "statistical stability" is shown to fail with rough data (see [3]). But no positive answer for a general class of random right-hand side seems to emerge. Although the two approaches are about slightly different problems, they seem related at least for some basic random processes on which we will focus in this article.

Our hope is that such an analysis about the propagation of random waves in a random medium should lead to results relying on dynamically stable hypotheses. We are led in this direction by the strategy followed by the second author with Z. Ammari in [1, 2] where they managed to give a general and robust class of initial data, dynamically stable, such that the quantum mean field dynamics can be followed.

About this very technical question a first attempt was tried by the first author in [5]. The idea was to exploit the link between gaussian random fields (and possibly other fields like the poissonian random fields) with quantum field theory. It rapidly appears that the asymptotic problem, of waves in a random medium in a gaussian random field in the kinetic regime, cannot be thought as an infinite semiclassical problem like the bosonic mean field problem. It has some similarities but the strength of the free wave propagator and the translation invariance lead to non quadratic and non "semiclassical" Wick quantized operators. For this reason the coherent state method presented in [5] led to an accurate Ansatz, only for $O(h^{1/2})$ macroscopic times, where $h > 0$ is the chosen small parameter, and the derivation of a linear Boltzmann equation was possible only by forcing the markovian nature of the asymptotic evolution by reinitializing on some intermediate time scale the random potential. It was not at all satisfactory. Actually the number estimates that we prove in this article confirm that a coherent state approach cannot work for those problems.

Another issue of this problem is the good understanding of the dispersive properties of the free wave propagator with the asymptotic behaviour of waves in a random medium. The different behaviours expected in small dimension, $d \leq 2$ for the Schrödinger equation in the kinetic regime compared to $d \geq 3$, are closely linked with the time integrability of the dispersion relation ($L^1 - L^\infty$ estimates). In the community of nonlinear PDE's, Strichartz estimates are known to be more robust and effective than the pointwise in time $L^1 - L^\infty$ estimate. With the endpoint Strichartz estimates proved by Keel and Tao in [21], those inequalities are now well adapted for linear critical problems. This article shows that they actually lead to very accurate and somehow surprising "number estimates" with some non trivial consequences.

Before giving the outline of this text, let us point out some limitations and features of the present analysis:

- We are not yet able to derive a full kinetic equation, except if one makes some connection with the existing results of [8]. The class of good initial data for which an asymptotic equation can be written is not yet identified.
- We work essentially with the Schrödinger equation in the presence of a gaussian random potential in the kinetic regime, as what we think to be the simplest, and richest model problem from the point of view of available structures.

- Once the two previous points are made clear, the interested reader will realize that several argument, especially the one making use of Strichartz estimates, have been written in a sufficiently general framework in order to be transposed in another framework.
- Some results like the possibility to define Wigner measures for all times, the localization in energy of the propagation phenomena, the class of potential corresponding to the scale invariant potential for Strichartz estimates, definitely bring a partial but accurate information.

Our main results are about accurate number estimates, stated in Proposition 4.3 in a rather general abstract setting and in Theorem 5.1 for the case of our model problem of the Schrödinger equation with a gaussian translation invariant potential in the kinetic regime and dimension $d \geq 3$.

Outline of the Article

(a) In Sect. 2 the link between gaussian Hilbert spaces and the bosonic Fock space is recalled and the equations in which we are interested are explicitely written.
(b) In Sect. 3 the translation invariance is used in order to make appear in a crucial way the center of mass variable, with respect to the position of the field variable. The expression of the creation and annihiliation operators are given explicitely in the center of mass and relative variables and finally L^p-estimates are carefully checked for those creation and annihilation operators under the suitable assumptions on the potential.
(c) Section 4 reviews the known results about endpoint Strichartz estimates, and gives consequences in connection with the L^p-estimate in the center of mass given in Sect. 3. Then a rather general fixed point is proved which combines endpoint Strichartz estimates with an adaptation of Cauchy-Kowalevski techniques.
(d) In Sect. 5, the general assumptions of Sect. 4 are checked in the framework of the Schrödinger equation with a gaussian random field in the kinetic regime and ambient dimension $d \geq 3$.
(e) Consequences and a priori information, for the asymptotic evolution of Wigner functions are given in Sect. 6, withouth computing them.
(f) Finally various approximation or stability results are deduced as consequences of the general estimates proved in Sects. 4, 5, and 6.

Before starting, be aware of the following assumed framework and conventions:

All our Hilbert spaces, real or complex, are separable. All measures are assumed sigma-finite. On a set $\mathcal{X}$ endowed with a sigma-set, a generic sigma-finite measure will be denoted $\mathbf{dx}$, while the normal calligraphy dx will be reserved for the Lebesgue measure on $\mathcal{X} = \mathbb{R}^d$. When $(\mathcal{X}, \mathbf{dx})$ and $(\mathcal{Y}, \mathbf{dy})$ are two sigma-finite measured spaces, the notation $L^p_x L^q_y$, $1 \leq p, q \leq +\infty$, is used for $L^p(\mathcal{X}, \mathbf{dx}; L^q(\mathcal{Y}, \mathbf{dy}))$. However a more general version of $L^p_x L^q_y$ will be introduced in Sect. 3.2.

2 Random Fields and Fock Space

2.1 *Gaussian Hilbert Space and Random Fields*

Let $\mathcal{G}$ be the stochastic gaussian measure (see e.g. [20]) on the Lebesgue measured space $(\mathbb{R}^d, \mathcal{L}, dy)$. This defines a real Hilbert gaussian space indexed by $L^2(\mathbb{R}^d, dy; \mathbb{R})$ which is generated, as a Hilbert space, by the centered real gaussian variables $X_A \sim N(0, |A|)$, with A measurable set of $\mathbb{R}^d$ and $|A| = \int_A dy$. By Minlös theorem (see [28]) the space $L^2(\Omega, \mathcal{G}; \mathbb{R})$ which contains powers of those gaussian processes can be realized with $\Omega = \mathcal{S}'(\mathbb{R}^d, dy; \mathbb{R})$.

Complex valued elements $F \in L^2(\Omega, \mathcal{G}; \mathbb{C})$ are written $F = \mathrm{Re}\, F + i\mathrm{Im}\, F$, $\mathrm{Re}\, F, \mathrm{Im}\, F \in L^2(\Omega, \mathcal{G}; \mathbb{R})$ handled by the $\mathbb{R}$-linearity of the decomposition.

Once the complexification is fixed in this order (see [20] for an accurate description of various complex structures of gaussian measures), the chaos decomposition of elements in $F \in L^2(\Omega, \mathcal{G}; \mathbb{C})$ can be written

$$F(\omega) = \bigoplus_{n=0}^{\infty} \int_{\mathbb{R}^{dn}} F_n(y_1, \dots, y_n) : X_{y_1} \cdots X_{y_n} : dy_1 \cdots dy_n\,, \tag{1}$$

where

- $F_n(y_{\sigma(1)}, \dots, y_{\sigma(n)}) = F_n(y_1, \dots, y_n)$ for all $\sigma \in \mathfrak{S}_n$ and complex valued functions are treated by the $\mathbb{R}$-linearity of the decomposition $F_n = \mathrm{Re}\,(F_n) + i\mathrm{Im}\, F_n$;
- the above symmetry can be written $F_n = S_n F_n$ where S_n is the symmetrizing orthogonal projection on $L^2(\mathbb{R}^{dn}, dy_1 \cdots dy_n; \mathbb{C})$ given by

$$(S_n F_n)(y_1, \dots, y_n) = \frac{1}{n!} \sum_{\sigma \in \mathfrak{S}_n} F_n(y_{\sigma(1)}, \dots, y_{\sigma(n)})\,; \tag{2}$$

- the family $(X_y)_{y \in \mathbb{R}^d}$ is made of jointly gaussian real centered random fields such that $\mathbb{E}(X_y X_{y'}) = \delta(y - y')$, which actually means

$$\mathbb{E}[(\int_{\mathbb{R}^d} f(y) X_y\, dy)(\int_{\mathbb{R}^d} g(y') X_{y'} dy')] = \int_{\mathbb{R}^d} f(y) g(y)\, dy$$

  for all $f, g \in \mathcal{S}(\mathbb{R}^d; \mathbb{C})$;[1]
- products or Wick products of singular random variables X_{y_j}, $j = 1 \dots J$, must be considered in their weak formulation as well;
- $: Y_1 \cdots Y_n :$ stands for the Wick product of the random variables $Y_1, \dots, Y_n$;

[1] We follow the general probabilistic convention which omits the ω argument with $X_y = X_y(\omega)$ e.g. in formula (1).

- with the assumed symmetry of the F_n components,

$$\mathbb{E}(|F|^2) = \int_\Omega |F(\omega)|^2 \, d\mathcal{G}(\omega)$$

$$= \sum_{n=0}^{\infty} n! \int_{\mathbb{R}^{dn}} |F_n(y_1, \dots, y_n)|^2 \, dy_1 \cdots dy_n = \sum_{n=0}^{\infty} n! \|F_n\|_{L^2}^2 \,. \tag{3}$$

A field is a random function of $x \in \mathbb{R}^d$ and we shall consider $F : \mathbb{R}^d \times \Omega \to \mathbb{C}$. A real gaussian centered translation invariant field can be written

$$\mathcal{V}(x, \omega) = \int_{\mathbb{R}^d} V(y - x) \, X_y \, dy \,.$$

An element $F \in L^2(\mathbb{R}_x^d \times \Omega, dx \otimes \mathcal{G}; \mathbb{C})$ has the chaos decomposition

$$F(x, \omega) = \bigoplus_{n=0}^{\infty} \int_{\mathbb{R}^{dn}} \tilde{F}_n(x, y_1, \dots, y_n) : X_{y_1} \cdots X_{y_n} : dy_1 \cdots dy_n \tag{4}$$

$$= \bigoplus_{n=0}^{\infty} \int_{\mathbb{R}^{dn}} F_n(x, y_1 - x, \dots, y_n - x) : X_{y_1} \cdots X_{y_n} : dy_1 \cdots dy_n \tag{5}$$

where $F_n(x, y_1, \dots, y_n) = \tilde{F}_n(x, y_1 + x, \dots, y_n + x)$ shares the same symmetry in $(y_1, \dots, y_n)$ as $\tilde{F}_n$ and

$$\|F\|_{L^2(\mathbb{R}_x^d \times \Omega)}^2 = \int_{\mathbb{R}^d} \mathbb{E}(|F(x, \cdot)|^2) \, dx$$

$$= \sum_{n=0}^{\infty} n! \|\tilde{F}_n\|_{L^2(\mathbb{R}^d \times \mathbb{R}^{dn})}^2 = \sum_{n=0}^{\infty} n! \|F_n\|_{L^2(\mathbb{R}^d \times \mathbb{R}^{dn})}^2 \,. \tag{6}$$

Assumptions on the real potential function V will be specified later but we can already compute the product $\mathcal{V}(x, \omega) \, F(x, \omega)$ by making use of Wick formula (see e.g. [20]-Theorem 3.15)

$$X_y \, : X_{y_1} \cdots X_{y_n} :$$

$$= \, : X_y X_{y_1} \cdots X_{y_n} : + \sum_{j=1}^{n} \delta(y - y_j) \, : X_{y_1} \cdots X_{y_{j-1}} \underbrace{X_{y_j}}_{\text{removed}} X_{y_{j+1}} \cdots X_{y_n} :$$

which leads to the chaos decomposition of $\mathcal{V}(x,\omega)F(x,\omega)$ as

$$\int_{\mathbb{R}^{d(n+1)}} \frac{1}{(n+1)!} \sum_{\sigma\in\mathfrak{S}_{n+1}} V(y_{\sigma(n+1)}-x)\, F_n(x, y_{\sigma(1)}-x, \dots, y_{\sigma(n)}-x)$$

$$: X_{y_1}\cdots X_{y_{n+1}}:\, dy_1\cdots dy_{n+1}$$

$$+ \int_{\mathbb{R}^{d(n-1)}} n\left[\int_{\mathbb{R}^d} V(y)\, F_n(x, y, y_1 - x, \dots, y_{n-1}-x)\, dy\right]$$

$$: X_{y_1}\cdots X_{y_{n-1}}:\, dy_1\cdots dy_{n-1}\,. \tag{7}$$

2.2 *The Fock Space Presentation*

The chaos decomposition (1) provides the isomorphism between $L^2(\Omega, \mathcal{G}; \mathbb{C})$ and the bosonic Fock space

$$\Gamma(L^2(\mathbb{R}^d, dy; \mathbb{C})) = \bigoplus_{n=0}^{\infty} (L^2(\mathbb{R}^d, dy; \mathbb{C}))^{\odot n}$$

where for a (real or complex) Hilbert space $\mathfrak{h}$, $\mathfrak{h}^{\odot n}$ is the symmetric Hilbert completed tensor product, equal to $\mathbb{C}$ (or $\mathbb{R}$) for $n = 0$, endowed with the norm such that

$$\|\varphi^{\otimes n}\|_{\mathfrak{h}^{\odot n}} = \|\varphi\|_{\mathfrak{h}}^n\,, \qquad \|f_n\|_{L^2(\mathbb{R}^d, dy; \mathbb{C})^{\odot n}} = \|f_n\|_{L^2(\mathbb{R}^{dn}, dy_1\cdots dy_n; \mathbb{C})}\,. \tag{8}$$

The above direct sum is also the Hilbert completed direct sum. Note that the Fock space norm (8) differs from the $\mathfrak{h}^{\odot n}$-norm chosen in [20] in adequation with Wick products by a factor $\sqrt{n!}$. The unitary operator from $L^2(\Omega, \mathcal{G}; \mathbb{C})$ to $\Gamma(L^2(\mathbb{R}^d_y, dy; \mathbb{C}))$ is thus given by

$$F \mapsto \bigoplus_{n=0}^{\infty} f_n\,, \qquad f_n = \sqrt{n!}\, F_n\,,$$

since

$$\|F\|^2_{L^2(\Omega,\mathcal{G};\mathbb{C})} = \sum_{n=0}^{\infty} n!\|F_n\|^2_{L^2(\mathbb{R}^{dn}, dy_1\cdots dy_n; \mathbb{C})} = \sum_{n=0}^{\infty} \|f_n\|^2_{L^2(\mathbb{R}^d, dy; \mathbb{C})^{\odot n}}\,.$$

The Fock space $\Gamma(\mathfrak{h})$ is endowed with densely defined Wick-quantized operators. For a monomial symbol $b(z) = \langle z^{\otimes q}, \tilde{b} z^{\otimes p} \rangle$ with $\tilde{b} \in \mathcal{L}(\mathfrak{h}^{\otimes p}; \mathfrak{h}^{\otimes q})$, the Wick quantization b^{Wick} is defined on $\bigoplus_{n\in\mathbb{N}}^{\mathrm{alg}} \mathfrak{h}^{\odot n}$ by

$$b^{\mathrm{Wick}} f_{n+p} = \frac{\sqrt{(n+p)!(n+q)!}}{n!} S_{n+q}(\tilde{b} \otimes \mathrm{Id}^{\otimes n}) f_{n+p}$$

where $S_m : \mathfrak{h}^{\otimes m} \to \mathfrak{h}^{\odot m}$ is the symmetrizing orthogonal projection given by

$$S_m(g_1 \otimes \cdots \otimes g_m) = \frac{1}{m!} \sum_{\sigma \in \mathfrak{S}_m} g_{\sigma(1)} \otimes \cdots \otimes g_{\sigma(m)} \tag{9}$$

already introduced in (2).

Basic examples in our case $\mathfrak{h} = L^2(\mathbb{R}^d, dy; \mathbb{C})$ are given by

$$a(g) = (\langle g, z \rangle)^{\mathrm{Wick}},$$

$$a(g) f_n(y_1, \ldots, y_{n-1}) = \sqrt{n} \int_{\mathbb{R}^d} \overline{g(y)} f_n(y_1, \ldots, y_{n-1}, y)\, dy,$$

$$a^*(f) = (\langle z, f \rangle)^{\mathrm{Wick}},$$

$$a^*(f) f_n(y_1, \ldots, y_{n+1}) = \frac{\sqrt{n+1}}{(n+1)!} \sum_{\sigma \in \mathfrak{S}_{n+1}} f(y_{\sigma(1)}) f_n(y_{\sigma(2)}, \ldots, y_{\sigma(n+1)}),$$

$$\phi(V) = (\sqrt{2}\mathrm{Re}\, \langle V, z \rangle)^{\mathrm{Wick}}, \quad \phi(V) = \frac{1}{\sqrt{2}}[a(V) + a^*(V)],$$

$$d\Gamma(A) = (\langle z, Az \rangle)^{\mathrm{Wick}}, \quad d\Gamma(A) = \sum_{k=0}^{n-1} \mathrm{Id}^{\otimes k} \otimes A \otimes \mathrm{Id}^{\otimes n-1-k}.$$

with

$$[a(g), a^*(f)] = a(g)a^*(f) - a^*(f)a(g) = \langle g, f \rangle \,\mathrm{Id},$$

Remember also that more generally, if $(A, D(A))$ generates a strongly continuous semigroup of contractions e^{tA}, $t \geq 0$, then $\Gamma(e^{tA}) f_n = [e^{tA}]^{\otimes n} f_n$ defines a strongly continuous semigroup of contractions $\Gamma(e^{tA})$ on $\Gamma(\mathfrak{h})$ with generator denoted by $(d\Gamma(A), D(d\Gamma(A)))$, which extends the above definition of $d\Gamma(A)$. In particular this makes sense for $A = -iB$ with $(B, D(B))$ self-adjoint on $\mathfrak{h}$ and $(d\Gamma(B), D(d\Gamma(B)))$ is a self-adjoint operator on $\Gamma(\mathfrak{h})$ when $(B, D(B))$ is self-adjoint on $\mathfrak{h}$.

According to (5) and (6), random $L^2(\mathbb{R}^d, dx; \mathbb{C})$ functions $F(x,\omega)$ can be written as elements f of $L^2(\mathbb{R}^d, dx; \mathbb{C}) \otimes \Gamma(L^2(\mathbb{R}^d, dy; \mathbb{C}))$,

$$F(x,\omega) \mapsto f(x, \cdot - x) = \bigoplus_{n\in\mathbb{N}} f_n(x, y_1 - x, \ldots, y_n - x)$$

$$\text{with } f_n \in L^2_{\text{sym}}(\mathbb{R}^d_x \times \mathbb{R}^{dn}, dxdy_1 \cdots dy_n; \mathbb{C}),$$

$$\|F\|^2_{L^2(\mathbb{R}^d\times\Omega, dx\otimes\mathcal{G})} = \sum_{n=0}^{\infty} \|f_n\|^2_{L^2(\mathbb{R}^d\times\mathbb{R}^{dn}, dxdy_1\cdots dy_n)},$$

and where L^2_{sym} refers to the exchange symmetry in the y-variables.

When $V \in L^2(\mathbb{R}^d, dy; \mathbb{R})$ and $\mathcal{V}(x,\omega) = \int_{\mathbb{R}^d} V(y-x) X_y \, dy$, the Wick product formula (7) for $\mathcal{V}(x,\omega)F(x,\omega)$ is transformed into

$$\mathcal{V}(x,\omega)F(x,\omega) \mapsto [a(V) + a^*(V)]f(x, \cdot - x) = [\sqrt{2}\phi(V)f](x, \cdot - x). \tag{10}$$

With the notation

$$D_y = \frac{1}{i}\partial_y = \begin{pmatrix} \frac{1}{i}\partial_{y^1} \\ \vdots \\ \frac{1}{i}\partial_{y^d} \end{pmatrix}$$

the operator $(x \cdot D_y, D(x \cdot D_y))$, with $x \cdot D_y = \sum_{k=1}^d x^k D_{y^k}$, is essentially self-adjoint on $\mathcal{S}(\mathbb{R}^d, dy; \mathbb{C})$ for all $x \in \mathbb{R}^d$. This defines a strongly continuous unitary representation of the additive group $(\mathbb{R}^d_x, +)$ on $L^2(\mathbb{R}^d_x) \otimes \Gamma(L^2(\mathbb{R}^d_y))$ given by

$$e^{-ix\cdot d\Gamma(D_y)}\Big(\bigoplus_{n=0}^{\infty} f_n(x, y_1, \ldots, y_n)\Big) = \bigoplus_{n=0}^{\infty} f_n(x, y_1 - x, \ldots, y_n - x).$$

Therefore the above unitary correspondence $F(x,\omega) \mapsto f(x, \cdot - x)$ gives a unitary correspondence

$$F \in L^2(\mathbb{R}^d \times \Omega, dx \otimes \mathcal{G}; \mathbb{C}) \mapsto f \in L^2(\mathbb{R}^d, dx; \mathbb{C}) \otimes \Gamma(L^2(\mathbb{R}^d, dy; \mathbb{C})), \tag{11}$$

while (10) becomes for $V \in L^2(\mathbb{R}^d, dy; \mathbb{R})$

$$\mathcal{V}F \mapsto \left[\sqrt{2}\Phi(V)f\right]. \tag{12}$$

We now translate a general pseudo-differential operator $a^{\text{Weyl}}(x, D_x) \otimes \text{Id}_{L^2(\Omega,\mathcal{G};\mathbb{C})}$ in the x-variable, under the above transformation (11).

When $\mathfrak{h}$ is a complex Hilbert space, we recall that $L^2(\mathbb{R}^d, dx; \mathbb{C}) \otimes \mathfrak{h}$ equals $L^2(\mathbb{R}^d, dx; \mathfrak{h})$ and

- the Fourier transform, with the normalization

$$Fu(\xi) = \int_{\mathbb{R}^d} e^{-i\xi\cdot x} u(x)\, dx\,, \quad F^{-1}v(x) = \int_{\mathbb{R}^d} e^{ix\cdot\xi} v(\xi)\, \frac{d\xi}{(2\pi)^d}\,,$$

  is unitary from $L^2(\mathbb{R}^d, dx; \mathfrak{h})$ to $L^2(\mathbb{R}^d, \frac{d\xi}{(2\pi)^d}; \mathfrak{h})$;
- $\mathcal{S}(\mathbb{R}^d; \mathfrak{h})$, $\mathcal{S}'(\mathbb{R}^d; \mathfrak{h})$ and the Fourier transform have the same properties as in the scalar case $\mathfrak{h} = \mathbb{C}$.

Be aware that the behavior of the Fourier transform when $\mathfrak{h}$ is a general Banach space is more tricky according to [25]. So when $\mathfrak{h}$ is a Hilbert space, we consider pseudo-differential operators in the x-variable of the form $a^{\mathrm{Weyl}}(x, D_x) = a^{\mathrm{Weyl}}(x, D_x) \otimes \mathrm{Id}_{\mathfrak{h}}$ for a symbol $a \in \mathcal{S}'(\mathbb{R}^{2d}_{x,\xi}; \mathbb{C})$ given by its Schwartz' kernel

$$[a^{\mathrm{Weyl}}(x, D_x)](x, y) = \int_{\mathbb{R}^d} e^{i(x-y)\cdot\xi} a\left(\frac{x+y}{2}, \xi\right) \frac{d\xi}{(2\pi)^d}\,.$$

When $\mathfrak{h} = \mathbb{C}$, $a^{\mathrm{Weyl}}(x, D_x)$ is a continuous endomorphism of $\mathcal{S}(\mathbb{R}^d_x; \mathbb{C})$ and $\mathcal{S}'(\mathbb{R}^d; \mathbb{C})$ with the formal adjoint $\overline{a}^{\mathrm{Weyl}}(x, D_x)$ and the alternative representations:

- When $v, u \in \mathcal{S}(\mathbb{R}^d; \mathbb{C})$,

$$\langle v\,, a^{\mathrm{Weyl}}(x, D_x)u\rangle = \int_{\mathbb{R}^{2d}} a(x, \xi)\, W[v, u](x, \xi)\, \frac{dx\, d\xi}{(2\pi)^d}$$

  where $W[v, u]$ is the Wigner function of the pair $[v, u]$ (or the Weyl symbol of $|u\rangle\langle v|$), given by

$$W[v, u](x, \xi) = \int_{\mathbb{R}^d} e^{i\xi\cdot s}\, u(x + \frac{s}{2})\, \overline{v}(x - \frac{s}{2})\, ds\,,$$

  and which belongs to $\mathcal{S}(\mathbb{R}^{2d}; \mathbb{C})$.
- By setting $[\![P, X]\!] = p_\xi\cdot x - p_x\cdot\xi$ for $P = (p_x, p_\xi)$, $X = (x, \xi)$ in $\mathbb{R}^{2d} = T^*\mathbb{R}^d$, and

$$\mathcal{F}a(P) = \int_{\mathbb{R}^{2d}} e^{i[\![P,X]\!]} a(X)\, \frac{dX}{(2\pi)^d}$$

  we have $a = \mathcal{F}(\mathcal{F}a)$ in $\mathcal{S}'(\mathbb{R}^{2d})$. When $\mathcal{F}a \in L^1(\mathbb{R}^{2d}; \mathbb{C})$,

$$a^{\mathrm{Weyl}}(x, D_x) = \int_{\mathbb{R}^{2d}} \mathcal{F}a(P)\, \tau_P\, \frac{dP}{(2\pi)^d}\,,$$

where $\tau_P = e^{i(p_\xi\cdot x - p_x\cdot D_x)} = [e^{i(p_\xi\cdot x - p_x\cdot \xi)}]^{\mathrm{Weyl}}(x, D_x)$ is the unitary phase translation

$$\tau_P\, u\,(x) = e^{ip_\xi\cdot(x-p_x/2)}u(x-p_x)\,.$$

In particular, the above integral is a $\mathcal{L}(L^2(\mathbb{R}^d, dx; \mathbb{C}))$-integral when $\mathcal{F}a \in L^1(\mathbb{R}^{2d}, dP; \mathbb{C})$ and a fortiori when $a \in \mathcal{S}(\mathbb{R}^{2d}; \mathbb{C})$.

With those two remarks, for a general $a \in \mathcal{S}'(\mathbb{R}^{2d}; \mathbb{C})$ the integral

$$a^{\mathrm{Weyl}}(x, D_x) = \int_{\mathbb{R}^{2d}} \mathcal{F}a(P)\, \underbrace{e^{i(p_\xi\cdot x - p_x\cdot D_x)}}_{=\tau_P}\, \frac{dP}{(2\pi)^d}$$

can be interpreted as the weak limit

$$a^{\mathrm{Weyl}}(x, D_x) = \mathrm{w}-\lim_{n\to\infty} \int_{\mathbb{R}^{2d}} \mathcal{F}a_n(P)\, e^{i(p_\xi\cdot x - p_x\cdot D_x)}\, \frac{dP}{(2\pi)^d}\,,$$

where $a_n \in \mathcal{S}(\mathbb{R}^{2d}; \mathbb{C})$ is any approximation of $a \in \mathcal{S}'(\mathbb{R}^{2d}; \mathbb{C})$.

While considering the $a^{\mathrm{Weyl}}(x, D_x) \otimes \mathrm{Id}_{\mathfrak{h}}$, the same construction makes sense after noticing that for $u, v \in \mathcal{S}(\mathbb{R}^d; \mathfrak{h})$, the Wigner transform $W[v, u]$ belongs to $\mathcal{S}(\mathbb{R}^{2d}_{x,\xi}; \mathcal{L}^1(\mathfrak{h}))$[2] and

$$\begin{aligned}\langle v\,,\, a^{\mathrm{Weyl}}(x, D_x)u\rangle &= \mathrm{Tr}\left[[a^{\mathrm{Weyl}}(x, D_x)\otimes \mathrm{Id}_{\mathfrak{h}}]\,|u\rangle\langle v|\right]\\ &= \int_{\mathbb{R}^{2d}} a(x,\xi)\,\mathrm{Tr}[W[v,u]](x,\xi)\,\frac{dxd\xi}{(2\pi)^d}\,.\end{aligned}$$

We apply this with $\mathfrak{h} = L^2(\Omega, \mathcal{G}; \mathbb{C})$ and $\mathfrak{h} = \Gamma(L^2(\mathbb{R}^d, dy; \mathbb{C}))$: We start from

$$\begin{aligned}a^{\mathrm{Weyl}}(x, D_x) &= a^{\mathrm{Weyl}}(x, D_x)\otimes \mathrm{Id}_{L^2(\Omega,\mathcal{G};\mathbb{C})}\\ &= \mathrm{w}-\lim_{n\to\infty}\int_{\mathbb{R}^{2d}} \mathcal{F}a_n(P) e^{i(p_\xi\cdot x - p_x\cdot D_x)}\,\frac{dP}{(2\pi)^d}\,,\end{aligned}$$

the correspondance

$$a^{\mathrm{Weyl}}(x, D_x)F \mapsto e^{ix\cdot d\Gamma(D_y)}a^{\mathrm{Weyl}}(x, D_x)e^{-ix\cdot d\Gamma(D_y)}f\,,$$

and

$$e^{ix\cdot\lambda}e^{i(p_\xi\cdot x - p_x\cdot D_x)}(e^{-ix\cdot\lambda}\times) = e^{i(p_x\cdot x - p_x\cdot(D_x-\lambda))}\quad \text{for all } \lambda\in\mathbb{R}^d$$

[2] $\mathcal{L}^p(\mathfrak{h})$ denotes the Schatten space of compact operators for $1 \le p \le +\infty$.

which gives by the functional calculus, the equality of unitary operators

$$e^{ix\cdot d\Gamma(D_y)}e^{i(p_\xi\cdot x-p_x\cdot D_x)}(e^{-ix\cdot d\Gamma(D_y)})=e^{i(p_x\cdot x-p_x\cdot(D_x-d\Gamma(D_y)))}\,.$$

We deduce that for $a\in\mathcal{S}'(\mathbb{R}^{2d};\mathbb{C})$, $a^{\mathrm{Weyl}}(x,D_x)F\in\mathcal{S}'(\mathbb{R}^d_x;L^2(\Omega,\mathcal{G};\mathbb{C}))$ is transformed into

$$a^{\mathrm{Weyl}}(x,D_x)\,F\mapsto a^{\mathrm{Weyl}}(x,D_x-d\Gamma(D_y))\,f\in\mathcal{S}'(\mathbb{R}^d;\Gamma(L^2(\mathbb{R}^d,dy;\mathbb{C})))\,. \tag{13}$$

with

$$a^{\mathrm{Weyl}}(x,D_x-d\Gamma(D_y))=\mathrm{w}-\lim_{n\to\infty}\int_{\mathbb{R}^{2d}}\mathcal{F}a_n(P)e^{i(p_\xi\cdot x-p_x\cdot(D_x-d\Gamma(D_y)))}\,\frac{dP}{(2\pi)^d}\,.$$

Let us continue by applying the Fourier transform in the x-variable with

$$F_xu(\xi)=\int_{\mathbb{R}^d}e^{-i\xi\cdot x}u(x)\,dx\,,\quad F_x^{-1}u(x)=\int_{\mathbb{R}^d}e^{ix\cdot\xi}u(\xi)\,\frac{d\xi}{(2\pi)^d}$$

and set for $f\in\mathcal{S}'(\mathbb{R}^d_x;\Gamma(L^2(\mathbb{R}^d,dy;\mathbb{C})))$

$$\hat{f}=F_xf\in\mathcal{S}'(\mathbb{R}^d;\Gamma(L^2(\mathbb{R}^d,dy;\mathbb{C})))\,.$$

With

$$F_x\,a^{\mathrm{Weyl}}(x,D_x)\,F_x^{-1}=a^{\mathrm{Weyl}}(-D_\xi,\xi)$$

where the functional calculus leads to $F_x\,a^{\mathrm{Weyl}}(x,D_x-d\Gamma(D_y))\,F_x^{-1}=a^{\mathrm{Weyl}}(-D_\xi,\xi-d\Gamma(D_y))$, we obtain the unitary correspondence

$$F\in L^2(\mathbb{R}^d\times\Omega,dx\otimes\mathcal{G};\mathbb{C})\mapsto\hat{f}=F_xf\in L^2\big(\mathbb{R}^d,\frac{d\xi}{(2\pi)^d};\Gamma(L^2(\mathbb{R}^d,dy;\mathbb{C}))\big)\,, \tag{14}$$

with

$$F(x,\omega)=\sum_{n=0}^{\infty}\int_{\mathbb{R}^{dn}}\frac{1}{\sqrt{n!}}f_n(x,y_1-x,\ldots,y_n-x):X_{y_1}\cdots X_{y_n}:\,dy_1\cdots dy_n\,, \tag{15}$$

and where (12) and (13) become

$$\mathcal{V} F \mapsto \sqrt{2}\phi(V)\, \hat{f}\,, \tag{16}$$

$$\underbrace{a^{\mathrm{Weyl}}(x, D_x)\, F}_{\in \mathcal{S}'(\mathbb{R}^d_\xi; L^2(\Omega, \mathcal{G}; \mathbb{C}))} \mapsto \underbrace{a^{\mathrm{Weyl}}(-D_\xi, \xi - d\Gamma(D_y))\, \hat{f}}_{\in \mathcal{S}'(\mathbb{R}^d_\xi; \Gamma(L^2(\mathbb{R}^d, dy; \mathbb{C})))}\,. \tag{17}$$

From this point of view, the Fock space and functional analysis presentation is simpler than sticking with the usual chaos decomposition (4) where Fourier transforms and pseudo-differential operators do not seem to have simple probabilistic interpretation.

Remark 2.1 As a final remark, all the above constructions can be tensorized with an additional separable Hilbert space $\mathfrak{h}' = L^2(Z, \mathbf{dz}; \mathbb{C})$.

2.3 *Our Problem*

We aim at studying the stochastic partial differential equation

$$\begin{cases} i\partial_t F = -\Delta_x F + \sqrt{h}\mathcal{V} F\,, \\ F(t=0) = F_0\,, \end{cases} \tag{18}$$

where

- $\mathcal{V}$ is the translation invariant gaussian random field

$$\mathcal{V}(x, \omega) = \int_{\mathbb{R}^d} V(y - x)\, X_y\, dy\,,$$

  with $V \in L^2(\mathbb{R}^d; \mathbb{R})$;
- the solution $F(t, x, \omega, z)$ is seeked in $C^0(\mathbb{R}; L^2(\mathbb{R}^d \times \Omega \times Z, dx \otimes \mathcal{G} \otimes \mathbf{dz}; \mathbb{C}))$;
- $h > 0$ is a small parameter which will tend to 0.

In particular we will consider the asymptotic behavior of quantities

$$\langle F(\frac{t}{h})\,,\, a^{\mathrm{Weyl}}(hx, D_x) F(\frac{t}{h})\rangle_{L^2(\mathbb{R}^d \times \Omega \times Z)} = \int_Z \mathbb{E}\left[\langle F(\frac{t}{h}, z)\,,\, a^{\mathrm{Weyl}}(hx, D_x) F(\frac{t}{h}, z)\rangle_{L^2(\mathbb{R}^d, dx)}\right] \mathbf{dz}(z) \tag{19}$$

for $a \in S(1, dx^2 + d\xi^2)$ and $t \in [0, T]$. Remember that the symbol class $S(1, dx^2 + d\xi^2)$ is the set of C^∞-functions on $\mathbb{R}^{2d}$ with all derivatives bounded on $\mathbb{R}^{2d}$.

Note that the variable $z \in Z$ does not appear in the equation. The dynamics is thus well defined when it is defined for $Z = \{z_0\}$ and $\mathbf{dz} = \delta_{z_0}$. A sufficient condition was provided in [5] by making use of Nelson commutator method.

Lemma 2.1 ***Proposition 4.4 in [5]:*** *Assume $V \in H^2(\mathbb{R}^d;\mathbb{R})$. Then the operator $-\Delta_x + \sqrt{h}\mathcal{V}$ is essentially self-adjoint on $\bigoplus_{n\in\mathbb{N}}^{\mathrm{alg}} \mathcal{S}(\mathbb{R}^d_x; (L^2(\mathbb{R}^d, dy;\mathbb{C}))^{\odot n})$ which is a dense subset of $L^2(\mathbb{R}^d_x, dx; L^2(\Omega, \mathcal{G};\mathbb{C})) = L^2(\mathbb{R}^d \times \Omega, dx \otimes \mathcal{G};\mathbb{C})$ by* (4).

Remark 2.2 A side corollary of our analysis says that the dynamics is well defined under the assumption $V \in L^{r'_\sigma}(\mathbb{R}^d;\mathbb{R})$ with $r'_\sigma = \frac{2d}{d+2}$ in dimension $d \geq 3$, See Sect. 7.4 at the end of the article.

Lemma 2.1 provides a natural self-adjoint realization of $-\Delta_x + \sqrt{h}\mathcal{V}$ in $\mathfrak{h} = L^2(\mathbb{R}^d \times \Omega \times Z, dx \otimes \mathcal{G} \otimes \mathbf{dz};\mathbb{C})$ and any initial datum $F_0 \in \mathfrak{h}$ defines a unique solution $F \in \mathcal{C}^0(\mathbb{R};\mathfrak{h})$.

There are various reasons for introducing an additional variable $z \in Z$, and this trick will be used repeatedly. One of them is the following: Starting with $Z = \{z_0\}$ and $\mathbf{dz} = \delta_{z_0}$, one may consider instead of $F(\frac{t}{h}) = U_{\mathcal{V}}(\frac{t}{h})F_0$ with $U_{\mathcal{V}}(t) = \exp(-it(-\Delta_x + \sqrt{h}\mathcal{V}))$, the evolution of a state

$$\varrho(\frac{t}{h}) = U_{\mathcal{V}}(\frac{t}{h})\varrho_0 U^*_{\mathcal{V}}(\frac{t}{h})$$

with $\varrho_0 \in \mathcal{L}^1(L^2(\mathbb{R}^d \times \Omega;\mathbb{C}))$, $\varrho_0 \geq 0$, $\mathrm{Tr}[\varrho_0] = 1$ possibly replacing $\|F_0\|_{L^2} = 1$. By writing $\varrho_0 = \varrho_0^{1/2}\varrho_0^{1/2}$ one gets

$$\varrho(\frac{t}{h}) = [U_{\mathcal{V}}(\frac{t}{h})\varrho_0^{1/2}][U_{\mathcal{V}}(\frac{t}{h})\varrho_0^{1/2}]^*$$

where $F(t) = U_{\mathcal{V}}(t)\varrho_0^{1/2}$ is the solution to (18) in

$$\mathcal{L}^2(L^2(\mathbb{R}^d \times \Omega, dx \times \mathcal{G};\mathbb{C})) \simeq L^2(\mathbb{R}^d \times \Omega \times Z, dx \otimes \mathcal{G} \otimes \mathbf{dz})$$
$$\text{with} \quad Z = \mathbb{R}^d \otimes \Omega, \quad \mathbf{dz} = dx \otimes \mathcal{G},$$

while the trace to be computed at time $\frac{t}{h}$ equals

$$\mathrm{Tr}\left[a^{\mathrm{Weyl}}(hx, D_x)\varrho(\frac{t}{h})\right] = \int_Z \mathbb{E}\left[\langle F(\frac{t}{h}, z), a^{\mathrm{Weyl}}(hx, D_x)F(\frac{t}{h}, z)\rangle_{L^2(\mathbb{R}^d, dx)}\right] \mathbf{dz}(z).$$

Thus considering the evolution of non negative trace class operators instead of projectors on wave functions, becomes the same problem by introducing the suitable additional parameter $z \in Z$.

The unitary correspondence (14), (15), with (16), (17) and Remark 2.1, transforms the dynamics (18) into

$$\begin{cases} i\partial_t \hat{f} = (\xi - d\Gamma(D_y))^2 \hat{f} + \sqrt{2h}\phi(V)\hat{f}\,, \\ \hat{f}(t=0) = \hat{f}_0\,, \end{cases} \tag{20}$$

and the quantity (19) into

$$\langle \hat{f}(\frac{t}{h})\,,\, a^{\mathrm{Weyl}}(-hD_\xi, \xi - d\Gamma(D_y))\, \hat{f}(\frac{t}{h})\rangle_{L^2(\mathbb{R}^d \times \Omega \times Z, \frac{d\xi}{(2\pi)^d}\otimes \mathcal{G} \otimes \mathbf{dz})}\,. \tag{21}$$

We will see that the variable $\xi \in \mathbb{R}^d$ and even some part Y' of the variable $Y = (y_1, \ldots, y_n)$, when the total number is fixed to n, can be taken as another parameter like $z \in Z$ for some points of the analysis. This leads to a parameter z'-dependent, $z' = (\xi, Y', z) \in \mathbb{R}^d \times \mathbb{R}^{dn'} \times Z$, analysis in $L^2(\mathbb{R}^{d(n-n')}, dY'')$. Those parameters appear in Sect. 3 by introducing the center of mass $Y'' = y_G = \frac{y_1 + \cdots + y_n}{n}$ and the relative coordinates $y'_j = y_j - y_G$, a general functional framework for parameter dependent Strichartz estimates and their consequences are presented in Sect. 4 and finally those are detailled in Sect. 5 for (20).

3 The Fock Space and the Center of Mass

According to (20) our stochastic dynamics has been translated in a parameter dependent dynamics in the Fock space. We shall consider an additional unitary transform using the center of mass and the relative variables

$$y_G^n = \frac{y_1 + \cdots + y_n}{n}\,, \quad y'_j = y_j - y_G^n$$

in the n-particles sector, $n \geq 1$. It trivializes the free dynamics when $\mathcal{V} \equiv 0$ or $V \equiv 0$. The expression of the interaction term $\sqrt{2h}\phi(V)$ becomes more tricky but various general estimates are given here.

3.1 The Unitary Transform Associated with the Center of Mass

We shall use the following notations for $n \geq 1$:

- A generic element of $\mathbb{R}^{dn}$ will be written

$$Y_n = (y_1, \ldots, y_n) \quad \text{with} \quad |Y_n|^2 = \sum_{j=1}^{n} |y_j|^2\,. \tag{22}$$

- The center of mass of $Y_n \in \mathbb{R}^{dn}$ will be written

$$y_G = y_G^n = \frac{y_1 + \cdots + y_n}{n} \tag{23}$$

and the relative coordinates $y'_j = y_j - y_G^n$ will be gathered into

$$Y'_n = (y'_1, \ldots, y'_n) = (y_1 - y_G^n, \ldots, y_n - y_G^n)\,. \tag{24}$$

The vector Y'_n actually belongs to the subspace $\mathcal{R}^n = \left\{Y_n \in \mathbb{R}^{dn}\,, \sum_{j=1}^n y_j = 0\right\}$ and we recall

$$|Y_n|^2 = n|y_G^n|^2 + \left|Y'_n\right|^2 = n|y_G^n|^2 + \sum_{j=1}^n |y'_j|^2\,. \tag{25}$$

With those notations the map $\mathbb{R}^{dn} \ni Y_n \mapsto (y_G^n, Y'_n) \in \mathbb{R}^d \times \mathcal{R}^n \subset \mathbb{R}^d \times \mathbb{R}^{dn}$ is a measurable map and the image measure of the Lebesgue measure $|dY_n| = \prod_{j=1}^n |dy_j|$ is nothing but

$$dy_G \otimes d\mu_n(Y'_n) = dy_G \otimes [n^d dy_1 \cdots dy_n \delta_0(y_1 + \cdots + y_n)]\,. \tag{26}$$

For $n \geq 2$ we can write $d\mu_n(Y'_n) = n^d \prod_{j \neq j_0} dy'_j$ for any fixed $j_0 \in \{\bar{1}, \ldots, n\}$ by taking the linear coordinates $(y'_j)_{j \neq j_0}$ on $\mathcal{R}^n$ where $y'_{j_0} = -\sum_{j \neq j_0} y'_j$. For $n = 1$, $\mathcal{R}^1 = \{0\}$ and integrating with respect to $Y'_1 = y'_1 \in \mathcal{R}^1$ is nothing but the evaluation at $y'_1 = 0$.

Definition 3.1 On $\sqcup_{n=1}^{\infty} \mathbb{R}^{dn}$ the measure μ carried by $\mathcal{R} = \sqcup_{n=1}^{\infty} \mathcal{R}^n$ is defined by

$$\begin{aligned} \int_{\mathcal{R}^n} g_n(Y')\, d\mu_n(Y') &= \int_{\mathbb{R}^{dn}} g_n(y_1, \ldots, y_n) \delta_0(y_1 + \cdots + y_n)\, n^d\, dy_1 \cdots dy_n \\ &\overset{n \geq 2}{=} \int_{\mathbb{R}^{d(n-1)}} g_n(y'_1, \ldots, y'_{n-1}, -\sum_{j=1}^{n-1} y'_j)\, n^d\, dy'_1 \cdots dy'_{n-1} \end{aligned}$$

for all g_n in $\mathcal{C}_c^0(\mathbb{R}^{dn})$.

For $1 \leq p < +\infty$, the space $L^p(\mathcal{R}, d\mu)$ is the direct sum $\bigoplus_{n=1}^{\infty} L^p(\mathcal{R}^n, d\mu_n)$ completed with respect to the norm

$$\left\| \bigoplus_{n=1}^{\infty} g_n \right\|_{L^p} = \left(\sum_{n=1}^{\infty} \|g_n\|_{L^p(\mathcal{R}^n, d\mu_n)}^p \right)^{1/p}.$$

The closed subspace of symmetric functions, $g_n(y'_{\sigma(1)}, \ldots, y'_{\sigma(n)}) = g_n(y'_1, \ldots, y'_n)$ for all $\sigma \in \mathfrak{S}_n$ and for all $n \geq 1$, is then denoted by $L^p_{\text{sym}}(\mathcal{R}, d\mu(Y'))$.

For $g_n \in L^2(\mathbb{R}^{dn} \times Z, dY_n \otimes \mathbf{dz}; \mathbb{C})$, $n \geq 1$, the function

$$g_{G,n}(y_G, Y'_n, z) = U_G g_n(y_G, Y'_n, z) = g_n(y_G + Y'_n, z) \tag{27}$$

belongs to $L^2(\mathbb{R}^d \times \mathcal{R}^n \times Z, dy_G \otimes d\mu_n \otimes \mathbf{dz}; \mathbb{C})$ with

$$\|U_G g_n\|_{L^2(\mathbb{R}^d \times \mathcal{R}^n \times Z, dy_G \otimes d\mu_n \otimes \mathbf{dz})} = \|g_n\|_{L^2(\mathbb{R}^{dn} \times Z, dY_n \otimes \mathbf{dz})}$$

$$\text{and} \quad g_n(Y_n, z) = (U_G^{-1} g_{G,n})(Y_n, z) = g_{G,n}(y_G^n, Y_n - y_G^n, z)\,.$$

Additionally

$$\begin{aligned} U_G : L^2(\mathbb{R}^d, dy)^{\odot n} &\to L^2(\mathbb{R}^d, dy_G; L^2_{\text{sym}}(\mathcal{R}^n, d\mu_n)) \\ &= L^2_{\text{sym}}(\mathcal{R}^n, d\mu_n; L^2(\mathbb{R}^d, dy_G)) \end{aligned}$$

is unitary and the same result holds for the parameter $z \in Z$ version.

Proposition 3.1 *The map U_G extended by $U_G g_0(z) = g_0(z)$ for $n = 0$, defines a unitary map*

$$\begin{aligned} U_G : L^2(Z, dz; \Gamma(L^2(\mathbb{R}^d, dy; \mathbb{C}))) \\ \to L^2(Z; \mathbb{C}) \oplus L^2_{\text{sym}}(Z \times \mathcal{R}, dz \otimes d\mu; L^2(\mathbb{R}^d, dy_G; \mathbb{C}))\,. \end{aligned} \tag{28}$$

When $d\Gamma_G(A) = U_G[d\Gamma(A) \otimes \mathrm{Id}_{L^2(Z,dz)}]U_G^{-1}$ for a self-adjoint operator $(A, D(A))$ in $L^2(\mathbb{R}^d_y, dy)$, the case $A = D_y$ gives

$$d\Gamma_G(D_y) = U_G\, d\Gamma(D_y)\, U_G^{-1} = D_{y_G}\,. \tag{29}$$

For any bounded measurable function ϕ on $\mathcal{R} \times Z$ the multiplication by $\phi(Y', z)\big|_{\mathcal{R}^n} = \phi_n(Y'_n, z)$ for $n \geq 1$, while $\phi_0 : Z \to \mathbb{C}$, commutes with $d\Gamma_G(D_y) = D_{y_G}$ according to

$$\begin{aligned} \forall t \in \mathbb{R}^d\,, \forall u \in L^2(Z, \mathbf{dz}; \mathbb{C}) \oplus L^2_{\text{sym}}(Z \times \mathcal{R}, \mathbf{dz} \otimes d\mu; L^2(\mathbb{R}^d, dy_G, \mathbb{C}))\,, \\ e^{it\cdot D_{y_G}}(\phi u) = \phi(e^{it\cdot D_{y_G}} u)\,. \end{aligned}$$

A particular case is $\phi_n(Y'_n, z) = \varphi(n)$ for a bounded function $\varphi : \mathbb{N} \to \mathbb{C}$.

Proof The unitarity of U_G comes at once from (27) and the componentwise unitarity already checked. For $d\Gamma_G(D_y) = D_{y_G}$, simply write

$$\begin{aligned}\partial_{y_G} g_{G,n}(y_G, Y_n') &= \partial_{y_G} g_n(y_G + y_1', \dots, y_G + y_n') \\ &= \sum_{j=1}^{n} (\partial_{y_j} g_n)(y_G + y_1', \dots, y_G + y_n') .\end{aligned}$$

The commutation statement comes from the separation of variables, y_G and (Y', z). □

Introducing the center of mass thus simplifies the free transport part of (20). It is not so for the interaction term $\sqrt{2h}\phi(V) = \sqrt{h}[a(V) + a^*(V)]$. An explicit and useful expression is nevertheless possible for

$$a_G(V) = U_G\, a(V)\, U_G^{-1} \quad \text{and} \quad a_G^*(V) = U_G\, a^*(V)\, U_G^{-1} . \tag{30}$$

Proposition 3.2 *The operator $a_G(V)$ and $a_G^*(V)$ for $V \in L^2(\mathbb{R}^d, dy; \mathbb{C})$ have the following action on $f_{G,n} \in L^2_{\text{sym}}(\mathcal{R}^n \times Z, d\mu_n \otimes \mathbf{dz}; L^2(\mathbb{R}^d, dy_G; \mathbb{C}))$ for $n \geq 1$ and $f_{G,0} \in L^2(Z, \mathbf{dz}; \mathbb{C})$ where we omit the transparent variable $z \in Z$:*

$$a_G(V) f_{G,0} = 0 , \quad [a_G(V) f_{G,1}] = \int_{\mathbb{R}^d} \overline{V(y_1)} f_{G,1}(y_1)\, dy_1 , \quad \textit{and} \tag{31}$$

$$[a_G(V) f_{G,n}](y_G, Y_{n-1}') = \sqrt{n} \int_{\mathbb{R}^d} \overline{V(y_G + y_n)} f_{G,n}(y_G + \frac{y_n}{n}, Y_n - \frac{y_n}{n})\, dy_n , \tag{32}$$

for all $n > 1$, with $Y_n = (y_1', \dots, y_{n-1}', y_n) \in \mathbb{R}^{dn}$, $Y_{n-1}' \in \mathcal{R}^{n-1}$, $Y_n - \frac{y_n}{n} \in \mathcal{R}^n$,

$$a_G^*(V) f_{G,0}(y_G) = V(y_G) f_{G,0} , \quad \textit{and} \tag{33}$$

$$\begin{aligned} &a_G^*(V) f_{G,n}(y_G, Y_{n+1}') \\ &\quad = \sqrt{n+1} S_{n+1}[V(y_G + y_{n+1}') f_{G,n}(y_G - \frac{y_{n+1}'}{n}, Y_n + \frac{y_{n+1}'}{n})] , \end{aligned} \tag{34}$$

for all $n > 0$, with $Y_n = (y_1', \dots, y_n') \in \mathbb{R}^{dn}$, $Y_{n+1}' \in \mathcal{R}^{n+1}$, $Y_n + \frac{y_{n+1}'}{n} \in \mathcal{R}^n$, and

$$S_{n+1}[v(y_{n+1}') u(y_1', \dots, y_{n+1}')] = \frac{1}{(n+1)!} \sum_{\sigma \in \mathfrak{S}_{n+1}} v(y_{\sigma(n+1)}') u(y_{\sigma(1)}', \dots, y_{\sigma(n+1)}') .$$

Proof Write for $n > 1$,

$$[a_G(V) f_{G,n}](y_G^{n-1}, Y'_{n-1}) = [a(V) U_G^{-1} f_{G,n}](Y'_{n-1} + y_G^{n-1})$$
$$= \sqrt{n} \int_{\mathbb{R}^d} \overline{V(\tilde{y}_n)} [U_G^{-1} f_{G,n}](Y'_{n-1} + y_G^{n-1}, \tilde{y}_n) \, d\tilde{y}_n .$$

By setting $\tilde{y}_n = y_G^{n-1} + y_n$ the formula $(U_G^{-1} g_{G,n})(\cdot) = g_{G,n}(y_G^n, \cdot - y_G^n)$ with

$$y_G^n = \frac{y_1 + \cdots + y_{n-1} + \tilde{y}_n}{n} = \frac{n-1}{n} y_G^{n-1} + \frac{\tilde{y}_n}{n} = y_G^{n-1} + \frac{y_n}{n}$$

leads to

$$[a_G(V) f_{G,n}](y_G^{n-1}, Y'_{n-1})$$
$$= \sqrt{n} \int_{\mathbb{R}^d} \overline{V(y_G^{n-1} + y_n)} f_{G,n}(y_G^{n-1} + \frac{y_n}{n}, Y'_{n-1} - \frac{y_n}{n}, y_n - \frac{y_n}{n}) \, dy_n$$
$$= \sqrt{n} \int_{\mathbb{R}^d} \overline{V(y_G^{n-1} + y_n)} f_{G,n}(y_G^{n-1} + \frac{y_n}{n}, Y_n - \frac{y_n}{n}) \, dy_n$$

with $Y_n = (y'_1, \ldots, y'_{n-1}, y_n)$.

The computation of $a_G^*(V) f_{G,n}$ is done by duality:

$$\langle a_G^*(V) f_{G,n-1} \, , \, g_{G,n} \rangle$$
$$= \langle f_{G,n-1} \, , \, a_G(V) g_{G,n} \rangle$$
$$= \int_{\mathbb{R}^d \times \mathcal{R}^{n-1}} \overline{f_{G,n-1}(y_G^{n-1}, Y'_{n-1})} \times$$
$$\left[\sqrt{n} \int_{\mathbb{R}^d} \overline{V(y_G^{n-1} + y_n)} g_{G,n}(y_G^{n-1} + \frac{y_n}{n}, Y_n - \frac{y_n}{n}) \, dy_n \right]$$
$$dy_G^{n-1} d\mu_{n-1}(Y'_{n-1}) .$$

Remember $Y_n = (y'_1, \ldots, y'_{n-1}, y_n)$ and $\tilde{Y}'_n = Y_n - \frac{y_n}{n} \in \mathcal{R}^n$. The change of variables

$$\tilde{Y}'_n = Y_n - \frac{y_n}{n}, \; y_G^n = y_G^{n-1} + \frac{y_n}{n}, \; y_G^{n-1} = y_G^n - \frac{\tilde{y}_n}{n-1}$$

$$Y'_{n-1} = \tilde{Y}_{n-1} + \frac{y_n}{n} = \tilde{Y}_{n-1} + \frac{\tilde{y}_n}{n-1},$$

with

$$\begin{aligned} dy_n dy_G^{n-1} d\mu^{n-1}(Y'_{n-1}) &= dy_G^{n-1}\delta_0(y'_1+\cdots+y'_{n-1})(n-1)^d dy'_1\cdots dy'_{n-1} \\ &= dy_G^n \frac{n^d}{(n-1)^d}(n-1)^d\delta_0(\tilde{y}'_1+\cdots+\tilde{y}'_n)d\tilde{y}'_1\cdots d\tilde{y}'_n \\ &= dy_G^n d\mu_n(\tilde{Y}'_n)\,, \end{aligned}$$

gives

$$\begin{aligned} &\langle a_G^*(V) f_{G,n-1}\,,\, g_{G,n}\rangle \\ &= \sqrt{n}\int_{\mathbb{R}^d\times\mathcal{R}^n} \overline{V(y_G^n+\tilde{y}'_n) f_{G,n-1}(y_G^n-\frac{\tilde{y}_n}{n-1},\tilde{Y}'_{n-1}+\frac{\tilde{y}'_n}{n-1})}\times \\ &\qquad\qquad g_{G,n}(y_G^n,\tilde{Y}'_n)\, dy_G^n\, d\mu_n(\tilde{Y}'_n)\,. \end{aligned}$$

Replacing n by $n+1$, while remembering that $a_G^*(V) f_{G,n}$ is symmetric in the variables $(y'_1,\ldots,y'_{n+1})$ yields

$$[a_G^*(V) f_{G,n}](y_G, Y'_{n+1}) = \sqrt{n+1}S_{n+1}[V(y_G+y'_{n+1}) f_{G,n}(y_G-\frac{y'_{n+1}}{n}, Y_n+\frac{y'_{n+1}}{n})]$$

with $Y_n=(y'_1,\ldots,y'_n)$. □

3.2 General $L_x^p L_y^q$ Spaces

When $(\mathcal{X},\mathbf{dx})$ and $(\mathcal{Y},\mathbf{dy})$ are sigma-finite measured spaces $L_x^pL_y^q$, $1\le p,q\le +\infty$, denotes the space $L_x^pL_y^q = L^p(\mathcal{X},\mathbf{dx}; L^q(\mathcal{Y},\mathbf{dy}))$. This shortened notation is especially useful when estimates are written in those spaces, like in Strichartz estimates (see Sect. 4). However the final space of the unitary map U_G in (28) shows already that the product space $\mathcal{X}\times\mathcal{Y}$ is too restrictive. Below is a convenient generalization.

Definition 3.2 Let $(\mathcal{X}_n,\mathbf{dx_n})_{n\in\mathcal{N}}$ and $(\mathcal{Y}_n,\mathbf{dy_n})_{n\in\mathcal{N}}$ be at most countable families ($\mathcal{N}\subset\mathbb{N}$) of sigma-finite measured spaces. Let $\mathcal{X}=\sqcup_{n\in\mathcal{N}}\mathcal{X}_n$ and $\mathcal{Y}=\sqcup_{n\in\mathcal{N}}\mathcal{Y}_n$ be endowed with the measures $\mathbf{dx}=\Sigma_{n\in\mathcal{N}}\mathbf{dx_n}$ and $\mathbf{dy}=\Sigma_{n\in\mathcal{N}}\mathbf{dy_n}$. In this framework, the space $L_x^pL_y^q$, $1\le p,q\le+\infty$, will denote the closed subspace of $L^p(\mathcal{X},\mathbf{dx}; L^q(\mathcal{Y},\mathbf{dy}))$ given by

$$\begin{aligned} L_x^pL_y^q = \Big\{ f\in L^p(\mathcal{X},\mathbf{dx}; L^q(\mathcal{Y},\mathbf{dy}))\,,\; & f(x,y) \\ &= \sum_{n\in\mathcal{N}} 1_{\mathcal{X}_n}(x)1_{\mathcal{Y}_n}(y)\, f(x,y) \text{ a.e.}\Big\}\,. \end{aligned}$$

The above definition is coherent with the specific product case, which is the particular case $\mathcal{N} = \{0\}$. The differences will be clear from the different frameworks when $(\mathcal{X}_n, \mathbf{dx_n})_{n\in\mathcal{N}}$ and $(\mathcal{Y}_n, \mathbf{dy_n})_{n\in\mathcal{N}}$ will be specified.

The two following properties of the product case are still valid in this extended framework:

- The dual of $L^p_x L^q_y$, $1 \le q, p < +\infty$ is $L^{p'}_x L^{q'}_y$ with $\frac{1}{q'} + \frac{1}{q} = 1$ and $\frac{1}{p'} + \frac{1}{p} = 1$.
- Minkowski's inequality says

$$\|f\|_{L^p_x L^q_y} \le \|f\|_{L^q_y L^p_x} \quad \text{for} \quad 1 \le q \le p \le +\infty. \tag{35}$$

Below are examples, associated with the decomposition associated with the introduction of the center of mass (23) and the relative coordinates (24), where those notations will be used

- $\mathcal{N} = \{n\}$, $n \ge 1$, $\mathcal{X}_n = \mathcal{R}_n \times Z'$, $\mathbf{dx_n} = d\mu_n \otimes \mathbf{dz'}$, $\mathcal{Y}_n = \mathbb{R}^d$, $\mathbf{dy_n} = dy_G$ and

$$L^p_{(Y'_n, z')} L^q_{y_G} = L^p_{\mathcal{X}_n} L^q_{y_G} = L^p(\mathcal{R}_n \times Z', d\mu_n \otimes \mathbf{dz'}; L^q(\mathbb{R}^d, dy_G)).$$

  The notation $L^p_{(Y'_n, z'),\mathrm{sym}} L^q_{y_G}$ will stand for the closed subspace of functions which are symmetric with respect to the variables $Y'_n \in \mathcal{R}_n$.
- $\mathcal{N} = \{0, 1\}$ with

$$\mathcal{X}_0 = Z', \quad \mathcal{X}_1 = \mathcal{R} \times Z' = (\sqcup_{n=1}^{\infty} \mathcal{R}_n) \times Z', \quad \mathbf{dx_0} = \mathbf{dz'}, \quad \mathbf{dx_1} = d\mu \otimes \mathbf{dz'},$$

$$\mathcal{Y}_0 = \{0\}, \quad \mathcal{Y}_1 = \mathbb{R}^d, \quad \mathbf{dy_0} = \delta_0, \quad \mathbf{dy_1} = dy_G,$$

  where

$$L^p_{(Y', z')} L^q_{y_G} = L^p(Z', \mathbf{dz'}) \oplus L^p(\mathcal{R} \times Z', d\mu \otimes \mathbf{dz'}; L^q(\mathbb{R}^d, dy_G)).$$

  With the same convention as above for $L^p_{(Y', z'),\mathrm{sym}} L^q_{y_G}$, which refers to the symmetry for the $Y' \in \mathcal{R}$ variable, the formula (28) becomes

$$U_G : L^2(Z', \mathbf{dz'}; \Gamma(L^2(\mathbb{R}^d, dy; \mathbb{C}))) \to L^2_{(Y', z'),\mathrm{sym}} L^2_{y_G}.$$

  The general spaces $L^2_{(Y', z'),\mathrm{sym}} L^p_{y_G}$, $1 \le p \le +\infty$, will be especially useful after Sect. 4.
- The previous example can be written with $\mathcal{N} = \mathbb{N}$ and

$$\mathcal{X}_0 = Z', \quad \mathcal{X}_{n>0} = \mathcal{R}_n \times Z', \quad \mathbf{dx_0} = \mathbf{dz'}, \quad \mathbf{dx_{n>0}} = d\mu_n \otimes \mathbf{dz'},$$

$$\mathcal{Y}_0 = \{0\}, \quad \mathcal{Y}_{n>0} = \mathbb{R}^d, \quad \mathbf{dy_0} = \delta_0, \quad \mathbf{dy_{n>0}} = dy_G.$$

3.3 L^p_{yG}-Estimates for $a_G(V)$ and $a_G^*(V)$

General L^p-esptimates, or more precisely $L^2_{(Y',z),\text{sym}} L^p_{yG}$-estimates, are proved in this paragraph for the operators $a_G(V)$ and $a_G^*(V)$. The use of the center of mass and the L^p_{yG} spaces, will be extremely useful for the application of Strichartz estimates in Sect. 4.

Let us start with a simple application of Young's inequality.

Lemma 3.1 *For any $q', p' \in [1,2]$ such that $q' \leq p'$, let $r' \in [1,2]$ be defined by $\frac{1}{r'} = \frac{1}{2} + \frac{1}{q'} - \frac{1}{p'}$. The inequality*

$$\|V(y_G + y')\varphi(y_G)\|_{L^2_{y'} L^{q'}_{y_G}} \leq \|V\|_{L^{r'}} \|\varphi\|_{L^{p'}},$$

holds for all $V \in L^{r'}(\mathbb{R}^d, dy; \mathbb{C})$ and all $\varphi \in L^{p'}(\mathbb{R}^d, dy; \mathbb{C})$.

Proof The conditions $\frac{1}{r'} + \frac{1}{p'} = \frac{1}{2} + \frac{1}{q'}$, $1 \leq q' \leq p' \leq 2$, ensure

$$\frac{1}{r'} = \frac{1}{2} + \frac{1}{q'} - \frac{1}{p'} \in [\frac{1}{2}, 1] \quad \text{and} \quad r' \in [1,2].$$

Young's inequality with $\frac{1}{\tilde{r}} + \frac{1}{\tilde{p}} = \frac{1}{\frac{2}{q'}} + 1$ and $\tilde{r}, \tilde{p}, \frac{2}{q'} \geq 1$ yields

$$\begin{aligned}\|V(y_G + y')\varphi(y_G)\|_{L^2_{y'} L^{q'}_{y_G}} &\leq \||V|(y_G - y')|\varphi|(y_G)\|_{L^2_{y'} L^{q'}_{y_G}} \\ &= \||V(-\cdot)|^{q'} * |\varphi|^{q'}\|^{1/q'}_{L^{2/q'}} \leq \||V|^{q'}\|^{1/q'}_{L^{\tilde{r}}} \||\varphi|^{q'}\|^{1/q'}_{L^{\tilde{p}}}.\end{aligned}$$

By taking $\tilde{p} = \frac{p'}{q'} \in [1,2]$ and $r' = \tilde{r} q'$ we obtain

$$\|V(y_G + y')\varphi(y_G)\|_{L^2_{y'} L^{q'}_{y_G}} \leq \|V\|_{L^{r'}} \|\varphi\|_{L^{p'}}.$$

□

The first result concerns the action of $a_G(V)$ and $a_G^*(V)$ on a fixed finite particles sector.

Proposition 3.3 *For any $p', q' \in [1,2]$ such that $q' \leq p'$, $2 \leq p \leq q \leq +\infty$, let $r' \in [1,2]$ be defined by $\frac{1}{r'} = \frac{1}{2} + \frac{1}{q'} - \frac{1}{p'}$ like in Lemma 3.1. For any $V \in L^{q'}(\mathbb{R}^d, dy; \mathbb{C}) \cap L^{r'}(\mathbb{R}^d, dy; \mathbb{C})$, the creation and annihilation operators satisfy* *the following estimates:*

$$\forall f_{G,0} \in L^2_z, \quad \|a^*_G(V) f_{G,0}\|_{L^2_z L^{q'}_{y_G}} \leq \|V\|_{L^{q'}} \|f_{G,0}\|_{L^2_z}, \tag{36}$$

$$\forall n > 0, \forall f_{G,n} \in L^2_{(Y'_n,z),\mathrm{sym}} L^{p'}_{y_G},$$

$$\|a^*_G(V)\, f_{G,n}\|_{L^2_{(Y'_{n+1},z)} L^{q'}_{y_G}} \leq \|V\|_{L^{r'}} \sqrt{n+1} \|f_{G,n}\|_{L^2_{(Y'_n,z)} L^{p'}_{y_G}}, \tag{37}$$

$$\forall f_{G,1} \in L^2_z L^q_{y_G}, \quad \|a_G(V) f_{G,1}\|_{L^2_z} \leq \|V\|_{L^{q'}} \|f_{G,1}\|_{L^2_z L^q_{y_G}}, \tag{38}$$

$$\forall n > 1, \forall f_{G,n} \in L^2_{(Y'_n,z),\mathrm{sym}} L^q_{y_G},$$

$$\|a_G(V)\, f_{G,n}\|_{L^2_{(Y'_{n-1},z)} L^p_{y_G}} \leq \|V\|_{L^{r'}} \sqrt{n} \|f_{G,n}\|_{L^2_{(Y'_n,z)} L^q_{y_G}}. \tag{39}$$

A notable case is when $q' = r'$ *and* $p' = p = 2$.

Proof The variable $z \in Z$ is actually a parameter which can be forgotten because our estimates are uniform w.r.t. $z \in Z$.

For (36) it suffices to notice $[a^*_G(V) f_{G,0}](y_G) = f_{G,0} \times V(y_G)$.

The estimate of $a^*_G(V) f_{G,n}$ for $n > 0$ relies on Lemma 3.1. We start from the expression (34)

$$(a_G(V)^* f_{G,n})(y_G, Y'_{n+1}) = \sqrt{n+1}\, S_{n+1}\, V(y_G + y'_{n+1})\, f_{G,n}(y_G - \frac{y'_{n+1}}{n}, Y_n + \frac{y'_{n+1}}{n})$$

with $Y'_{n+1} = (y'_1, \ldots, y'_n, y'_{n+1}) \in \mathcal{R}^{n+1}$, $Y_n = (y'_1, \ldots, y'_n) \in \mathbb{R}^{dn}$, $Y_n + \frac{y'_{n+1}}{n} \in \mathcal{R}^n$. The symmetrization S_{n+1} simply takes the average of $n+1$-terms which have all the same form as

$$\sqrt{n+1} V(y_G + y'_{n+1})\, f_{G,n}(y_G - \frac{y'_{n+1}}{n}, Y_n + \frac{y'_{n+1}}{n}),$$

after circular permutation of the variables y'_j which does not change the $L^2_{Y'_{n+1}} L^{q'}_{y_G}$-norm. We can therefore forget the symmetrization S_{n+1} for proving the upper bound (37). When $n > 1$ integrations must be performed with respect to the independent variables $(y'_2, \ldots, y'_n) \in \mathbb{R}^{d(n-1)}$. Remember that $(y'_2, \ldots, y'_n, y'_{n+1})$ are coordinates on $\mathcal{R}^{n+1}$ such that $y'_1 = -y'_2 \cdots - y'_n - y'_{n+1}$, $d\mu_{n+1}(Y'_{n+1}) = (n+1)^d dy'_2 \cdots dy'_{n+1}$ and that the quantity

$$\left\| V(y_G + y'_{n+1}) f_{G,n}(y_G - \frac{y'_{n+1}}{n}, Y_n + \frac{y'_{n+1}}{n}) \right\|_{L^2_{Y'_{n+1}} L^{q'}_{y_G}}$$

equals

$$(n+1)^{\frac{d}{2}} \| V(y_G + y'_{n+1})$$
$$f_{G,n}(y_G - \frac{y'_{n+1}}{n}, Y_n + \frac{y'_{n+1}}{n}) \|_{L^2(\mathbb{R}^d, dy'_{n+1}; L^2(\mathbb{R}^{d(n-1)}, dy'_2 \cdots dy'_n; L^{q'}_{y_G}))}.$$

When $y'_{n+1} \in \mathbb{R}^d$ is fixed, setting $y'_1 = -\sum_{j=2}^{n}(y'_j + \frac{y'_{n+1}}{n})$ and $Y'_n = Y_n + \frac{y'_{n+1}}{n}$, provides the coordinates $(y'_2 + \frac{y'_{n+1}}{n}, \ldots, y'_n + \frac{y'_{n+1}}{n})$ on $\mathcal{R}^n$ with $d\mu_n(Y'_n) = n^{d/2} dy'_2 \cdots dy'_n$, and then $\| V(y_G + y'_{n+1}) f_{G,n}(y_G - \frac{y'_{n+1}}{n}, Y_n + \frac{y'_{n+1}}{n}) \|_{L^2(\mathbb{R}^{d(n-1)}, dy'_2 \cdots dy'_n; L^{q'}_{y_G})}$ equals

$$n^{-d/2} \| V(y_G + y'_{n+1}) f_{G,n}(y_G - \frac{y'_{n+1}}{n}, Y'_n) \|_{L^2_{Y'_n} L^{q'}_{y_G}} =$$
$$n^{-d/2} \| V(\tilde{y}_G + \frac{n+1}{n} y'_{n+1}) f_{G,n}(\tilde{y}_G, Y'_n) \|_{L^2_{Y'_n} L^{q'}_{\tilde{y}_G}}.$$

We deduce

$$\begin{aligned}
&\left\| V(y_G + y'_{n+1}) f_{G,n}(y_G - \frac{y'_{n+1}}{n}, Y_n + \frac{y'_{n+1}}{n}) \right\|_{L^2_{Y'_{n+1}} L^{q'}_{y_G}} \\
&= \frac{(n+1)^{d/2}}{n^{d/2}} \left\| \| V(\tilde{y}_G + \frac{n+1}{n} y'_{n+1}) f_{G,n}(\tilde{y}_G, Y'_n) \|_{L^2_{Y'_n} L^{q'}_{\tilde{y}_G}} \right\|_{L^2(\mathbb{R}^d, dy'_{n+1})} \\
&= \left\| \| V(\tilde{y}_G + y') f_{G,n}(\tilde{y}_G, Y'_n) \|_{L^2_{Y'_n} L^{q'}_{\tilde{y}_G}} \right\|_{L^2(\mathbb{R}^d, dy')} \\
&= \left\| \| V(\tilde{y}_G + y') f_{G,n}(\tilde{y}_G, Y'_n) \|_{L^2(\mathbb{R}^d, dy'; L^{q'}_{\tilde{y}_G})} \right\|_{L^2_{Y'_n}},
\end{aligned}$$

after using the change of variable $y' = \frac{n+1}{n} y'_{n+1}$ in $\mathbb{R}^d$ for the third line and $L^2_{y'} L^2_{Y'_n} = L^2_{Y'_n} L^2_{y'}$ for the last one. We now use Lemma 3.1 with

$$\| V(\tilde{y}_G + y') f_{G,n}(\tilde{y}_G, Y'_n) \|_{L^2(\mathbb{R}^d, dy'; L^{q'}_{\tilde{y}_G})} \leq \| V \|_{L^{r'}} \| f_{G,n}(\tilde{y}_G, Y'_n) \|_{L^{p'}_{\tilde{y}_G}}$$

for almost all $Y_n' \in \mathcal{R}^n$ and $1 \leq q' \leq p' \leq 2$, $\frac{1}{r'} = \frac{1}{2} + \frac{1}{q'} - \frac{1}{p'}$. Integrating w.r.t. $Y_n' \in \mathcal{R}^n$ gives

$$\left\| V(y_G + y_{n+1}') f_{G,n}(y_G - \frac{y_{n+1}'}{n}, Y_n + \frac{y_{n+1}'}{n}) \right\|_{L^2_{Y_{n+1}'} L^{q'}_{y_G}} \leq \|V\|_{L^{r'}} \|f_{G,n}\|_{L^2_{Y_n'} L^{p'}_{y_G}}.$$

By multiplying by $\sqrt{n+1}$ and with the symmetrization S_{n+1}, we have proved (37).

The estimates with $a_G(V)$ follow by duality using

$$\|a_G(V) f_{G,n+1}\|_{L^2_{Y_{n+1}'} L^p_{y_G}} = \sup_{\|f_{G,n}\|_{L^2_{Y_n'} L^{p'}_{y_G}} = 1} |\langle a_G(V) f_{G,n+1}, f_{G,n} \rangle| .$$

Indeed, if $\|f_{G,n}\|_{L^2_{Y_n'} L^{p'}_{y_G}} = 1$, then

$$\begin{aligned} |\langle a_G(V) f_{G,n+1}, f_{G,n} \rangle| &= |\langle f_{G,n+1}, a_G^*(V) f_{G,n} \rangle| \\ &\leq \|f_{G,n+1}\|_{L^2_{Y_{n+1}'} L^q_{y_G}} \|a_G^*(V) f_{G,n}\|_{L^2_{Y_n'} L^{q'}_{y_G}} \\ &\leq \begin{cases} \|V\|_{L^{q'}} \|f_{G,1}\|_{L^q_{y_G}} & \text{when } n = 0, \\ \|V\|_{L^{r'}} \sqrt{n+1} \|f_{G,n+1}\|_{L^2_{Y_{n+1}'} L^q_{y_G}}, & \text{when } n > 0, \end{cases} \end{aligned}$$

which implies the bounds (38) and (39). □

Remark 3.1 Instead of Young's inequality one could use the more general Brascamp-Lieb inequality (see[4, 22]). This would not change the result (up to multiplicative constants). One may wonder whether it is possible to improve Lebesgue's exponent, in particular the integrability by reaching exponents $p < 2$ in (39) by strengthening the assumptions on V. Actually it is not. Take $V \in \mathcal{S}(\mathbb{R}^d)$ and $\varphi \in L^2(\mathbb{R}^d, dy; \mathbb{C})$, then $a(V)\varphi^{\otimes n} = \sqrt{n}\langle V, \varphi \rangle \varphi^{\otimes n-1}$ and $a_G(V) U_G^{-1}(\varphi^{\otimes n})$ cannot be put in $L^2_{z, Y_{n-1}'} L^p_{y_G}$ with $p < 2$ in general.

Proposition 3.4 *Take $\alpha, \alpha' \in \mathbb{R}$, $\alpha < \alpha'$ and for $1 \leq q' \leq p' \leq 2$, $2 \leq p \leq q \leq +\infty$, and let $r' \in [1, 2]$ be defined by $\frac{1}{r'} = \frac{1}{2} + \frac{1}{q'} - \frac{1}{p'}$. For any $V \in L^{r'}(\mathbb{R}^d) \cap L^{q'}(\mathbb{R}^d)$, the following estimates hold*

$$\forall f \in e^{-\alpha' N} L^2_{z, Y', \text{sym}} L^{p'}_{y_G},$$

$$\|e^{\alpha N} a_G^*(V) f\|_{L^2_{z,Y'} L^{q'}_{y_G}} \leq \frac{\max(\|V\|_{L^{r'}}, \|V\|_{L^{q'}}) e^{\alpha'}}{2\sqrt{\alpha' - \alpha}} \|e^{\alpha' N} f\|_{L^2_{z,Y'} L^{p'}_{y_G}}, \tag{40}$$

$$\forall f \in e^{-\alpha' N} L^2_{z,Y',\mathrm{sym}} L^q_{y_G}\,,$$

$$\|e^{\alpha N} a_G(V) f\|_{L^2_{z,Y'} L^p_{y_G}} \le \frac{\max(\|V\|_{L^{r'}}, \|V\|_{L^{q'}}) e^{-\alpha}}{2\sqrt{\alpha' - \alpha}} \|e^{\alpha' N} f\|_{L^2_{z,Y'} L^q_{y_G}}\,. \tag{41}$$

Again, a notable case is when $q' = r'$ and $p = p' = 2$.

Proof By writing

$$e^{\alpha N} a_G^*(V) e^{-\alpha' N} (\bigoplus_{n=0}^{\infty} f_{G,n}) = \bigoplus_{n=0}^{\infty} e^{\alpha(n+1) - \alpha' n} a_G^*(V) f_{G,n}\,,$$

and
$$e^{\alpha N} a_G(V) e^{-\alpha' N} (\bigoplus_{n=0}^{\infty} f_{G,n}) = \bigoplus_{n=1}^{\infty} e^{\alpha(n-1) - \alpha' n} a_G(V) f_{G,n}\,,$$

Proposition 3.3 tells us that it suffices to bound

$$\sup_{n \in \mathbb{N}} \sqrt{n+1} e^{-(\alpha' - \alpha)(n+1)} e^{\alpha'} \le \frac{e^{\alpha'}}{\sqrt{2e}\sqrt{\alpha' - \alpha}} \le \frac{e^{\alpha'}}{2\sqrt{\alpha' - \alpha}}\,,$$

and
$$\sup_{n \in \mathbb{N}} \sqrt{n} e^{-(\alpha' - \alpha) n} e^{-\alpha} \le \frac{e^{-\alpha}}{\sqrt{2e}\sqrt{\alpha' - \alpha}} \le \frac{e^{-\alpha}}{2\sqrt{\alpha' - \alpha}}\,.$$

□

4 Strichartz Estimates in the Center of Mass Variable

Here we review the celebrated results of Keel and Tao in [21] and adapt them to our framework. We shall use like those authors the short notations

- $a(z) \lesssim b(z)$ for the uniform inequality

$$\forall z \in Z\,, \quad a(z) \le C b(z)\,,$$

  where C is a constant which depends only on the following data: the dimension d or the free one particle evolution on $\mathbb{R}^d$;
- for $1 \le p, q \le +\infty$, various uses of the general notation $L^p_x L^q_y$ introduced in Definition 3.2 will be specified;
- except in specified cases, L^p_x is used for $2 \le p \le +\infty$ while $L^{p'}_x$ is used for $1 \le p' \le 2$.

4.1 Endpoint Strichartz Estimates

Keel and Tao's results about endpoint Strichartz estimates (see [21]) written with uniform inequalities, obviously induce a parameter dependent version which will be needed. They start with a time-dependent operator $U(t) : \mathfrak{h}_{\text{in}} \to L^2_x = L^2(X, \mathbf{dx}; \mathbb{C})$ where $t \in \mathbb{R}$ and $\mathfrak{h}_{\text{in}}$ is a (separable) Hilbert space of initial data. We rather consider a parameter dependent operator $U(t, z_1) : \mathfrak{h}_{\text{in}} \to L^2_x$ defined for $(t, z_1) \in \mathbb{R} \times Z_1$ such that

$$\|U(t, z_1)f\|_{L^2_x} \lesssim \|f\|_{\mathfrak{h}_{\text{in}}}, \tag{42}$$

$$\|U(t, z_1)U^*(s, z)g\|_{L^\infty_x} \lesssim \frac{\|g\|_{L^1_x}}{|t-s|^\sigma} \quad \text{for all } t \neq s, \tag{43}$$

while $U^*(t, z_1)$ may be defined only on a dense set of L^1_x.

On the measured space $(Z_1, \mathbf{dz_1})$ the map $(t, z_1) \mapsto U(t, z_1)f \in L^2_x$ is assumed measurable for all $f \in \mathfrak{h}_{\text{in}}$ and $U(t) : L^w(Z_1, \mathbf{dz_1}; \mathfrak{h}_{\text{in}}) \to L^w_{z_1} L^2_x$, where $L^w_{z_1} L^2_x = L^w(Z_1, \mathbf{dz_1}; L^2(X, \mathbf{dx}))$ here, is defined by pointwise multiplication $(U(t)f)(z_1) = U(t, z_1)f(z_1)$.

The set of sharp σ-admissible space-time exponents is given by

$$q, r \geq 2 \quad \frac{1}{q} + \frac{\sigma}{r} = \frac{\sigma}{2},$$

and the dual exponents are denoted by q', r', $\frac{1}{q} + \frac{1}{q'} = 1$, $\frac{1}{r} + \frac{1}{r'} = 1$ with $1 \leq q', r' \leq 2$, $\frac{1}{q'} + \frac{\sigma}{r'} = \frac{\sigma+2}{2}$.

We will consider cases where $\sigma > 1$ and the endpoint Strichartz estimates for $P = (2, \frac{2\sigma}{\sigma-1})$ holds true. The results for sharp σ-admissible pairs (q, r) and $(\tilde{q}, \tilde{r})$ are:

- the homogeneous estimate

$$\|U(t)f\|_{L^w_{z_1} L^q_t L^r_x} \lesssim \|f\|_{L^w(Z_1, \mathbf{dz_1}; \mathfrak{h}_{\text{in}})}; \tag{44}$$

- the inhomogeneous estimate

$$\|\int U(s)^* F(s)\, ds\|_{L^w(Z_1, dz_1; \mathfrak{h}_{\text{in}})} \lesssim \|F\|_{L^w_{z_1} L^{\tilde{q}'}_t L^{\tilde{r}'}_x}; \tag{45}$$

- the retarded estimate

$$\|\int_{s<t} U(t)U(s)^* F(s)\, ds\|_{L^w_{z_1} L^q_t L^r_x} \lesssim \|F\|_{L^w_{z_1} L^{\tilde{q}'}_t L^{\tilde{r}'}_x}, \tag{46}$$

where $s < t$ can be replaced by $s > t$.

Keel and Tao's results are written in [21] with $Z_1 = \{z_0\}$ and $\mathbf{dz_1} = \delta_{z_0}$, but the uniform inequalities with respect to $z_1 \in Z_1$ can be integrated afterwards for data in $L^w_{z_1}$.

By requiring $\sigma > 1$, the endpoint estimate allows to take $q = \tilde{q}' = 2$ with the endpoint exponents $r_\sigma = \frac{2\sigma}{\sigma-1}$ and $r'_\sigma = \frac{2\sigma}{\sigma+1}$. This is a very convenient framework for fixed point and bootstrap method in our linear setting.

Below are the typical inequalities which will be used. In our applications like in Sect. 3.3, the vaccuum sector plays a separate role and it is convenient to use the general Definition 3.2 for $L^w_z L^q_x$

$$\mathcal{N} = \{0, 1\}\,, \quad Z = Z_0 \sqcup Z_1$$
$$\text{and } \mathcal{X}_0 = \{0\}\,, \quad \mathcal{X}_1 = X\,, \quad \mathbf{dx_0} = \delta_0\,, \quad \mathbf{dx_1} = \mathbf{dx}\,.$$

In particular the spaces $L^2_z L^q_x$ for $1 \le q \le \infty$ equal

$$L^2_z L^q_x = L^2(Z_0, \mathbf{dz_0}) \oplus L^2(Z_1, \mathbf{dz_1}; L^2(X, \mathbf{dx})) = \underbrace{L^2_{z_0}}_{\text{vacuum}} \oplus L^2_{z_1} L^2_x\,. \tag{47}$$

At this level the action of the dynamics $U(t)U(s)^*$ is considered only on the $L^w_{z_1} L^q_x$ component.

Proposition 4.1 *Consider $L^2_z L^q_x = L^2_{z_0} \oplus L^2_{z_1} L^2_x$ like in (47) and according to Definition 3.2.*

Assume that there is a dense Banach space D in $L^2_{z_1,x} = L^2_{z_1} L^2_x$ such that $D \subset L^2_{z_1} L^{r_\sigma}_x$ and $U(t)U(s)^* u \in L^2_{z_1} L^{r_\sigma}_x$ is measurable with respect to $t, s \in \mathbb{R}$ for all $u \in D$ with the uniform estimate $\|U(t)U(s)^* u\|_{L^2_{z_1} L^{r_\sigma}_x} \lesssim \|u\|_D$ for almost all $t, s \in \mathbb{R}$.

Assume that the bounded operator $B^*_{t,s} : L^2_z L^2_x \to L^2_{z_1} L^{r'_\sigma}_x$ and its adjoint $B_{t,s} : L^2_{z_1} L^{r_\sigma}_x \to L^2_z L^2_x$ are strongly measurable with respect to $(t, s) \in [0, T] \times [0, T]$ with the assumption

$$\sup_{t \in [0,T]} \int_0^T \|B^*_{t,s}\|^2 \, ds < +\infty\,, \quad \|B^*_{t,s}\| = \|B^*_{t,s}\|_{L^2_{z_1} L^{r'_\sigma}_x \leftarrow L^2_z L^2_x}\,, \tag{48}$$

$$\textit{resp.} \quad \sup_{s \in [0,T]} \int_0^T \|B_{t,s}\|^2 dt < +\infty\,, \quad \|B_{t,s}\| = \|B_{t,s}\|_{L^2_z L^2_x \leftarrow L^2_{z_1} L^{r_\sigma}_x}\,. \tag{49}$$

The operator A^*_T (resp. A_T) defined by

$$[A^*_T f](t) = 1_{Z_1}(z) \int_0^T U(t)U(s)^* B^*_{t,s} f(s) \, ds\,, \tag{50}$$

$$\textit{resp.} \quad [A_T f](t) = \int_0^T B_{t,s} U(t)U(s)^* 1_{Z_1}(z) f(s) \, ds\,, \tag{51}$$

acts continuously on $L^\infty([0,T];L^2_zL^2_x)$ *(resp. extends as a continuous operator on* $L^1([0,T];L^2_zL^2_x)$*) with*

$$Ran\,A_T^* \subset L^\infty([0,T];L^2_{z_1,x})\,,\quad Ker(A_T)\supset L^1([0,T];L^2_{z_0})\,, \tag{52}$$

$$\|(A_T^*)^n\|_{\mathcal{L}(L^\infty([0,T];L^2_zL^2_x))} \lesssim \left(\sup_{t_{n+1}\in[0,T]}\int_{[0,T]^n}\|B^*_{t_{n+1},t_n}\|^2\ldots\|B^*_{t_2,t_1}\|^2\,dt_1\cdots dt_n\right)^{1/2}, \tag{53}$$

$$\text{resp.}\quad \|(A_T)^n\|_{\mathcal{L}(L^1([0,T];L^2_zL^2_x))} \lesssim \left(\sup_{t_0\in[0,T]}\int_{[0,T]^n}\|B_{t_n,t_{n-1}}\|^2\ldots\|B_{t_1,t_0}\|^2\,dt_1\cdots dt_n\right)^{1/2}, \tag{54}$$

for all non zero $n\in\mathbb{N}$.

When $B^\sharp_{t,s}=B^\sharp_{t,s}1_{s<t}$ *or* $B^\sharp_{t,s}=B^\sharp_{t,s}1_{s>t}$ *(*$B^\sharp=B^*$ *resp.* $B^\sharp=B$*), the domain of integration* $[0,T]^n$ *can be replaced by the corresponding n-dimensional simplex* $0<t_1<\ldots<t_n<T$ *or* $T>t_1\ldots>t_n>0$.

Remark 4.1 The dense subspace D is introduced in order to get a dense domain of $L^1([0,T];L^2_{z_1,x})$ where A_T is well defined by its integral formula. The extension to the whole space $L^1([0,T];L^2_{z_1,x})$ is proved by using the fact that $L^\infty([0,T];L^2_{z_1,x})$ is the dual of $L^1([0,T];L^2_{z_1,x})$ and it cannot be done in the other way.

Examples where the dense subset D is easy to construct are when $L^2(X,\mathbf{dx};\mathbb{C})=L^2(\mathbb{R}^d,dx;\mathbb{C})$ and $U(t)U(s)^*:H^\mu(\mathbb{R}^d;\mathbb{C})\to H^\mu(\mathbb{R}^d;\mathbb{C})$ are measurable and uniformly bounded w.r.t. $t,s\in\mathbb{R}$ for some $\mu>d/2$. In this simple case, the set D can be $L^2(Z_1,\mathbf{dz_1};H^\mu(\mathbb{R}^d;\mathbb{C}))$ with $\mu>\frac{d}{2}$.

Proof Let us start with A_T^*. When $f\in L^\infty([0,T],dt;L^2_zL^2_x)$ the function $1_{[0,T]}f$ belongs to $L^2_zL^2_{t,x}$ and, for almost all $t_0\in[0,T]$, the function $(z,s,x)\mapsto B^*_{t_0,s}1_{[0,T]}(s)f(s)$ belongs to $L^2_{z_1}L^2_sL^{r'_\sigma}_x$. The inhomogeneous endpoint Strichartz estimate implies for almost all $t_0\in[0,T]$

$$\begin{aligned}\|A_T^*f(t_0)\|^2_{L^2_zL^2_x} &\lesssim \int_0^T \|B^*_{t_0,s}f(s)\|^2_{L^2_{z_1}L^{r'_\sigma}_x}\,ds\\ &\lesssim \left(\int_0^T\|B^*_{t_0,s}\|^2\,ds\right)\|f\|^2_{L^\infty([0,T];L^2_zL^2_x)}\,.\end{aligned} \tag{55}$$

This proves firstly that A_T^* acts continuously on $L^\infty([0,T]; L_z^2L_x^2)$. The property $\mathrm{Ran}\, A_T^* \subset L^\infty([0,T]; L^2_{z_1,x})$ comes from the assumption $B_{t,s}^* : L_z^2L_x^2 \to L^2_{z_1}L_x^{r'_\sigma}$ and the redundant multiplication by $1_{Z_1}(z)$ in (50). Secondly iterating (55) with $(t_0, s) = (t_{n+1}, t_n)$ leads to (53).

Consider now $A_T f$ when $f = 1_{Z_1}(z) f \in L^1([0,T]; L^2_{z_1,x})$. For f in the dense subspace $L^1([0,T]; D)$ of $L^1([0,T]; L^2_{z_1,x})$, our assumptions ensure that $A_T f$ belongs to $L^\infty([0,T]; L_z^2L_x^2) \subset L^1([0,T]; L_z^2L_x^2)$ with

$$\|A_T f\|_{L^1([0,T];L_z^2L_x^2)} \lesssim C_T \|f\|_{L^1([0,T];D)}\,.$$

With

$$\begin{aligned}\int_0^T \langle v(t)\,,\, A_T f(t)\rangle\, dt &= \int_0^T \langle 1_{Z_1}(z) \int_0^T U(s)U^*(t)B_{t,s}^* v(t)\, dt\,,\, f(s)\rangle\, ds\\ &= \int_0^T \langle (\tilde{A}_T^* v)(s)\,,\, f(s)\rangle\, ds\,,\end{aligned}$$

where $B_{t,s}^*$ has simply been replaced by $B_{s,t}^*$ in

$$\tilde{A}_T^* v(t) = 1_{Z_1}(z) \int_0^T U(t)U(s)^* B_{s,t}^* v(s)\, ds\,,$$

we obtain

$$\begin{aligned}&\forall v \in L^\infty([0,T]; L_z^2L_x^2)\,,\\ &\quad |\langle v\,,\, A_T f\rangle| \lesssim \left(\int_0^T \|B_{s,t}\|^2\, ds\right)^{1/2} \|v\|_{L^\infty([0,T];L_z^2L_x^2)} \|f\|_{L^1([0,T];L^2_{z_1,x})}\,,\end{aligned}$$

while $L^\infty([0,T]; L_z^2L_x^2) = (L^1([0,T]; L_z^2L_x^2))'$.

This proves that A_T extends as a continuous operator from $L^1([0,T]; L^2_{z_1,x})$ to $L^1([0,T]; L_z^2L_x^2)$ and the formula contains the extension by 0 on $L^1([0,T]; L^2_{z_0})$, with $L^1([0,T]; L_z^2L_x^2) = L^1([0,T]; L^2_{z_0}) \oplus L^1([0,T]; L^2_{z_1,x})$. Its adjoint is $\tilde{A}_T^* : L^\infty([0,T]; L_z^2L_x^2) \to L^\infty([0,T]; L_z^2L_x^2)$. The estimate (53) for $\tilde{A}_T^*$ with $(\|B_{t,s}^*\|, t_k)$ replaced by $(\|B_{s,t}^*\| = \|B_{s,t}\|, t_{n+1-k})$ yields (54). □

Note that when $B_{t,s}^\sharp = B_{t,s}^\sharp 1_{t>s}$ or $B_{t,s}^\sharp = B_{t,s}^\sharp 1_{t<s}$ with $\|B_{t,s}^\sharp\| \leq \beta$, the upper bounds of (53) and (54) are below

$$\left(\frac{(\beta^2 T)^n}{n!}\right)^{1/2} \lesssim \left(\frac{e\beta^2 T}{n}\right)^{n/2}\,.$$

This gives a hint of times scales with respect to β, e.g. when $\beta^2 T \leq C$ here, where iterative methods lead to convergent series or the associated fixed point methods can be used. We will use some refined versions of the scaling rule $\beta^2 T \leq C$. Although the L^p_t spaces estimates are written with $p = +\infty$ and $p = 1$, this scaling really relies on the endpoint Strichartz estimate with $p = 2$.

We complete our general corollaries of endpoint Strichartz estimates with a result which combines the action of operators like $B_{t,s}$ and $B^*_{t,s}$ in Proposition 4.1.

Proposition 4.2 *Let $\mathcal{I}, \mathcal{J}$ be at most countable families of disjoint finite intervals, and set $U\mathcal{I} = \sqcup_{I\in\mathcal{I}} I$ and $U\mathcal{J} = \sqcup_{J\in\mathcal{J}} J$. For a given $\varphi_\infty \in L^\infty(U\mathcal{J}; L^2_z L^2_x)$ consider*

$$\varphi_{1,I}(t) = 1_I(t) \sum_{J\in\mathcal{J}} \int_0^t B_{1,IJ} U(t) U(s)^* B^*_{2,IJ}(s) \varphi_{\infty,J}(s)\, ds$$

with $\varphi_{\infty,J}(s) = \varphi_\infty(s) 1_J(s)$,

and $\|B_{1,IJ}\|_{L^2_z L^2_x \leftarrow L^2_{z_1} L^{r_\sigma}_x} \leq \beta_{1,IJ}$, $\sup_{s\in J} \|B^*_{2,IJ}(s)\|_{L^2_{z_1} L^{r'_\sigma}_x \leftarrow L^2_z L^2_x} \leq \beta_{2,IJ}$,

where $B_{1,IJ} : L^2_{z_1} L^{r_\sigma}_x \to L^2_z L^2_x$ does not depend on $(t,s) \in I \times J$ while $B^*_{2,IJ}(s) : L^2_z L^2_x \to L^2_{z_1} L^{r'_\sigma}_x$ does not depend on the time variable $t \in I$ and is strongly measurable with respect to $s \in J$. Then the function $\varphi_1 = \sum_{I\in\mathcal{I}} \varphi_{1,I}$ belongs to $L^1(U\mathcal{I}, dt; L^2_z L^2_x)$ with

$$\|\varphi_1\|_{L^1(U\mathcal{I},dt;L^2_z L^2_x)} \lesssim \left[\sum_{\substack{I\in\mathcal{I},\, J\in\mathcal{J} \\ \inf J < \sup I}} |I|^{1/2} \beta_{1,IJ} \beta_{2,IJ} |J|^{1/2} \right] \|\varphi_\infty\|_{L^\infty(U\mathcal{J},dt;L^2_z L^2_x)},$$

as soon as $\left[\sum_{I\in\mathcal{I},J\in\mathcal{J}} 1_{]0,+\infty[}(\sup I - \inf J) |I|^{1/2} \beta_{1,IJ} \beta_{2,IJ} |J|^{1/2}\right] < +\infty$.

Proof Every term of $\varphi_{1,I}$ can be written

$$\psi_{IJ}(t) = B_{1,IJ} \int_0^t U(t) U(s)^* \phi_{2,IJ}(s)\, ds$$

where $\phi_{2,IJ} = B^*_{2,IJ}(\cdot)\varphi_{\infty,J}(\cdot) \in L^2(\mathbb{R}; L^2_{z_1} L^{r'_\sigma}_x)$ satisfies $\phi_{2,IJ} = 0$ if $\inf J \geq \sup I$, and

$$\begin{aligned}\|\phi_{2,IJ}\|_{L^2(\mathbb{R},dt;L^2_{z_1} L^{r'_\sigma}_x)} &\leq |J|^{1/2} \beta_{2,IJ} \|\varphi_{\infty,J}\|_{L^\infty(J,dt;L^2_z L^2_x)} \\ &\leq |J|^{1/2} \beta_{2,IJ} \|\varphi_\infty\|_{L^\infty(U\mathcal{J},dt;L^2_z L^2_x)}.\end{aligned}$$

The retarded endpoint Strichartz estimate with $\|B_{1,IJ}\|_{L^2_zL^2_x\leftarrow L^2_zL^{r_\sigma}_x}\leq\beta_{1,IJ}$ implies

$$\|\psi_{IJ}\|_{L^2(I,dt;L^2_zL^2_x)}\lesssim 1_{]0,+\infty[}(\sup I-\inf J)\beta_{1,IJ}\beta_{2,IJ}|J|^{1/2}\|\varphi_\infty\|_{L^\infty(UJ,dt;L^2_zL^2_x)}$$

and therefore

$$\|\psi_{IJ}\|_{L^1(I,dt;L^2_zL^2_x)}\lesssim\left[1_{]0,+\infty[}(\sup I-\inf J)|I|^{1/2}\beta_{1,IJ}\beta_{2,IJ}|J|^{1/2}\right]\|\varphi_\infty\|_{L^\infty(UJ,dt;L^2_zL^2_x)}\,.$$

The finiteness of $\sum_{I\in\mathcal{I},J\in\mathcal{J}}\left[1_{]0,+\infty[}(\sup I-\inf J)|I|^{1/2}\beta_{1,IJ}\beta_{2,IJ}|J|^{1/2}\right]$ ensures that $\varphi_{1,I}=\sum_{J\in\mathcal{J}}\psi_{IJ}$ belongs to $L^1(I,dt;L^2_zL^2_x)$ and finally

$$\begin{aligned}\|\varphi_1\|_{L^1(UI,dt;L^2_zL^2_x)}&=\sum_{I\in\mathcal{I}}\|\varphi_{1,I}\|_{L^1(I,dt;L^2_zL^2_x)}\\&\lesssim\sum_{I\in\mathcal{I},J\in\mathcal{J}}\left[1_{]0,+\infty[}(\sup I-\inf J)|I|^{1/2}\beta_{1,IJ}\beta_{2,IJ}|J|^{1/2}\right]\\&\quad\|\varphi_\infty\|_{L^\infty(UJ,dt;L^2_zL^2_x)}\,.\end{aligned}$$

□

4.2 Fixed Point in Weighted Spaces

In this section, we apply the general framework of Strichartz estimates for evolution equations in the spaces

$$\begin{aligned}F^2&=L^2(Z',\mathbf{dz'};\Gamma(L^2(\mathbb{R}^d,dy;\mathbb{C})))\\&=L^2(Z',\mathbf{dz'};\mathbb{C})\oplus L^2_{\mathrm{sym}}(\mathcal{R}\times Z',d\mu\otimes\mathbf{dz'};L^2(\mathbb{R}^d,dy_G;\mathbb{C}))\,.\end{aligned}$$

The measured space of parameters $(Z',\mathbf{dz'})$ will be specified later and by following the notations of Definition 3.2 and (47) for the application of Strichartz estimates, we write

$$\begin{aligned}&Z_0=Z'\,,\quad Z_1=\mathcal{R}\times Z'\,,\quad \mathbf{dz_0}=\mathbf{dz'}\,,\quad \mathbf{dz_1}=\mu\otimes\mathbf{dz'}\,,\\&X_0=\{0\}\,,\quad X_1=\mathbb{R}^d\,,\quad \mathbf{dx_0}=\delta_0\,,\quad \mathbf{dx}=dy_G\,,\\&F_2=L^2_{z,\mathrm{sym}}L^2_{y_G}=L^2_{z_0}\oplus L^2_{z_1,\mathrm{sym}}L^2_{y_G}=L^2_{z_0}\oplus L^2_{(Y',z'),\mathrm{sym}}L^2_{y_G}\,,\end{aligned}\tag{56}$$

where the second variable $x \in \mathcal{X} = X_0 \sqcup X_1$ has been replaced by y_G in order to recall its link with the center of mass on the non vacuum sector.

We will use the $L^p_{y_G}$,$1 \leq p \leq +\infty$, version

$$L^2_{z,\mathrm{sym}} L^p_{y_G} = L^2_{z_0} \oplus L^2_{(Y',z'),\mathrm{sym}} L^p_{y_G} \qquad \text{with } z_1 = (Y', z')\,.$$

In all the above identities the subscript $_{\mathrm{sym}}$ refers to the symmetry for the relative variable $Y' \in \mathcal{R}$. Because the symmetry is preserved by all our defined operators, this subscript will be forgotten when we write estimates.

Only the useful conditions on the "free dynamics" $U(t)$, or more precisely $U(t)U(s)^* : F^2 \to F^2$ will be specified. Those will be checked for our model later in Sect. 5. The free dynamics or more precisely $U(t)U(s)^* : F^2 \to F^2$ is assumed to preserve the number of particles

$$[U(t)U(s)^*, N] = 0$$

with the following decomposition:

$$U(t)U(s)^* = (K_0(t,z')\overline{K_0(s,z')}\times_{z'}) \oplus (U_1(t,Y',z')U_1^*(s,Y',z')\times_{(Y',z')}) \tag{57}$$

$$\text{in} \quad F^2 = \underbrace{L^2(Z', \mathbf{dz'}; \mathbb{C})}_{=L^2_{z_0} \quad \text{(vacuum)}} \oplus \underbrace{L^2_{\mathrm{sym}}(\mathcal{R} \times Z', d\mu \otimes \mathbf{dz'}; L^2(\mathbb{R}^d, dy_G; \mathbb{C}))}_{=L^2_{(Y',z'),\mathrm{sym}} L^2_{y_G}}\,, \tag{58}$$

where $\times_{z'}$ or $\times_{(Y',z')}$ stands for the pointwise multiplication. So the operator $U_1(t,Y',z')U_1^*(s,Y',z')$ is a one particle operator acting in the y_G-variable, parametrized by $z_1 = (Y',z')$ and we add the following conditions which make the results of Sect. 4.1 relevant:

- The measured space $(X_1, \mathbf{dx_1})$ is nothing but $(\mathbb{R}^d, dy_G)$ in the center of mass variable and the $z_1 = (Y',z')$-dependent one particle operators $U_1(t,z_1) : \mathfrak{h}_{\mathrm{in}} \to L^2(\mathbb{R}^d, dy_G; \mathbb{C})$ and its adjoint are assumed to satisfy the estimate (42)(43) with $\sigma > 1$. Remember $r'_\sigma = \frac{2\sigma}{\sigma+1}$ and $r_\sigma = \frac{2\sigma}{\sigma-1}$.
- The additional assumption of Proposition 4.1 concerned with the dense subset D is also assumed for $U_1(t,z_1)$.
- The vacuum component K_0 belongs to $L^\infty(\mathbb{R} \times Z', dt \otimes \mathbf{dz'}; \mathbb{C})$.

The interaction terms will be

$$B^*_{t,s} = c_1(t,s)e^{\alpha(t,s)N}\sqrt{h}a^*_G(V_1)e^{-\alpha'(t,s)N}$$

$$\text{and} \quad B_{t,s} = c_2(t,s)\sqrt{h}e^{\alpha(t,s)N}a_G(V_2)e^{-\alpha'(t,s)N}$$

with $V_1, V_2 \in L^{r'_\sigma}(\mathbb{R}^d, dy; \mathbb{C})$ (complex valued V are allowed here) and where c_1, c_2, α and α' are real measurable functions of $(t,s) \in [0,T]^2$ with $\alpha - \alpha' < 0$.

Those will be specified further and we shall check the estimates (48)(49). Because $Z_0 = Z'$ corresponds to the vacuum sector, $N = 0$, on which $a_G(V)$ vanishes while the range of $a_G(V)^*$ lies in the non vacuum sector $N \geq 1$, the range $B_{t,s}^*$ lies naturally in $L^2_{z_1} L^{r'_\sigma}_{yG}$, $z_1 = (Y', z')$, once the proper estimates are checked while it adjoints $B_{t,s}$ sends $L^2_{z_1} L^{r_\sigma}_{yG}$ into $L^2_z L^2_{yG}$ and is naturally extended by 0 on the vacuum sector $L^2_{z_0}$.

We will consider the following system

$$u^h_\infty(t) = -i \int_0^t U(t)U^*(s)\big(\sqrt{h}a^*_G(V_1)u^h_\infty(s) + \sqrt{h}u^h_2(s) + u^h_1(s)\big)\,ds + f^h_\infty(t)\,, \tag{59}$$

$$u^h_2(t) = -i \int_0^t a_G(V_2)U(t)U(s)^*\sqrt{h}u^h_2(s)\,ds \qquad + f^h_2(t)\,, \tag{60}$$

$$u^h_1(t) = -i \int_0^t a_G(V_2)U(t)U(s)^*\big(ha^*_G(V_1)u^h_\infty(s) + \sqrt{h}u^h_1(s)\big)\,ds \qquad + f^h_1(t)\,. \tag{61}$$

written shortly as

$$\forall q \in \{\infty, 2, 1\}, \quad u^h_q = \sum_{p \in \{\infty,2,1\}} L_{qp}(u^h_p) + f^h_q \tag{62}$$

or

$$\begin{pmatrix} u^h_\infty \\ u^h_2 \\ u^h_1 \end{pmatrix} = L \begin{pmatrix} u^h_\infty \\ u^h_2 \\ u^h_1 \end{pmatrix} + \begin{pmatrix} f^h_\infty \\ f^h_2 \\ f^h_1 \end{pmatrix}, \quad L = \begin{pmatrix} L_{\infty\infty} & L_{\infty 2} & L_{\infty 1} \\ 0 & L_{22} & 0 \\ L_{1\infty} & 0 & L_{11} \end{pmatrix}. \tag{63}$$

This system will be studied in spaces with the number weight $e^{\alpha N}$ and we will use the following functional spaces.

Definition 4.3 For $T > 0$, $h \in]0, h_0[$, I^h_T denotes the interval $I^h_T =]-T/h, T/h[$. Fix $\alpha_0, \alpha_1 \in \mathbb{R}$, $\alpha_0 < \alpha_1$ and set $M_{\alpha 01} = \frac{\max(e^{\alpha_1}, e^{-\alpha_0})}{2} \geq 1/2$. Assume $V_1, V_2 \in L^{r'_\sigma}(\mathbb{R}^d, dy; \mathbb{C})$ with $\max(\|V_1\|_{L^{r'_\sigma}}, \|V_2\|_{L^{r'_\sigma}}) < C_V$. For a parameter $\gamma > 0$ and $\alpha \in [\alpha_0, \alpha_1[$ set

$$T_\alpha = \gamma(\alpha_1 - \alpha)\,.$$

The space $\mathcal{E}^h_{\alpha_0,\alpha_1,\gamma}$ is the set of $(e^{-\alpha_0 N}L^2_{z,\mathrm{sym}}L^2_{yG})^3$-valued measurable functions

$$I^h_{T_{\alpha_0}} \ni t \mapsto \begin{pmatrix} u_\infty(t) \\ u_2(t) \\ u_1(t) \end{pmatrix}$$

such that for all α in $[\alpha_0, \alpha_1[$,

$$\begin{aligned} |t|^{-1/2}u_\infty &\in L^\infty(I^h_{T_\alpha}, dt; e^{-\alpha N}L^2_z L^2_{yG})\,, \\ u_2 &\in L^2_{\mathrm{loc}}(I^h_{T_\alpha}, dt; e^{-\alpha N}L^2_z L^2_{yG})\,, \\ |t|^{-1/2}u_1 &\in L^1_{\mathrm{loc}}(I^h_{T_\alpha}, dt; e^{-\alpha N}L^2_z L^2_{yG})\,. \end{aligned}$$

and $M(u_\infty, u_2, u_1) < +\infty$ with

$$M(u_\infty, u_2, u_1) = M_\infty(u_\infty) + M_2(u_2) + M_1(u_1)\,, \tag{64}$$

$$M_\infty(u_\infty) = \sup_{\alpha_0 \le \alpha < \alpha_1} \left\| \left(\frac{T_\alpha - |ht|}{|ht|}\right)^{1/2} e^{\alpha N} u_\infty \right\|_{L^\infty(I^h_{T_\alpha}; L^2_z L^2_{yG})}\,, \tag{65}$$

$$M_2(u_2) = \frac{1}{M_{\alpha 01} C_V \gamma^{1/2}} \sup_{\substack{\alpha_0 \le \alpha < \alpha_1 \\ \tau \in [0, T_\alpha[}} \sqrt{T_\alpha - \tau} \left\| e^{\alpha N} u_2 \right\|_{L^2(I^h_\tau; L^2_z L^2_{yG})}\,, \tag{66}$$

$$M_1(u_1) = \frac{1}{M_{\alpha 01} C_V \gamma^{1/2}} \sup_{\substack{\alpha_0 \le \alpha < \alpha_1 \\ \tau \in [0, T_\alpha[}} \sqrt{T_\alpha - \tau} \left\| \frac{e^{\alpha N} u_1}{\sqrt{|ht|}} \right\|_{L^1(I^h_\tau; L^2_z L^2_{yG})}\,. \tag{67}$$

Endowed with the norm $M(u_\infty, u_2, u_1)$, $\mathcal{E}^h_{\alpha_0,\alpha_1,\gamma}$ is a Banach space for all $h \in\ ]0, h_0[$. The α-dependent time domain $I^h_{T_\alpha}$ where weighted L^∞_t, L^2_t and L^1_t norms are evaluated is illustrated in Fig. 1.

The constants $C_V > 0$ and $M_{\alpha 01} = \max(e^{\alpha_1}, e^{-\alpha_0})/2 \ge 1/2$ were chosen so that Proposition 3.4 applied with $q' = r'_\sigma$ and $p' = 2$, gives

$$\begin{aligned} \|e^{\alpha N} a^*_G(V) e^{-\alpha' N}\varphi\|_{L^2_{z_1} L^{r'_\sigma}_{yG}} &\le \frac{C_V e^{\alpha'}}{2\sqrt{\alpha' - \alpha}} \|\varphi\|_{L^2_z L^2_{yG}} \le \frac{M_{\alpha 01} C_V}{\sqrt{\alpha' - \alpha}} \|\varphi\|_{L^2_z L^2_{yG}}\,, \\ \|e^{\alpha N} a_G(V) e^{-\alpha' N}\varphi\|_{L^2_z L^2_{yG}} &\le \frac{C_V e^{-\alpha}}{2\sqrt{\alpha' - \alpha}} \|\varphi\|_{L^2_{z_1} L^{r_\sigma}_{yG}} \le \frac{M_{\alpha 01} C_V}{\sqrt{\alpha' - \alpha}} \|\varphi\|_{L^2_z L^{r_\sigma}_{yG}}\,, \end{aligned}$$

for all $\alpha, \alpha' \in [\alpha_0, \alpha_1[$, $\alpha < \alpha'$.

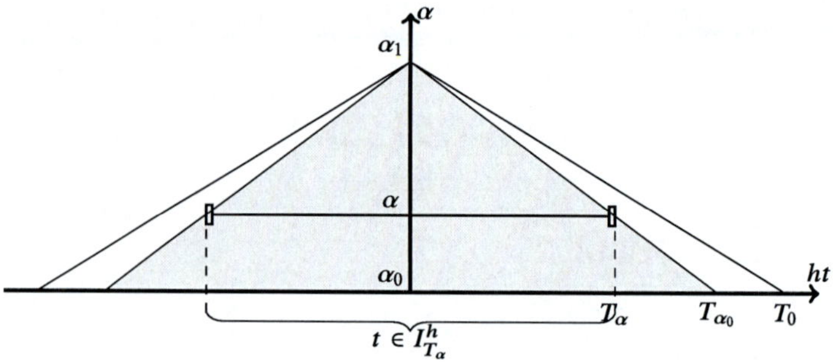

Fig. 1 The time interval $I^h_{T_\alpha} = \left]-\frac{\gamma(\alpha_1-\alpha)}{h}, \frac{\gamma(\alpha_1-\alpha)}{h}\right[$ according to α

Finally the normalization of (66) and (67) was chosen in order to make the contraction statement simple.

Proposition 4.3 *Assume that the free dynamics $U_1(t, z_1) : \mathfrak{h}_{in} \to L^2(\mathbb{R}^d, dy_G; \mathbb{C})$ satisfies (42)(43) (uniformly w.r.t. $z \in Z$) with $\sigma > 1$ and the additional existence of the dense subset D assumed in Proposition 4.1.*

Let $h_0 > 0$, $\alpha_0, \alpha_1 \in \mathbb{R}$, $\alpha_0 < \alpha_1$ and $V_1, V_2 \in L^{r'_\sigma}(\mathbb{R}^d, dy; \mathbb{C})$ be fixed. The positive constants $M_{\alpha 01}, C_V$, the space $\mathcal{E}^h_{\alpha_0,\alpha_1,\gamma}$ and its norm M are the ones of Definition 4.3. By choosing the parameter $\gamma > 0$ small enough the linear operator L given by (63) is a contraction of the Banach space $(\mathcal{E}^h_{\alpha_0,\alpha_1,\gamma}, M)$ for all $h \in]0, h_0[$ and the system (63), explicitely written (59)(60)(61), admits a unique solution for any $(f^h_\infty, f^h_2, f^h_1) \in \mathcal{E}^h_{\alpha_0,\alpha_1,\gamma}$.

More precisely there exists a constant $C_{d,U}$ determined by the dimension d and the free dynamics U, given by the pair K_0 and U_1, such that

$$\forall h \in]0, h_0[\,, \quad \|L\|_{\mathcal{L}(\mathcal{E}^h_{\alpha_0,\alpha_1,\gamma})} \leq C_{d,U} M_{\alpha 01} C_V \gamma^{1/2}\,.$$

Taking e.g. $\gamma = \frac{1}{2C^2_{d,U} M^2_{\alpha 01} C^2_V}$ ensures $\|L\|_{\mathcal{L}(\mathcal{E}^h_{\alpha_0,\alpha_1,\gamma})} \leq \frac{1}{2}$ so that the solution to (63) satisfies

$$M(u^h_\infty, u^h_2, u^h_1) \leq 2M(f^h_\infty, f^h_2, f^h_1)\,.$$

Proof The non-vanishing entries of $L\begin{pmatrix} u_\infty \\ u_2 \\ u_1 \end{pmatrix}$, namely

$$L_{\infty\infty}(u_\infty)\,, \quad L_{\infty 2}(u_2)\,, \quad L_{\infty 1}(u_1)\,, \quad L_{22}(u_2)\,, \quad L_{11}(u_1) \quad \text{and} \quad L_{1\infty}(u_\infty)$$

will be considered separately in this order of increasing difficulty. Additionally the symmetry $t \mapsto -t$ allows us to restrict the analysis to $t \geq 0$, that is $t \in [0, \frac{T_\alpha}{h}[$ for $\alpha \in [\alpha_0, \alpha_1[$. Accordingly I^h_T is, in this proof, the restricted interval $[0, \frac{T}{h}[$.

We use like in Sect. 4 the symbol $\lesssim$ for inequalities with constants which depend only on the dimension d and the free dynamics U.

$\mathbf{L}_{\infty\infty}(\mathbf{u}_\infty)$ For this term and up to the square root and the parameter $h\in]0,h_0[$, we follow exactly the method of [23] for Cauchy-Kowalevski theorem. Write for $t\in]0,T_\alpha/h[$, $ht\in]0,T_\alpha[$, $\alpha<\alpha_1-\frac{ht}{\gamma}$, and

$$\left(\frac{T_\alpha-ht}{ht}\right)^{1/2} e^{\alpha N}L_{\infty\infty}(u_\infty)(t)$$
$$= -i\int_0^{T_\alpha/h} U(t)U(s)^* B^*_{t,s}\left(\frac{T_{\alpha_s}-hs}{hs}\right)^{1/2} e^{\alpha_s N}u_\infty(s)\,ds$$

with

$$B^*_{t,s}=1_{s<t}\left(\frac{T_\alpha-ht}{ht}\right)^{1/2} e^{\alpha N}\sqrt{h}a^*_G(V)e^{-\alpha_s N}\left(\frac{hs}{T_{\alpha_s}-hs}\right)^{1/2}, \tag{68}$$

and $\alpha<\alpha_s<\alpha_1-\frac{hs}{\gamma}$. Hence $hs<T_{\alpha_s}$ and

$$\left(\frac{T_{\alpha_s}-hs}{hs}\right)^{1/2} \|e^{\alpha_s N}u_\infty(s)\|_{L^2_zL^2{yG}}\leq M_\infty(u_\infty) \tag{69}$$

while $\alpha<\alpha_s$ implies that $\|B^*_{t,s}\|=\|B^*_{t,s}\|_{L^2_zL^{r'_\sigma}_{yG}\leftarrow L^2_zL^2_{yG}}$ satisfies

$$\|B^*_{t,s}\|^2\leq h1_{s<t}\frac{M^2_{\alpha 01}C^2_V}{(\alpha_s-\alpha)}\frac{(T_\alpha-ht)(hs)}{ht(T_{\alpha_s}-hs)}=h1_{s'<t'}\frac{M^2_{\alpha 01}C^2_V}{(\alpha_{s'/h}-\alpha)}\frac{(T_\alpha-t')s'}{t'(T_{\alpha_{s'/h}}-s')},$$

by setting $s'=hs$, $t'=ht$. By choosing

$$\alpha_s=\frac{\alpha_1+\alpha-hs/\gamma}{2}=\frac{\alpha_1+\alpha-s'/\gamma}{2},$$

we obtain

$$\gamma(\alpha_s-\alpha)=\frac{\gamma(\alpha_1-\alpha)-s'}{2}=\frac{T_\alpha-s'}{2},$$
$$T_{\alpha_{s'/h}}=\gamma(\alpha_1-\alpha_{s'/h})=\frac{\gamma(\alpha_1-\alpha)+s'}{2},\quad T_{\alpha_{s'/h}}-s'=\frac{T_\alpha-s'}{2},$$

and

$$\frac{(T_\alpha-t')s'}{(\alpha_{s'/h}-\alpha)t'(T_{\alpha_{s'/h}}-s')}=4\gamma\frac{(T_\alpha-t)s'}{t'(T_\alpha-s')^2}.$$

This yields

$$\int_0^{T_\alpha/h} \|B^*_{t,s}\|^2\, ds \le 4\gamma M^2_{\alpha 01} C_V^2 \frac{T_\alpha - t'}{t'} \int_0^{t'} \frac{s'}{(T_\alpha - s')^2}\, ds' \le 4\gamma M^2_{\alpha 01} C_V^2\,. \tag{70}$$

The inequalities (69) and (70) combined with the inequality (53) with $n = 1$ of Proposition 4.1 imply

$$\left\| \left(\frac{T_\alpha - ht}{ht}\right)^{1/2} e^{\alpha N} L_{\infty\infty}(u_\infty) \right\|_{L^\infty([0,T_\alpha/h];L^2_z L^2_{y_G})} \lesssim 2\gamma^{1/2} M_{\alpha 01} C_V M_\infty(u_\infty)\,. \tag{71}$$

$\mathbf{L_{\infty 2}(u_2)}$ The Cauchy-Schwarz inequality applied to

$$\sqrt{\frac{T_\alpha - ht}{ht}} e^{\alpha N} L_{\infty 2}(u_2)(t) = -i\sqrt{T_\alpha - ht}\frac{1}{\sqrt{t}} \int_0^t U(t)U(s)^* e^{\alpha N} u_2(s)\, ds\,,$$

imply

$$\begin{aligned}
\left\| \sqrt{\frac{T_\alpha - ht}{ht}} e^{\alpha N} L_{\infty 2}(u_2)(t) \right\|_{L^2_z L^2_{y_G}} &\le \sqrt{T_\alpha - ht}\frac{1}{\sqrt{t}} \|e^{\alpha N} u_2(s)\|_{L^2_z L^1([0,t];L^2_{y_G})} \\
&\le \sqrt{T_\alpha - ht}\|e^{\alpha N} u_2(s)\|_{L^2([0,t];L^2_z L^2_{y_G})} \\
&\le \sup_{\tau \in]0,T_\alpha[} \sqrt{T_\alpha - \tau}\|e^{\alpha N} u_2(s)\|_{L^2([0,\tau/h];L^2_z L^2_{y_G})}\,.
\end{aligned}$$

Taking the supremum over $\alpha \in [\alpha_0, \alpha_1[$ yields

$$M_\infty(L_{\infty 2}(u_2)) \lesssim M_{\alpha 01} C_V \gamma^{1/2} M_2(u_2). \tag{72}$$

$\mathbf{L_{\infty 1}(u_1)}$ The expression

$$\sqrt{\frac{T_\alpha - ht}{ht}} e^{\alpha N} L_{\infty 1}(u_1)(t) = -i\sqrt{T_\alpha - ht} \int_0^t \frac{\sqrt{hs}}{\sqrt{ht}} \Big[U(t)U(s)^* e^{\alpha N} \frac{1}{\sqrt{hs}} u_1(s) \Big]\, ds\,,$$

gives

$$\begin{aligned}\left\|\sqrt{\frac{T_\alpha - ht}{ht}}\, e^{\alpha N} L_{\infty 1}(u_1)(t)\right\|_{L^2_z L^2_{y_G}} &\le \sqrt{T_\alpha - ht}\,\|e^{\alpha N}\frac{u_1(s)}{\sqrt{hs}}\|_{L^1([0,t];L^2_z L^2_{y_G})}\\ &\le \sup_{\tau\in]0,T_\alpha[} \sqrt{T_\alpha - \tau}\,\|\frac{u_1(s)}{\sqrt{hs}}\|_{L^1([0,\frac{\tau}{h}];L^2_z L^2_{y_G})}\\ &\le M_{\alpha 01} C_V \gamma^{1/2} M_1(u_1)\end{aligned}$$

and

$$\left\|\sqrt{\frac{T_\alpha - ht}{ht}}\, e^{\alpha N} L_{\infty 1}(u_1)\right\|_{L^\infty([0,\frac{T_\alpha}{h}];L^2_z L^2_{y_G})} \le M_{\alpha 01} C_V \gamma^{1/2} M_1(u_1)\,. \tag{73}$$

The entries $L_{22}(u_2)$, $L_{11}(u_1)$ and finally $L_{1\infty}(u_\infty)$ require some additional techniques. The proof, done in several steps for each of them, relies on a dyadic partition of the interval $[0, T_\alpha[$ around T_α . In the two cases of $L_{22}(u_2)$ and $L_{11}(u_1)$, the norms $M_2(\varphi)$ and $M_1(\varphi)$ are transformed into equivalent norms corresponding to this dyadic partition, the proof being given in Lemma 4.1 below. Finally the entry $L_{1\infty}(u_\infty)$ is treated via dyadic partitions around T_α and 0 and happens to be a direct application of Proposition 4.2.

Splitting the Interval [0,T[Fix $\alpha \in [\alpha_0, \alpha_1[$ and therefore $T = T_\alpha$. The intervals J^n_T are defined for $n \in \mathbb{N}$ by

$$\begin{aligned} &J^n_T = T + 2^{-n}[-T, -T/2[= [(1-2^{-n})T, (1-2^{-n-1})T[\,,\\ &J^{\le n_0}_T = \bigcup_{n\le n_0} J^n_T \text{ for } n_0 \in \mathbb{N}, \end{aligned}$$

so that $[0, T] = \cup_{n\in\mathbb{N}} J^n_T = J^{\le n_0}_T \cup (\cup_{n>n_0} J^n_T)$, see Figure 2.

With the exponents

$$\alpha'_0 = \frac{\alpha_1 + 6\alpha}{7} \quad \text{and} \quad \alpha'_n = \frac{\alpha_1 + (2^{n+2}-1)\alpha}{2^{n+2}} \text{ for } n \ge 1$$

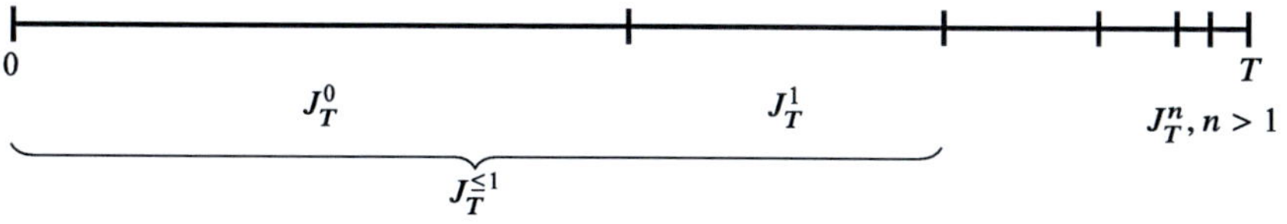

Fig. 2 The time intervals J^n_T , $n \in \mathbb{N}$, with length $\frac{T}{2^{n+1}}$

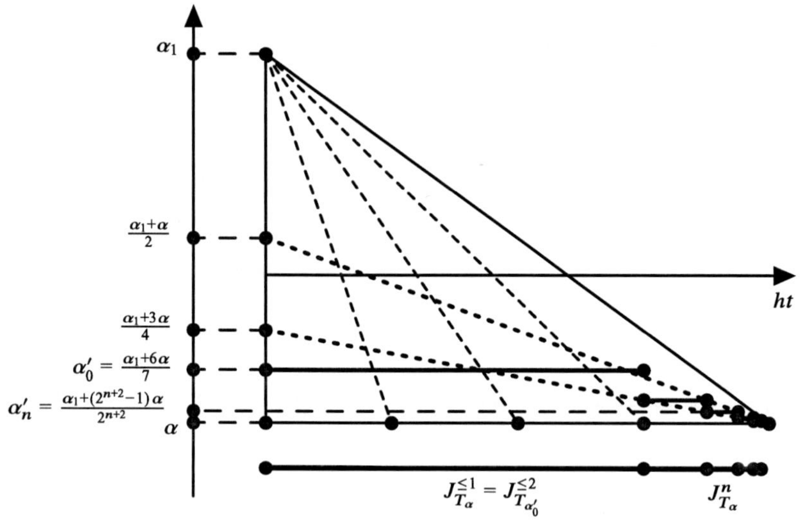

Fig. 3 The exponent α'_0 is determined by $\frac{7}{8}T_{\alpha'_0} = \frac{3}{4}T_\alpha$ while for $n > 1$, α'_n is determined by $T_{\alpha'_n} = (1 - \frac{2^{-n-1}}{2})T_\alpha = (1 - 2^{-n-2})T_\alpha$.

we note that

$$J^{\leq 2}_{T_{\alpha'_0}} = \frac{7}{8}T_{\alpha'_0} = \frac{7}{8}\frac{6}{7}T_\alpha = \frac{3}{4}T_\alpha = J^{\leq 1}_{T_\alpha}\,,$$

$$\text{and for } n \geq 1 \quad T_{\alpha'_n} = T_\alpha - \frac{1}{2}\frac{T_\alpha}{2^{n+1}} = (1 - 2^{-n-2})T_\alpha\,.$$

By taking $\delta_n = \frac{T_\alpha}{2^{n+2}}$ and $2\delta_n = \frac{T_\alpha}{2^{n+1}}$ for $n > 1$, we obtain in particular

$$J^n_{T_\alpha} = [T_\alpha - 4\delta_n, T_\alpha - 2\delta_n[= [T_{\alpha'_n} - 3\delta_n, T_{\alpha'_n} - \delta_n[\quad \text{with } \delta_n \leq \frac{T_{\alpha'_n}}{12} \ (n > 1)$$

as summarized in Fig. 3.

The equivalence of norms

$$\kappa_2^{-1} N_{2,1}(\varphi) \leq N_{2,i}(\varphi) \leq \kappa_2 N_{2,1}(\varphi)\,, \quad 2 \leq i \leq 4\,, \tag{74}$$

for some universal constant $\kappa_2 > 1$ is proved in Lemma 4.1 for

$$N_{2,1}(\varphi) = \sup_{\tau \in [0,T[} \sqrt{T-\tau}\, \|\varphi\|_{L^2([0,\frac{\tau}{h}]; L^2_z L^2_{y_G})}\,, \tag{75}$$

$$N_{2,2}(\varphi) = \sqrt{T}\, \|\varphi\|_{L^2(J^{\leq 1}_{T/h}; L^2_z L^2_{y_G})} + \sup_{\delta \in]0,T/8]} \sqrt{\delta}\, \|\varphi\|_{L^2(h^{-1}[T-2\delta, T-\delta]; L^2_z L^2_{y_G})}\,, \tag{76}$$

$$N_{2,3}(\varphi) = \sqrt{T} \sup_{n\in\mathbb{N}} 2^{-n/2} \|\varphi\|_{L^2(J^n_{T/h};L^2_z L^2_{y_G})}, \tag{77}$$

$$N_{2,4}(\varphi) = \sqrt{T} \|\varphi\|_{L^2(J^{\leq 2}_{T/h};L^2_z L^2_{y_G})} + \sup_{\delta\in]0,T/12]} \sqrt{\delta} \|\varphi\|_{L^2(h^{-1}[T-3\delta,T-\delta];L^2_z L^2_{y_G})}. \tag{78}$$

$\mathbf{L_{22}(u_2)}$ For $\alpha \in [\alpha_0, \alpha_1[$, we seek an upper bound of $N_{2,1}(\varphi)$ (with $T = T_\alpha$) for

$$\varphi(t) = e^{\alpha N} L_{22}(u_2)(t) = -i \int_0^t e^{\alpha N}\sqrt{h} a_G(V_2) U(t) U(s)^* u_2(s)\, ds\,.$$

By the equivalence of norms $N_{2,1}$ and $N_{2,3}$ this is the same as finding an upper bound for

$$\sqrt{T_\alpha}\, 2^{-n/2} \|\varphi\|_{L^2(h^{-1}J^n_{T_\alpha};L^2_z L^2_{y_G})}$$

uniformly in both $\alpha \in [\alpha_0, \alpha_1[$ and $n \geq 0$, or equivalently for

$$\sqrt{T_\alpha}\|\varphi\|_{L^2(h^{-1}J^{\leq 1}_{T_\alpha};L^2_z L^2_{y_G})} \quad \text{and} \quad \sqrt{T_\alpha} 2^{-n/2} \|\varphi\|_{L^2(h^{-1}J^n_{T_\alpha};L^2_z L^2_{y_G})} \ (n > 1)\,,$$

with the same uniformity.

For $t \in h^{-1}J^{\leq 1}_{T_\alpha}$ we write

$$\sqrt{T_\alpha}\varphi(t) = -i\sqrt{T_\alpha} e^{\alpha N} a_G(V) e^{-\alpha_0' N} \int_{s<t} U(t)U(s)^* \sqrt{h}\, w_1(s)\, ds$$

with

$$w_1(s) = e^{\alpha_0' N} 1_{h^{-1}J^{\leq 1}_{T_\alpha}}(s) u_2(s) = e^{\alpha_0' N} 1_{h^{-1}J^{\leq 2}_{T_{\alpha_0'}}}(s)\, u_2(s)\,.$$

Then Proposition 3.4, the retarded Strichartz estimate (46) and the Cauchy-Schwarz inequality yield

$$\begin{aligned}
&\sqrt{T_\alpha}\|\varphi\|_{L^2(h^{-1}J^{\leq 1}_{T_\alpha};L^2_z L^2_{y_G})} \\
&\quad\lesssim \sqrt{T_\alpha}\frac{C_V M_{\alpha 01}}{\sqrt{\alpha_0' - \alpha}} \|\int_{s<t} U(t)U(s)^*\sqrt{h} w_1(s)\, ds\|_{L^2_z L^2_t(h^{-1}J^{\leq 1}_{T_\alpha};L^{r_\sigma}_{y_G})} \\
&\quad\lesssim C_V M_{\alpha 01}\sqrt{\gamma}\|\sqrt{h} w_1\|_{L^2_z L^1(h^{-1}J^{\leq 2}_{T_{\alpha_0'}};L^2_{y_G})}
\end{aligned}$$

$$\lesssim C_V M_{\alpha 01}\sqrt{\gamma}\sqrt{T_{\alpha'_0}}\|w_1\|_{L^2_z L^2_t(h^{-1}J^{\leq 2}_{T_{\alpha'_0}};L^2_{y_G})}$$

$$\lesssim C_V M_{\alpha 01}\sqrt{\gamma}\sqrt{T_{\alpha'_0}}\|e^{\alpha'_0}u_2\|_{L^2_t(h^{-1}J^{\leq 2}_{T_{\alpha'_0}};L^2_z L^2_{y_G})}\,.$$

The equivalence between the norms $N_{2,1}$ and $N_{2,4}$ implies

$$\sqrt{T_\alpha}\|\varphi\|_{L^2(h^{-1}J^{\leq 1}_{T_\alpha};L^2_z L^2_{y_G})} \lesssim C_V^2 M_{\alpha 01}^2 \gamma M_2(u_2)\,. \tag{79}$$

For $t\in h^{-1}J^n_{T_\alpha}$, $n>1$, write

$$\sqrt{T_\alpha}\,2^{-n/2}\varphi(t) = -\,i\sqrt{T_\alpha}2^{-n/2}e^{\alpha N}a_G(V)e^{-\alpha'_0 N}\int_{s<t}U(t)U(s)^*\sqrt{h}\,w_1(s)\,ds$$
$$\underbrace{-\,i\sqrt{T_\alpha}2^{-n/2}\sum_{m=2}^{n}e^{\alpha N}a_G(V_2)e^{-\alpha'_m N}\int_{s<t}U(t)U(s)^*\sqrt{h}\,w_m(s)\,ds}_{=\tilde{\varphi}}$$

with for $m\geq 2$

$$w_m(s) = 1_{h^{-1}J^m_{T_\alpha}}(s)e^{\alpha'_m N}u_2(s) = 1_{h^{-1}\left[T_{\alpha'_m}-3\delta_m,T_{\alpha'_m}-\delta_m\right]}(s)e^{\alpha'_m N}u_2(s)\,.$$

The first term is actually estimate as we did for (79) with the additional factor $2^{-n/2}\leq 1$. It suffices to consider the application of Proposition 3.4, the retarded Strichartz estimate (46) and the Cauchy-Schwarz inequality to

$$\sqrt{T_\alpha}\,2^{-n/2}\|\tilde{\varphi}\|_{L^2(h^{-1}J^n_{T_\alpha};L^2_z L^2_{y_G})}$$
$$\lesssim \sqrt{T_\alpha}\,2^{-n/2}\sum_{m=2}^{n}\frac{C_V\,M_{\alpha 01}}{\sqrt{\alpha'_m-\alpha}}\left\|\int_{s<t}U(t)U(s)^*\sqrt{h}w_m(s)\,ds\right\|_{L^2_z L^2_t(h^{-1}J^n_{T_\alpha};L^{r_\sigma}_{y_G})}$$
$$\lesssim C_V\,M_{\alpha 01}\sqrt{\gamma}\,2^{-n/2}\sum_{m=2}^{n}2^{m/2}\left\|\sqrt{h}w_m\right\|_{L^2_z L^1_t(h^{-1}J^m_{T_\alpha};L^2_{y_G})}$$
$$\lesssim C_V\,M_{\alpha 01}\sqrt{\gamma}\,2^{-n/2}\sum_{m=2}^{n}\sqrt{T_\alpha}\,\|w_m\|_{L^2_z L^2_t(h^{-1}J^m_{T_\alpha};L^2_{y_G})}\,. \tag{80}$$

Thanks to the equivalence of the norms $N_{2,1}$ and $N_{2,4}$ (with $T = T_{\alpha'_m}$), we obtain for $m \geq 2$

$$\begin{aligned}
&\sqrt{T_\alpha}\, \|w_m\|_{L^2_z L^2_t(h^{-1}J^m_{T_\alpha};L^2_{yG})} \\
&= 2^{\frac{m+2}{2}} \sqrt{\delta_m} \left\| e^{\alpha'_m N} u_2(s) \right\|_{L^2_t(h^{-1}\left[T_{\alpha'_m}-3\delta_m, T_{\alpha'_m}-\delta_m\right]; L^2_z L^2_{yG})} \\
&\lesssim 2^{m/2} C_V\, M_{\alpha 01} \sqrt{\gamma} M_2(u_2)\,.
\end{aligned} \tag{81}$$

Putting together (80) and (81) gives

$$\begin{aligned}
\sqrt{T_\alpha}\, 2^{-n/2} \|\varphi\|_{L^2(h^{-1}J^n_{T_\alpha};L^2_z L^2_{yG})} &\lesssim 2^{-n/2} \sum_{m=0}^{n} 2^{m/2} C_V^2\, M^2_{\alpha 01}\, \gamma\, M_2(u_2) \\
&\lesssim C_V^2\, M^2_{\alpha 01}\, \gamma\, M_2(u_2)
\end{aligned}$$

which, combined with (79) and the normalization of $M_2(L_{22}(u_2))$, yields

$$M_2(L_{22}(u_2)) \lesssim C_V\, M_{\alpha 01} \sqrt{\gamma}\, M_2(u_2)\,. \tag{82}$$

The estimate of $L_{11}(u_1)$ starts with the same decomposition of the interval $[0, T/h]$ with the norms

$$N_{1,1}(\varphi) = \sup_{\tau\in[0,T[} \sqrt{T-\tau} \left\| \frac{\varphi(t)}{\sqrt{ht}} \right\|_{L^1([0,\frac{\tau}{h}];L^2_z L^2_{yG})}\,, \tag{83}$$

$$N_{1,2}(\varphi) = \left\| \sqrt{\frac{T}{ht}}\varphi \right\|_{L^1(J^{\leq 1}_{T/h};L^2_z L^2_{yG})} + \sup_{\delta\in]0,\frac{T}{8}]} \sqrt{\frac{\delta}{T}}\, \|\varphi\|_{L^1(h^{-1}[T-2\delta,T-\delta];L^2_z L^2_{yG})}\,, \tag{84}$$

$$N_{1,3}(\varphi) = \left\| \sqrt{\frac{T}{ht}}\varphi \right\|_{L^1(J^{\leq 1}_{T/h};L^2_z L^2_{yG})} + \sup_{n>1} 2^{-n/2}\, \|\varphi\|_{L^1(J^n_{T/h};L^2_z L^2_{yG})}\,, \tag{85}$$

$$N_{1,4}(\varphi) = \left\| \sqrt{\frac{T}{ht}}\varphi \right\|_{L^1(J^{\leq 2}_{T/h};L^2_z L^2_{yG})} + \sup_{\delta\in]0,\frac{T}{12}]} \sqrt{\frac{\delta}{T}}\, \|\varphi\|_{L^1(h^{-1}[T-3\delta,T-\delta];L^2_z L^2_{yG})}\,. \tag{86}$$

Those norms are equivalent according to

$$\kappa_1^{-1} N_{1,1}(\varphi) \leq N_{1,i}(\varphi) \leq \kappa_1 N_{1,1}(\varphi)\,, \quad 2 \leq i \leq 4 \tag{87}$$

with a universal constant $\kappa_1 > 1$. See Lemma 4.1 for the proof.

$\mathbf{L_{11}(u_1)}$-Step 1, Decomposition of $L_{11}(u_1)$ For $\alpha \in [\alpha_0, \alpha_1[$, we seek an upper bound of $N_{1,1}(\varphi)$ for

$$\varphi(t) = e^{\alpha N} L_{11}(u_1)(t) = -i \int_0^t e^{\alpha N} \sqrt{h} a_G(V_2) U(t) U(s)^* u_1(s)\, ds\,.$$

By the equivalence of norms $N_{1,1}$ and $N_{1,3}$ this is the same as finding a uniform upper bound for

$$\left\| \left(\frac{T_\alpha}{ht}\right)^{1/2} \varphi \right\|_{L^1(h^{-1}J_{T_\alpha}^{\leq 1}; L_z^2 L_{y_G}^2)} \quad \text{and} \quad 2^{-n/2} \|\varphi\|_{L^1(h^{-1}J_{T_\alpha}^n; L_z^2 L_{y_G}^2)} \text{ for } n > 1\,.$$

Setting $\psi_1(t) = \left(\frac{T_\alpha}{ht}\right)^{1/2} 1_{h^{-1}J_{T_\alpha}^{\leq 1}}(t)\varphi(t)$ and, for $n > 1$, $\psi_n(t) = 2^{-n/2} 1_{h^{-1}J_{T_\alpha}^n}(t)\varphi(t)$ gives

$$\psi_1(t) = -i \int_0^t \left(\frac{T_\alpha}{ht}\right)^{1/2} e^{\alpha N} \sqrt{h} a_G(V_2) U(t) U(s)^* 1_{h^{-1}J_{T_\alpha}^{\leq 1}}(s) u_1(s)\, ds\,,$$
$$t \in h^{-1} J_{T_\alpha}^{\leq 1}\,,$$

and, for $n > 1$,

$$\begin{aligned}\psi_n(t) = &-i \int_0^t 2^{-n/2} e^{\alpha N} \sqrt{h} a_G(V_2) U(t) U(s)^* 1_{h^{-1}J_{T_\alpha}^{\leq 1}}(s) u_1(s)\, ds \\ &- i \sum_{1 < m \leq n} \int_0^t \sqrt{h} 2^{-n/2} e^{\alpha N} a_G(V_2) U(t) U(s)^* 1_{h^{-1}J_{T_\alpha}^m}(s) u_1(s)\, ds \\ &t \in h^{-1} J_{T_\alpha}^n\,.\end{aligned}$$

This allows to rewrite the above decomposition as

$$\psi_1(t) \overset{t \leq \frac{3T_\alpha}{4h}}{=} -i \int_0^{\frac{3T_\alpha}{4h}} \underbrace{1_{[0,t]}(s) \sqrt{\frac{T_\alpha}{T_{\alpha'_0}}} \sqrt{\frac{hs}{ht}} e^{\alpha N} \sqrt{h} a_G(V_2) e^{-\alpha'_0 N}}_{B_{11}(t,s)} U(t) U(s)^* w_1(s)\, ds\,,$$

and, for $n > 1$,

$$\psi_n(t) \overset{t\in h^{-1}J^n_{T_\alpha}}{=}$$

$$-i\int_0^{\frac{T_\alpha}{h}} \underbrace{1_{[0,\frac{3T_\alpha}{4h}]}(s)2^{-n/2}e^{\alpha N}\sqrt{h}a_G(V_2)e^{-\alpha'_0 N}\sqrt{\frac{hs}{T_{\alpha'_0}}}}_{B_{n1}(t,s)}U(t)U(s)^*w_1(s)\,ds$$

$$-i\sum_{m=2}^{n}\int_0^{\frac{T_\alpha}{h}} \underbrace{1_{[0,t]\cap h^{-1}J^m_{T_\alpha}}(s)2^{-\frac{n-m}{2}}e^{\alpha N}\sqrt{h}a_G(V_2)e^{-\alpha'_m N}}_{B_{nm}(t,s)}U(t)U(s)^*w_m(s)\,ds\,,$$

with $w_1(s) = 1_{h^{-1}J^{\leq 2}_{T_{\alpha'_0}}}(s)\left(\frac{T_{\alpha'_0}}{hs}\right)^{1/2}e^{\alpha'_0 N}u_1(s)$ and

$$w_m(s) \overset{m\geq 1}{=} 2^{-m/2}1_{h^{-1}J^m_{T_\alpha}}(s)e^{\alpha'_m N}u_1(s) = 2^{-m/2}1_{\left[\frac{T_{\alpha'_m}-3\delta_m}{h},\frac{T_{\alpha'_m}-\delta_m}{h}\right]}(s)e^{\alpha'_m N}u_1(s)\,.$$

Proposition 4.1 tells us

$$\|\psi_1\|_{L^1(h^{-1}J^{\leq 1}_{T_\alpha};L^2_zL^2_{y_G})} \lesssim \left(\sup_{s\in[0,\frac{3T_\alpha}{4h}]}\int_0^{\frac{3T_\alpha}{4h}}\|B_{11}(t,s)\|^2\,dt\right)^{1/2}$$
$$\|w_1\|_{L^1(h^{-1}J^{\leq 2}_{T_{\alpha'_0}};L^2_zL^2_{y_G})}\,,$$

and, for $n > 1$,

$$\|\psi_n\|_{L^1(h^{-1}J^n_{T_\alpha};L^2_zL^2_{y_G})}$$

$$\lesssim \left(\sup_{s\in[0,\frac{3T_\alpha}{4h}]}\int_{h^{-1}J^n_{T_\alpha}}\|B_{n1}(t,s)\|^2\,dt\right)^{1/2}\|w_1\|_{L^1(h^{-1}J^{\leq 2}_{T_{\alpha'_0}};L^2_zL^2_{y_G})}$$

$$+\sum_{m=2}^{n}\left(\sup_{s\in h^{-1}J^m_{T_\alpha}}\int_{h^{-1}J^n_{T_\alpha}}\|B_{nm}(t,s)\|^2\,dt\right)^{1/2}\|w_m\|_{L^1([\frac{T_{\alpha'_m}-3\delta_m}{h},\frac{T_{\alpha'_m}-\delta_m}{h}];L^2_zL^2_{y_G})}\,.$$

From the comparison between the norms $N_{1,1}$ and $N_{1,4}$ we know

$$\|w_1\|_{L^1(h^{-1}J^{\leq 2}_{T_{\alpha'_0}};L^2_zL^2_{y_G})} \lesssim \sup_{\tau\in[0,T_{\alpha'_0}[} \sqrt{T_{\alpha'_0}-\tau}\left\|\frac{e^{\alpha'_0 N}u_1}{\sqrt{ht}}\right\|_{L^1([0,\frac{\tau}{h}];L^2_zL^2_{y_G})}$$
$$\lesssim M_{\alpha 01}C_V\gamma^{1/2}M_1(u_1)\,,$$

while for $m>1$,

$$\|w_m\|_{L^1([\frac{T_{\alpha'_m}-3\delta_m}{h},\frac{T_{\alpha'_m}-\delta_m}{h}];L^2_zL^2_{y_G})}$$
$$\lesssim \left(\frac{T_{\alpha'_m}}{\delta_m}\right)^{1/2} 2^{-m/2} \sup_{\tau\in[0,T_{\alpha'_m}[} \sqrt{T_{\alpha'_m}-\tau}\,\|\frac{e^{\alpha'_m N}u_1}{\sqrt{ht}}\|_{L^1([0,\frac{\tau}{h}];L^2_zL^2_{y_G})}$$
$$\lesssim \left(\frac{T_\alpha}{T_\alpha 2^{-m-2}}\right)^{1/2} 2^{-m/2}M_{\alpha 01}C_V\gamma^{1/2}M_1(u_1)$$
$$\lesssim M_{\alpha 01}C_V\gamma^{1/2}M_1(u_1)\,.$$

We have proved

$$\frac{\sup_{\tau\in[0,T_\alpha[}\sqrt{T_\alpha-\tau}\left\|\frac{e^{\alpha N}L_{11}(u_1)}{\sqrt{ht}}\right\|_{L^1([0,\frac{\tau}{h}];L^2_zL^2_{y_G})}}{M_{\alpha 01}C_V\gamma^{1/2}M_1(u_1)}$$
$$\lesssim \left(\sup_{s\in[0,\frac{3T_\alpha}{4h}]}\int_0^{\frac{3T_\alpha}{4h}}\|B_{11}(t,s)\|^2\,dt\right)^{1/2} + \sup_{n\geq 1}\left(\sup_{s\in[0,\frac{3T_\alpha}{4h}]}\int_{h^{-1}J^n_{T_\alpha}}\|B_{n1}(t,s)\|^2\,dt\right)^{1/2}$$
$$+\sup_{n>1}\sum_{m=2}^{n}\left(\sup_{s\in h^{-1}J^m_{T_\alpha}}\int_{h^{-1}J^n_{T_\alpha}}\|B_{nm}(t,s)\|^2\,dt\right)^{1/2}. \tag{88}$$

It remains to estimate every term of the above right-hand side.
$\mathbf{L_{11}(u_1)}$-Step 2, Estimate for B_{11} The expression

$$B_{11}(t,s) = 1_{[0,t]}(s)\left(\frac{T_\alpha}{T_{\alpha'_0}}\right)^{1/2}\left(\frac{hs}{ht}\right)^{1/2} e^{\alpha N}\sqrt{h}a_G(V_2)e^{-\alpha'_0 N}$$

implies, with $T_\alpha=\gamma(\alpha_1-\alpha)=7\gamma(\alpha'_0-\alpha)$,

$$\|B_{11}(t,s)\|^2 \leq \frac{T_\alpha}{T_{\alpha'_0}}\frac{7\gamma}{T_\alpha}M^2_{\alpha 01}C^2_V h 1_{[0,t]}(s)\frac{hs}{ht} \leq 7\gamma M^2_{\alpha 01}C^2_V\frac{4hs}{3T_\alpha}\frac{1_{[0,t]}(s)}{t}\,.$$

We obtain

$$\int_0^{\frac{3T_\alpha}{4h}} \|B_{11}(t,s)\|^2\,dt \leq 7\gamma M_{\alpha 01}^2 C_V^2 \frac{4hs}{3T_\alpha}\ln(\frac{3T_\alpha}{4hs}).$$

and

$$\left(\sup_{s\in[0,\frac{3T_\alpha}{4h}]} \int_0^{\frac{3T_\alpha}{4h}} \|B_{11}(t,s)\|^2\,dt\right)^{1/2} \lesssim \gamma^{1/2} M_{\alpha 01} C_V\,. \tag{89}$$

$L_{11}(u_1)$-Step 3, Estimate for B_{n1}, $n>1$ From

$$B_{n1}(t,s) = 1_{[0,\frac{3T_\alpha}{4h}]}(s) 2^{-n/2} e^{\alpha N}\sqrt{h} a_G(V_2) e^{-\alpha_0' N}\left(\frac{hs}{T_{\alpha_0'}}\right)^{1/2}$$

we deduce with $\alpha_0' - \alpha = \frac{\alpha_1-\alpha}{7} = \frac{T_\alpha}{7\gamma}$ and $T_{\alpha_0'} = \frac{6T_\alpha}{7}$,

$$\begin{aligned}\|B_{n1}(t,s)\|^2 &\leq 1_{[0,\frac{3T_\alpha}{4h}]}(s) 2^{-n}\frac{hM_{\alpha 01}^2 C_V^2}{(\alpha_0'-\alpha)}\left(\frac{3T_\alpha/4}{T_{\alpha_0'}}\right)\\ &\leq 1_{[0,\frac{3T_\alpha}{4h}]}(s)\frac{2^{-n}7\gamma h}{T_\alpha} M_{\alpha 01}^2 C_V^2\,.\end{aligned}$$

With $J_{T_\alpha}^n = [(1-2^{-n})T_\alpha, (1-2^{-n-1})T_\alpha[$ for $n>1$ we obtain

$$\int_{h^{-1}J_{T_\alpha}^n} \|B_{n1}(t,s)\|^2\,dt \leq 2^{-n-1}T_\alpha \times \frac{2^{-n}7\gamma}{T_\alpha} M_{\alpha 01}^2 C_V^2 \leq \frac{7\gamma}{4} M_{\alpha 01}^2 C_V^2\,.$$

and

$$\sup_{n>1}\left(\sup_{s\in[0,\frac{3T_\alpha}{4h}]} \int_{h^{-1}J_{T_\alpha}^n} \|B_{n1}(t,s)\|^2\,dt\right)^{1/2} \lesssim \gamma^{1/2} M_{\alpha 01} C_V\,. \tag{90}$$

$L_{11}(u_1)$-Step 4, Estimate for the B_{nm}'s, $n,m>1$ From

$$B_{nm}(t,s) = 1_{[0,t]\cap h^{-1}J_{T_\alpha}^m}(s) 2^{-(n-m)/2} e^{\alpha N}\sqrt{h} a_G(V_2) e^{-\alpha_m' N}$$

and $\alpha'_m - \alpha = 2^{-(m+2)}(\alpha_1 - \alpha) = \frac{2^{-(m+2)}T_\alpha}{\gamma}$, we deduce

$$\|B_{nm}(t,s)\|^2 \leq 1_{[0,t]\cap h^{-1}J^m_{T_\alpha}}(s) 2^{-(n-m)} \frac{h2^{m+2}\gamma}{T_\alpha} M^2_{\alpha 01} C^2_V .$$

Using again that the length of $J^n_{T_\alpha}$ is $2^{-(n+1)}T_\alpha$, we get

$$\sup_{s\in h^{-1}J^m_{T_\alpha}} \int_{h^{-1}J^n_{T_\alpha}} \|B_{nm}(t,s)\|^2 \, dt \leq 2\gamma 2^{-2(n-m)} M^2_{\alpha 01} C^2_V$$

and

$$\sup_{n\geq 1} \sum_{m=1}^{n} \left(\sup_{s\in h^{-1}J^m_{T_\alpha}} \int_{h^{-1}J^n_{T_\alpha}} \|B_{nm}(t,s)\|^2 \, dt \right)^{1/2} \lesssim \gamma^{1/2} M_{\alpha 01} C_V . \tag{91}$$

$\mathbf{L}_{\mathbf{1}\infty}(\mathbf{u}_\infty)$-Step 1, Decomposition of $\mathbf{L}_{\mathbf{1}\infty}(\mathbf{u}_\infty)$

Compared with the decomposition of $\mathbf{L_{22}(u_2)}$ and $\mathbf{L_{11}(u_1)}$, an additional dyadic decomposition has to be done around 0 in order to absorb the weight $\frac{1}{\sqrt{ht}}$ properly and to use Proposition 4.2. Decompose now $[0, T] = \cup_{n\in\mathbb{Z}} J^n_T$ where J^0_T is now the interval $[T/4, T/2[$ and $J^n_T = 2^n J^0_T$ for $n < 0$, according to figure 4. In particular, the interval previously denoted by J^0_T is now $J^{\leq 0}_T$ while $J^{\leq n_0}_T$ is not changed for $n_0 > 0$.

We seek an upper bound of $N_{1,1}(\varphi)$ for

$$\varphi(t) = e^{\alpha N} L_{1\infty}(u_\infty)(t) = -h \int_0^t e^{\alpha N} a_G(V_2) U(t) U(s)^* a_G(V_1)^* u_\infty(s) \, ds .$$

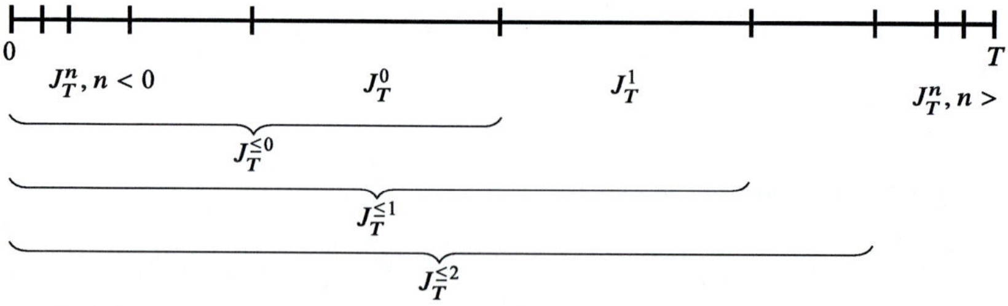

Fig. 4 The time intervals J^n_T, $n \in \mathbb{Z}$.

By the equivalence of norms $N_{1,1}$ and $N_{1,3}$ this is equivalent to proving a uniform upper bound for

$$\left\|\left(\frac{T_\alpha}{ht}\right)^{1/2}\varphi\right\|_{L^1(h^{-1}J_{T_\alpha}^{\leq 1};L_z^2L_{yG}^2)} \quad \text{and} \quad 2^{-n/2}\|\varphi\|_{L^1(h^{-1}J_{T_\alpha}^n;L_z^2L_{yG}^2)} \text{ for } n>1\,.$$

But the dyadic decomposition around 0 says

$$\begin{aligned}\left\|\left(\frac{T_\alpha}{ht}\right)^{1/2}\varphi\right\|_{L^1(h^{-1}J_{T_\alpha}^{\leq 1};L_z^2L_{yG}^2)} &\leq 2\sum_{n\leq 1}\|2^{-\frac{n+1}{2}}\varphi\|_{L^1(h^{-1}J_{T_\alpha}^n;L_z^2L_{yG}^2)}\\ &= 2\|\sum_{n\leq 1}2^{-\frac{n+1}{2}}1_{h^{-1}J_{T_\alpha}^n}(t)\varphi\|_{L^1(h^{-1}J_{T_\alpha}^{\leq 1};L_z^2L_{yG}^2)}\,.\end{aligned}$$

$L_{1\infty}(u_\infty)$-Step 2, Estimate on $h^{-1}J_{T_\alpha}^{n\leq 1}$

We write $\varphi_1=\sum_{n\leq 1}2^{-\frac{n+1}{2}}1_{h^{-1}J_{T_\alpha}^n}(t)e^{\alpha N}L_{1\infty}(u_\infty)=\sum_{n\leq 1}\varphi_{1,n}(t)$ where

$$\begin{aligned}\varphi_{1,n}(t) &= -h\sum_{m=-\infty}^{1}2^{-\frac{n+1}{2}}1_{h^{-1}J_{T_\alpha}^n}(t)\int_0^t e^{\alpha N}a_G(V_2)e^{-\frac{\alpha+\alpha_0'}{2}N}\times\\ &\qquad U(t)U(s)^*e^{\frac{\alpha+\alpha_0'}{2}N}a_G^*(V_1)e^{-\alpha_0'N}1_{h^{-1}J_{T_\alpha}^m}(s)e^{\alpha_0'N}u_\infty^h(s)\,ds\\ &= -h\sum_{m=-\infty}^{1}1_{h^{-1}J_{T_\alpha}^n}\int_0^t B_{1n}U(t)U(s)^*B_{2m}^*(s)\varphi_{\infty,m}(s)\,ds\end{aligned}$$

with

$$\begin{aligned}&B_{1n}=2^{-\frac{n+1}{2}}e^{\alpha N}a_G(V_2)e^{-\frac{\alpha+\alpha_0'}{2}N}\,,\\ &\|B_{1n}\|_{L_z^2L_{yG}^2\leftarrow L_z^2L_{yG}^{r_\sigma}}\lesssim\frac{M_{\alpha 01}\|V_2\|_{L^{r_\sigma'}}}{\sqrt{\alpha_0'-\alpha}}2^{-n/2}\lesssim\frac{M_{\alpha 01}C_V\gamma^{1/2}}{T_{\alpha_1}^{1/2}}2^{-n/2}\,,\\ &B_{2m}^*=e^{\frac{\alpha+\alpha_0'}{2}N}a_G^*(V_1)e^{-\alpha_0'N}1_{h^{-1}J_{T_\alpha}^m}(s)\frac{\sqrt{hs}}{\sqrt{T_\alpha-hs}}\,,\\ &\|B_{2m}^*\|_{L_z^2L_{yG}^{r_\sigma'}\leftarrow L_z^2L_{yG}^2}\lesssim\frac{M_{\alpha 01}\|V_1\|_{L^{r_\sigma'}}}{\sqrt{\alpha_0'-\alpha}}2^{m/2}\lesssim\frac{M_{\alpha 01}C_V\gamma^{1/2}}{T_\alpha^{1/2}}2^{m/2}\,,\\ &\varphi_{\infty,m}(s)=1_{h^{-1}J_{T_\alpha}^m}(s)\varphi_\infty(s)\,,\quad \varphi_\infty(s)=e^{\alpha_0'N}\frac{\sqrt{T_\alpha-hs}}{\sqrt{hs}}u_\infty(s)\,.\end{aligned}$$

By noticing

$$|h^{-1}J^n_{T_\alpha}| \leq T_\alpha h^{-1}2^n$$

the upper bound of Proposition 4.2 gives

$$\begin{aligned}
&\|\varphi_1\|_{L^1(h^{-1}J^{\leq 1}_{T_\alpha};L^2_z L^2_{yG})} \\
&\lesssim \left[\sum_{-\infty\leq m\leq n\leq 1} 2^{n/2}(M_{\alpha 01}C_V\gamma^{1/2}2^{-n/2})(M_{\alpha 01}C_V\gamma^{1/2}2^{m/2})\right] \\
&\qquad \times \|\varphi_\infty\|_{L^\infty(h^{-1}J^{\leq 1}_{T_\alpha};L^2_z L^2_{yG})} \\
&\lesssim M^2_{\alpha 01}C^2_V\gamma M_\infty(u_\infty)\,.
\end{aligned}$$

We proved

$$\|\frac{1}{\sqrt{ht}}e^{\alpha N}L_{1\infty}(u_\infty)\|_{L^1(h^{-1}J^{\leq 1}_{T_\alpha};L^2_z L^2_{yG})} \lesssim M_{\alpha 01}C^2_V\gamma M_\infty(u_\infty)\,. \tag{92}$$

$\mathbf{L_{1\infty}(u_\infty)}$-Step 3, Estimate on $\mathbf{h^{-1}J^n_{T_\alpha}}$, n>1

Write $\varphi_1(t) = 2^{-n/2}1_{h^{-1}J^n_{T_\alpha}}(t)e^{\alpha N}L_{1\infty}(u_\infty)$, where

$$\begin{aligned}
\varphi_1(t) = &-h\sum_{m=-\infty}^{1} 2^{-\frac{n}{2}}1_{h^{-1}J^n_{T_\alpha}}(t)\int_0^t e^{\alpha N}a_G(V_2)e^{-\frac{\alpha+\alpha'_0}{2}N}U(t)U(s)^*\times \\
&\qquad e^{\frac{\alpha+\alpha'_0}{2}N}a^*_G(V_1)e^{-\alpha'_0 N}1_{h^{-1}J^m_{T_\alpha}}(s)e^{\alpha'_0 N}u^h_\infty(s)\,ds \\
&-h\sum_{m=2}^{n} 2^{-\frac{n}{2}}1_{h^{-1}J^n_{T_\alpha}}(t)\int_0^t e^{\alpha N}a_G(V_2)e^{-\frac{\alpha+\alpha'_m}{2}N}U(t)U(s)^*\times \\
&\qquad e^{\frac{\alpha+\alpha'_m}{2}N}a^*_G(V_1)e^{-\alpha'_m N}1_{h^{-1}J^m_{T_\alpha}}(s)e^{\alpha'_m N}u^h_\infty(s)\,ds \\
= &-h\sum_{m=-\infty}^{1} 1_{h^{-1}J^n_{T_\alpha}}(t)\int_0^t B_{1n}U(t)U(s)^*B^*_{2m}(s)\varphi_{\infty,m}(s)\,ds \\
&-h\sum_{m=2}^{n} 1_{h^{-1}J^n_{T_\alpha}}(t)\int_0^t B_{1nm}U(t)U(s)^*B^*_{2nm}(s)\varphi_{\infty,m}(s)\,ds\,.
\end{aligned}$$

The family $\mathcal{I}$ of Proposition 4.2 is made here of the single interval $h^{-1}J^n_{T_\alpha}$ while the family $\mathcal{J} = \left\{h^{-1}J^m_{T_\alpha}, m \leq n\right\}$ is splitted in two parts $m \leq 1$ and $2 \leq m \leq n$. In the last two lines the notations correspond to

$$B_{1n} = 2^{-\frac{n}{2}} e^{\alpha N} a_G(V_2) e^{-\frac{\alpha+\alpha'_0}{2}N},$$

$$\|B_{1n}\|_{L^2_z L^2_{y_G} \leftarrow L^2_z L^{r_\sigma}_{y_G}} \lesssim \frac{M_{\alpha 01}\|V_2\|_{L^{r'_\sigma}}}{\sqrt{\alpha'_0 - \alpha}} 2^{-n/2} \lesssim \frac{M_{\alpha 01} C_V \gamma^{1/2}}{T_{\alpha_1}^{1/2}} 2^{-n/2},$$

$$m \leq 1 \quad B^*_{2m} = e^{\frac{\alpha+\alpha'_0}{2}N} a^*_G(V_1) e^{-\alpha'_0 N} 1_{h^{-1}J^m_{T_\alpha}}(s) \frac{\sqrt{hs}}{\sqrt{T_\alpha - hs}},$$

$$m \leq 1 \quad \|B^*_{2m}\|_{L^2_z L^{r'_\sigma}_{y_G} \leftarrow L^2_z L^2_{y_G}} \lesssim \frac{M_{\alpha 01}\|V_1\|_{L^{r'_\sigma}}}{\sqrt{\alpha'_0 - \alpha}} 2^{m/2} \lesssim \frac{M_{\alpha 01} C_V \gamma^{1/2}}{T_\alpha^{1/2}} 2^{m/2},$$

$$m \geq 2 \quad B_{1nm} = 2^{-\frac{n}{2}} e^{\alpha N} a_G(V_2) e^{-\frac{\alpha+\alpha'_m}{2}N},$$

$$m \geq 2 \quad \|B_{1nm}\|_{L^2_z L^2_{y_G} \leftarrow L^2_z L^{r_\sigma}_{y_G}} \lesssim \frac{M_{\alpha 01}\|V_2\|_{L^{r'_\sigma}}}{\sqrt{\alpha'_m - \alpha}} 2^{-n/2} \lesssim \frac{M_{\alpha 01} C_V \gamma^{1/2}}{T_{\alpha_1}^{1/2}} 2^{m/2-n/2},$$

$$m \geq 2 \quad B^*_{2nm} = e^{\frac{\alpha+\alpha'_m}{2}N} a^*_G(V_1) e^{-\alpha'_m N} 1_{h^{-1}J^m_{T_\alpha}}(s) \frac{\sqrt{hs}}{\sqrt{T_\alpha - hs}},$$

$$m \geq 2 \quad \|B^*_{2mn}\|_{L^2_z L^{r'_\sigma}_{y_G} \leftarrow L^2_z L^2_{y_G}} \lesssim \frac{M_{\alpha 01}\|V_1\|_{L^{r'_\sigma}}}{\sqrt{\alpha'_m - \alpha}} 2^{m/2} \lesssim \frac{M_{\alpha 01} C_V \gamma^{1/2}}{T_\alpha^{1/2}} 2^{m},$$

$$\varphi_m(s) = 1_{h^{-1}J^m_{T_\alpha}}(s) \varphi_\infty(s),$$

$$\varphi_\infty(s) = 1_{h^{-1}J^{\leq 1}_{T_\alpha}}(s) e^{\alpha'_0 N} \frac{\sqrt{T_\alpha - hs}}{\sqrt{hs}} u_\infty(s)$$
$$+ \sum_{m=2}^{\infty} e^{\alpha'_m N} 1_{h^{-1}J^m_{T_\alpha}}(s) \frac{\sqrt{T_\alpha - hs}}{\sqrt{hs}} u_\infty(s).$$

The size of the intervals are estimated respectively by $|h^{-1}J^n_{T^\alpha}| \lesssim h^{-1}2^{-n}T_\alpha$ and

$$|h^{-1}J^m_{T_\alpha}| \lesssim h^{-1}2^m T_\alpha \quad \text{for } m \leq 1 \quad , \quad |h^{-1}J^m_{T_\alpha}| \lesssim h^{-1}2^{-m}T_\alpha \quad \text{for} \quad m \geq 2.$$

Proposition 4.2 gives

$$\begin{aligned}
&\frac{\|\varphi_1\|_{L^1(h^{-1}J^n_{T\alpha};L^2_zL^2_{y_G})}}{\|\varphi_\infty\|_{L^\infty(h^{-1}J^{\leq n}_{T\alpha};L^2_zL^2_{y_G})}}\\
&\lesssim \left[\sum_{-\infty\leq m\leq 1} 2^{-n/2}(M_{\alpha 01}C_V\gamma^{1/2}2^{-n/2})(M_{\alpha 01}C_V\gamma^{1/2}2^{m/2})2^{m/2}\right]\\
&\quad+\left[\sum_{m=2}^{n} 2^{-n/2}(M_{\alpha 01}C_V\gamma^{1/2}2^{m/2-n/2})(M_{\alpha 01}C_V\gamma^{1/2}2^{m})2^{-m/2}\right]\\
&\lesssim M_{\alpha 01}C_V^2\gamma\,.
\end{aligned}$$

With

$$\|\varphi_\infty\|_{L^\infty(h^{-1}J^{\leq n}_{T\alpha};L^2_zL^2_{y_G})}\leq\|\varphi_\infty\|_{L^\infty([0,T_\alpha/h[;L^2_zL^2_{y_G})}\leq M_\infty(u_\infty)\,,$$

we have proved

$$\sup_{n\geq 1}2^{-n/2}\|e^{\alpha N}L_{1\infty}(u_\infty)\|_{L^1(h^{-1}J^n_{T\alpha};L^2_zL^2_{y_G})}\lesssim M_{\alpha 01}C_V^2\gamma M_\infty(u_\infty)\,. \tag{93}$$

Conclusion From (71), (72) and (73) we deduce

$$M_\infty(L_{\infty\infty}(u_\infty)+L_{\infty 2}(u_2)+L_{\infty 1}(u_1))\lesssim\gamma^{1/2}M_{\alpha 01}C_V M(u_\infty,u_2,u_1)\,. \tag{94}$$

Combining (88), (89), (90), (91), and taking the supremum over $\alpha\in[\alpha_0,\alpha_1[$ yields

$$M_1(L_{11}(u_1))\lesssim\gamma^{1/2}M_{\alpha 01}C_V M_1(u_1)\,,$$

while (82) says

$$M_2(L_{22}(u_2))\lesssim C_V M_{\alpha 01}\sqrt{\gamma}M_2(u_2)\,.$$

Finally the upper bounds (92),(93) combined, firstly with the equivalence of norms N_{11} and N_{31}, and secondly the normalization of (67) of M_1 yields

$$M_1(L_{1\infty}(u_\infty))\lesssim\gamma^{1/2}M_{\alpha 01}C_V M(u_\infty,u_2,u_1)\,.$$

The sum of all those inequalities is

$$M(L(u_\infty,u_2,u_1))\lesssim\gamma^{1/2}M_{\alpha 01}C_V M(u_\infty,u_2,u_1)\,,$$

which means that there exists a constant $C_{d,U}$ determined by the dimension d and the free dynamics U such that

$$\|L\|_{\mathcal{L}(\mathcal{E}^h_{\alpha_0,\alpha_1,\gamma})} \leq C_{d,U} M_{\alpha 01} C_V \gamma^{1/2}\,.$$

□

Lemma 4.1 *The norms $N_{p,1}$, $N_{p,2}$, $N_{p,3}$, $N_{p,4}$ defined in (75)(76)(77)(78) for $p = 2$ (resp. (83)(84)(85)(86) for $p = 1$) are equivalent according to (74) (resp. (87)).*

Proof We forget the notation $L^2_z L^2_{y_G}$ because it is a time integration issue and it can be done with any Banach space valued functions.

With the Definition (77) of $N_{2,3}(\varphi)$, the equality

$$\|\varphi\|_{L^2(h^{-1}J_T^{\leq 1})} = \left(\|\varphi\|^2_{L^2(h^{-1}J_T^0)} + \|\varphi\|^2_{L^2(h^{-1}J_T^1)}\right)^{1/2}$$

allows to replace $N_{2,3}(\varphi)$ by the equivalent norm

$$\sqrt{T}\|\varphi\|_{L^2(h^{-1}J_T^{\leq 1})} + \sqrt{T}\sup_{n>1} 2^{-n/2}\|\varphi\|_{L^2(h^{-1}J_T^n)}$$

For $p = 1$, the inequality

$$\forall t \in [\frac{3T}{4h}, \frac{T}{h}[, \quad \frac{1}{T} \leq \frac{1}{ht} \leq \frac{4}{3T}$$

allows to replace the second term of the definitions (84) (85)(86) of $N_{1,2}$, $N_{1,3}$ and $N_{1,4}$, respectively by

$$\sup_{\delta\in]0,T/8]} \sqrt{\delta}\left\|\frac{\varphi}{\sqrt{ht}}\right\|_{L^1(h^{-1}[T-2\delta,T-\delta[)}\,,$$

$$\sup_{n>1}\sqrt{T}2^{-n/2}\left\|\frac{\varphi}{\sqrt{ht}}\right\|_{L^1(h^{-1}J_T^n)}\,,$$

$$\sup_{\delta\in]0,T/12]} \sqrt{\delta}\left\|\frac{\varphi}{\sqrt{ht}}\right\|_{L^1(h^{-1}[T-3\delta,T-\delta[)}\,.$$

Additionally the $\sup_{\tau\in[0,T[}$ in the definitions (75)(83) can be replaced by $\sup_{\tau\in[3T/4,T[}$. We are thus led to compare the norms, for $p = 1, 2$,

$$N_{p,1,T,h}(\varphi) = \sup_{\tau\in[3T/4,T[} \sqrt{T-\tau}\left\|\frac{\varphi}{(ht)^{\frac{1}{p}-\frac{1}{2}}}\right\|_{L^p([0,\frac{\tau}{h}])},$$

$$N_{p,2,T,h}(\varphi) = \sqrt{T}\left\|\frac{\varphi}{(ht)^{\frac{1}{p}-\frac{1}{2}}}\right\|_{L^p(J^{\leq 1}_{T/h})} + \sup_{\delta\in]0,\frac{T}{8}]}\sqrt{\delta}\left\|\frac{\varphi}{(ht)^{\frac{1}{p}-\frac{1}{2}}}\right\|_{L^p(h^{-1}[T-2\delta,T-\delta])},$$

$$N_{p,3,T,h}(\varphi) = \sqrt{T}\left\|\frac{\varphi}{(ht)^{\frac{1}{p}-\frac{1}{2}}}\right\|_{L^p(J^{\leq 1}_{T/h})} + \sqrt{T}\sup_{n>1} 2^{-n/2}\left\|\frac{\varphi}{(ht)^{\frac{1}{p}-\frac{1}{2}}}\right\|_{L^p(h^{-1}J^n_T)},$$

$$N_{p,4,T,h}(\varphi) = \sqrt{T}\left\|\frac{\varphi}{(ht)^{\frac{1}{p}-\frac{1}{2}}}\right\|_{L^p(J^{\leq 2}_{T/h})} + \sup_{\delta\in]0,\frac{T}{12}]}\sqrt{\delta}\left\|\frac{\varphi}{(ht)^{\frac{1}{p}-\frac{1}{2}}}\right\|_{L^p(h^{-1}[T-3\delta,T-\delta])}.$$

The elementary homogeneity of those expressions gives

$$N_{p,i,T,h}(\varphi) = \frac{T}{h^{1/p}} N_{p,i,1,1}(\tilde{\varphi}) \quad \text{with } \tilde{\varphi}(t) = \varphi(ht) \quad \text{for } p = 1, 2 \quad \text{and } 1 \leq i \leq 4,$$

and it suffices to consider the case $T = h = 1$ while setting $\psi = \frac{\tilde{\varphi}}{t^{1/p-1/2}}$.

For $\tau \in [3/4, 1[$ the identity

$$\|\psi\|_{L^p([0,\tau])} = \left(\|\psi\|^p_{L^p(J^{\leq 1}_1)} + \|\psi\|^p_{L^p([3/4,\tau])}\right)^{1/p}$$

reduces the comparison of $N_{p,1,1,1}(\tilde{\varphi})$, $N_{p,2,1,1}(\tilde{\varphi})$ and $N_{p,3,1,1}(\tilde{\varphi})$ to the comparison of

$$A_1(\psi) = \sup_{\tau\in[3/4,1[} \sqrt{1-\tau}\, \|\psi\|_{L^p([3/4,\tau])},$$

$$A_2(\psi) = \sup_{\delta\in]0,1/8]} \sqrt{\delta}\, \|\psi\|_{L^p[1-2\delta,1-\delta])},$$

$$A_3(\psi) = \sup_{n>1} 2^{-n/2}\, \|\psi\|_{L^p(J^n_1)}.$$

Taking $\tau = 1-\delta$, $\delta \leq 1/8$, in $A_1(\psi)$ and $\delta = 2^{-n-1}$, $n > 1$, in $A_2(\psi)$ gives

$$A_3(\psi) \leq \sqrt{2}A_2(\psi) \leq \sqrt{2}A_1(\psi).$$

For $\tau \in [3/4, 1[$ there exists $n_\tau > 1$ such that $\tau \in [1-2^{-n_\tau}, 1-2^{-n_\tau-1}[= J_1^{n_\tau}$ and

$$\|\psi\|^p_{L^p([3/4,\tau])} = \sum_{n=2}^{n_\tau} \|\psi\|^p_{L^p(J_1^n)} \leq \sum_{n=2}^{n_\tau} 2^{np/2}(2^{-n/2}\|\psi\|_{L^p(J_1^n)})^p$$

$$\leq \frac{2^{p(n_\tau+1)/2}}{2^{p/2}-1} A_3(\psi)^p\,.$$

The inequality

$$(1-2^{-n_\tau}) \leq \tau \quad \text{or} \quad \sqrt{1-\tau} \leq 2^{-n_\tau/2}\,,$$

while taking the supremum over $\tau \in [3/4, 1[$, implies

$$A_1(\psi) \leq \frac{\sqrt{2}}{(2^{p/2}-1)^{1/p}} A_3(\psi)\,.$$

We have proved the equivalence

$$\kappa_{p,1}^{-1} N_{p,1}(\varphi) \leq N_{p,i}(\varphi) \leq \kappa_{p,1} N_{p,1}(\varphi) \quad \text{for } p=1,2,\ i=2,3\,,$$

with a universal constant $\kappa_{p,1} > 1$.

It now suffices to compare $N_{p,2}$ and $N_{p,4}$ or equivalently $N_{p,2,1,1}(\tilde{\varphi})$ and $N_{p,4,1,1}(\tilde{\varphi})$ written with $\psi = \frac{\tilde{\varphi}}{t^{1/p-1/2}}$

$$N_{p,2,1,1}(\tilde{\varphi}) = \|\psi\|_{L^p(J_1^{\leq 1})} + \sup_{\delta \in]0,1/8]} \sqrt{\delta}\|\psi\|_{L^p([1-2\delta,1-\delta])} =: B_2(\psi)\,,$$

$$N_{p,4,1,1}(\tilde{\varphi}) = \|\psi\|_{L^p(J_1^{\leq 2})} + \sup_{\delta \in]0,1/12]} \sqrt{\delta}\|\psi\|_{L^p([1-3\delta,1-\delta])} =: B_4(\psi)\,.$$

For the first terms of $B_2(\psi)$ and $B_4(\psi)$,

$$\|\psi\|^p_{L^p(J_1^{\leq 1})} \leq \|\psi\|^p_{L^p(J_1^{\leq 2})} = \|\psi\|^p_{L^p(J_1^{\leq 1})} + \|\psi\|^p_{L^p(J_1^2)}$$

gives

$$\|\psi\|_{L^p(J_1^{\leq 1})} \leq \|\psi\|_{L^p(J_1^{\leq 2})} \leq \|\psi\|_{L^p(J_1^{\leq 1})} + \sup_{\delta \in]0,1/8]} \|\psi\|_{L^p([1-2\delta,1-\delta])}\,.$$

For the second terms of $B_2(\psi)$ and $B_4(\psi)$,

$$\begin{aligned}\sqrt{\delta}\|\psi\|_{L^p([1-3\delta,1-3\delta/2])} &\leq \sqrt{\delta}\|\psi\|_{L^p([1-3\delta,1-\delta])}\\ &\leq \sqrt{\delta}\|\psi\|_{L^p([1-3\delta,1-3\delta/2])} + \sqrt{\delta}\|\psi\|_{L^p([1-2\delta,1-\delta])}\end{aligned}$$

leads to

$$\begin{aligned}(2/3)^{1/2} \sup_{\delta\in]0,1/8]} \sqrt{\delta}\|\psi\|_{L^p([1-3\delta,1-\delta])} &\leq \sup_{\delta\in]0,1/12]} \sqrt{\delta}\|\psi\|_{L^p([1-2\delta,1-\delta])}\\ &\leq 2 \sup_{\delta\in]0,1/8]} \sqrt{\delta}\|\psi\|_{L^p([1-3\delta,1-\delta])}\,.\end{aligned}$$

Adding the two terms yields the equivalences

$$\begin{aligned}\kappa_{p,2}^{-1}B_2(\psi) \leq B_4(\psi) \leq \kappa_{p,2}B_2(\psi)\\ \text{and } \kappa_{p,2}^{-1}N_{2,2}(\varphi) \leq N_{2,4}(\varphi) \leq \kappa_{p,2}N_{2,2}(\varphi)\end{aligned}$$

for a universal constant $\kappa_{p,2} > 1$. The proof ends by taking $\kappa_p = \kappa_{p,1}\kappa_{p,2} > 1$. □

5 Consequences of Strichartz Estimates for Our Model Problem

The general results of Sect. 4 are applied to our model problem presented in Sect. 2.3.

5.1 Validity of the General Hypotheses and Main Result

Let us consider (20)

$$\begin{cases} i\partial_t \hat{f}^h = (\xi - d\Gamma(D_y))^2 \hat{f}^h + \sqrt{h}[a(V) + a^*(V)]\hat{f}\,, \\ \hat{f}^h(t=0) = \hat{f}_0^h\,, \end{cases} \tag{95}$$

where $\hat{f}^h(t) \in L^2(\mathbb{R}^d \times Z'', \frac{d\xi}{(2\pi)^d} \otimes \mathbf{dz''}; \Gamma(L^2(\mathbb{R}^d, dy; \mathbb{C})))$, ξ is the Fourier variable of $x \in \mathbb{R}^d$ and $z'' \in Z''$ is a parameter, e.g. $L^2(Z'', \mathbf{dz''}) = L^2(\mathbb{R}^d, \frac{d\xi}{(2\pi)^d}; \mathbb{C}) \otimes \Gamma(L^2(\mathbb{R}^d, dy; \mathbb{C}))$ when we want to handle the evolution of Hilbert-Schmidt operators on $L^2(\mathbb{R}^d, \frac{d\xi}{(2\pi)^d}; \mathbb{C}) \otimes \Gamma(L^2(\mathbb{R}^d, dy; \mathbb{C}))$ as described in the end of

Sect. 2.3. Our complete parameter is thus

$$z' = (\xi, z') \in \mathbb{R}^d \times Z'' = Z'$$

and remember the writing introduced in Definition 3.2 and specified in (47) and (58)

$$L^2(\mathbb{R}^d \times Z'', \frac{d\xi}{(2\pi)^d} \otimes \mathbf{dz}; \Gamma(L^2(\mathbb{R}^d, dy; \mathbb{C}))) = L^2_{z,\mathrm{sym}} L^2_{y_G} = \underbrace{L^2_{z_0}}_{\text{vacuum}} \oplus L^2_{z_1,\mathrm{sym}} L^2_{y_G}$$

with $Z_0 = Z'$, $Z_1 = \mathcal{R} \otimes Z'$, where the subscript ${}_{\mathrm{sym}}$ refers to the symmetry for the relative coordinate variable $Y' \in \mathcal{R}$.

Using the center of mass variable (see Sect. 3) by setting $t \mapsto u^h_G(t) = U_G^{-1} \hat{f}^h(t)$, (95) becomes

$$\begin{cases} i\partial_t u^h_G = (\xi - D_{y_G})^2 u^h_G + \sqrt{h}[a^*_G(V) + a_G(V)]u^h_G\,, \\ u^h_G(t=0) = u^h_{G,0}\,. \end{cases} \tag{96}$$

In this context, the free dynamics $U(t)$ involved in (57) acts simply on $L^2_z L^2_{y_G}$ and equals

$$U(t) = K_0(t, z') \oplus U_1(t, Y', z') = (e^{-it|\xi|^2} \times_{z'}) \oplus (e^{-it(\xi - D_{y_G})^2} \times_{(Y', z')})$$

where we recall $z_0 = z' \in Z'$ and $z_1 = (Y', z') \in \mathcal{R} \times Z'$. Because $\|e^{\pm i\xi\cdot y}\varphi\|_{L^p_y} = \|\varphi\|_{L^p_y}$ for all $1 \le p \le +\infty$ and $e^{-itD_y^2} = e^{it\Delta_y}$ satisfies

$$\|e^{it\Delta_y} f\|_{L^2_y} \le \|f\|_{L^2_y}\,,$$

$$\|e^{it\Delta_y}(e^{is\Delta_y})^* g\|_{L^\infty_y} = \|e^{i(t-s)\Delta_y} g\|_{L^\infty_y} \le \frac{\|g\|_{L^1_y}}{(4\pi)^{d/2}|t-s|^{d/2}} \quad t \ne s\,,$$

the assumption (42)(43) are satisfied for $U_1(t, z_1)$, $z_1 = (Y', z')$, as soon as $d \ge 3$ with $\sigma = \frac{d}{2} > 1$, uniformly with respect to $z_1 \in \mathcal{R} \times Z'$.

The dense subset D in $L^2_{z_1} L^2_{y_G}$ such that $D \subset L^2(Z_1, \mathbf{dz_1}; L^{r_\sigma}(\mathbb{R}^d, dy_G; \mathbb{C}))$, with $r^\sigma = \frac{2d}{d-2}$ and $d \ge 3$ here, is simply $D = L^2(Z_1, dz_1; H^\mu(\mathbb{R}^d))$ with $\mu > d/2$. Remember that the dense subset D was introduced in Proposition 4.1 for the dense *a priori* definition of the operator A_T on $L^1([0, T]; L^2_{z,x})$ (see Remark 4.1 and the proof of Proposition 4.1).

Below are reviewed assumptions on V:

1. If $V \in L^{\frac{2d}{d+2}}(\mathbb{R}^d, dy; \mathbb{R})$, the assumptions of Proposition 4.3 are satisfied with $C_V = 1 + \|V\|_{L^{\frac{2d}{d+2}}} > 0$ and $r'_\sigma = \frac{2\sigma}{\sigma+1} = \frac{2d}{d+2}$.

2. If $V \in H^2(\mathbb{R}^d;\mathbb{R})$ then (95) (or (96)) defines a unitary dynamics with a rather well understood domain of its generator in $L^2(\mathbb{R}^d \times Z'', \frac{d\xi}{(2\pi)^d} \otimes$ **dz"**; $\Gamma(L^2(\mathbb{R}^d, dy;\mathbb{C}))) \simeq L^2_z L^2_{yG}$.

We will always assume $V \in L^{r'_\sigma}$ in the sequel, and depending on the statement we might assume that $V \in H^2$ or not.

If $V \in H^2(\mathbb{R}^d;\mathbb{R})$, the unique solution $t \mapsto u^h_G(t) = U_G^{-1}\hat{f}^h(t) \in C^0(\mathbb{R}; L^2_z L^2_{yG})$ to (96) satisfies

$$u^h_G(t) = U(t)u^h_{G,0} - i\int_0^t U(t)U(s)^*\sqrt{h}[a^*_G(V) + a_G(V)]u^h_G(s)\,ds\,, \tag{97}$$

and we will now seek for a solution of this equation using the fixed point method developed in Sect. 4.2, for $V \in L^{r'_\sigma}$ but not necessarily $V \in H^2(\mathbb{R}^d;\mathbb{R})$.

If $(u^h_\infty, u^h_2, u^h_1)$ solves

$$u^h_\infty(t) = -i\int_0^t U(t)U^*(s)\left(\sqrt{h}a^*_G(V)u^h_\infty(s) + \sqrt{h}u^h_2(s) + u^h_1(s)\right)ds + f^h_\infty(t)\,, \tag{98}$$

$$u^h_2(t) = -i\int_0^t a_G(V)U(t)U(s)^*\sqrt{h}u^h_2(s)\,ds \qquad + f^h_2(t)\,, \tag{99}$$

$$u^h_1(t) = -i\int_0^t a_G(V)U(t)U(s)^*\left(ha^*_G(V)u^h_\infty(s) + \sqrt{h}u^h_1(s)\right)ds\,, \tag{100}$$

with

$$f^h_\infty(t) = -i\int_0^t U(t)U(s)^*a^*_G(V)\sqrt{h}U(s)u^h_{G,0}\,ds\,, \tag{101}$$

$$f^h_2(t) = -i\,a_G(V)\int_0^t U(t)U(s)^*a^*_G(V)\sqrt{h}U(s)u^h_{G,0}\,ds + a_G(V)U(t)u^h_{G,0}\,, \tag{102}$$

written shortly as

$$\begin{pmatrix} u^h_\infty \\ u^h_2 \\ u^h_1 \end{pmatrix} = L\begin{pmatrix} u^h_\infty \\ u^h_2 \\ u^h_1 \end{pmatrix} + \begin{pmatrix} f^h_\infty \\ f^h_2 \\ 0 \end{pmatrix}, \quad L = \begin{pmatrix} L_{\infty\infty} & L_{\infty 2} & L_{\infty 1} \\ 0 & L_{22} & 0 \\ L_{1\infty} & 0 & L_{11} \end{pmatrix}, \tag{103}$$

then $u^h_G(t) = u_\infty(t) + U(t)u^h_{G,0}$ will yield a solution to (97).

Actually, with $u_G^h(t) = u_\infty^h(t) + U(t)u_{G,0}^h$, applying $a_G(V)$ to (98) on the one hand, and summing $\sqrt{h}\times$(99) and (100) on the other hand yields $\sqrt{h}a_G(V)u_G^h = u_1^h + \sqrt{h}u_2^h$, which inserted in (98) provides (97).

Theorem 5.1 *Assume $d \geq 3$ and $V \in L^{\frac{2d}{d+2}}(\mathbb{R}^d, dy; \mathbb{R})$ with*

$$\|V\|_{L^{\frac{2d}{d+2}}} < C_V\,.$$

Assume that there exists $\alpha_1 > 0$ and $C_{\alpha_1} > 0$ such that

$$\forall h \in]0, h_0[\,, \quad \|e^{\alpha_1 N} u_{G,0}^h\|_{L_z^2 L_{y_G}^2} \leq C_{\alpha_1}\,.$$

There exists a constant $C_d > 0$ depending on the dimension $d \geq 3$, such that when $\gamma > 0$ is chosen such that

$$2\|L\|_{\mathcal{L}(\mathcal{E}^h_{-\alpha_1,\alpha_1,\gamma})} \leq C_d e^{\alpha_1} C_V \gamma^{1/2} \leq 1\,,$$

the function $u_G^h(t) = u_\infty^h(t) + U(t)u_{G,0}^h$ with $(u_\infty^h, u_2^h, u_1^h)$ the unique solution to (103) in $(\mathcal{E}^h_{-\alpha_1,\alpha_1,\gamma}, M)$ satisfies

$$\forall t \in I_{T_\alpha}^h\,, \quad \left\|e^{\alpha N}[u_G^h(t) - U(t)u_{G,0}^h]\right\|_{L_z^2 L_{y_G}^2} \leq C_d C_V e^{\alpha_1} C_{\alpha_1} \gamma^{1/2} \frac{\sqrt{|ht|}}{\sqrt{T_\alpha - |ht|}}\,, \tag{104}$$

with

$$T_\alpha = \gamma(\alpha_1 - \alpha) \tag{105}$$

for all $\alpha \in [0, \alpha_1[$ and all $h \in]0, h_0[$.

If, moreover, $V \in H^2(\mathbb{R}^d; \mathbb{R})$, then u_G^h is the only solution to (96) in $C^0(I_{T_0}^h; L_z^2 L_{y_G}^2)$.

Proof We take $\alpha_0 = -\alpha_1$ where $\alpha_1 > 0$ is fixed. The constant $M_{\alpha 01}$ of Definition 4.3 is nothing but

$$M_{\alpha 01} = \frac{e^{\alpha_1}}{2}\,.$$

Accordingly to Definition 4.3, for a fixed $\gamma > 0$ the time scale T_α is given by $T_\alpha = \gamma(\alpha_1 - \alpha)$ for all $\alpha \in [-\alpha_1, \alpha_1[$. Proposition 4.3 tells us that the condition

$$C_{d,U}\frac{e^{\alpha_1}}{2} C_V \gamma^{1/2} \leq \frac{1}{2}$$

where $C_{d,U} = C_d$ is determined by the dimension $d \geq 3$ here, ensures that the operator L is a contraction in $\mathcal{E}^h_{-\alpha_1,\alpha_1,\gamma}$ for all $h \in]0, h_0[$:

$$\|L\|_{\mathcal{L}(\mathcal{E}^h_{-\alpha_1,\alpha_1,\gamma})} \leq \frac{1}{2}.$$

If $M(f^h_\infty, f^h_2, 0) < \infty$, then the system (103) admits a unique solution in $\mathcal{E}^h_{-\alpha_1,\alpha_1,\gamma}$ for all $h \in]0, h_0[$ with

$$M(u^h_\infty, u^h_2, u^h_1) \leq 2M(f^h_\infty, f^h_2, 0).$$

It remains to check two things:

- the right-hand side $(f^h_\infty, f^h_2, 0)$ given by (101)(102) belongs to $\mathcal{E}^h_{-\alpha_1,\alpha_1,\gamma}$ and to estimate $M(f^h_\infty, f^h_2, 0)$;
- the unique solution $(u^h_\infty, u^h_2, u^h_1)$ to (103) yields after setting $u^h_G(t) = u^h_\infty(t) + U(t)u^h_{G,0}$ the unique solution to (18) in $C^0(]-\frac{T_0}{h}, \frac{T_0}{h}[; L^2_{z,y_G})$.

The first step is simpler than what we did for Proposition 4.3. Let us start with

$$\begin{aligned} e^{\alpha N} f^h_\infty(t) &= -i \int_0^t U(t)U(s)^* e^{\alpha N}\sqrt{h} a^*_G(V) e^{-\alpha_1 N} U(s) e^{\alpha_1 N} u^h_{G,0}\, ds \\ &= \int_0^t U(t)U(s)^* F(s)\, ds \end{aligned}$$

with $F(s) = -i 1_{[0,t]}(s) e^{\alpha N}\sqrt{h} a^*_G(V) e^{-\alpha_1 N} U(s) e^{\alpha_1 N} u^h_{G,0}$. By Proposition 3.4 we know that

$$\|F\|_{L^2_s L^2_{z_1} L^{r'_\sigma}_{y_G}} \leq \frac{C_V e^{\alpha_1}}{2\sqrt{\alpha_1 - \alpha}} |ht|^{1/2} \|e^{\alpha_1 N} u^h_{G,0}\|_{L^2_z L^2_{y_G}} \leq \frac{C_V e^{\alpha_1} \gamma^{1/2}}{\sqrt{T_\alpha}} C_{\alpha_1} |ht|^{1/2}.$$

A direct application of the retarded endpoint Strichartz estimate (46) yields

$$\left(\frac{T_\alpha - |ht|}{|ht|}\right)^{1/2} \|e^{\alpha N} f^h_\infty(t)\|_{L^2_z L^2_{y_G}} \lesssim C_V e^{\alpha_1} C_{\alpha_1} \gamma^{1/2}.$$

and by taking the supremum over $|ht| < T_\alpha$,

$$M(f^h_\infty, 0) \lesssim C_V e^{\alpha_1} C_{\alpha_1} \gamma^{1/2}. \tag{106}$$

For

$$f^h_{2,1}(t) = -i\, a_G(V) \int_0^t U(t)U(s)^* a^*_G(V)\sqrt{h} U(s) u^h_{G,0}\, ds,$$

the Proposition 3.4 and the retarded Strichartz estimate (44) give

$$\sqrt{T_\alpha-\tau}\|e^{\alpha N}f^h_{2,1}\|_{L^2_t(I^h_\tau;L^2_zL^2_{y_G})}$$
$$\lesssim \sqrt{T_\alpha-\tau}\frac{C_Ve^{\alpha_1}}{\sqrt{\alpha_1-\alpha}}\left\|\int_0^t U(t)U(s)^*e^{\frac{\alpha+\alpha_1}{2}N}a^*_G(V)\sqrt{h}U(s)u^h_{G,0}\,ds\right\|_{L^2_{z_1}L^2_t(I^h_\tau;L^{r_\sigma}_{y_G})}$$
$$\lesssim \sqrt{T_\alpha-\tau}\frac{C_Ve^{\alpha_1}}{\sqrt{\alpha_1-\alpha}}\left\|e^{\frac{\alpha+\alpha_1}{2}N}a^*_G(V)\sqrt{h}U(s)u^h_{G,0}\right\|_{L^2_{z_1}L^2_s(I^h_\tau;L^{r'_\sigma}_{y_G})},$$

where here $r'_\sigma=\frac{2d}{d+2}$ and $r_\sigma=\frac{2d}{d-2}$.

Then using Proposition 3.4 again, the square integrability of 1 on I^h_τ and the boundedness of $U(s)$ in the L^2 norm,

$$\sqrt{T_\alpha-\tau}\|e^{\alpha N}f^h_{2,1}\|_{L^2_t(I^h_\tau;L^2_zL^2_{y_G})}$$
$$\lesssim \sqrt{T_\alpha-\tau}\frac{C_V^2e^{2\alpha_1}}{\alpha_1-\alpha}\left\|e^{\alpha_1N}\sqrt{h}U(s)u^h_{G,0}\right\|_{L^2_zL^2_s(I^h_\tau;L^2_{y_G})}$$
$$\lesssim C_V^2e^{2\alpha_1}\gamma\frac{\sqrt{T_\alpha-\tau}\sqrt{\tau}}{T_\alpha}\left\|e^{\alpha_1N}U(s)u^h_{G,0}\right\|_{L^\infty_s(I^h_\tau;L^2_zL^2_{y_G})}$$
$$\lesssim C_V^2e^{2\alpha_1}\gamma\left\|e^{\alpha_1N}u^h_{G,0}\right\|_{L^2_zL^2_{y_G}}$$

By taking the supremum w.r.t. $\alpha\in[-\alpha_1,\alpha_1[$ and dividing by $C_Ve^{\alpha_1}\gamma^{1/2}/2$ we obtain

$$M_2(f^h_{2,1})\lesssim C_Ve^{\alpha_1}C_{\alpha_1}\gamma^{1/2}. \tag{107}$$

It remains to control

$$f^h_{2,2}(t)=a_G(V)U(t)u^h_{G,0}.$$

For $-\alpha_1\le\alpha<\alpha_1$ and $0\le\tau<T_\alpha$, Proposition 3.4 and the homogeneous Strichartz estimate (44) yield

$$\sqrt{T_\alpha-\tau}\|e^{\alpha N}a_G(V)U(t)u^h_{G,0}\|_{L^2(I^h_\tau;L^2_zL^2_{y_G})}$$
$$\lesssim \sqrt{T_\alpha-\tau}\frac{C_V\,e^{\alpha_1}}{\sqrt{\alpha_1-\alpha}}\|U(t)e^{\alpha_1N}u^h_{G,0}\|_{L^2_{z_1}L^2(I^h_\tau;L^{r_\sigma}_{y_G})}$$
$$\lesssim C_V\,e^{\alpha_1}\sqrt{\gamma}\|e^{\alpha_1N}u^h_{G,0}\|_{L^2_zL^2_{y_G}}. \tag{108}$$

Taking the supremum over $\tau \in [0, T_\alpha[$, $\alpha \in [-\alpha, \alpha_1[$ and dividing by $C_V e^{\alpha_1}\gamma^{1/2}/2$ gives

$$M(0, f^h_{2,2}, 0) \lesssim C_{\alpha_1}$$

It can be improved by rewriting the system

$$\begin{pmatrix} u^h_\infty \\ u^h_2 \\ u^h_1 \end{pmatrix} = (\mathrm{Id} - L)^{-1} \begin{pmatrix} f^h_\infty \\ f^h_{2,1} + f^h_{2,2} \\ 0 \end{pmatrix} = \begin{pmatrix} 0 \\ f^h_{2,2} \\ 0 \end{pmatrix} + (\mathrm{Id} - L)^{-1} \begin{pmatrix} f^h_\infty \\ f^h_{2,1} \\ 0 \end{pmatrix}$$
$$+ (\mathrm{Id} - L)^{-1} L \begin{pmatrix} 0 \\ f^h_{2,2} \\ 0 \end{pmatrix}$$

from which we deduce

$$M(u^h_\infty, u^h_2 - f^h_{2,2}, u^h_1) \lesssim C_V e^{\alpha_1}\sqrt{\gamma}\left[M(f^h_\infty, f^h_{2,1}, 0) + M(0, f^h_{2,2}, 0)\right].$$

The inequalities (106), (107) and (108) prove that $(f^h_\infty, f^h_2, 0) \in \mathcal{E}^h_{-\alpha_1,\alpha_1,\gamma}$ and thus

$$M(u^h_\infty, u^h_2 - f^h_{2,2}, u^h_1) \leq 2M(f^h_\infty, f^h_2, 0) \lesssim C_V e^{\alpha_1} C_{\alpha_1}\gamma^{1/2}.$$

By possibly enlarging the constant $C_d > 0$, the above inequality becomes

$$M(u^h_\infty, u^h_2 - f^h_{2,2}, u^h_1) \leq C_d C_V C_{\alpha_1} e^{\alpha_1}\gamma^{1/2}.$$

We have finished the proof as soon as we can identify

$$u^h_\infty(t) = u^h_G(t) - U(t)u^h_{G,0}$$

for $t \in I^h_{T_\alpha}$ and $\alpha \in [0, \alpha_1[$. For $t \in I^h_{T_0}$, the function $u^h_G(t) = u^h_\infty(t) + U(t)u^h_{G,0}$ belongs to $C^0(I^h_{T_0}; L^2_z L^2_{y_G})$ and satisfies (97) which is equivalent to (96). By the existence and uniqueness for (97) or (96) in $C^0(I^h_{T_0}; L^2_z L^2_{y_G})$ when $V \in H^2(\mathbb{R}^d; \mathbb{R})$, u^h_G is the unique solution to (97) or (96) in $C^0(I^h_{T_0}; L^2_z L^2_{y_G})$. □

5.2 Consequences of Theorem 5.1

Let us work now with a general initial time t_0, specified later, and consider (96)

$$\begin{cases} i\partial_t u_G^h = (\xi - D_{y_G})^2 u_G^h + \sqrt{h}[a_G^*(V) + a_G(V)]u_G^h \,, \\ u_G^h(t = t_0) = u_{G,t_0}^h \,, \end{cases} \tag{109}$$

with the solution $u_G^h(t) = u_G^h(t' + t_0) = U(t')u_{G,t_0}^h + u_\infty^h$ in the framework of Theorem 5.1. For simplicity and because we work definitely in the framework of (109) we use here $U(t)U(s)^* = U(t-s)$. Remember that $(u_\infty^h, u_2^h, u_1^h)$ solves

$$\begin{pmatrix} u_\infty^h \\ u_2^h \\ u_1^h \end{pmatrix} = \underbrace{\begin{pmatrix} L_{\infty\infty} & L_{\infty 2} & L_{\infty 1} \\ 0 & L_{22} & 0 \\ L_{1\infty} & 0 & L_{11} \end{pmatrix}}_{=L} \begin{pmatrix} u_\infty^h \\ u_2^h \\ u_1^h \end{pmatrix} + \begin{pmatrix} f_\infty^h \\ f_2^h \\ 0 \end{pmatrix} \tag{110}$$

with

$$L_{\infty\infty}(\varphi)(t') = -i\int_0^{t'} U(t'-s)\sqrt{h}a_G^*(V)\varphi(s)\, ds \,,$$

$$L_{\infty 1}(\varphi)(t') = -i\int_0^{t'} U(t'-s)\varphi(s)\, ds \,, \quad L_{\infty 2}(\varphi)(t') = \sqrt{h}L_{\infty 1}(\varphi)(t') \,,$$

$$L_{qq}(\varphi)(t') = -i\int_0^{t'} a_G(V)U(t'-s)\sqrt{h}\varphi(s)\, ds \,, \quad q \in \{2, 1\} \,,$$

$$L_{1\infty}(\varphi)(t') = -ih\int_0^{t'} a_G(V)U(t'-s)a_G^*(V)\varphi(s)\, ds \,,$$

and

$$f_\infty^h(t') = -i\int_0^{t'} U(t'-s)a_G^*(V)\sqrt{h}U(s)u_{G,t_0}^h\, ds \,,$$

$$f_2^h(t') = \underbrace{-ia_G(V)\int_0^{t'} U(t'-s)a_G^*(V)\sqrt{h}U(s)u_{G,t_0}^h\, ds}_{f_{2,1}^h(t')} + \underbrace{a_G(V)U(t')u_{G,t_0}^h}_{f_{2,2}^h(t')} \,.$$

Theorem 5.1 provides a framework in which L is a contraction and we will use it twice while inverting

$$\begin{pmatrix} u^h_\infty \\ u^h_2 \\ u^h_1 \end{pmatrix} = (\mathrm{Id}-L)^{-1}\begin{pmatrix} f^h_\infty \\ f^h_2 \\ 0 \end{pmatrix} = \begin{pmatrix} 0 \\ f^h_{2,2} \\ 0 \end{pmatrix} + (\mathrm{Id}-L)^{-1}\begin{pmatrix} f^h_\infty \\ f^h_{2,1} \\ 0 \end{pmatrix} + (\mathrm{Id}-L)^{-1}L\begin{pmatrix} 0 \\ f^h_{2,2} \\ 0 \end{pmatrix}$$

and then using the Neumann expansion $(\mathrm{Id}-L)^{-1} = \sum_{k=0}^{\infty} L^k$ for different values of t_0 and of the parameter γ in Theorem 5.1. The following result is an easy consequence of Theorem 5.1.

Proposition 5.1 *Assume that the initial datum $u^h_{G,0}$ for $t_0 = 0$ in (109) satisfies the uniform bound $\|e^{2\alpha_1 N}u^h_{G,0}\|_{L^2_z L^2_{y_G}} \leq C_{\alpha_1}$ for all $h \in]0, h_0[$. Then there exists $\hat{T}_{\alpha_1} > 0$ and $\tilde{C}_{\alpha_1} > 0$, $\delta_{\alpha_1} > 0$, such that*

(a) The following weighted estimate

$$\|e^{\alpha_1 N}u^h_G(t)\|_{L^2_z L^2_{y_G}} \leq \tilde{C}_{\alpha_1}$$

holds true for all $t \in I^h_{\hat{T}_{\alpha_1}} =]-\frac{\hat{T}_{\alpha_1}}{h}, \frac{\hat{T}_{\alpha_1}}{h}[$ and all $h \in]0, h_0[$.

(b) For $t_0 \in I^h_{\hat{T}_{\alpha_1}}$, $u^h_G(t_0 + \delta/h)$ admits in $e^{-\frac{\alpha_1}{2}N}L^2_z L^2_{y_G}$ the following asymptotic expansion in terms of $\delta \in [-\delta_1, \delta_1]$,

$$\begin{aligned} u^h_G(t_0+\delta/h) &= \underbrace{U(\delta/h)u^h_G(t_0)}_{O(1)} \\ &\quad \underbrace{-\, i\sqrt{h}\int_0^{\delta/h} U(\delta/h - s)\,[a^*_G + a_G](V)\,U(s)u^h_G(t_0)\,ds}_{O(|\delta|^{1/2})} \\ &\underbrace{-h\int_0^{\frac{\delta}{h}}\int_0^s U(\frac{\delta}{h}-s)\,[a^*_G+a_G](V)\,U(s-s')\,[a^*_G+a_G](V)\,U(s')u^h_G(t_0)\,ds'\,ds}_{O(|\delta|)} \\ &\quad + O(|\delta|^{3/2}) \end{aligned}$$

where $v(h,\delta) = O(|\delta|^{k/2})$, $k = 0, 1, 2, 3$, means $\|e^{\frac{\alpha_1}{2}N}v(h,\delta)\|_{L^2_z L^2_{y_G}} \leq \tilde{C}_{\alpha_1}|\delta|^{k/2}$ uniformly with respect to $h \in]0, h_0[$ and $t_0 \in I^h_{\hat{T}_{\alpha_1}}$.

Proof

(a) Fix $\alpha_1 > 0$ and apply Theorem 5.1 with α_1 replaced by $2\alpha_1$. There exists $\gamma = \gamma_1 > 0$, determined by α_1, $C_{12}(V)$ and the dimension $d \geq 3$, such that the operator L is a contraction in $\mathcal{E}^h_{-2\alpha_1,2\alpha_1,\gamma_1}$. The system (110) for $t_0 = 0$ admits a unique solution with the norm M in $\mathcal{E}^h_{-2\alpha_1,2\alpha_1,\gamma_1}$ estimated by

$$M(u^h_\infty, u^h_2, u^h_1) \lesssim C_{\alpha_1} \tag{111}$$

and the solution u^h_G to (18) equals

$$u^h_G(t) = U(t)u^h_{G,0} + u^h_\infty(t)\,.$$

With $T_{\alpha_1} = \gamma_1(2\alpha_1 - \alpha_1) = \gamma_1\alpha_1$, the estimate (111) says in particular

$$\forall t \in I^h_{T_{\alpha_1}}\,, \quad \|e^{\alpha_1 N}u^h_\infty(t)\|_{L^2_z L^2_{y_G}} \lesssim C_{\alpha_1}\frac{\sqrt{|ht|}}{\sqrt{T_{\alpha_1} - |ht|}}\,.$$

By taking $\hat{T}_{\alpha_1} = \frac{T_{\alpha_1}}{2}$ with $|ht| \leq \frac{T_{\alpha_1}}{2}$ when $t \in I_{\hat{T}^h_{\alpha_1}}$ and with

$$\|e^{\alpha_1 N}U(t)u^h_{G,0}\|_{L^2_z L^2_{y_G}} \leq \|e^{\alpha_1}u^h_{G,0}\|_{L^2_z L^2_{y_G}} \leq C_{\alpha_1}\,,$$

we finally obtain

$$\forall t \in I^h_{\hat{T}_{\alpha_1}}\,, \quad \|e^{\alpha_1 N}u^h_G(t)\|_{L^2_z L^2_{y_G}} \lesssim \tilde{C}_{\alpha_1}\,,$$

for $\tilde{C}_{\alpha_1}$ large enough.

[(**b**)] With **a)** the initial datum $u^h_{G,t_0} = u^h_G(t_0)$ of (109) fulfils the assumptions of Theorem 4.1 after time translation $t' = t - t_0$ and where $t' \in I^h_T$ means $t \in t_0 + I^h_T$. For any $\gamma > 0$ small enough and by setting $T_\alpha = \gamma(\alpha_1 - \alpha)$ for $\alpha \in [0, \alpha_1]$ we know that the system (110) satisfies

$$\|L\|_{\mathcal{L}(\mathcal{E}^h_{-\alpha_1,\alpha_1,\gamma})} \lesssim \gamma^{1/2}\,, \quad M(f^h_\infty, f^h_{2,1}, 0) \lesssim C_{\alpha_1}\gamma^{1/2}\,, \quad M(0, f^h_{2,2}, 0) \lesssim C_{\alpha_1}\,,$$

while $u^h_G(t' + t_0) = U(t')u^h_G(t_0) + u^h_\infty(t')$ for $t' \in I^h_{T_\alpha}$.

In particular

$$\begin{pmatrix} u^h_\infty \\ u^h_2 \\ u^h_1 \end{pmatrix} = \begin{pmatrix} 0 \\ f^h_{2,2} \\ 0 \end{pmatrix} + (\mathrm{Id} - L)^{-1}\begin{pmatrix} f^h_\infty \\ f^h_{2,1} \\ 0 \end{pmatrix} + (\mathrm{Id} - L)^{-1}L\begin{pmatrix} 0 \\ f^h_{2,2} \\ 0 \end{pmatrix}$$

leads to

$$\begin{pmatrix} u_\infty^h \\ u_2^h \\ u_1^h \end{pmatrix} = \begin{pmatrix} f_\infty^h + L_{\infty 2}(f_{2,2}^h) + L_{\infty\infty}(f_\infty^h + L_{\infty 2}(f_{2,2}^h)) + L_{\infty 2}(f_{2,1}^h + L_{22}(f_{2,2}^h)) \\ f_{2,2}^h + f_{2,1}^h + L_{22}(f_{2,2}^h) + L_{22}(f_{2,1}^h + L_{22}(f_{2,2}^h)) \\ L_{1\infty}(f_\infty^h + L_{\infty 2}(f_{2,2}^h)) \end{pmatrix} + O(\gamma^{3/2})$$

in $\mathcal{E}^h_{-\alpha_1,\alpha_1,\gamma}$. By using the first line with $\alpha = \frac{\alpha_1}{2}$, and by setting

$$v^h(t') = U(t')u_G^h(t_0) + [f_\infty^h(t') + L_{\infty 2}(f_{2,2}^h)](t') + L_{\infty\infty}[f_\infty^h + L_{\infty 2}(f_{2,2}^h)](t') + L_{\infty 2}[f_{2,1}^h + L_{22}(f_{2,2}^h)](t')$$

the difference $u_G^h(t_0 + t') - v^h(t')$ satisfies satisfies

$$\forall t' \in I^h_{T_{\frac{\alpha_1}{2}}}, \quad \|e^{\frac{\alpha_1}{2}N}[u_G^h(t_0 + t') - v^h(t')]\|_{L_z^2 L_{y_G}^2} \lesssim \gamma^{3/2} \frac{\sqrt{|ht'|}}{\sqrt{T_0 - |ht'|}},$$

where $T_{\frac{\alpha_1}{2}} = \frac{\gamma\alpha_1}{2}$. For $\delta = \pm\frac{T_{\frac{\alpha_1}{2}}}{2} = \pm\frac{\gamma\alpha_1}{4}$ we obtain

$$\|e^{\frac{\alpha_1}{2}N}[u_G^h(t_0 + \delta/h) - v^h(\delta/h)]\|_{L_z^2 L_{y_G}^2} = O(|\delta|^{3/2}).$$

It now suffices to specify all the terms of $v^h(\delta/h)$:

- The first one is nothing but $U(\delta/h)u_G^h(t_0)$ with an $O(1)$-norm.
- The second term

$$f_\infty^h(\delta/h) + L_{\infty 2}(f_{2,2}^h)(\delta/h) = -i\int_0^{\delta/h} U(\delta/h - s)\sqrt{h}[a_G^*(V) + a_G(V)] U(s)u_G^h(t_0)\, ds$$

  has an $O(\delta^{1/2})$-norm.
- All the other terms have an $O(\delta)$-norm and equal

$$L_{\infty\infty}(f_\infty^h)(\delta/h) = -h\int_0^{\delta/h}\int_0^s U(\delta/h - s)a_G^*(V)U(s - s')a_G^*(V)U(s')u_G^h(t_0)\, ds'ds,$$

$$L_{\infty\infty}(L_{\infty 2}(f^h_{2,2}))(\delta/h) =$$
$$-h\int_0^{\delta/h}\int_0^s U(\delta/h-s)a^*_G(V)U(s-s')a_G(V)U(s')u^h_G(t_0)\,ds'ds\,,$$
$$L_{\infty 2}(f^h_{2,1})(\delta/h) =$$
$$-h\int_0^{\delta/h}\int_0^s U(\delta/h-s)a_G(V)U(s-s')a^*_G(V)U(s')u^h_G(t_0)\,ds'ds\,,$$
$$L_{\infty 2}(L_{22}(f^h_{2,2}))(\delta/h) =$$
$$-h\int_0^{\delta/h}\int_0^s U(\delta/h-s)a_G(V)U(s-s')a_G(V)U(s')u^h_G(t_0)\,ds'ds\,.$$

This ends the proof. □

6 Semiclassical Measures

We will check here that semiclassical (or Wigner) measures for our model problem presented in Sect. 2.3 can be defined simultaneously for all macroscopic times $t \in]-\hat{T}_{\alpha_1}, \hat{T}_{\alpha_1}[$.

6.1 Framework

Below are reviewed a few properties of semiclassical measures or Wigner measures. We refer the reader e.g. to [6, 14–16, 19, 27] for various presentations of those now well known objects.

(a) The Anti-Wick quantization on $\mathbb{R}^d$ is defined by

$$a^{\text{A-Wick}}(hx, D_x) = \int_{T^*\mathbb{R}^d} a(X)\,|\varphi^h_X\rangle\langle\varphi^h_X|\,\frac{dX}{(2\pi h)^d}$$

is defined for any $a \in L^\infty(T^*\mathbb{R}^d, dx; \mathbb{C})$ with

$$\varphi^h_{X_0}(x) = \frac{h^{d/4}}{\pi^{d/4}}e^{i\xi_0\cdot(x-\frac{x_0}{2h})}e^{-\frac{h(x-\frac{x_0}{h})^2}{2}}\,, \qquad X_0 = (x_0, \xi_0) \in T^*\mathbb{R}^d\,.$$

It is a non negative quantization for which

$$(a \geq 0) \Rightarrow (a^{\text{A-Wick}}(hx, D_x) \geq 0)$$

and

$$\|a^{\text{A-Wick}}(hx, D_x)\|_{\mathcal{L}(L^2(\mathbb{R}^d,dx;\mathbb{C}))} \le \|a\|_{L^\infty}\,.$$

A natural separable subspace of $L^\infty(T^*\mathbb{R}^d;\mathbb{C})$ is

$$\begin{aligned} C_0^0(T^*\mathbb{R}^d;\mathbb{C}) &= \left\{a \in C^0(T^*\mathbb{R}^d;\mathbb{C})\,,\ \lim_{X\to\infty} a(X) = 0\right\} \\ \text{resp.}\quad C^0(T^*\mathbb{R}^d \sqcup \{\infty\}\,;\mathbb{C}) &= C_0^0(T^*\mathbb{R}^d;\mathbb{C}) \oplus \mathbb{C}\,1 \\ &= \left\{a \in C^0(T^*\mathbb{R}^d;\mathbb{C})\,,\ \lim_{X\to\infty} a(X) \in \mathbb{C}\right\}, \end{aligned}$$

endowed with the C^0 norm, of which the dual is the space $\mathcal{M}_b(T^*\mathbb{R}^d;\mathbb{C})$ (resp. $\mathcal{M}_b(T^*\mathbb{R}^d \sqcup \{\infty\}\,;\mathbb{C})$) of bounded Radon measures on $T^*\mathbb{R}^d$ (resp. $T^*\mathbb{R}^d \sqcup \{\infty\}$).

(b) For a bounded family $(\varrho_h)_{h\in]0,h_0[}$ of normal states $\varrho_h \in \mathcal{L}^1(L^2(\mathbb{R}^d,dx;\mathbb{C}))$, $\varrho_h \ge 0$, $\mathrm{Tr}[\varrho_h] = 1$, the set of semiclassical measures on $T^*\mathbb{R}^d$ (resp. $T^*\mathbb{R}^d \sqcup \{\infty\}$) is defined as the weak* limit point in $\mathcal{M}_b(T^*\mathbb{R}^d;\mathbb{R}_+)$ (resp. $\mathcal{M}_b(T^*\mathbb{R}^d \sqcup \{\infty\}\,;\mathbb{R}_+)$) of $\frac{\sigma^{\text{Wick}}(\varrho_h)}{(2\pi h)^d}$ with

$$\sigma^{\text{Wick}}(\varrho_h)(X) = \langle \varphi_X^h\,,\, \varrho_h \varphi_X^h\rangle_{L^2(\mathbb{R}^d)} = \mathrm{Tr}\left[\varrho_h |\varphi_X^h\rangle\langle\varphi_X^h|\right].$$

This is extended by linearity for any bounded family $(\varrho_h)_{h\in]0,h_0[}$ in the space $\mathcal{L}^1(L^2(\mathbb{R}^d,dx;\mathbb{C}))$.

The set of semiclassical measures is denoted by

$$\mathcal{M}(\varrho_h, h \in]0,h_0[)\,,$$

and when h is restricted to a set $\mathcal{E} \subset]0,h_0[$, $0 \in \overline{\mathcal{E}}$, we use

$$\mathcal{M}(\varrho_h, h \in \mathcal{E})\,.$$

After recalling

$$\int_{T^*\mathbb{R}^d} a(X)\,\sigma^{\text{Wick}}(\varrho_h)(X)\,\frac{dX}{(2\pi h)^d} = \mathrm{Tr}\left[a^{\text{A-Wick}}(hx,D_x)\,\varrho_h\right],$$

semiclassical measures $\mu \in \mathcal{M}(\varrho_h, h \in]0, h_0[)$ are characterized by the existence of a sequence $(h_k)_{k\in\mathbb{N}^*}$, $h_k \in \mathcal{E}$ such that

$$\lim_{k\to\infty} h_k = 0\,,$$

$$\lim_{k\to\infty} \mathrm{Tr}\left[a^{\mathrm{A-Wick}}(h_k x, D_x)\,\varrho_{h_k}\right] = \int_{T^*\mathbb{R}^d} a(X)\,d\mu(X)\,, \quad \forall a \in \mathcal{D}\,,$$

$$\lim_{k\to\infty} \mathrm{Tr}\left[\varrho_{h_k}\right] = \mu(T^*\mathbb{R}^d \sqcup \{\infty\}) = \mu(T^*\mathbb{R}^d) + \mu(\infty)\,,$$

where $\mathcal{D}$ is any dense set of $\mathcal{C}_0^0(T^*\mathbb{R}^d;\mathbb{C})$.

(c) After choosing $\mathcal{D} = \mathcal{C}_0^\infty(T^*\mathbb{R}^d;\mathbb{C})$ and by recalling $\|a^{\mathrm{A-Wick}}(hx, D_x) - a^{\mathrm{Weyl}}(hx, D_x)\| = O(h)$, for any $a \in S(1, dx^2 + d\xi^2) \supset \mathcal{C}_0^\infty(T^*\mathbb{R}^d;\mathbb{C})$, semiclassical measures are characterized by

$$\forall a \in \mathcal{C}_0^\infty(T^*\mathbb{R}^d;\mathbb{C})\,, \quad \lim_{k\to\infty} \mathrm{Tr}\left[a^{\mathrm{Weyl}}(h_k x, D_x)\,\varrho_{h_k}\right] = \int_{T^*\mathbb{R}^d} a(X)\,d\mu(X)\,,$$

or

$$\forall P \in T^*\mathbb{R}^d\,, \quad \lim_{k\to\infty} \mathrm{Tr}\left[\tau_P^{h_k}\varrho_{h_k}\right] = \int_{T^*\mathbb{R}^d} e^{i(p_\xi\cdot x - p_x\cdot\xi)}\,d\mu(x,\xi)\,,$$

with

$$\tau_P^h = (e^{i(p_\xi\cdot x - p_x\cdot\xi)})^{\mathrm{Weyl}}(hx, D_x) = e^{i(p_\xi\cdot(hx) - p_x\cdot D_x)}\,, \quad P = (p_x, p_\xi)\,.$$

The compactification $T^*\mathbb{R}^d \sqcup \{\infty\}$ is just a way to count the mass of $(\varrho_{h_k})_{k\in\mathbb{N}^*}$ which is not caught by the compactly supported obervables $a \in \mathcal{C}_0^\infty(T^*\mathbb{R}^d;\mathbb{C})$.

(d) Semiclassical measures are transformed by the dual action of semiclassical Fourier integral operator on $a^{\mathrm{Weyl}}(hx, D_x)$, $a \in \mathcal{C}_0^\infty(T^*\mathbb{R}^d;\mathbb{C})$.

(e) When $\mathcal{M}(\varrho_{h,1}, h \in \mathcal{E}) = \{\mu_1\}$ and $\mathcal{M}(\varrho_{h,2}, h \in \mathcal{E}) = \{\mu_2\}$ the total variation between μ_1 and μ_2 is estimated by

$$|\mu_2 - \mu_1|(\underbrace{T^*\mathbb{R}^d}_{\text{or } T^*\mathbb{R}^d\sqcup\{\infty\}}) \le 4 \liminf_{h\to 0} \|\varrho_{h,1} - \varrho_{h,2}\|_{\mathcal{L}^1}\,.$$

(f) When (Λ, d_Λ) is a metric space and $(\varrho_h(\lambda))_{h\in]0,h_0[,\lambda\in\Lambda}$ is a bounded family in $\mathcal{L}^1(L^2(\mathbb{R}^d, dx;\mathbb{C}))$, semiclassical measures can be defined simultaneously for all $\lambda \in \Lambda$, if for any sequence $(h_n)_{n\in\mathbb{N}}$, $\lim_{n\to\infty} h_n = 0^+$, there exists a subsequence $(h_{n_k})_{k\in\mathbb{N}}$ such that

$$\forall \lambda \in \Lambda\,, \exists \mu_\lambda \in \mathcal{M}_b(T^*\mathbb{R}^d \sqcup \{\infty\})\,,$$

$$\lim_{k\to\infty} \mathrm{Tr}\left[a^{\mathrm{A-Wick}}(h_{n_k} x, D_x)\,\varrho_{h_{n_k}}(\lambda)\right] = \int_{T^*\mathbb{R}^d\sqcup\{\infty\}} a(X)\,d\mu_\lambda(X)\,.$$

By assuming (Λ, d_Λ) separable, sufficient conditions for this property are either

- For all given $a \in C_0^\infty(T^*\mathbb{R}^d; \mathbb{C})$, $\mathrm{Tr}\left[a^{\mathrm{Weyl}}(hx, D_x)\,\varrho_h(\lambda)\right]$ is an equicontinuous family of continuous functions from Λ to $\mathbb{C}$, or
- The map $(P, \lambda) \mapsto \mathrm{Tr}\left[\tau_P^h\,\varrho_h(\lambda)\right]$ is an equicontinuous family of continuous functions from $T^*\mathbb{R}^d \times \Lambda$ to $\mathbb{C}$.

For the first characterization, apply a diagonal extraction process for a dense countable subset of (Λ, d_Λ) (and a dense countable subset of $C_0^0(T^*\mathbb{R}^d)$ lying in $C_0^\infty(T^*\mathbb{R}^d; \mathbb{C})$) and then apply the various characterisations of elements of $\mathcal{M}(\varrho_h(\lambda), h \in \mathcal{E})$.

Like in our problem, semiclassical measures can be defined for bounded families $\varrho_h \in \mathcal{L}^1(L^2(\mathbb{R}^d \times Z', dx \otimes \mathbf{dz'}; \mathbb{C}))$ after using observables $a^{\mathrm{Weyl}}(hx, D_x) \otimes \mathrm{Id}_{L^2_z}$.

When $(\varrho_h)_{h\in]0,h_0[}$ is a family of states, $\varrho_h \geq 0$ and $\mathrm{Tr}[\varrho_h] = 1$, the relationship with the study of pure states can be done in two ways:

- Firstly by writing a general state as a convex combination of pure states, provided that this convex decomposition is explicit enough to follow the behaviour as $h \to 0^+$.
- Secondly by writing $\varrho_h = \varrho_h^{1/2}\varrho_h^{1/2}$ and taking $\Psi_h = \varrho_h^{1/2} \in \mathcal{L}^2(L^2(\mathbb{R}^d \times Z', dx \otimes \mathbf{dz'}; \mathbb{C})) \sim L^2(\mathbb{R}^d \times Z' \times \hat{Z}, dx \otimes \mathbf{dz'} \otimes \mathbf{d\hat{z}}; \mathbb{C})$ where $\hat{Z}$ is another copy of $\mathbb{R}^d \times Z'$ with $\mathbf{d\hat{z}} = dx \otimes \mathbf{dz'}$. Then

$$\mathrm{Tr}\left[(a^{\mathrm{Weyl}}(hx, D_x) \otimes \mathrm{Id}_{L^2_{z'}})\varrho_h\right] = \langle \Psi_h\,,\, (a^{\mathrm{Weyl}}(hx, D_x) \otimes \mathrm{Id}_{L^2_{z',\hat{z}}})\Psi_h \rangle\,.$$

6.2 Equicontinuity

The following result, which is the first useful information about semiclassical measures, before computing them, comes from the equicontinuity directly deduced from Proposition 5.1. The unitary transforms introduced in Sects. 2.3 and 3 in order to transform (18) into (96) and $a^{\mathrm{Weyl}}(hx, D_x) \otimes \mathrm{Id}$ into $a^{\mathrm{Weyl}}(-hD_\xi, \xi - D_{y_G})$ are not recalled here and the results are directly formulated for the initial problem (18) and the semiclassical observables $a^{\mathrm{Weyl}}(hx, D_x) \otimes \mathrm{Id}$.

Proposition 6.1 *Assume*

$$V \in L^{r'_\sigma}(\mathbb{R}^d, dx; \mathbb{R}) \cap H^2(\mathbb{R}^d; \mathbb{R})\,, \quad r'_\sigma = \frac{2d}{d+2}\,, \quad d \geq 3\,,$$

and let $U_{\mathcal{V}}(t) = e^{-it(-\Delta_x + \sqrt{h}\mathcal{V})}$ *like in Sect. 2.3.*

Assume that there exists $\alpha_1 > 0$ *such that* $\varrho_h(0) \in \mathcal{L}^1(L^2(\mathbb{R}^d \times \Omega, dx \otimes \mathcal{G}; \mathbb{C}))$, $\varrho_h(0) \geq 0$, $\mathrm{Tr}[\varrho_h(0)] = 1$ *satisfies*

$$\exists C_{\alpha_1} > 0\,,\ \forall h \in]0, h_0[\,,\quad \mathrm{Tr}\left[e^{\alpha_1 N}\varrho_h(0)e^{\alpha_1 N}\right] \leq C_{\alpha_1}\,.$$

Then there exists $\hat{T}_{\alpha_1} > 0$ *such that elements of* $\mathcal{M}(\varrho_h(t), h \in]0, h_0[)$ *can be defined simultaneously for all macroscopic times* $t \in]-\hat{T}_{\alpha_1}, \hat{T}_{\alpha_1}[$ *when* $\varrho_h(t) = U_{\mathcal{V}}(\frac{t}{h})\,\varrho_h(0)\,U_{\mathcal{V}}^*(\frac{t}{h})$.

Proof When $U(s) = e^{-is(-\Delta_x)}$ denotes the free unitary transform, the time evolved observable $U^*(\frac{s}{h})[a^{\mathrm{Weyl}}(hx, D_x) \otimes \mathrm{Id}_{L^2_\omega}]U(\frac{s}{h})$ equals exactly $a^{\mathrm{Weyl}}(hx, D_x, s) \otimes \mathrm{Id}_{L^2_\omega}$ with

$$a(x, \xi, s) = a(x + 2\xi s, \xi)\,.$$

It is clearly equicontinuous in $h \in]0, h_0[$ with respect to $s \in [-\hat{T}_{\alpha_1}, \hat{T}_{\alpha_1}]$ in $\mathcal{L}(L^2_{x,\omega})$ for any given $a \in C_0^\infty(T^*\mathbb{R}^d; \mathbb{C})$:

$$\|a^{\mathrm{Weyl}}(hx, D_x, s) - a^{\mathrm{Weyl}}(hx, D_x, 0)\|_{\mathcal{L}(L^2_{x,\omega})} \leq C_a|s|\,.$$

We drop the tensorization with Id_{L^2}. With

$$\begin{aligned}\mathrm{Tr}&\left[a^{\mathrm{Weyl}}(hx, D_x)\varrho_h(t+\delta)\right] - \mathrm{Tr}\left[a^{\mathrm{Weyl}}(hx, D_x)\varrho_h(t)\right]\\ &= \mathrm{Tr}\left[a^{\mathrm{Weyl}}(hx, D_x, \delta)U^*(\frac{\delta}{h})U_{\mathcal{V}}(\frac{\delta}{h})\varrho_h(t)U_{\mathcal{V}}^*(\frac{\delta}{h})U(\frac{\delta}{h})\right]\\ &\qquad\qquad\qquad - \mathrm{Tr}\left[a^{\mathrm{Weyl}}(hx, D_x, 0)\varrho_h(t)\right]\end{aligned}$$

it thus suffices to check, uniformly with respect to $(h, t) \in]0, h_0[\times]-\hat{T}_{\alpha_1}, \hat{T}_{\alpha_1}[$, the estimate

$$\|U^*(\frac{\delta}{h})U_{\mathcal{V}}(\frac{\delta}{h})\varrho_h(t)U_{\mathcal{V}}^*(\frac{\delta}{h})U(\frac{\delta}{h}) - \varrho_h(t)\|_{\mathcal{L}^1} = o_{\delta\to 0}(1)\,. \tag{112}$$

We now use the decomposition $\varrho_h(0) = \varrho_h(0)^{1/2}\varrho_h(0)^{1/2}$ and consider the evolution

$$U_{\mathcal{V}}(\frac{t}{h})\varrho_h(0)^{1/2} \in \mathcal{L}^2(L^2(\mathbb{R}^d \times \Omega, dx \otimes \mathcal{G}; \mathbb{C})) \sim L^2(\mathbb{R}^d \times \Omega \times \hat{Z}, dx \otimes \mathcal{G} \otimes \mathbf{d\hat{z}}; \mathbb{C})$$

with $\hat{Z} = \mathbb{R}^d \times \Omega$, $\mathbf{d\hat{z}} = dx \otimes \mathcal{G}$.

The estimate (112) is done as soon as

$$\|U^*(\frac{\delta}{h})U_{\mathcal{V}}(\frac{\delta}{h})[U_{\mathcal{V}}(\frac{t}{h})\varrho_h(0)^{1/2}] - [U_{\mathcal{V}}(\frac{t}{h})\varrho_h(0)^{1/2}]\|_{L^2_{x,\omega,\hat{z}}} = o_{\delta\to 0}(1)$$

uniformly with respect to $(h,t) \in]0,h_0[\times] - \hat{T}_{\alpha_1}, \hat{T}_{\alpha_1}[$.

This problem is now translated in a problem in

$$\underbrace{L^2(\mathbb{R}^d \times \hat{Z}, \frac{d\xi}{(2\pi)^d} \otimes \mathbf{d\hat{z}}; \mathbb{C})}_{\text{vacuum}} \oplus L^2_{\text{sym}}(\mathbb{R}^d_{y_G} \times Z_1; dy_G \otimes \mathbf{dz_1}; \mathbb{C})$$

by the unitary transform U_G associated with the center of mass y_G of Sect. 3, the translation invariance and its Fourier variable $\xi \in \mathbb{R}^d$ and the relative coordinates $Y' \in \mathcal{R}$. The variable $z_1 \in Z_1$ is nothing but $z_1 = (\xi, Y', \hat{z}) \in \mathbb{R}^d \times \mathcal{R} \times \hat{Z}$ with $\mathbf{dz_1} = \frac{d\xi}{(2\pi)^d} \otimes \mu \otimes \mathbf{d\hat{z}}$. The subscript $_{\text{sym}}$ refers to the symmetry in the variable $Y' \in \mathcal{R}$. All the assumptions of Theorem 5.1 have been checked in Sect. 5. In particular we can use Proposition 5.1-b) with

$$u^h_G(\frac{t}{h}) = U_{\mathcal{V}}(\frac{t}{h})\varrho_h(0)^{1/2} \quad \text{and} \quad \frac{t}{h} \in I^h_{\hat{T}_{\alpha_1}}.$$

It says in particular

$$u^h_G(\frac{t}{h} + \frac{\delta}{h}) = U(\frac{\delta}{h})u^h_G(\frac{t}{h}) + O(|\delta|^{1/2}),$$

uniformly with respect to $(h, \frac{t}{h}) \in]0,h_0[\times I^h_{\hat{T}_{\alpha_1}}$, and therefore

$$\|U^*(\frac{\delta}{h})U_{\mathcal{V}}(\frac{\delta}{h})[U_{\mathcal{V}}(\frac{t}{h})\varrho_h(0)^{1/2}] - [U_{\mathcal{V}}(\frac{t}{h})\varrho_h(0)^{1/2}]\|_{L^2_{x,\omega,\hat{z}}} = O_{\delta\to 0}(|\delta|^{1/2})$$

uniformly with respect to $(h,t) \in]0,h_0[\times] - \hat{T}_{\alpha_1}, \hat{T}_{\alpha_1}[$.

This ends the proof. □

7 Approximations

With our number estimates stated in Sect. 5, various approximations can be considered for the general class of initial data $(\varrho_h(0))_{h\in]0,h_0[}$, $\varrho_h(0) \in \mathcal{L}^1(L^2(\mathbb{R}^d \times \Omega, dx \otimes \mathcal{G}; \mathbb{C}))$, $\varrho_h(0) \geq 0$, $\mathrm{Tr}[\varrho_h(0)] = 1$ under the sole additional assumption $\mathrm{Tr}[e^{\alpha_1 N}\varrho_h(0)e^{\alpha_1 N}] \leq C_{\alpha_1}$. Before computing the evolution of the semiclassical measures $(\mu_t)_{t\in]-\hat{T}_{\alpha_1},\hat{T}_{\alpha_1}[}$ given by Proposition 6.1 (this will be done in a subsequent article), it provides useful a priori information for them.

7.1 Truncation with Respect to the Number Operator N

For $\varepsilon > 0$, let $\chi_\varepsilon : [0, +\infty) \to [0, 1]$ be a decaying function such that

$$\forall k \in \mathbb{N}, \forall \varepsilon \in]0, 1[, \exists C_{k,\varepsilon} > 0, \qquad \sup_{s\in[0,+\infty)} s^k \chi_\varepsilon(s) \leq C_{k,\varepsilon}, \tag{113}$$

$$\forall \alpha_1 > 0, \exists C_{\alpha_1} > 0, \forall \varepsilon \in]0, 1[, \qquad \sup_{s\in[0,+\infty)} e^{-\alpha_1 s}(1 - \chi_\varepsilon(s)) \leq C_{\alpha_1} \times \varepsilon. \tag{114}$$

Examples are

$$\chi_\varepsilon(s) = 1_{[0,\varepsilon^{-1}]}(s) \quad \text{and} \quad \chi_\varepsilon(s) = e^{-\varepsilon s}.$$

Then the operators

$$a_{G,\varepsilon}(V) = \chi_\varepsilon(N) a_G(V) \chi_\varepsilon(N), \quad a^*_{G,\varepsilon}(V) = \chi_\varepsilon(N) a^*_G(V) \chi_\varepsilon(N)$$

are bounded operators on

$$F^2 = L^2(Z', \mathbf{dz'}; \Gamma(L^2(\mathbb{R}^d, dy; \mathbb{C}))) = L^2_{z,\mathrm{sym}} = L^2_{z_0} \oplus L^2_{z_1,\mathrm{sym}} L^2_{y_G}$$

according to (56) and $\sqrt{h}(a_{G,\varepsilon}(V) + a^*_{G,\varepsilon}(V))$ is an $O_\varepsilon(\sqrt{h})$ bounded self-adjoint perturbation of $(\xi - D_{y_G})^2$. Additionally for $\varepsilon > 0$ the estimates of Proposition 3.3 hold true when $a_G(V)$ and $a^*_G(V)$ are replaced by $a_{G,\varepsilon}(V)$ and $a^*_{G,\varepsilon}(V)$. Actually, (39) with $n > 1$ and (37) with $n > 0$ become

$$\begin{aligned} \|a_{G,\varepsilon}(V) f_{G,n}\|_{L^2_{z',Y'_{n-1}} L^p_{y_G}} &\leq \|V\|_{L^{r'}} \chi_\varepsilon(n-1)^2 \sqrt{n} \|f_{G,n}\|_{L^2_{z',Y'_n} L^q_{y_G}} \\ &\leq C_\varepsilon \|V\|_{L^{r'}} \|f_{G,n}\|_{L^2_{z',Y'_n} L^q_{y_G}} \end{aligned} \tag{115}$$

$$\begin{aligned} \|a^*_{G,\varepsilon}(V) f_{G,n}\|_{L^2_{z',Y'_{n+1}} L^{q'}_{y_G}} &\leq \|V\|_{L^{r'}} \chi_\varepsilon(n)^2 \sqrt{n+1} \|f_{G,n}\|_{L^2_{z',Y'_n} L^{p'}_{y_G}} \\ &\leq C_\varepsilon \|V\|_{L^{r'}} \|f_{G,n}\|_{L^2_{z',Y'_n} L^{p'}_{y_G}} \end{aligned} \tag{116}$$

when $V \in L^{q'}(\mathbb{R}^d; \mathbb{C}) \cap L^{r'}(\mathbb{R}^d; \mathbb{C})$, $\frac{1}{r'} = \frac{1}{2} + \frac{1}{q'} - \frac{1}{p'}$, $p', q' \in [1, 2]$. All the analysis can thus be carried out with $a_G(V)$ and $a^*_G(V)$ replaced by $a_{G,\varepsilon}(V)$ and $a^*_{G,\varepsilon}(V)$, either with estimates which are uniform in $\varepsilon \in]0, 1[$, or by replacing the N-dependent estimates by constants C_ε depending on $\varepsilon \in]0, 1[$.

In particular the solution $v^h_{G,\varepsilon}$ to

$$\begin{cases} i\partial_t v^h_{G,\varepsilon} = (\xi - D_{y_G})^2 v^h_{G,\varepsilon} + \sqrt{h}[a^*_{G,\varepsilon}(V) + a_{G,\varepsilon}(V)]v^h_{G,\varepsilon} \\ v^h_{G,\varepsilon}(t=0) = v^h_{G,\varepsilon,0} = u^h_{G,0}\,, \end{cases} \tag{117}$$

satisfies the same properties as the solution u^h_G to (96) stated in Theorem 5.1 and Proposition 5.1, uniformly with respect to $\varepsilon \in]0, 1[$.

Proposition 7.1 *Assume* $\|e^{2\alpha_1 N} u^h_{G,0}\|_{L^2_z L^2_{y_G}} \leq C_{\alpha_1}$ *for all* $h \in]0, h_0[$ *like in Proposition 5.1. There exists* $\hat{C}_{\alpha_1} > 0$ *and* $\hat{T}_{\alpha_1} > 0$ *such that the solutions* u^h_G *to (96) and* $v^h_{G,\varepsilon}$ *to (117) for* $\varepsilon \in]0, 1[$*, satisfy*

$$\|u^h_G(t) - v^h_{G,\varepsilon}(t)\|_{L^2_z L^2_{y_G}} \leq \hat{C}_{\alpha_1}\varepsilon$$

for all $t \in I^h_{\hat{T}_\alpha} =]-\frac{\hat{T}_{\alpha_1}}{h}, \frac{\hat{T}_{\alpha_1}}{h}[$.

Additionally the statement **b)** *of Proposition 5.1 holds true when* u^h_G, $a_G(V)$, $a^*_G(V)$ *are replaced by* $v^h_{G,\varepsilon}$, $a_{G,\varepsilon}(V)$, $a^*_{G,\varepsilon}(V)$.

Proof The statements **a)** and **b)** of Proposition 5.1 hold true uniformly with respect to $\varepsilon \in]0, 1[$ for $v^h_{G,\varepsilon}$ as a consequence of the previous arguments.

In particular $v^h_{G,\varepsilon}(t) = U(\frac{t}{h})u^h_{G,0} + v^h_{\infty,\varepsilon}$ where $(v^h_{\infty,\varepsilon}, v^h_{2,\varepsilon}, v^h_{1,\varepsilon})$ solves the system

$$\begin{pmatrix} v^h_{\infty,\varepsilon} \\ v^h_{2,\varepsilon} \\ v^h_{1,\varepsilon} \end{pmatrix} = L_\varepsilon \begin{pmatrix} v^h_{\infty,\varepsilon} \\ v^h_{2,\varepsilon} \\ v^h_{1,\varepsilon} \end{pmatrix} + \begin{pmatrix} f^h_{\infty,\varepsilon} \\ f^h_{2,\varepsilon} \\ 0 \end{pmatrix}, \quad L_\varepsilon = \begin{pmatrix} L_{\infty\infty,\varepsilon} & L_{\infty 2,\varepsilon} & L_{\infty 1,\varepsilon} \\ 0 & L_{22,\varepsilon} & 0 \\ L_{1\infty,\varepsilon} & 0 & L_{11,\varepsilon} \end{pmatrix}, \tag{118}$$

with

$$f^h_{\infty,\varepsilon}(t) = -i\int_0^t U(t)U(s)^* a^*_{G,\varepsilon}(V)\sqrt{h}U(s)u^h_{G,0}\,ds\,, \tag{119}$$

$$f^h_{2,\varepsilon}(t) = -i\,a_{G,\varepsilon}(V)\int_0^t U(t)U(s)^* a^*_{G,\varepsilon}(V)\sqrt{h}U(s)u^h_{G,0}\,ds + a_{G,\varepsilon}(V)U(t)u^h_{G,0}\,, \tag{120}$$

and where the entries L_ε are the same as the ones of L with $a_G(V)$ and $a^*_G(V)$ replaced by $a_{G,\varepsilon}(V)$ and $a^*_{G,\varepsilon}(V)$. When $\chi_\varepsilon(s) = e^{-\varepsilon s}$, one recovers the system for u^h_G by taking $\varepsilon = 0$.

We start now with the equation for u_G^h

$$u_G^h(t) = U(\frac{t}{h})u_{G,0}^h - i\sqrt{h}\int_0^{\frac{t}{h}} U(t-s)[a_G(V) + a_G^*(V)]u_G^h(s)\, ds\,,$$

which implies

$$\chi_\varepsilon(N)u_G^h(t) =$$
$$U(\frac{t}{h})\chi_\varepsilon(N)u_{G,0}^h - i\sqrt{h}\int_0^{\frac{t}{h}} U(\frac{t}{h}-s)\chi_\varepsilon(N)[a_G(V) + a_G^*(V)]\chi_\varepsilon(N)^2 u_G^h(s)\, ds$$
$$- i\sqrt{h}\chi_\varepsilon(N)\int_0^{\frac{t}{h}} U(\frac{t}{h}-s)[a_G(V) + a_G^*(V)](1-\chi_\varepsilon^2(N))u_G^h(s)\, ds\,.$$

The function $w_{G,\varepsilon}^h(t) = \chi_\varepsilon(N)u_G^h(t)$ solves

$$w_{G,\varepsilon}^h(t) = U(\frac{t}{h})\chi_\varepsilon(N)u_{G,0}^h$$
$$- i\sqrt{h}\int_0^{\frac{t}{h}} U(t-s)[a_{G,\varepsilon}(V) + a_{G,\varepsilon}^*(V)]w_{G,\varepsilon}^h(s)\, ds + g_{\infty,\varepsilon}^h \tag{121}$$

with

$$g_{\infty,\varepsilon}^h = -i\sqrt{h}\chi_\varepsilon(N)\int_0^{\frac{t}{h}} U(t-s)[a_G(V) + a_G^*(V)](1-\chi_\varepsilon^2(N))u_G^h(s)\, ds\,. \tag{122}$$

The system for $(w_{\infty,\varepsilon}^h, w_{2,\varepsilon}^h, w_{1,\varepsilon}^h)$ after decomposing $w_{G,\varepsilon}^h(t) = U(\frac{t}{h})\chi_\varepsilon(N)u_{G,0}^h + w_{\infty,\varepsilon}^h(t)$ is

$$\begin{pmatrix} w_{\infty,\varepsilon}^h \\ w_{2,\varepsilon}^h \\ w_{1,\varepsilon}^h \end{pmatrix} = L_\varepsilon \begin{pmatrix} w_{\infty,\varepsilon}^h \\ w_{2,\varepsilon}^h \\ w_{1,\varepsilon}^h \end{pmatrix} + \begin{pmatrix} \tilde{f}_{\infty,\varepsilon}^h \\ \tilde{f}_{2,\varepsilon}^h \\ 0 \end{pmatrix} + \begin{pmatrix} g_{\infty,\varepsilon}^h \\ 0 \\ 0 \end{pmatrix},$$

where $\tilde{f}^h_{\infty,\varepsilon}$ and $\tilde{f}^h_{2,\varepsilon}$ have the same expressions as (119)(120) with $u^h_{G,0}$ replaced by $\chi_\varepsilon(N)u^h_{G,0}$. By taking the difference with (118), and because $\|L_\varepsilon\|_{\mathcal{L}(\mathcal{E}_{0,\alpha_1,\gamma})} \leq 1/2$ for $\gamma > 0$ small enough, the proof is done as soon as the three norms

$$\|u^h_G(t) - \chi_\varepsilon(N)u^h_G(t)\|_{L^2_z L^2_{y_G}} \tag{123}$$

$$M(\tilde{f}^h_{\infty,\varepsilon} - f^h_{\infty,\varepsilon}, \tilde{f}^h_{2,\varepsilon} - f^h_{2,\varepsilon}, 0) \tag{124}$$

$$M(g^h_{\infty,\varepsilon}, 0, 0)\,, \tag{125}$$

are bounded by $\hat{C}_{\alpha_1}\varepsilon$.

Because the time interval is restricted to $I^h_{\hat{T}_{\alpha_1}}$ with $\hat{T}_{\alpha_1} < T_{\alpha_1}$, the weight $\sqrt{T_{\alpha_1} - |ht|}$ or $\sqrt{T_{\alpha_1} - \tau}$ used in Definition 4.3 or in Proposition 4.3 can be forgotten now (simply multiply $f^h_{q,\varepsilon}, \tilde{f}^h_{q,\varepsilon}, q \in \{\infty, 2\}$ and $g^h_{\infty,\varepsilon}$ by $1_{I_{h^{-1}\hat{T}_{\alpha_1}}}(t)$).

The estimate of (123) is obvious since

$$\|(1 - \chi_\varepsilon(N))u^h_G(t)\|_{L^2_z L^2_{y_G}} \leq \underbrace{\sup_{s\geq 0} |(1 - \chi_\varepsilon(s))e^{-\alpha_1 s}|}_{O(\varepsilon)} \times \underbrace{\|e^{\alpha_1 N}u^h_G(t)\|_{L^2_z L^2_{y_G}}}_{\leq \tilde{C}_{\alpha_1}}\,.$$

The estimate of (124) is very similar. Actually in the proof of Theorem 5.1 we checked $M(f^h_\infty, f^h_2, 0) \lesssim \|e^{\alpha_1 N}u^h_{G,0}\|_{L^2_z L^2_{y_G}}$. It gives now

$$M(\tilde{f}^h_{\infty,\varepsilon} - f^h_{\infty,\varepsilon}, \tilde{f}^h_{2,\varepsilon} - f^h_{2,\varepsilon}, 0) \lesssim \|e^{\alpha_1 N}(\chi_\varepsilon(N) - 1)u^h_{G,0}\|_{L^2_z L^2_{y_G}} \leq \hat{C}_{\alpha_1}\varepsilon\,.$$

For (125) let us first decompose $g^h_{\infty,\varepsilon}$ as

$$g^h_{\infty,\varepsilon} = g^h_{\infty,1,\varepsilon} + g^h_{\infty,2,\varepsilon}$$

$$\text{with } g^h_{\infty,1,\varepsilon} = -i\sqrt{h}\chi_\varepsilon(N)\int_0^{\frac{t}{h}} U(t-s)\, a^*_G(V)\,(1 - \chi^2_\varepsilon(N))\, u^h_G(s)\, ds$$

$$\text{and } g^h_{\infty,2,\varepsilon} = -i\sqrt{h}\chi_\varepsilon(N)\int_0^{\frac{t}{h}} U(t-s)\, a_G(V)\,(1 - \chi^2_\varepsilon(N))\, u^h_G(s)\, ds\,.$$

The estimate of $g^h_{\infty,1,\varepsilon}$ follows the method for the bound of $M(f^h_\infty, 0, 0)$ in the proof of Theorem 5.1, where we simply used the uniform bound in time for $\|U(s)e^{\alpha_1 N}u^h_{G,0}\|_{L^2_z L^2_{y_G}}$. With

$$\sup_t \|(1 - \chi^2_\varepsilon(N))u^h_G(t)\|_{L^2_z L^2_{y_G}} \leq \underbrace{\sup_{s\geq 0} |(1 - \chi^2_\varepsilon(s))e^{-\alpha_1 s}|}_{O(\varepsilon)} \times \underbrace{\|e^{\alpha_1 N}u^h_G(t)\|_{L^2_z L^2_{y_G}}}_{\leq \tilde{C}_{\alpha_1}}\,,$$

this gives

$$M(g^h_{\infty,1,\varepsilon},0,0)\leq \hat{C}_{\alpha_1}\varepsilon\,.$$

For $g^h_{\infty,2,\varepsilon}$, remember firstly that the assumption is $\|e^{2\alpha_1 N}u^h_{G,0}\|_{L^2_zL^2_{y_G}}\leq C_{\alpha_1}$ and by possibly reducing $\hat{T}_{\alpha_1}$, we may assume $\|e^{\frac{3\alpha_1}{2}N}u_{G,h}(t)\|_{L^2_zL^2_{y_G}}\leq \tilde{C}_{\alpha_1}$. We now use the obvious relation $a_G(V)\phi(N)=\phi(N+1)a_G(V)$ and write

$$g^h_{\infty,2,\varepsilon}=-i\chi_\varepsilon(N)e^{-\frac{\alpha_1}{2}(N+1)}(1-\chi^2_\varepsilon(N+1))e^{\frac{\alpha_1}{2}(N+1)}$$
$$\int_0^{\frac{t}{h}}U(t-s)\sqrt{h}a_G(V)u^h_G(s)\,ds\,.$$

Remember that the equivalent system (96) says $\sqrt{h}a_G(V)u^h_G(t)=u^h_1(t)+\sqrt{h}u^h_2(t)$ with $M(0,u^h_2,u^h_1)\lesssim C_{\alpha_1}$. The above equality becomes

$$g^h_{\infty,2,\varepsilon}(t)=\chi_\varepsilon(N)(1-\chi^2_\varepsilon(N+1))e^{-\frac{\alpha_1}{2}(N+1)}e^{\frac{\alpha_1}{2}(N+1)}[L_{\infty 1}(u^h_1)+L_{\infty 2}(u^h_2)]\,.$$

The bounds for $L_{\infty 1}$ and $L_{\infty 2}$ in the Theorem 5.1, lead to

$$\||ht|^{-1/2}e^{\frac{\alpha_1}{2}(N+1)}[L_{\infty 1,\varepsilon}(u^h_1)+L_{\infty 2,\varepsilon}(u^h_2)](t)\|_{L^\infty(I^h_{\hat{T}_{\alpha_1}};L^2_zL^2_{y_G})}\lesssim C_{\alpha_1}\,.$$

With

$$\|\chi_\varepsilon(N)(1-\chi^2_\varepsilon(N+1))e^{-\frac{\alpha_1}{2}(N+1)}\|_{\mathcal{L}(L^2_zL^2_{y_G})}\leq\sup_{s\geq 0}|(1-\chi^2_\varepsilon(s))e^{-\frac{\alpha_1}{2}s}|=O(\varepsilon)\,,$$

this proves

$$M(g^h_{\infty,2,\varepsilon},0,0)\leq\hat{C}_{\alpha_1}\varepsilon\,.$$

□

Let us go back to our initial problem and let us compare the evolution of states for the dynamics $U(\frac{t}{h})=e^{-it(-\Delta_x+\sqrt{h}\mathcal{V})}$ for $\varepsilon=0$ and the case $\varepsilon>0$ where $\chi_\varepsilon(N)\mathcal{V}\chi_\varepsilon(N)$ is a bounded self-adjoint perturbation of $-\Delta_x$. Set in particular

$$U_{\mathcal{V},\varepsilon}=e^{-it(-\Delta_x+\sqrt{h}\mathcal{V}_\varepsilon)}\quad\text{with}\quad\mathcal{V}_\varepsilon=\chi_\varepsilon(N)\mathcal{V}\chi_\varepsilon(N)\,.\tag{126}$$

Proposition 7.2 *Assume like in Proposition 6.1*

$$V \in L^{r'_\sigma}(\mathbb{R}^d, dx; \mathbb{R}) \cap H^2(\mathbb{R}^d; \mathbb{R}), \quad r'_\sigma = \frac{2d}{d+2}, \quad d \geq 3,$$

and assume that there exists $\alpha_1 > 0$ *such that* $\varrho_h(0) \in \mathcal{L}^1(L^2(\mathbb{R}^d \times \Omega, dx \otimes \mathcal{G}; \mathbb{C}))$, $\varrho_h(0) \geq 0$, $\mathrm{Tr}[\varrho_h(0)] = 1$ *satisfies*

$$\exists C_{\alpha_1} > 0, \ \forall h \in]0, h_0[, \quad \mathrm{Tr}\left[e^{\alpha_1 N} \varrho_h(0) e^{\alpha_1 N}\right] \leq C_{\alpha_1}.$$

Call $\varrho_h(t) = U_{\mathcal{V}}(\frac{t}{h})\varrho_h(0)U^*_{\mathcal{V}}(\frac{t}{h})$ *and* $\varrho_{h,\varepsilon}(t) = U_{\mathcal{V},\varepsilon}(\frac{t}{h})\varrho_h(0)U^*_{\mathcal{V},\varepsilon}(\frac{t}{h})$. *When the subset* $\mathcal{E} \subset]0, h_0[$, $0 \in \overline{\mathcal{E}}$, *is chosen such that*

$$\forall t \in]-\hat{T}_{\alpha_1}, \hat{T}_{\alpha_1}[, \ \mathcal{M}(\varrho_h(t), h \in \mathcal{E}) = \{\mu_t\} \text{ and } \mathcal{M}(\varrho_{h,\varepsilon}(t), h \in \mathcal{E}) = \left\{\mu_{t,\varepsilon}\right\}.$$

Then the total variation of $\mu_t - \mu_{t,\varepsilon}$ *is estimated by*

$$\forall t \in]-\hat{T}_{\alpha_1}, \hat{T}_{\alpha_1}[, \quad |\mu_t - \mu_{t,\varepsilon}| \underbrace{(T^*\mathbb{R}^d)}_{\text{or } T^*\mathbb{R}^d \sqcup \{\infty\}} \leq C'_{\alpha_1}\varepsilon,$$

for some constant $C'_{\alpha_1} > 0$ *determined by* $\alpha_1 > 0$.

Proof From

$$\varrho_h(t) - \varrho_{h,\varepsilon}(t) = \left[U_{\mathcal{V}}(\frac{t}{h})\varrho_h(0)^{1/2} - U_{\mathcal{V},\varepsilon}(\frac{t}{h})\varrho_h(0)^{1/2}\right]\varrho_h(0)^{1/2}U^*_{\mathcal{V}}(\frac{t}{h})$$
$$+ U_{\mathcal{V},\varepsilon}(\frac{t}{h})\varrho_h(0)^{1/2}[\varrho_h(0)^{1/2}U^*_{\mathcal{V}}(\frac{t}{h}) - \varrho_h(0)^{1/2}U^*_{\mathcal{V},\varepsilon}(\frac{t}{h})]$$

we deduce

$$|\mu(t) - \mu_\varepsilon(t)|(T^*\mathbb{R}^d \cup \{\infty\}) \leq 4 \liminf_{h\in\mathcal{E}, h\to 0} \|\varrho_h(t) - \varrho_{h,\varepsilon}(t)\|_{\mathcal{L}^1}$$
$$\leq 8 \liminf_{h\in\mathcal{E}, h\to 0} \|\Psi^h(t) - \Psi^h_\varepsilon(t)\|_{L^2_{x,\omega,\hat{z}}}$$

with $\Psi^h_\varepsilon(t) = U_{\mathcal{V},\varepsilon}(\frac{t}{h})\varrho_h(0)^{1/2} \in \mathcal{L}^2(L^2(\mathbb{R}^d \times \Omega, dx \otimes \mathcal{G}; \mathbb{C})) \sim L^2(\mathbb{R}^d \times \Omega \times \hat{Z}, dx \otimes \mathcal{G} \otimes \mathbf{d\hat{z}}; \mathbb{C})$ with $\hat{Z} = \mathbb{R}^d \times \Omega$, $\mathbf{d\hat{z}} = dx \otimes \mathcal{G}$.

But Proposition 7.1 implies

$$\forall t \in]-\hat{T}_{\alpha_1}, \hat{T}_{\alpha_1}[, \quad \|\Psi^h(t) - \Psi^h_\varepsilon(t)\|_{L^2_{x,\omega,\hat{z}}} \leq \hat{C}_{\alpha_1}\varepsilon.$$

□

7.2 Asymptotic Conservation of Energy

The result of this paragraph is a consequence of the approximation of the $U_{\mathcal{V}}$ dynamics by the one of $U_{\mathcal{V}_\varepsilon}$ in terms of wave functions in Proposition 7.1, states and semiclassical measures in Proposition 7.2.

Proposition 7.3 *Assume like in Proposition 6.1*

$$V \in L^{r'_\sigma}(\mathbb{R}^d, dx; \mathbb{R}) \cap H^2(\mathbb{R}^d; \mathbb{R}), \quad r'_\sigma = \frac{2d}{d+2}, \quad d \geq 3,$$

and assume that there exists $\alpha_1 > 0$ *such that* $\varrho_h(0) \in \mathcal{L}^1(L^2(\mathbb{R}^d \times \Omega, dx \otimes \mathcal{G}; \mathbb{C}))$, $\varrho_h(0) \geq 0$, $\mathrm{Tr}[\varrho_h(0)] = 1$ *satisfies*

$$\exists C_{\alpha_1} > 0, \forall h \in]0, h_0[, \quad \mathrm{Tr}\left[e^{\alpha_1 N} \varrho_h(0) e^{\alpha_1 N}\right] \leq C_{\alpha_1}.$$

Call $\varrho_h(t) = U_{\mathcal{V}}(\frac{t}{h})\varrho_h(0)U^*_{\mathcal{V}}(\frac{t}{h})$ *and let the subset* $\mathcal{E} \subset]0, h_0[$, $0 \in \overline{\mathcal{E}}$, *be such that*

$$\forall t \in] - \hat{T}_{\alpha_1}, \hat{T}_{\alpha_1}[, \quad \mathcal{M}(\varrho_h(t), h \in \mathcal{E}) = \{\mu_t\}$$

with the additional assumption at time $t = 0$,

$$\mathrm{supp}\, \mu_0 \subset \left\{(x, \xi) \in T^*\mathbb{R}^d, \ |\xi|^2 \in F\right\} \tag{127}$$

where F *is a closed subset of* $\mathbb{R}$. *Then for all* $t \in] - \hat{T}_{\alpha_1}, \hat{T}_{\alpha_1}[$, *the support of* μ_t *restricted to* $T^*\mathbb{R}^d$ *satisfies*

$$\mathrm{supp}\, \mu_t\big|_{T^*\mathbb{R}^d} \subset \left\{(x, \xi) \in T^*\mathbb{R}^d, \ |\xi|^2 \in F\right\}.$$

Proof For $\varepsilon > 0$ and $z \in \mathbb{C} \setminus \mathbb{R}$ the resolvent estimate

$$\|[z + \Delta_x]^{-1} - [z - (-\Delta_x + \sqrt{h}\mathcal{V}_\varepsilon)]^{-1}\|_{\mathcal{L}(L^2_{x,\omega})} \leq \frac{C_\varepsilon \sqrt{h}}{|\mathrm{Im}\, z|^2}$$

with $\mathcal{V}_\varepsilon = \chi_\varepsilon(N)\mathcal{V}\chi_\varepsilon(N) \in \mathcal{L}(L^2_{x,\omega})$ as in (126) combined with Helffer-Sjöstrand formula [17] gives

$$\forall \varepsilon > 0, \forall \chi \in C_0^\infty(\mathbb{R}; \mathbb{C}), \exists C_{\chi,\varepsilon} > 0,$$
$$\|\chi(-\Delta_x) - \chi(-\Delta_x + \sqrt{h}\mathcal{V}_\varepsilon)\|_{\mathcal{L}(L^2_{x,\omega})} \leq C_{\chi,\varepsilon}\sqrt{h}.$$

The semiclassical calculus then implies

$$\Big\|\chi(-\Delta_x+\sqrt{h}\mathcal{V}_\varepsilon)\,a^{\mathrm{Weyl}}(hx,D_x)\,\chi(-\Delta_x+\sqrt{h}\mathcal{V}_\varepsilon) - [\chi^2(|\xi|^2)a]^{\mathrm{Weyl}}(hx,D_x)\Big\|_{\mathcal{L}(L^2_{x,\omega})} = O_{a,\chi,\varepsilon}(\sqrt{h})$$

for all $a\in C_0^\infty(T^*\mathbb{R}^d;\mathbb{C})$ and all $\chi\in C_0^\infty(\mathbb{R};\mathbb{C})$.

Hence, the assumption (127) implies

$$\forall\chi\in C_0^\infty(\mathbb{R}\setminus F;[0,1])\,,\quad \lim_{h\in\mathcal{E},h\to0}\|\chi(-\Delta_x+\sqrt{h}\mathcal{V}_\varepsilon)\,\varrho_h(0)\,\chi(-\Delta_x+\sqrt{h}\mathcal{V}_\varepsilon)\|_{\mathcal{L}^1(L^2_{x,\omega})}=0\,,$$

and therefore

$$\forall\chi\in C_0^\infty(\mathbb{R}\setminus F;[0,1])\,,\ \forall t\in]-\hat{T}_{\alpha_1},\hat{T}_{\alpha_1}[\,,\quad \lim_{h\in\mathcal{E},h\to0}\|\chi(-\Delta_x+\sqrt{h}\mathcal{V}_\varepsilon)\,\varrho_{h,\varepsilon}(t)\,\chi(-\Delta_x+\sqrt{h}\mathcal{V}_\varepsilon)\|_{\mathcal{L}^1(L^2_{x,\omega})}=0\,,$$

with $\varrho_{h,\varepsilon}(t)=U_{\mathcal{V}_\varepsilon}(\frac{t}{h})\varrho_h(0)U^*_{\mathcal{V}_\varepsilon}(\frac{t}{h})$ and $U_{\mathcal{V}_\varepsilon}(t)=e^{-it(-\Delta_x+\sqrt{h}\mathcal{V}_\varepsilon)}$.

When $\mathcal{E}'\subset\mathcal{E}$, $0\in\overline{\mathcal{E}'}$, is such that

$$\mathcal{M}(\varrho_{h,\varepsilon}(t)\,,h\in\mathcal{E}')=\left\{\mu_{t,\varepsilon}\right\}\,,$$

Proposition 7.2 tells us

$$|\mu_t-\mu_{t,\varepsilon}|(T^*\mathbb{R}^d)\le C'_{\alpha_1}\varepsilon\,,$$

while

$$\int_{T^*\mathbb{R}^d}a(x,\xi)\,|\chi|^2(|\xi|^2)\,d\mu_{t,\varepsilon}(x,\xi) = \lim_{h\in\mathcal{E}',h\to0}\mathrm{Tr}\left[\chi(-\Delta_x+\sqrt{h}\mathcal{V}_\varepsilon)\,a^{\mathrm{Weyl}}(hx,D_x)\,\chi(-\Delta_x+\sqrt{h}\mathcal{V}_\varepsilon)\,\varrho_{h,\varepsilon}(t)\right]=0\,,$$

for $a\in C_0^\infty(T^*\mathbb{R}^d;\mathbb{C})$ and $\chi\in C_0^\infty(\mathbb{R}\setminus F;[0,1])$. We deduce

$$\forall a\in C_0^\infty(T^*\mathbb{R}^d;\mathbb{C})\,,\ \forall\chi\in C_0^\infty(\mathbb{R}\setminus F;[0,1])\,,\ \forall t\in]-\hat{T}_{\alpha_1},\hat{T}_{\alpha_1}[\,,\quad \int_{T^*\mathbb{R}^d}a(x,\xi)\,\chi^2(|\xi|^2)\,d\mu_t(x,\xi)=0\,,$$

which yields the result. □

7.3 Changing V

The formulation of Theorem 5.1 $u_G^h(t) = U_{\mathcal{V}}(t)u_{G,0}^h = U(\frac{t}{h})u_{G,0}^h + u_\infty^h(t)$ where $(u_q^h)_{q\in\{\infty,2,1\}}$ is a solution of a fixed point problem, solved in Proposition 4.3, where only $\|V\|_{L^{r'_\sigma}}$, $r'_\sigma = \frac{2d}{d+2}$, is used, allows to consider perturbations of V, which can be done separately in the the terms $a_G(V)$ and $a_G^*(V)$ and with complex valued perturbations.

Remember that our state $\varrho_h(t) = U_{\mathcal{V}}(\frac{t}{h})\varrho_h(0)U_{\mathcal{V}}^*(\frac{t}{h})$ is written

$$\varrho_h(t) = [U_{\mathcal{V}}(\frac{t}{h})\varrho_h(0)^{1/2}][\varrho_h(0)^{1/2}U_{\mathcal{V}}^*(\frac{t}{h})],$$

and the link with the fixed point problem is done after setting

$$U(t)u_{G,0}^h + u_\infty^h(t) = u_G^h(t) = U_{\mathcal{V}}(\frac{t}{h})\varrho_h(0)^{1/2} \quad \text{in } \mathcal{L}^2(L^2_{x,\omega}) \sim L^2_{z,y_G},$$

where the last identification is done via the unitary transform U_G of Sect. 3, omitted here and explained in the proof of Proposition 6.1.

A generalization is done by writing for a pair $\tilde{\mathcal{V}} = (V_1, V_2) \in L^{r'_\sigma}(\mathbb{R}^d, dy; \mathbb{C})^2$,

$$\varrho_{h,\tilde{\mathcal{V}}}(t) = u_{G,\tilde{\mathcal{V}}}^h(\frac{t}{h})[u_{G,\tilde{\mathcal{V}}}^h(\frac{t}{h})]^* \in \mathcal{L}^1(L^2_{x,\omega}), \tag{128}$$

where $u_{G,\tilde{\mathcal{V}}}^h(t) = U(t)\varrho_h(0)^{1/2} + u_{\infty,\tilde{\mathcal{V}}}^h(t)$ and $(u_{q,\tilde{\mathcal{V}}}^h)_{q\in\{\infty,2,1\}}$ solves the fixed point problem (59)(60)(61) with $f_1^h(t) = 0$ and f_∞^h and f_2^h given by

$$f_\infty^h(t) = f_{\infty,\tilde{\mathcal{V}}}^h(t) = -i\int_0^t U(t)U(s)^* a_G^*(V_1)\sqrt{h}U(s)u_{G,0}^h\, ds\,, \tag{129}$$

$$f_2^h(t) = f_{2,\tilde{\mathcal{V}}}(t) = -i\, a_G(V_2)\int_0^t U(t)U(s)^* a_G^*(V_1)\sqrt{h}U(s)u_{G,0}^h\, ds + a_G(V_2)U(t)u_{G,0}^h\,. \tag{130}$$

This fixed point problem will be written

$$\begin{pmatrix} u_{\infty,\tilde{\mathcal{V}}}^h \\ u_{2,\tilde{\mathcal{V}}}^h \\ u_{1,\tilde{\mathcal{V}}}^h \end{pmatrix} = L_{\tilde{\mathcal{V}}}\begin{pmatrix} u_{\infty,\tilde{\mathcal{V}}}^h \\ u_{2,\tilde{\mathcal{V}}}^h \\ u_{1,\tilde{\mathcal{V}}}^h \end{pmatrix} + \begin{pmatrix} f_{\infty,\tilde{\mathcal{V}}}^h \\ f_{2,\tilde{\mathcal{V}}}^h \\ 0 \end{pmatrix}. \tag{131}$$

Proposition 7.4 *For two pairs* $\tilde{\mathcal{V}}_k = (V_{1,k}, V_{2,k}) \in L^{r'_\sigma}(\mathbb{R}^d, dy; \mathbb{C})^2$, *for* $\|e^{\alpha_1 N} u^h_{G,0}\| \leq C_{\alpha_1}$ *and by choosing* $\hat{T}_{\alpha_1} > 0$ *small enough, the two solutions to* (131) *with the right-hand sides given by* (129)(130) *satisfy*

$$\forall t \in] - \hat{T}_{\alpha_1}, \hat{T}_{\alpha_1}[,$$
$$\|u^h_{\infty,\tilde{\mathcal{V}}_2}(\frac{t}{h}) - u^h_{\infty,\tilde{\mathcal{V}}_1}(\frac{t}{h})\|_{L^2_{z,y_G}} \leq C\left[\|V_{1,2} - V_{1,1}\|_{L^{r'_\sigma}} + \|V_{2,2} - V_{2,1}\|_{L^{r'_\sigma}}\right]$$

for some constant $C > 0$ *given by* $\alpha_1 > 0$, C_{α_1}, *the dimension* d, *and* $\max_{i,j} \|V_{i,j}\|_{L^{r'_\sigma}}$.

Proof It suffices to notice that the difference $v^h = u^h_{\tilde{\mathcal{V}}_2} - u^h_{\tilde{\mathcal{V}}_1}$ with $u^h_{\tilde{\mathcal{V}}_k} = (u^h_{q,\tilde{\mathcal{V}}_k})_{q\in\{\infty,2,1\}}$, $k = 1, 2$, solves

$$v^h - L_{\tilde{\mathcal{V}}_1}(v^h) = (L_{\tilde{\mathcal{V}}_2} - L_{\tilde{\mathcal{V}}_1})(u^h_{\tilde{\mathcal{V}}_2}) + \begin{pmatrix} f^h_{\infty,\tilde{\mathcal{V}}_2} - f^h_{\infty,\tilde{\mathcal{V}}_1} \\ f^h_{2,\tilde{\mathcal{V}}_2} - f^h_{2,\tilde{\mathcal{V}}_1} \\ 0 \end{pmatrix}.$$

Estimates for all the terms of the right-hand side have essentially been proved for Proposition 4.3 and for Theorem 5.1. Although they are written for $V_1 = V_2$ real-valued in Theorem 5.1 the generalization is straightforward (like in Proposition 4.3) and upper bounds are proportional the $L^{r'_\sigma}$ of the potential which is either $(V_{1,2} - V_{1,1})$ or $(V_{2,2} - V_{2,1})$.

The time interval $] - T_{\alpha_1}, T_{\alpha_1}[=] - 2\hat{T}_{\alpha_1}, 2\hat{T}_{\alpha_1}[$ is actually chosen like in Proposition 4.3 such that $\|L_{\tilde{\mathcal{V}}_1}\|_{\mathcal{L}(\mathcal{E}_{\alpha_1,-\alpha_1,\gamma})} \leq \frac{1}{2}$ and this ends the proof. □

For a general pair $\tilde{\mathcal{V}} = (V_1, V_2) \in L^{r'_\sigma}(\mathbb{R}^d, dy; \mathbb{C})^2$, the trace-class operator $\varrho^h_{\tilde{\mathcal{V}}}(t)$ is no more a state and neither self-adjoint. However it remains uniformly bounded in $\mathcal{L}^1(L^2_{x,\omega})$ and complex-valued semiclassical measures $\mu_{\tilde{\mathcal{V}}}(t)$ make sense for $t \in] - \hat{T}_{\alpha_1}, \hat{T}_{\alpha_1}[$. Moreover the results of Proposition 5.1 and Proposition 6.1 can be adapted mutatis mutandis for such a general pair, so that semiclassical measures (extraction process) can be defined simultaneously for all $t \in]\hat{T}_{\alpha_1}, \hat{T}_{\alpha_1}[$.

The above comparison result can be translated in terms of trace-class operators and asymptotically for semiclassical measures.

Proposition 7.5 *Assume*

$$V \in L^{r'_\sigma}(\mathbb{R}^d, dx; \mathbb{R}) \cap H^2(\mathbb{R}^d; \mathbb{R}), \; V_1, V_2 \in L^{r'_\sigma}(\mathbb{R}^d, dx; \mathbb{C}),$$
$$r'_\sigma = \frac{2d}{d+2}, \; d \geq 3,$$

and assume that there exists $\alpha_1 > 0$ *such that* $\varrho_h(0) \in \mathcal{L}^1(L^2(\mathbb{R}^d \times \Omega, dx \otimes \mathcal{G}; \mathbb{C}))$, $\varrho_h(0) \geq 0$, $\mathrm{Tr}[\varrho_h(0)] = 1$ *satisfies*

$$\exists C_{\alpha_1} > 0\,,\ \forall h \in]0, h_0[\,, \quad \mathrm{Tr}\left[e^{\alpha_1 N} \varrho_h(0) e^{\alpha_1 N}\right] \leq C_{\alpha_1}\,.$$

Let $\varrho_h(t) = U_{\mathcal{V}}(\frac{t}{h})\varrho(0)U^*_{\mathcal{V}}(\frac{t}{h})$ *and let* $\varrho_{h,\tilde{\mathcal{V}}}(t)$ *be defined by* (128). *Then*

$$\exists C > 0\,,\ \forall t \in]-\hat{T}_{\alpha_1}, \hat{T}_{\alpha_1}[\,,$$
$$\|\varrho_h(t) - \varrho_{h,\tilde{\mathcal{V}}}(t)\|_{\mathcal{L}^1(L^2_{x,\omega})} \leq C\left[\|V_1 - V\|_{L^{r'_\sigma}} + \|V_2 - V\|_{L^{r'_\sigma}}\right].$$

When the subset $\mathcal{E} \subset]0, h_0[$, $0 \in \overline{\mathcal{E}}$, *is chosen such that*

$$\forall t \in]-\hat{T}_{\alpha_1}, \hat{T}_{\alpha_1}[\,,\ \mathcal{M}(\varrho_h(t),\ h \in \mathcal{E}) = \{\mu_t\} \ \text{and}\ \mathcal{M}(\varrho_{h,\tilde{\mathcal{V}}}(t),\ h \in \mathcal{E}) = \left\{\mu_{t,\tilde{\mathcal{V}}}\right\}.$$

Then the total variation of $\mu_t - \mu_{t,\tilde{\mathcal{V}}}$ *is estimated by*

$$\exists C' > 0\,,\ \forall t \in]-\hat{T}_{\alpha_1}, \hat{T}_{\alpha_1}[\,,$$
$$|\mu_t - \mu_{t,\tilde{\mathcal{V}}}| \underbrace{(T^*\mathbb{R}^d)}_{\text{or } T^*\mathbb{R}^d \sqcup \{\infty\}} \leq C'\left[\|V_1 - V\|_{L^{r'_\sigma}} + \|V_2 - V\|_{L^{r'_\sigma}}\right].$$

Proof It suffices to write

$$\varrho_{h,\tilde{\mathcal{V}}}(t) - \varrho_h(t)$$
$$= \left[u^h_{G,\tilde{\mathcal{V}}}(\frac{t}{h}) - u^h_G(\frac{t}{h})\right][u^h_{G,\tilde{\mathcal{V}}}(\frac{t}{h})]^* + [u^h_G(\frac{t}{h})]\left[u^h_{G,\tilde{\mathcal{V}}}(\frac{t}{h}) - u^h_G(\frac{t}{h})\right]^*$$

and to remember that Hilbert-Schmidt norms correspond to L^2_{z,y_G}-norms estimated in Proposition 7.4. □

7.4 *Quantum Dynamics with Low Regularity*

We conclude with an easy application of Proposition 7.4 which says that the dynamics $(U_{\mathcal{V}}(t))_{t\in\mathbb{R}}$ is actually well defined under the sole assumption

$$V \in L^{r'_\sigma}(\mathbb{R}^d; \mathbb{R})\,, \quad r'_\sigma = \frac{2d}{d+2} \quad d \geq 3\,, \tag{132}$$

with good approximations when $V_n \in L^{r'_\sigma}(\mathbb{R}^d; \mathbb{R}) \cap H^2(\mathbb{R}^d; \mathbb{R})$ satisfies $\lim_{n\to\infty} \|V_n - V\|_{L^{r'_\sigma}} = 0$.

Proposition 7.6 *Let V belong to $L^{r'_\sigma}(\mathbb{R}^d;\mathbb{R})$ and let $(V_n)_{n\in\mathbb{N}}$ be a sequence in $L^{r'_\sigma}(\mathbb{R}^d;\mathbb{R})\cap H^2(\mathbb{R}^d;\mathbb{R})$ such that $\lim_{n\to\infty}\|V-V_n\|_{L^{r'_\sigma}}=0$. Then for any $t\in\mathbb{R}$ the unitary operator $U_{\mathcal{V}_n}(t)$ converges strongly to a unitary operator $U_{\mathcal{V}}(t)$.*

Therefore $(U_{\mathcal{V}}(t))_{t\in\mathbb{R}}$ is a strongly continous unitary group in $L^2(\mathbb{R}^d\times Z'', dx\otimes dz''; \Gamma(L^2(\mathbb{R}^d, dy;\mathbb{C})))=L^2_{z,\mathrm{sym}}L^2_{y_G}$ with a self-adjoint generator denoted $(-\Delta_x+\sqrt{h}\mathcal{V}, D(-\Delta_x+\sqrt{h}\mathcal{V}))$.

The convergence $(-\Delta_x+\sqrt{h}\mathcal{V}_n, D(-\Delta_x+\sqrt{h}\mathcal{V}_n))$ to $(-\Delta_x+\sqrt{h}\mathcal{V}, D(-\Delta_x+\sqrt{h}\mathcal{V}))$ holds in the strong resolvent sense.

Remark 7.1 Although the dynamics $(U_{\mathcal{V}}(t))_{t\in\mathbb{R}}$ and its self-adjoint generator $(-\Delta_x+\sqrt{h}\mathcal{V}, D(-\Delta_x+\sqrt{h}\mathcal{V}))$ is well defined for $V\in L^{r'_\sigma}(\mathbb{R}^d;\mathbb{R})$, we have no information on the domain $D(-\Delta_x+\sqrt{h}\mathcal{V})$. The approximation process by $V_n\in L^{r'_\sigma}(\mathbb{R}^d;\mathbb{R})\cap H^2(\mathbb{R}^d;\mathbb{R})$ for which a core of $\Delta_x+\sqrt{h}\mathcal{V}_n$ is given by Proposition 4.4 in [5] recalled in Lemma 2.1, provides a substitute for the analysis.

It could be interesting to see if this Schrödinger type approach relying on endpoint Strichartz estimates could be applied to other quantum field theoretic problem and whether it would bring additional information of tools as compared with the euclidean approach (see [28] and refences therein).

Proof Actually we can work here with $h=1$. The convergence of

$$U_{\mathcal{V}_n}(t)u_{G,0}=U(t)u_{G,0}+u_{\infty,\mathcal{V}_n}(t)$$

is deduced from the convergence (see Proposition 7.4) of $u_{\infty,\mathcal{V}_n}(t)$ to $u_{\infty,\mathcal{V}}(t)$ when $e^{\alpha_1 N}u_{G,0}\in L^2_{z,\mathrm{sym}}L^2_{y_G}$ for some $\alpha_1>0$.

From $\|U_{\mathcal{V}_n}(t)u_{G,0}\|_{L^2_{z,\mathrm{sym}}L^2_{y_G}}=\|u_{G,0}\|_{L^2_{z,\mathrm{sym}}L^2_{y_G}}$ we deduce $\|U_{\mathcal{V}}(t)u_{G,0}\|_{L^2_{z,\mathrm{sym}}L^2_{y_G}}=\|u_{G,0}\|_{L^2_{z,\mathrm{sym}}L^2_{y_G}}$. This finally provides the extension of $U_{\mathcal{V}}(t)u_{G,0}$ for any $u_{G,0}$ in $L^2_{z,\mathrm{sym}}L^2_{y_G}$ with the convergence of $U_{\mathcal{V}_n}(t)u_{G,0}$ to $U_{\mathcal{V}}(t)u_{G,0}$, because the space $e^{-\alpha_1 N}L^2_{z,\mathrm{sym}}L^2_{y_G}$ is dense in $L^2_{z,\mathrm{sym}}L^2_{y_G}$. Passing from the strong convergence of unitary groups to the strong resolvent convergence of generators is standard. □

Acknowledgments The authors thank Zied Ammari for beneficial first discussions about this project. The research of S.B. was partly done during a CNRS sabbatical semester.

References

1. Ammari, Z., Nier, F.: Mean field propagation of Wigner measures and BBGKY hierarchies of general bosonic states. J. Math. Pures Appl. **9**(6), 585–626 (2011)
2. Ammari, Z., Nier, F.: Mean field propagation of infinite-dimensional Wigner measures with a singular two-body interaction potential. Ann. Sc. Norm. Super. Pisa, Cl. Sci. **5**(14), 155–220 (2015)

3. Bal, G.: On the self-averaging of wave energy in random media. Multiscale Model. Simul. **2**(3), 398–420 (2004)
4. Brascamp, H.J., Lieb, E.: Best constants in Young fs inequality, its converse and its generalization to more than three functions. Adv. Math. **20**, 151–173 (1976)
5. Breteaux, S.: A geometric derivation of the Boltzmann equation for a particle interacting with a Gaussian random field, using a Fock space approach. Ann. Inst. Fourier **64**(3), 1031–1076 (2014)
6. Colin de Verdière, Y.: Ergodicité et fonctions propres du Laplacien. Commun. Math. Phys. **102**, 497–502 (1985)
7. Duerinckx, M., Shirley, C.: A new spectral analysis of stationnary random Schrödinger operators. J. Math. Phys. **62**, 0721006 (2021)
8. Erdös, L., Yau, H.T.: Linear Boltzmann equation as a weak coupling limit of random Schrödinger equation. Commun. Pure Appl. Math. **53**(6), 667–735 (2000)
9. Erdös, L., Yau, H.T., Salmhofer, M.: Quantum diffusion of the random Schrödinger evolution in the scaling limit. Acta Math. **200**(2), 211–277 (2008)
10. Erdös, L., Yau, H.T., Salmhofer, M.: Quantum diffusion of the random Schrödinger evolution in the scaling limit II. Commun. Math. Phys. **271**(1), 1–53 (2007)
11. Fouque, J.P., Garnier, J., Papanicolaou, G., Solna, K.: Wave Propagation and Time Reversal in Randomly Layered Media. Stochastic Modelling and Applied Probability, vol. 56. Springer, Berlin (2007)
12. Gallavotti, G.: Rigorous theory of the Boltzmann equation in the Lorentz model. Nota internan no. 358. Physics department "la Sapienza", Roma (1972), available on mparc@math.utexas.edu# 93–304
13. Garnier, J.: Multiscale analysis of wave propagation in random media. In: Proceedings of the International Congress of Mathematicians 2018 Rio de Janeiro, vol. 3, pp. 2865–2890 (2018)
14. Gérard, P.: Mesures semi-classiques et ondes de Bloch. In: Séeminaire sur les Équations aux Dérivées Partielles, 1990–1991, Exp. No. XVI, 19pp., École Polytech., Palaiseau (1991)
15. Gérard, P., Markowich, P.A., Mauser, N.J., Poupaud, F.: Homogenization limits andWigner transforms. Commun. Pure Appl. Math. **50**(4), 323–379 (1997)
16. Helffer, B., Martinez, A., Robert, D.: Ergodicité et limite semi-classique. Commun. Math. Phys. **109**(2), 313–326 (1987)
17. Helffer, B., Sjöstrand, J.: Equation de Harper. Lecture Notes in Physics, vol. 345, pp. 118–197. Springer, Berlin (1989)
18. Hernandez, F.: Quantum diffusion via an approximate semigroup property (2022). arXiv:2206.12998v1
19. Lions, P.L., Paul, T.: Sur les mesures de Wigner. Rev. Mat. Iberoamericana **9**(3), 553–618 (1993)
20. Janson, S.: Gaussian Hilbert Spaces. Cambridge Tracts in Mathematics, vol. 129. Cambridge University Press, Cambridge (1997)
21. Keel, M., Tao, T.: Endpoint Strichartz estimates. Am. J. Math. **120**, 955–980 (1998)
22. Lieb, E.: Gaussian kernels have only gaussian maximizers. Invent. Math. **102**, 179–208 (1990)
23. Nirenberg, L.: An abstract form of the nonlinear Cauchy-Kowalewski theorem J. Differ. Geom. **6**, 561–576 (1972)
24. Papanicolaou, G.: Waves in one-dimensional random media. In: Ecole d'été de Probabilités de Saint-Flour. Lecture Notes in Mathematics, pp. 205–275. Springer, Berlin (1988)
25. Peetre, J.: Sur la transformation de Fourier à valeurs vectorielles. Rendiconti del Seminario Matematico della Università di Padova **42**, 15–26 (1969)
26. Ryzhik, L., Papanicolaou, G., Keller, J.B.: Transport equations for elastic and other waves in random media. Wave Motion **24**(4), 327–370 (1996)
27. Schnirelman, A.: Ergodic properties of eigenfunctions. Uspehi Mat. Nauk. **29**, 181–182 (1974)
28. Simon, B.: The $P(\Phi)_2$ Euclidean (Quantum) Field Theory. Princeton Series in Physics, Princeton University Press, Princeton (1974)
29. Spohn, H.: Derivation of the transport equation for electrons moving through random impurities. J. Stat. Phys. **17**(6), 385–412 (1977)

On the Semiclassical Regularity of Thermal Equilibria

Jacky J. Chong, Laurent Lafleche, and Chiara Saffirio

1 Introduction

We consider a system of N non-interacting fermions in a harmonic trap and study the semiclassical structure of its one-particle density matrix associated with the corresponding Gibbs (equilibrium) state of the trapped gas at temperature $T > 0$. More precisely, given a family of one-particle normalized nonnegative trace class operators $\boldsymbol{\rho}_\beta$ acting on $L^2(\mathbb{R}^d)$ and indexed by the parameter β, proportional to T^{-1}, we consider the following quantities

$$\nabla_x \boldsymbol{\rho}_\beta := \left[\nabla, \boldsymbol{\rho}_\beta\right] \quad \text{and} \quad \nabla_\xi \boldsymbol{\rho}_\beta := \left[\tfrac{x}{i\hbar}, \boldsymbol{\rho}_\beta\right], \tag{1}$$

which we refer to as the quantum gradients of $\boldsymbol{\rho}_\beta$, and show that for any $p \in [2, \infty]$,

$$\boldsymbol{\rho}_\beta, \sqrt{\boldsymbol{\rho}_\beta} \in \mathcal{W}^{1,p}(\mathsf{m})$$

uniformly in $\hbar$, where $\mathcal{W}^{1,p}(\mathsf{m})$ are the semiclassical analogue of weighted Sobolev spaces equipped with the norm

$$\|\boldsymbol{\rho}\|^p_{\mathcal{W}^{1,p}(\mathsf{m})} := \|\boldsymbol{\rho}\,\mathsf{m}\|^p_{\mathcal{L}^p} + \|\nabla_x \boldsymbol{\rho}\,\mathsf{m}\|^p_{\mathcal{L}^p} + \left\|\nabla_\xi \boldsymbol{\rho}\,\mathsf{m}\right\|^p_{\mathcal{L}^p}$$

J. J. Chong · L. Lafleche
Department of Mathematics, The University of Texas at Austin, Austin, TX, USA
e-mail: jwchong@math.utexas.edu; lafleche@math.utexas.edu; chiara.saffirio@unibas.ch

C. Saffirio (✉)
Department of Mathematics and Computer Science, University of Basel, Basel, Switzerland

M. Correggi, M. Falconi (eds.), *Quantum Mathematics I*,
Springer INdAM Series 57, https://doi.org/10.1007/978-981-99-5894-8_2

and $\mathsf{m} := 1 + |\boldsymbol{p}|^n$. Here $\boldsymbol{p} = -i\hbar\nabla$ is the momentum operator and $\mathcal{L}^p$ are the semiclassical versions of the Lebesgue spaces endowed with the rescaled Schatten norms

$$\|\boldsymbol{\rho}\|_{\mathcal{L}^p} := h^{\frac{d}{p}} \mathrm{Tr}\big(|\boldsymbol{\rho}|^p\big)^{\frac{1}{p}}$$

where $h = 2\pi\hbar$.

The quantities (1) play key roles in obtaining semiclassical estimates, such as in the problems of the derivation of the Vlasov equation from quantum mechanics and of the Hartree and the Hartree–Fock equations from many-body quantum mechanics. To the authors' knowledge, they first appeared in [2], where the Hartree equation was derived in the mean-field regime for arbitrary long times from a system of N fermions interacting through a smooth potential, under the assumption that the initial state $\boldsymbol{\rho}^{\mathrm{in}}$ of the Hartree dynamics is a pure state, i.e. $(\boldsymbol{\rho}^{\mathrm{in}})^2 = \boldsymbol{\rho}^{\mathrm{in}}$, and satisfies the following semiclassical structures

$$\left\|\nabla_x \boldsymbol{\rho}^{\mathrm{in}}\right\|_{\mathcal{L}^1} \leq C \quad \text{and} \quad \left\|\nabla_\xi \boldsymbol{\rho}^{\mathrm{in}}\right\|_{\mathcal{L}^1} \leq C. \tag{2}$$

Similarly, bounds on semiclassical Schatten norms of commutators are used in [9] to obtain uniform in $\hbar$ mean-field estimates, in [13] in the context of the semiclassical limit, in [11] in the context of the convergence of numerical schemes, in [10] to get estimates on the quantum Wasserstein distance.

When the interaction is the Coulomb potential only partial results are available on a time scale of order one (see [14]) and they rely on a semiclassical structure expressed in terms of L^p norms of the diagonal of the operator $|\nabla_\xi \boldsymbol{\rho}|$, with $p > 3$ (see also [13]).

In [8] and [1], the authors obtained semiclassical bounds of the form (2) for pure states in the non-interacting case, which in turn provided explicit examples of initial pure states that satisfy the assumptions in [2] on the semiclassical structures. In this paper, we extend this analysis to the finite positive temperature case, where the one-particle reduced density matrix $\boldsymbol{\rho}$ is no longer a projection. When dealing with mixed states the analogue relevant quantities considered in [3, 6] are

$$\left\|\nabla_x \sqrt{\boldsymbol{\rho}^{\mathrm{in}}}\right\|_{\mathcal{L}^p} \leq C \quad \text{and} \quad \left\|\nabla_\xi \sqrt{\boldsymbol{\rho}^{\mathrm{in}}}\right\|_{\mathcal{L}^p} \leq C. \tag{3}$$

In [3], the interaction potential is integrable, its Fourier transform has bounded moments of order two, and Eq. (3) is required to hold true for the the initial state for $p = 1$. In [6], singular interactions of the form $|x|^{-a}$ with $a \leq 1$ are considered and Eq. (3) is required to hold for $p \in [2, \infty]$.

In this work, we focus on quasi-free Gibbs states, which is an important class of examples as they model the equilibria of an ideal gas at finite positive temperature. It is well-known that the one-particle reduced density matrix associated to a Gibbs state can be computed explicitly (see Proposition 2). Using the explicit expression (4), we prove bounds on the weighted $\mathcal{L}^p$ norms of the quantum gradients (1) with $\boldsymbol{\rho}_\beta$ given by the one-particle reduced density matrix associated to the Gibbs state of a system of non-interacting fermions confined by a harmonic potential, and on its partition function. Our main result is the following.

Theorem 1 *Let* $H = \frac{|p|^2+|x|^2}{2}$ *be the d-dimensional harmonic oscillator Hamiltonian and define the family of density matrix operators given by*

$$\boldsymbol{\rho}_\beta := \frac{1}{N h^d}\left(\mathbf{1} + e^{\beta(H-\mu)}\right)^{-1} \tag{4}$$

where μ *depends on* β, N *and* h *and is chosen so that* $h^d \mathrm{Tr}(\boldsymbol{\rho}_\beta) = 1$ *holds. Then for any* $p \geq 2$, $\boldsymbol{\rho}_\beta \in \mathcal{W}^{1,p}(m)$ *and* $\sqrt{\boldsymbol{\rho}_\beta} \in \mathcal{W}^{1,p}(m)$ *uniformly with respect to* $\hbar \in (0, 1)$. *More explicitly, let*

$$Z_\beta := h^d \mathrm{Tr}(e^{-\beta H}), \quad \text{and} \quad Z_\mu := \lambda\, e^{-\beta\mu} \tag{5}$$

with $\lambda := N h^d$, *then the following holds. For any fixed* $\beta \in \mathbb{R}_+$, $\hbar \in (0, 1)$ *and for* $p \in [1, \infty]$, *we have the bound*

$$\left\|\nabla \boldsymbol{\rho}_\beta\right\|_{\mathcal{L}^p} \leq C_{d,p}\, \frac{\beta^{\frac{1}{2}-\frac{d}{p}}}{Z_\mu}\, \frac{\max\left(2\sqrt{2}, \beta\hbar\right)^{\frac{1}{2}-\frac{1}{p}}}{(\theta(\beta\hbar))^{\frac{1}{p}}} \tag{6}$$

where $\theta(x) = \mathrm{th}(x)/x$ *with* $\mathrm{th}(x) = \frac{e^x - e^{-x}}{e^x + e^{-x}}$, *and*

$$C_{\lambda,\beta}^{-1}\, Z_\beta \leq Z_\mu \leq Z_\beta \tag{7}$$

with $C_{\lambda,\beta} = 2$ *if* $\mu \leq d\hbar/2$ *and* $C_{\lambda,\beta} = 1 + e^{\beta\lambda^{1/d}/\pi}$ *if* $\mu \geq d\hbar/2$.

Remark 1 As a corollary of our result, we exhibit a class of initial states satisfying the regularity assumptions of [6, Theorem 3.1].

Remark 2 Our proof in the case $p = \infty$ is rather general and in particular can be adapted without difficulty to other external trapping potentials $V(x)$, that is to the case of an Hamiltonian of the form $H = \frac{|p|^2}{2} + V(x)$.

Remark 3 As indicated in [6, Figure 1], there are three scaling regimes to keep in mind when studying mean-field dynamics of quantum systems (both bosonic and fermionic), namely, $Nh^d \to 0$, $Nh^d \to \infty$, and Nh^d converges to some constant (in fact, $Nh^d = 1$ which we called the critical scaling). We refer to the limit $Nh^d \to 0$ as the lower density scaling regime, which means that the density growth in N is slower than in the critical scaling regime. Similarly, we refer to the limit $Nh^d \to \infty$ as the higher density scaling regime.

Let $\beta > 0$ be fixed. In the lower density scaling regime $Nh^d \to 0$, we have that $e^{\beta\mu} \to 0$ or equivalently $\mu \to -\infty$, which is an immediate consequence of (7) in the above theorem and the fact that Z_β is bounded (cf. Sect. 2.2). It is not difficult to see that in this case $\rho_\beta = \left(Nh^d + Z_\mu e^{\beta H}\right)^{-1}$ will converge to $Z_\beta^{-1} e^{-\beta H}$. In other words, in the semiclassical limit $\hbar \to 0$, ρ_β will converge to the Maxwell–Boltzmann distribution. This result should not be surprising to the reader since in the lower density scaling regime $Nh^d \to 0$, the Fermi gas behaves like a classical ideal gas.

In the case of the higher density scaling regime $Nh^d \to \infty$, we recall that as in [6, Equation (16)], $\lambda = Nh^d \leq \frac{1}{\|\rho\|_{\mathcal{L}^\infty}} =: C_\infty^{-1}$. This shows that $\|\rho\|_{\mathcal{L}^\infty} \to 0$ as $\lambda \to \infty$. From a more physical perspective, a high density Fermi gas exhibits quantum "degeneracy" behavior, consequence of the Pauli exclusion principle.

Finally, in the critical scaling case, we see that ρ_β does not converge to the Maxwell–Boltzmann distribution but instead to the Fermi–Dirac distribution. The divergence from classical behavior is attributed to the fact that at the critical scaling we are in fact modeling the degeneracy behavior of the Fermi gas. See e.g. [5, Chapter 17.5].

Remark 4 In the semiclassical limit $\hbar \to 0$, then our main Inequality (6) implies

$$\left\|\nabla \rho_\beta\right\|_{\mathcal{L}^p} \leq C_{d,p,\lambda}\, \beta^{\frac{1}{2}+\frac{d}{p'}} + o(1)$$

where $o(1)$ converges to 0 when $\hbar \to 0$. This is optimal in the sense that it is consistent with the behavior of the classical norm $\left\|\nabla\left(Z_\beta^{-1} e^{-\beta|z|^2}\right)\right\|_{L^p(\mathbb{R}^{2d})}$ in terms of β provided we are away from the critical scaling regime. It is interesting to observe that the behavior of the right-hand-side of Inequality (6) changes at low temperature and when $\hbar$ is not negligible (i.e. when $\beta\hbar$ is large). Moreover, we do not capture the correct behavior when the state does not converge to the Maxwell–Boltzmann distribution.

1.1 Second Quantization Formalism

To introduce quasi-free Gibbs states, we will need the second quantized description of a many-body system of non-interacting fermions in an external trap. Here we give a brief review of the Fock space formalism. For a more comprehensive treatment, the reader could consult [4, 15].

The fermionic (antisymmetric) Fock space over the one-particle Hilbert space $\mathfrak{h} = L^2(\mathbb{R}^d)$ is defined to be the closure of

$$\mathcal{F} := \bigoplus_{n=0}^{\infty} \mathfrak{h}^{\wedge n} = \mathbb{C} \oplus \mathfrak{h} \oplus (\mathfrak{h} \wedge \mathfrak{h}) \oplus \cdots$$

with respect to the norm induced by the standard associated inner product. Here the n-th sector of $\mathcal{F}$ given by $\mathfrak{h}^{\wedge n}$ denotes the subspace of $\mathfrak{h}^n = L^2(\mathbb{R}^{dn})$ containing functions that satisfy the antisymmetry property

$$\psi(x_1,\ldots,x_n)=\operatorname{sgn}(\sigma)\,\psi(x_{\sigma(1)},\ldots,x_{\sigma(n)})$$

for any permutation σ. An important vector to note is the vacuum vector $\Omega=(1,0,0,\ldots)\in\mathcal{F}$ which describes a pure state with no particles.

The operator-valued distribution a_x and a_x^* on $\mathcal{F}$ are defined by their actions on the nth sector of $\mathcal{F}$ as follows. Let $\psi\in\mathfrak{h}^{\wedge n}$, then

$$(a_x\psi)(x_1,\ldots,x_{n-1}):=\sqrt{n}\,\psi(x,x_1,\ldots,x_{n-1}),$$

$$(a_x^*\psi)(x_1,\ldots,x_{n+1}):=\sum_{j=1}^{n+1}\frac{(-1)^{j-1}}{\sqrt{n+1}}\,\delta_x(x_j)\,\psi(x_1,\ldots,x\!\!/_j,\ldots,x_{n+1}).$$

Furthermore, for each $\varphi\in\mathfrak{h}$, we associate to it the annihilation operator and its adjoint the creation operator, defined by

$$a(\varphi)=\int_{\mathbb{R}^d}\overline{\varphi(x)}\,a_x\,\mathrm{d}x\quad\text{ and }\quad a^*(\varphi)=\int_{\mathbb{R}^d}\varphi(x)\,a_x^*\,\mathrm{d}x.$$

It can be shown that creation and annihilation operators are bounded on $\mathcal{F}$ with operator norm $\|a(\varphi)\|_\infty=\|a^*(\varphi)\|_\infty=\|\varphi\|_{\mathfrak{h}}$ and that they satisfy the canonical anticommutation relations, i.e.

$$\{a(\varphi),a(\phi)\}=\{a^*(\varphi),a^*(\phi)\}=0\quad\text{ and }\quad\{a^*(\varphi),a(\phi)\}=\langle\phi,\varphi\rangle_{\mathfrak{h}}\,\mathbf{1}$$

hold for all $(\varphi,\phi)\in\mathfrak{h}^2$ where $\{\mathsf{A},\mathsf{B}\}:=\mathsf{AB}+\mathsf{BA}$ is the standard anticommutator of the operators A and B.

In this work, we are interested in trapped non-interacting Fermionic systems modeled by the one-particle Hamilton operator

$$H=\tfrac{1}{2}\,|\boldsymbol{p}|^2+V(x),\tag{8}$$

where $\boldsymbol{p}=-i\hbar\nabla_x$ and V is an external trapping potential. The second quantization of the one-particle operator H on $\mathcal{F}$ is defined to be the operator H whose action on the nth sector is given by

$$(\mathsf{H}\Psi)\,(x_1,\ldots,x_n)=\sum_{j=1}^{n}(-\tfrac{\hbar^2}{2}\Delta_{x_j}+V(x_j))\Psi(x_1,\ldots,x_n).$$

Another useful operator on $\mathcal{F}$ is the number operator

$$\mathcal{N}=\bigoplus_{n=0}^{\infty}n\,\mathbf{1}_{\mathfrak{h}^{\wedge n}}.$$

1.2 Harmonic Oscillators

To simplify the presentation, we choose the external potential V to be the harmonic trapping potential, that is,

$$H = \tfrac{1}{2}\,|p|^2 + \tfrac{1}{2}\,|x|^2 = \sum_{i=1}^{d} \tfrac{1}{2}\,p_i^2 + \tfrac{1}{2}\,x_i^2 = \sum_{i=1}^{d} H_i. \tag{9}$$

Since H is the d-dimensional quantum harmonic oscillator, it is well-known that it emits a discrete spectrum where the eigenvalues and eigenvectors are given by

$$E_n = \left(|n|_1 + \tfrac{d}{2}\right)\hbar \qquad \text{for } n = (n_1, \ldots, n_d) \in \mathbb{N}_0^d,$$

$$u_{E_n} = \phi_{n_1} \otimes \cdots \otimes \phi_{n_d} \qquad \text{with } H_i \phi_{n_i} = \left(n_i + \tfrac{1}{2}\right)\hbar\,\phi_{n_i}.$$

where $|n|_1 = n_1 + \ldots + n_d$ and the functions ϕ_{n_i} are the standard one-variable Hermite functions

$$\phi_{n_i}(x_i) = c_{n_i}\, e^{-x^2/2} P_{n_i}(x_i) \quad \text{with} \quad c_{n_i}^{-1} = 2^{\frac{n_i}{2}}\,(n_i!)^{\frac{1}{2}}\,(\pi\hbar)^{\frac{1}{4}},$$

with $P_{n_i}(x)$ the Hermite polynomials. Moreover, for each eigenvalue E_n the corresponding multiplicity is given by $g_{|n|_1,d} = \binom{|n|_1+d-1}{d-1}$. In particular, by the spectral decomposition of H, we could now rewrite H as

$$\mathsf{H} = \sum_{n\in\mathbb{Z}_0^d} E_n\, a^*(u_{E_n})\, a(u_{E_n}). \tag{10}$$

1.3 Thermal States

We consider quantum states that are given by density operators ω on $\mathcal{F}$, that is, the expectation of an observable A is given by $\langle \mathsf{A}\rangle_\omega = \mathrm{Tr}_{\mathcal{F}}(\mathsf{A}\,\omega)$ for every $\mathsf{A} \in \mathcal{B}(\mathcal{F})$, where $\mathcal{B}(\mathcal{F})$ denotes the space of bounded operators on the Fock space. The Gibbs equilibrium of a trapped non-interacting Fermi gas associated to some positive temperature $T > 0$ is defined to be the unique minimizer of the Gibbs free energy functional

$$\mathcal{F}(\omega) = \mathrm{Tr}_{\mathcal{F}}(\mathsf{H}\,\omega) - k_{\mathrm{B}}\,T\,\mathcal{S}(\omega) \quad \text{with} \quad \mathcal{S}(\omega) = -\mathrm{Tr}_{\mathcal{F}}(\omega \ln \omega)$$

in $\mathcal{X}_N = \{\omega \in \mathcal{B}(\mathcal{F}) \mid \omega \geq 0,\ \mathrm{Tr}_{\mathcal{F}}(\omega) = 1,\ \mathrm{Tr}_{\mathcal{F}}(\mathcal{N}\omega) = N\}$ where $\mathcal{S}$ is the von Neumann entropy. It can be checked that the Gibbs state associated to the temperature T is given by the normalized positive trace class operator

$$\omega_N = \frac{1}{\mathcal{Z}_N} e^{-\beta\left(\mathsf{H}-\mu\mathcal{N}\right)} \tag{11}$$

where $\beta = 1/(k_{\mathrm{B}}T)$ and the chemical potential μ is chosen so that $\mathrm{Tr}(\mathcal{N}\omega_N) = N$. Here $\mathcal{Z}_N$ is the grand canonical partition function

$$\mathcal{Z}_N = 1 + \sum_{n=1}^{\infty} e^{n\beta\mu} \mathrm{Tr}_{\mathfrak{h}^{\wedge n}} \exp\left(-\beta \sum_{j=1}^{dn} H_j\right) = \prod_{n\in\mathbb{Z}_0^d} (1 + e^{-\beta(E_n-\mu)}).$$

1.4 Quasi-Free States

Here we will only give a rudimentary introduction to the tools necessary for the subsequent sections. A more comprehensive exposition of the following content can be found in [4, 15]. The state ω is said to be quasi-free if for all $n \in \mathbb{N}$, we have that

$$\mathrm{Tr}_{\mathcal{F}}\big(a^{\sharp_1}(f_1)\cdots a^{\sharp_{2n+1}}(f_{2n+1})\omega\big) = 0,$$

$$\mathrm{Tr}_{\mathcal{F}}\big(a^{\sharp_1}(f_1)\cdots a^{\sharp_{2n}}(f_{2n})\omega\big) = \sum_{\sigma}(-1)^{\sigma}\prod_{j=1}^{n}\mathrm{Tr}\big(a^{\sharp_{\sigma(j)}}(f_{\sigma(j)})a^{\sharp_{\sigma(j+n)}}(f_{\sigma(j+n)})\omega\big),$$

where $a^{\sharp_j}$ stands for either a or a^*, and the sum is taken over all permutations σ satisfying $\sigma(1) < \sigma(2) < \ldots < \sigma(n)$, and $\sigma(j) < \sigma(j+n)$, for all $j \in \{1, \ldots, n\}$. The definition indicates that quasi-free states are determined by the one-particle reduced density matrix operator defined via its integral kernel

$$\rho(x, y) = \frac{1}{Nh^d}\mathrm{Tr}_{\mathcal{F}}\left(a_y^* \, a_x \, \omega\right), \tag{12}$$

and the antisymmetric pairing function defined by

$$\alpha(x, y) = \frac{1}{Nh^d}\mathrm{Tr}_{\mathcal{F}}\left(a_y \, a_x \, \omega\right). \tag{13}$$

More compactly, we introduce the generalized one-particle density matrix $\boldsymbol{\Gamma} : \mathfrak{h} \oplus \mathfrak{h}^* \to \mathfrak{h} \oplus \mathfrak{h}^*$ defined by

$$\boldsymbol{\Gamma} = \begin{bmatrix} \boldsymbol{\rho} & \boldsymbol{\alpha} \\ -\mathbb{J}\boldsymbol{\alpha}\mathbb{J} & \mathbf{1} - \mathbb{J}\boldsymbol{\rho}\mathbb{J}^* \end{bmatrix} \tag{14}$$

where $\mathbb{J} : \mathfrak{h} \to \mathfrak{h}^*$ is the map $\mathbb{J}(\phi) = \langle \phi, \cdot \rangle$ and $\mathbb{J}^*$ is its adjoint operator. This observation is summarized by the following proposition (see e.g. [15, Appendix G]).

Proposition 1 *Let* $\boldsymbol{\Gamma} : \mathfrak{h} \oplus \mathfrak{h}^* \to \mathfrak{h} \oplus \mathfrak{h}^*$ *be an operator of the form* (14). *Then* $\boldsymbol{\Gamma}$ *is the generalized one-particle density matrix of some quasi-free state with finite particle number if and only if* $\boldsymbol{\Gamma} \geq 0$ *and* $\mathrm{Tr}\boldsymbol{\rho} < \infty$.

In the case of $\boldsymbol{\omega}_N$, it can be readily shown that its generalized one-particle density matrix is given by

$$\boldsymbol{\Gamma}_\beta = \begin{bmatrix} \boldsymbol{\rho}_\beta & 0 \\ 0 & \mathbf{1} - \mathbb{J}\boldsymbol{\rho}_\beta\mathbb{J}^* \end{bmatrix}$$

where $\boldsymbol{\alpha} = 0$, which we refer to as the gauge-invariant condition. The following proposition, whose proof can be found for example in [4, Proposition 5.2.23], makes the link between the Fock space Gibbs state (11) and the one body Gibbs state considered in our main theorem.

Proposition 2 *Let* $\boldsymbol{\omega}_N$ *denote the Gibbs state* (11) *with* $\mathrm{Tr}\left(e^{-\beta H}\right) < \infty$. *Then* $\boldsymbol{\omega}_N$ *is quasi-free and its one-particle reduced density matrix operator is given by Eq.* (4).

2 Semiclassical Thermal Regularity

In this section, we give the proof of the main theorem.

2.1 Preliminaries

For a trapped non-interacting Fermi gas with Hamiltonian (8), let us write $G = e^{-\beta H}$ and $\lambda = Nh^d$. Consider the quantities

$$g_\beta = Z_\beta^{-1}\, e^{-\beta H}, \tag{15a}$$

$$\boldsymbol{\rho}_\beta = \lambda^{-1}\left(\mathbf{1} + e^{\beta(H-\mu)}\right)^{-1} = G_\mu\left(\mathbf{1} + \lambda\, G_\mu\right)^{-1}, \tag{15b}$$

where $G_\mu = \lambda^{-1}\, e^{-\beta(H-\mu)} = Z_\mu^{-1}\, G$ and Z_β and Z_μ are given in Eq. (5). The quantum gradients of these operators are given by the following lemma.

Lemma 1 *Let ∇ be a quantum gradient. Then we have*

$$\nabla g_\beta = -\frac{\beta}{Z_\beta}\int_0^1 G^{1-s}\,(\nabla H)\,G^s\,\mathrm{d}s, \tag{16a}$$

$$\nabla \rho_\beta = -\beta\int_0^1 G_\mu^{1-s}(\mathbf{1}+\lambda G_\mu)^{-1}\,(\nabla H)\,(\mathbf{1}+\lambda G_\mu)^{-1} G_\mu^s\,\mathrm{d}s. \tag{16b}$$

Proof Notice that for any operators A and B, we have that

$$\partial_t\left[A, e^{tB}\right] = \left[A, B\,e^{tB}\right] = [A, B]\,e^{tB} + B\left[A, e^{tB}\right].$$

Therefore, $\partial_t\big(e^{-tB}\left[A, e^{tB}\right]\big) = e^{-tB}\,[A, B]\,e^{tB}$ and we obtain the Duhamel-like formula

$$\left[A, e^{tB}\right] = \int_0^t e^{(t-s)B}\,[A, B]\,e^{sB}\,\mathrm{d}s. \tag{17}$$

In particular, taking $t = 1$ and $B = -\beta H$, we obtain the identity

$$\nabla G = -\beta\int_0^1 G^{1-s}\,(\nabla H)\,G^s\,\mathrm{d}s$$

from which we deduce Identity (16a). Then observe that if A is an invertible operator, by the Leibniz rule for commutators, $0 = \nabla\big(AA^{-1}\big) = (\nabla A)\,A^{-1} + A\nabla\big(A^{-1}\big)$ and so

$$\nabla\big(A^{-1}\big) = -A^{-1}\,(\nabla A)\,A^{-1}. \tag{18}$$

In particular, since $\nabla\,\big(\mathbf{1}+\lambda G_\mu\big) = \lambda\nabla G_\mu$, we deduce that

$$\begin{aligned}\nabla\rho_\beta &= \big(\nabla G_\mu\big)\,\big(\mathbf{1}+\lambda G_\mu\big)^{-1} - G_\mu\,\big(\mathbf{1}+\lambda G_\mu\big)^{-1}\lambda\,\big(\nabla G_\mu\big)\,\big(\mathbf{1}+\lambda G_\mu\big)^{-1}\\ &= \big(\mathbf{1}+\lambda G_\mu\big)^{-1}\,\big(\nabla G_\mu\big)\,\big(\mathbf{1}+\lambda G_\mu\big)^{-1}\end{aligned}$$

which leads to Formula (16b). □

Taking the square root of G changes β by $\beta/2$. Hence it follows that

$$\nabla\sqrt{g_\beta} = -\frac{\beta}{2\sqrt{Z_\beta}}\int_0^1 G^{(1-s)/2}\,(\nabla H)\,G^{s/2}\,\mathrm{d}s.$$

On the other hand, it is more difficult to compute explicitly $\nabla\sqrt{\rho_\beta}$. The following lemma allows us to bound $\sqrt{\rho_\beta}$ by reducing the problem to estimating G.

Lemma 2 *Let* m *be a self-adjoint operator. Then*

$$\|\nabla\sqrt{\rho_\beta}\|_{\mathcal{L}^p(\mathsf{m})} \le \left(\|\nabla\sqrt{G_\mu}\|_{\mathcal{L}^p(\mathsf{m})} + \frac{\lambda}{2}\|\sqrt{\rho_\beta}\|_{\mathcal{L}^q(\mathsf{m})}\|\nabla G_\mu\|_{\mathcal{L}^r}\right)$$

for any $1 \le p, q, r \le \infty$ *such that* $\frac{1}{p} = \frac{1}{q} + \frac{1}{r}$.

Proof By Identity (18), we have that

$$\nabla\sqrt{\rho_\beta} = \left(\nabla\sqrt{G_\mu} - \sqrt{\rho_\beta}\left(\nabla\sqrt{\mathbf{1}+\lambda G_\mu}\right)\right)(\mathbf{1}+\lambda G_\mu)^{-\frac{1}{2}}. \tag{19}$$

To bound $\nabla\sqrt{\mathbf{1}+\lambda G_\mu}$, we proceed as in [6, Lemma 7.1]. Let $A := \lambda G_\mu$. Since $0 \le A \le 2c := \lambda Z_\mu^{-1}$, we deduce that $\|A - c\|_{\mathcal{L}^\infty} \le c$ and so the following series is absolutely convergent

$$\sqrt{\mathbf{1}+A} = \sqrt{1+c}\sqrt{\mathbf{1}+\tfrac{1}{1+c}(A-c)} = \sum_{n=0}^{\infty}\binom{1/2}{n}\left(\tfrac{1}{1+c}\right)^{n-\frac{1}{2}}(A-c)^n.$$

Since it follows from the Jacobi identity that

$$\nabla(A-c)^n = \sum_{k=1}^{n}(A-c)^{k-1}(\nabla A)(A-c)^{n-k},$$

then we deduce the estimate

$$\begin{aligned}\left\|\nabla\sqrt{\mathbf{1}+A}\right\|_{\mathcal{L}^p} &\le \sum_{n=1}^{\infty}\left|\binom{1/2}{n}\right|\left(\tfrac{1}{c+1}\right)^{n-\frac{1}{2}} n\,\|\nabla A\|_{\mathcal{L}^p}\|A-c\|_{\mathcal{L}^\infty}^{n-1}\\ &\le \frac{1}{2\sqrt{c+1}}\sum_{n=1}^{\infty}\binom{-1/2}{n-1}\left(\tfrac{-c}{c+1}\right)^{n-1}\|\nabla A\|_{\mathcal{L}^p} = \frac{1}{2}\|\nabla A\|_{\mathcal{L}^p}.\end{aligned}$$

We conclude by applying the fact that $\left\|\nabla\sqrt{\rho_\beta}\,\mathsf{m}\right\|_{\mathcal{L}^p} = \left\|\mathsf{m}\,\nabla\sqrt{\rho_\beta}\right\|_{\mathcal{L}^p}$, which follows from taking the adjoint, and the triangle inequality to Formula (19). This yields the desired result. □

2.2 Bounds on the Inverse Fugacity

In the rest of this paper, we assume H is the harmonic oscillator Hamiltonian $H = \frac{|p|^2+|x|^2}{2}$. In this case of the partition function in g_β has the closed form

$$Z_\beta = \left(\frac{2\pi}{\beta}\right)^d \frac{1}{\mathrm{shc}(\frac{\beta\hbar}{2})^d} \tag{20}$$

where $\mathrm{shc}(x) = \mathrm{sh}(x)/x$ denotes the hyperbolic sinc function. In particular $Z_\beta \leq (2\pi/\beta)^d$ and $Z_\beta \sim (2\pi/\beta)^d$ as $\beta\hbar \to 0$. The function Z_μ, that one could call the inverse fugacity, can be compared to the partition function Z_β as proved in the following proposition.

Proposition 3 *Let $Z_\mu = \lambda\, e^{-\beta\mu}$ with $\lambda = Nh^d$. Then the following inequality holds*

$$C_{\lambda,\beta}^{-1}\, Z_\beta \leq Z_\mu \leq Z_\beta \tag{21}$$

with $C_{\lambda,\beta} = 2$ if $\mu \leq d\hbar/2$ and $C_{\lambda,\beta} = 1 + e^{\beta\lambda^{1/d}/2\pi}$ if $\mu \geq d\hbar/2$.

Proof Let $c_\mu = e^{\beta\mu} = Nh^d/Z_\mu$. Notice that

$$1 = h^d \mathrm{Tr}(\boldsymbol{\rho}_\beta) = \frac{1}{N}\mathrm{Tr}\Big(\big(\mathbf{1} + c_\mu^{-1} e^{\beta H}\big)^{-1}\Big) \leq \frac{c_\mu}{N}\mathrm{Tr}\Big(e^{-\beta H}\Big),$$

which implies the second inequality in Formula (21).

Next, define the function $g : \mathbb{R}_+ \to \mathbb{R}_+$ by $g(r) = \big(1 + c_\mu^{-1}\, e^{\beta r}\big)^{-1}$, then for any $R > 0$, it holds $g(r) \geq \big(1 + c_\mu^{-1}\, e^{\beta r}\big)^{-1}\, \mathbb{1}_{r \leq R}$, which leads to

$$1 = \frac{1}{N}\mathrm{Tr}\,(g(H)) \geq \frac{1}{N\Big(1 + c_\mu^{-1}\, e^{\beta R}\Big)}\, \mathrm{Tr}\left(\mathbb{1}_{H \leq R}\right).$$

Recalling the property of the harmonic oscillator given in Sect. 1.2, one sees that the trace of the characteristic function is nothing but

$$\mathrm{Tr}\big(\mathbb{1}_{H \leq R}\big) = \left|\left\{n \in \mathbb{N}_0^d : \left(|n|_1 + \tfrac{d}{2}\right)\hbar \leq R\right\}\right|.$$

Since $|n|_1 \geq \sup_j n_j =: |n|_\infty$, then this can be crudely estimated by

$$\mathrm{Tr}\big(\mathbb{1}_{H \leq R}\big) \geq \left|\left\{n \in \mathbb{N}_0^d : |n|_\infty \leq \tfrac{R}{\hbar} - \tfrac{d}{2}\right\}\right| = \left(\left\lfloor \tfrac{R}{\hbar} - \tfrac{d}{2}\right\rfloor + 1\right)^d. \tag{22}$$

In particular, since $\mu \geq d\hbar/2 > 0$, then taking $R = \mu = \frac{\ln c_\mu}{\beta}$ yields $\frac{\mu}{\hbar} - \frac{d}{2} \leq \left(N^{1/d} + 1\right)$ and so since $N \geq 1$,

$$\mu \leq 2\, N^{1/d} \hbar + \frac{d\hbar}{2}. \tag{23}$$

Now, let us obtain upper bounds for c_μ. First, observe that we have the lower bounds

$$1 = \frac{c_\mu}{N} \mathrm{Tr}\Big(\big(\mathbf{1} + c_\mu e^{-\beta H}\big)^{-1} e^{-\beta H}\Big) \geq \frac{Z_\beta\, c_\mu}{N h^d \left(1 + c_\mu e^{-d\beta\hbar/2}\right)}. \tag{24}$$

If $c_\mu \leq e^{d\beta\hbar/2}$ (i.e. $\mu \leq d\hbar/2$) then it follows from Inequality (24) that

$$c_\mu \leq 2\, N h^d Z_\beta^{-1}.$$

On the other hand, if $c_\mu \geq e^{d\beta\hbar/2} > 1$ (i.e. $\mu \geq d\hbar/2$), then it follows from Inequality (23) and Inequality (24) that we have the bound

$$c_\mu \leq \left(1 + c_\mu\, e^{-d\beta\hbar/2}\right) N h^d Z_\beta^{-1} \leq \left(1 + e^{2\beta\hbar N^{1/d}}\right) N h^d Z_\beta^{-1} \tag{25}$$

This completes our proof of the proposition. □

2.3 $\mathcal{L}^\infty$ Bounds

This section is devoted to the proof of the following proposition.

Proposition 4 *Let $\beta > 0$ and $H = \frac{|\boldsymbol{p}|^2 + |x|^2}{2}$, then we have the estimates*

$$\left\|\nabla g_\beta\right\|_{\mathcal{L}^\infty} \leq \frac{2}{Z_\beta} \max\left(\sqrt{\beta}, \beta\sqrt{\hbar}\right),$$

$$\left\|\nabla \boldsymbol{\rho}_\beta\right\|_{\mathcal{L}^\infty} \leq \frac{2}{Z_\mu} \max\left(\sqrt{\beta}, \beta\sqrt{\hbar}\right).$$

Prior to giving the proof of the above proposition, let us make the following observation. Since $\nabla_x H = x$ and $\nabla_\xi H = \boldsymbol{p}$, we deduce that $\nabla_z H = \boldsymbol{z} := (x, \boldsymbol{p})$ where $z = (x, \xi)$. By Identity (16a), we have that

$$\nabla \boldsymbol{\rho}_\beta = -\beta \int_0^1 (\mathbf{1} + \lambda G_\mu)^{-1} G_\mu^{1-s}\, \boldsymbol{z}\, G_\mu^s (\mathbf{1} + \lambda G_\mu)^{-1}\, \mathrm{d}s$$

which implies the estimate

$$\|\nabla \boldsymbol{\rho}_\beta\|_{\mathcal{L}^p} \leq \beta \int_0^1 \left\|G_\mu^{1-s} z\, G_\mu^s\right\|_{\mathcal{L}^p} \mathrm{d}s = \beta\, Z_\mu^{-1} \int_0^1 \left\|G^{1-s} z\, G^s\right\|_{\mathcal{L}^p} \mathrm{d}s.$$

Since $\|G\|_{\mathcal{L}^\infty} \leq 1$, then we have the estimate

$$\begin{aligned} \|\nabla \boldsymbol{\rho}_\beta\|_{\mathcal{L}^p} &\leq \beta\, Z_\mu^{-1} \int_0^{1/2} \left\|G^{1-s} z\right\|_{\mathcal{L}^p} \mathrm{d}s + \beta\, Z_\mu^{-1} \int_{1/2}^1 \|z\, G^s\|_{\mathcal{L}^p} \mathrm{d}s \\ &\leq 2\,\beta\, Z_\mu^{-1} \int_{1/2}^1 \|z\, G^s\|_{\mathcal{L}^p} \mathrm{d}s. \end{aligned} \tag{26}$$

Hence to estimate $\nabla \boldsymbol{\rho}_\beta$ and $\nabla \boldsymbol{g}_\beta$, it remains to estimate the value of $\|x\, e^{-\beta s H}\|_{\mathcal{L}^p}$ and $\|\boldsymbol{p}\, e^{-\beta s H}\|_{\mathcal{L}^p}$ for $s \in [\frac{1}{2}, 1]$. Let start with the case $p = \infty$.

Lemma 3 *Let $\beta > 0$, then we have the estimate*

$$\left\||x|^n\, e^{-\beta H}\right\|_{\mathcal{L}^\infty}^{2/n} = \left\||\boldsymbol{p}|^n\, e^{-\beta H}\right\|_{\mathcal{L}^\infty}^{2/n} \leq n \max\left(\tfrac{2}{\beta}, \sqrt{2}\,\hbar\right).$$

Remark 5 In the classical case, it is not difficult to prove that the maximum of the function $x \mapsto |x|^n\, e^{-\beta|x|^2}$ is $\left(\frac{n}{2e\beta}\right)^{n/2}$, and more generally,

$$\left\||x|^n\, e^{-\beta|x|^2}\right\|_{L^p}^p = \omega_d\, \Gamma(\tfrac{d+np}{2})\, (\beta p)^{-\frac{d+np}{2}}$$

where ω_d is the volume of the d-dimensional unit ball and Γ is the gamma function.

Proof of Lemma 3 It is sufficient to prove a bound on the first quantity because H is symmetric in x and $\boldsymbol{p}$. Let $\varphi \in L^2$ and $\psi = e^{-tH}\varphi$. Notice that for any $t \geq 0$, $\|\psi\|_{L^2} \leq \|\varphi\|_{L^2}$. Let $y := \left\||x|^n\, \psi\right\|_{L^2}^{2/n}$. Since $2\,\partial_t \psi = -\left(|x|^2 - \hbar^2\Delta\right)\psi$, then integrating by parts yields

$$\begin{aligned} \frac{\mathrm{d}}{\mathrm{d}t} y^n &= -\int_{\mathbb{R}^d} |\psi|^2\, |x|^{2(n+1)}\, \mathrm{d}x - \hbar^2 \mathrm{Re}\left(\int_{\mathbb{R}^d} \nabla\left(\overline{\psi}\, |x|^{2n}\right) \cdot \nabla\psi\, \mathrm{d}x\right) \\ &\leq -\int_{\mathbb{R}^d} |\psi|^2\, |x|^{2(n+1)} + \hbar^2\, |\nabla\psi|^2\, |x|^{2n}\, \mathrm{d}x + 2\,n\,\hbar^2 \int_{\mathbb{R}^d} |\psi|\, |x|^{2n-1}\, |\nabla\psi|\, \mathrm{d}x. \end{aligned}$$

Applying Young's inequality for the product, we get

$$\begin{aligned} \frac{\mathrm{d}}{\mathrm{d}t} y^n &\leq -\int_{\mathbb{R}^d} |\psi|^2\, |x|^{2(n+1)}\, \mathrm{d}x + (n\hbar)^2 \int_{\mathbb{R}^d} |\psi|^2\, |x|^{2(n-1)}\, \mathrm{d}x \\ &\leq -c^{-1}\, y^{n+1} + (n\hbar)^2\, c\, y^{n-1} \end{aligned}$$

where $c = \|\varphi\|_{L^2}^{2/n}$. This yields the differential inequality

$$y' \leq -\frac{1}{n\,c}\left(y^2 - (n\hbar\,c)^2\right). \tag{27}$$

This ordinary differential equation has a fixed point at $y = n\hbar\,c$. If initially, $y \leq \hbar\,c$, then $y' \geq 0$ but y remains smaller than $n\hbar\,c$. If not, then at any time $y' < 0$ and $\hbar\,c < y(t) < y(0)$. If initially $y > \sqrt{2}\,n\,\hbar\,c$, then as long as it remains true, it holds

$$y' \leq -\frac{1}{2\,n\,c}\,y^2$$

which implies

$$y(t) \leq \max\left(\frac{2}{t}, \sqrt{2}\,\hbar\right) n\,c$$

and proves the result by taking $t = \beta$. □

To complete the proof of Proposition 4, we use the above lemma to get that for any $s \in [\frac{1}{2}, 1]$,

$$\left\|x\,e^{-\beta s H}\right\|_{\mathcal{L}^\infty} \leq 2\max\left(\tfrac{1}{\sqrt{\beta}}, \sqrt{\hbar}\right).$$

Finally, we conclude using Inequality (26).

2.4 $\mathcal{L}^p$ Bounds for $2 \leq p < \infty$

Proposition 5 *Let $\beta > 0$ and $\hbar \in (0, 1)$. Suppose $p \in [2, \infty]$ then there exists $C_{d,p} > 0$ such that*

$$\left\|\nabla g_\beta\right\|_{\mathcal{L}^p} C_{d,p}\,\frac{\beta^{\frac{1}{2}-\frac{d}{p}}}{Z_\beta}\,\frac{\max\left(2\sqrt{2}, \beta\hbar\right)^{\frac{1}{2}-\frac{1}{p}}}{\theta(\beta\hbar)^{\frac{1}{p}}},$$

$$\left\|\nabla \rho_\beta\right\|_{\mathcal{L}^p} \leq C_{d,p}\,\frac{\beta^{\frac{1}{2}-\frac{d}{p}}}{Z_\mu}\,\frac{\max\left(2\sqrt{2}, \beta\hbar\right)^{\frac{1}{2}-\frac{1}{p}}}{\theta(\beta\hbar)^{\frac{1}{p}}}.$$

where where $\theta(x) = \mathrm{th}(x)/x$ with $\mathrm{th}(x) = \frac{e^x - e^{-x}}{e^x + e^{-x}}$.

Lemma 4 *Let $\beta > 0$ and $n > -d$. Then*

$$h^d \mathrm{Tr}\big(|x|^n\, g_\beta\big) = h^d \mathrm{Tr}\big(|\boldsymbol{p}|^n\, g_\beta\big) = \frac{C_{d,n}}{(\beta\,\theta(\beta\hbar/2)/2)^{n/2}}$$

where $C_{d,n} = \Gamma(\frac{d+n}{2})/\,\Gamma(\frac{d}{2})$.

Proof Let $H_\circ = \frac{x^2+|\boldsymbol{p}|^2}{2}$ be the one-dimensional Hamiltonian of the harmonic oscillator, and ψ_n be its eigenvalues verifying

$$H_\circ\psi_n = \left(n+\tfrac{1}{2}\right)\hbar\,\psi_n.$$

The Wigner transform of the corresponding density operator $|\psi_n\rangle\langle\psi_n|$, is classically given by (see e.g. [12, Sect. 5.04] or [7, Theorem 1.105])

$$f_n(z) = 2\,(-1)^n\, e^{-|z|^2/\hbar}\, L_n\!\left(\frac{2\,|z|^2}{\hbar}\right)$$

where $z = (x,\xi)$ and L_n is the Laguerre polynomial of order n defined by

$$L_n(z) = \frac{e^x}{n!}\,\partial_x^n\big(x^n e^{-x}\big) = \sum_{k=0}^{n}\binom{n}{k}\frac{(-x)^k}{k!}$$

By the formula of the generating function of the Laguerre polynomials, we deduce that

$$\sum_{n=0}^{\infty} t^n f_n(z) = \frac{2}{1+t}\, e^{-\frac{|z|^2}{\hbar}\frac{1-t}{1+t}}$$

Taking $t = e^{-\beta\hbar}$, we obtain the Wigner transform of the thermal state. Since in dimension d it is factorized, it yields to the following formula for the Wigner transform of $Z_\beta^{-1}\, e^{-\beta H}$

$$f_\beta(z) = \left(\frac{\beta\,\theta(\beta\hbar/2)}{2\pi}\right)^d e^{-\beta|z|^2\theta(\beta\hbar/2)}$$

Its spatial moments are given by

$$h^d \mathrm{Tr}\big(|x|^n\, g_\beta\big) = \iint_{\mathbb{R}^{2d}} f_\beta(z)\,|x|^n\,\mathrm{d}z = \left(\frac{\beta\,\theta(\beta\hbar/2)}{2\pi}\right)^{d/2}\int_{\mathbb{R}^d} |x|^n\, e^{-\beta|x|^2\theta(\beta\hbar/2)}\,\mathrm{d}x$$

which yields the result. □

Proof of Proposition 5 Since $h^d \mathrm{Tr}\big(\big||x|^n\, G\big|^2\big) = Z_{2\beta}\, h^d \mathrm{Tr}\big(|x|^{2n}\, g_{2\beta}\big)$, then, by Lemma 4, we have the identity

$$\big\| |x|^n\, G \big\|_{\mathcal{L}^2}^2 = \big\| |p|^n\, G \big\|_{\mathcal{L}^2}^2 = \frac{C_{d,2n}\, Z_{2\beta}}{(\beta\, \theta(\beta\hbar))^n}$$

whenever $n > -d/2$. Now applying linear interpolation of Schatten norms and Lemma 3, we obtain the intermediate Schatten norm bounds

$$\big\| |x|^n\, G \big\|_{\mathcal{L}^p} = \big\| |p|^n\, G \big\|_{\mathcal{L}^p} \leq C_{d,n,p}\, Z_{2\beta}^{\frac{1}{p}}\, \frac{\max\left(\frac{2}{\beta}, \sqrt{2}\,\hbar\right)^{n\left(\frac{1}{2}-\frac{1}{p}\right)}}{(\beta\, \theta(\beta\hbar))^{\frac{n}{p}}}$$

where $C_{d,n,p} = C_{d,2n}^{\frac{1}{p}}\, n^{n\left(\frac{1}{2}-\frac{1}{p}\right)}$. Finally, by Identity (16a), we arrive at

$$\begin{aligned}
\|\nabla g_\beta\|_{\mathcal{L}^p} &\leq \frac{2\beta}{Z_\beta} \int_{1/2}^{1} \|z\, G^s\|_{\mathcal{L}^p}\, ds \\
&\leq C_{d,1,p}\, \frac{2\beta}{Z_\beta} \sup_{s\in[\frac{1}{2},1]} Z_{2s\beta}^{\frac{1}{p}}\, \frac{\max\left(\frac{2}{s\beta}, \sqrt{2}\,\hbar\right)^{\frac{1}{2}-\frac{1}{p}}}{(s\beta\, \theta(s\beta\hbar))^{\frac{1}{p}}} \\
&\leq C_{d,p}\, \frac{\beta^{\frac{1}{2}-\frac{d}{p}}}{Z_\beta}\, \frac{\max\left(2\sqrt{2}, \beta\hbar\right)^{\frac{1}{2}-\frac{1}{p}}}{(\theta(\beta\hbar))^{\frac{1}{p}}}
\end{aligned}$$

where we used the fact that $Z_\beta \leq (2\pi/\beta)^d$ and $C_{d,p} = 2^{\frac{5}{4}+\frac{2d+1}{p}}\, C_{d,1,p}\, \pi^{\frac{d}{p}}$. Similarly, Inequality (26) implies the same bound for ρ_β with Z_β replaced by Z_μ.

□

Acknowledgments J.C. was supported by the NSF through the RTG grant DMS- RTG 184031. C.S. acknowledges the support of the Swiss National Science Foundation through the Eccellenza project PCEFP2_181153 and of the NCCR SwissMAP.

References

1. Benedikter, N.: Effective dynamics of interacting fermions from semiclassical theory to the random phase approximation. J. Math. Phys. **63**(8), 081101 (2022)
2. Benedikter, N., Porta, M., Schlein, B.: Mean-field evolution of fermionic systems. Commun. Math. Phys. **331**(3), 1087–1131 (2014)
3. Benedikter, N., Jakšić, V., Porta, M., Saffirio, C., Schlein, B.: Mean-field evolution of fermionic mixed states. Commun. Pure Appl. Math. **69**(12), 2250–2303 (2016)

4. Bratteli, O., Robinson, D.W.: Operator Algebras and Quantum Statistical Mechanics II. Springer-Verlag, Berlin/Heidelberg (1981)
5. Chapman, S., Cowling, T.G.: The Mathematical Theory of Non-uniform Gases: An Account of the Kinetic Theory of Viscosity, Thermal Conduction and Diffusion in Gases, 3rd edn. Cambridge Mathematical Library Series. Cambridge University Press, Cambridge.New York (1990)
6. Chong, J.J., Lafleche, L., Saffirio, C.: From Many-Body Quantum Dynamics to the Hartree–Fock and Vlasov Equations with Singular Potentials, pp. 1–74 (2021). arXiv:2103.10946
7. Folland, G.B.: Harmonic Analysis in Phase Space. The Annals of Mathematics Studies, vol. 122 Princeton University Press, Princeton (1989)
8. Fournais, S., Mikkelsen, S.: An optimal semiclassical bound on commutators of spectral projections with position and momentum operators. Lett. Math. Phys. **110**(12), 3343–3373 (2020)
9. Golse, F., Paul, T.: Empirical measures and quantum mechanics: application to the mean-field limit. Commun. Math. Phys. **369**(3), 1021–1053 (2019)
10. Golse, F., Paul, T.: Semiclassical evolution with low regularity. J. Math. Pures Appl. **151**, 257–311 (2021)
11. Golse, F., Jin, S., Paul, T.: On the convergence of time splitting methods for quantum dynamics in the semiclassical regime. Found. Comput. Math. **21**(3), 613–647 (2021)
12. Groenewold, H.J.: On the principles of elementary quantum mechanics. Physica **12**(7), 405–460 (1946)
13. Lafleche, L, Saffirio, C.: Strong semiclassical limit from Hartree and Hartree–Fock to Vlasov–Poisson equation. Anal. PDE **16**(4), 891–926 (2023). https://doi.org/10.2140/apde.2023.16.891
14. Porta, M., Rademacher, S., Saffirio, C., Schlein, B.: Mean field evolution of fermions with coulomb interaction. J. Stat. Phys. **166**(6), 1345–1364 (2017)
15. Solovej, J.P.: Many Body Quantum Mechanics. Course Notes. University of Vienna, Wien (2014). https://web.math.ku.dk/~solovej/MANYBODY/mbnotes-ptn-5-3-14.pdf

Quasi-Classical Spin Boson Models

Michele Correggi, Marco Falconi, and Marco Merkli

1 Introduction and Main Result

The spin-boson model describes the interaction between a bosonic scalar field, playing the role of environment or reservoir, and a 'small' quantum system, whose spin degrees of freedom are the only relevant ones. It has been widely studied in the mathematical physics and physics literature, from various standpoints. The spin-boson model is one of *the* paradigmatic examples of an open quantum system. It is used to investigate general open system phenomena such as decoherence, entanglement, thermalization, to test the validity of markovian approximations and to analyze non-markovian behavior. We cannot attempt to give an exhaustive list of references of the model. We point the mathematically interested reader to the following inconclusive list of works, [1, 6–10, 22], as well as to references therein contained.

On the more physical side, the spin-boson model is used to describe atom-radiation interaction in quantum optics, qubit-noise coupling in quantum information and computation, environment induced transport phenomena and chemical processes in quantum chemistry. Some of these aspects can be found in the references [23–26, 29–37, 40].

For the purpose of this paper, in which we focus on mathematical aspects, we assume that the reader is familiar with the basic mathematical tools of free quantum fields, namely Fock spaces, second quantization, creation/annihilation operators,

M. Correggi (✉) · M. Falconi
Dipartimento di Matematica, Politecnico di Milano, Milano, Italy
e-mail: marco.falconi@polimi.it

M. Merkli
Department of Mathematics and Statistics, Memorial University of Newfoundland, St. John's, NL, Canada
e-mail: merkli@mun.ca

M. Correggi, M. Falconi (eds.), *Quantum Mathematics I*,
Springer INdAM Series 57, https://doi.org/10.1007/978-981-99-5894-8_3

etc.; if not, they may refer, e.g., to [16, 18]. Let us denote by $\mathscr{H}$ the Hilbert space of the spin system, and by $\mathfrak{h}$ the Hilbert space of a single bosonic excitation. We denote by G_ε the second quantization functor,[1] where $0 < \varepsilon \ll 1$ is a scale parameter. The spin-boson Hamiltonian has the general form

$$H_\varepsilon = \mathfrak{S} \otimes 1 + \nu(\varepsilon)\, 1 \otimes \mathrm{d}G_\varepsilon(\omega) + \mathfrak{s} \otimes \varphi_\varepsilon(g) \ ,$$

as an operator on $\mathscr{H} \otimes G^{\mathrm{s}}_\varepsilon(\mathfrak{h})$. Here, $\mathfrak{S}, \mathfrak{s} \in \mathscr{B}(\mathscr{H})$ are self-adjoint, $\nu(\varepsilon)$ is either $\nu(\varepsilon) = 1$ or $\nu(\varepsilon) = \frac{1}{\varepsilon}$, ω is a positive—with possibly unbounded inverse—operator on $\mathfrak{h}$, and $g \in \mathfrak{h}$. The bipartition of the total Hilbert space $\mathscr{H} \otimes G^{\mathrm{s}}_\varepsilon(\mathfrak{h})$ reflects the separation of the total physical system into two subsystems. Commonly, especially in the physics literature, the Hilbert space $\mathscr{H}$ is finite-dimensional. For instance, $\mathscr{H}$ has dimension 2^N in the case of N spins $1/2$, or N qubits. One of the most studied cases is $N = 2$, hence the name "spin-boson" model. A further characteristic of the spin-boson model is that the interaction operator is of the simple product form $\mathfrak{s} \otimes \varphi_\varepsilon(g)$, or a finite sum of such terms. This simplifies the (rigorous) analysis of the model. Nevertheless, other models in which the interaction term is more complicated, are also of interest. For instance, in the Nelson, the Pauli-Fierz or the polaron model, the interaction operator is of the form $\int_{\mathbb{R}^3} \mathfrak{s}(k) \otimes \varphi_\varepsilon(k)\mathrm{d}^3k$. While these models are also treatable with the methods explained here (see [15]), we focus in the present manuscript, for ease of presentation, on the simple form of the interaction as in H_ε above.

We shall consider a more general setup though. The guiding principle is that we want to describe two *qualitatively unequal* interacting parts. The 'spin' part which is 'small' and the boson part (or field, reservoir, environment) which is 'large'. A quantification of what small versus large means can be implemented in different ways, depending on the physical reality being modeled. For instance, finite dimensional ($\mathscr{H}$) versus infinite dimensional ($G^{\mathrm{s}}_\varepsilon(\mathfrak{h})$) Hilbert spaces, or Hamiltonians with discrete spectrum ($\mathfrak{S}$) versus Hamiltonians with continuous

[1] We use a somewhat unorthodox notation for the second quantization functor. We denote by $G^{\mathrm{s}}_\varepsilon(\mathfrak{h}) = \bigoplus_{n\in\mathbb{N}} \mathfrak{h}_n$ the symmetric Fock space over $\mathfrak{h}$ in which the canonical creation and annihilation operators have ε-dependent commutation relations:

$$[a_\varepsilon(f), a^*_\varepsilon(g)] = \varepsilon\langle f, g\rangle_{\mathfrak{h}};\, .$$

The second quantization of an operator A on $\mathfrak{h}$ is written thus as

$$\mathrm{d}G_\varepsilon(A) = \sum_{i,j=0}^{\infty} A_{ij} a^*_{\varepsilon,i} a_{\varepsilon,j} \ ,$$

with $A_{ij} = \langle e_i, Ae_j\rangle_{\mathfrak{h}}$, and $a^\sharp_{\varepsilon,k} = a^\sharp_\varepsilon(e_k)$, with $\{e_k\}_{k\in\mathbb{N}}$ an O.N.B. of $\mathfrak{h}$. The quasi-classical parameter ε clearly plays the role of a semiclassical parameter for the (Segal) field $\varphi_\varepsilon(f) = a^*_\varepsilon(f) + a_\varepsilon(f)$: as $\varepsilon \to 0$, the field becomes a classical commutative observable [see 15, for a gentler and more detailed introduction to the quasi-classical scaling], and [11–14] for other recent papers concerning the quasi-classical regime.

spectrum $(\mathrm{d}\mathcal{G}_\varepsilon(\omega))$. In the quasi-classical setup we are discussing here, the field is large in the sense that it is in a state which contains many more particles (or excitations) than the spin system does. This is formalized by saying that the (average) number of particles of the spin system is fixed, while that number in the field state is $\propto 1/\varepsilon \gg 1$—constituting the quasi-classical limit.

We will soon explain how the choice of $\nu(\varepsilon)$ affects the quasi-classical limit $\varepsilon \to 0$. There are further possible generalizations of the model, namely by taking $\mathfrak{s}$ and g to be vector-valued, or by taking $\mathfrak{S}$ to be only bounded from below, or by taking $g \notin \mathfrak{h}$. Self-adjointness for the latter case has been recently studied in [27]. For our purposes, these generalizations do not present serious obstacles as long as H_ε can be defined as a self-adjoint operator, however for the sake of clarity we keep the setting as described above.

Proposition 1 (Self-Adjointness of H_ε**)** *For all* $g \in \mathfrak{h}$*,* H_ε *is essentially self-adjoint on*[2] $D\big(\mathrm{d}\mathcal{G}_\varepsilon(\omega)\big) \cap C_0^\infty\big(\mathrm{d}\mathcal{G}_\varepsilon(1)\big)$*. In addition, if both* $g \in \mathfrak{h}$ *and* $\omega^{-1/2}g \in \mathfrak{h}$*, then* H_ε *is self-adjoint on* $D\big(\mathrm{d}\mathcal{G}_\varepsilon(\omega)\big)$ *and bounded from below.*

Proof The essential self-adjointness is proved in [20], for a general class of operators describing the interaction between matter and radiation; self-adjointness and boundedness from below with the additional assumption $\omega^{-1/2}g \in \mathfrak{h}$ is an easy consequence of the Kato-Rellich theorem on relatively bounded perturbations of self-adjoint operators. □

Remark 1 (Form Factors) There are form factors $g \in \mathfrak{h}$ with $\omega^{-1/2}g \notin \mathfrak{h}$ such that H_ε is unbounded from below (even though it is still self-adjoint). This is analogous to what happens for the van Hove model, and it is caused by some infrared singularity: in physical models, $\omega^{-1/2}$ is unbounded (and thus it could happen that $\omega^{-1/2}g \notin \mathfrak{h}$) only if the field is massless [see 17, for further details]. For our purposes, uniqueness of the quantum dynamics (i.e. essential self-adjointness of H_ε) is enough.

1.1 Main Result

Our goal is to characterize explicitly the dynamics of quantum states, in the limit $\varepsilon \to 0$. In order to do that, let us define quantum states as density matrices

$$\Gamma_\varepsilon \in \mathfrak{L}^1_{+,1}\big(\mathscr{H} \otimes \mathcal{G}^{\mathrm{s}}_\varepsilon(\mathfrak{h})\big) ,$$

[2] We denote by $C_0^\infty\big(\mathrm{d}\mathcal{G}_\varepsilon(1)\big)$ the Fock space vectors with a finite number of particles (*i.e.*, for which the k-particle components are all zero for $k > \underline{k}$, for some $\underline{k} \in \mathbb{N}$).

where $\mathfrak{L}^1$ is the trace ideal, and $\mathfrak{L}^1_{+,1}$ stands for elements in the positive cone, with trace one. A time-evolved state is then given by

$$\Gamma_\varepsilon(t) = e^{-itH_\varepsilon}\Gamma_\varepsilon e^{itH_\varepsilon} .$$

To be more precise, the question we will answer in this note is the following:

> Knowing the behavior of the initial state Γ_ε as $\varepsilon \to 0$, what is the behavior of $\Gamma_\varepsilon(t)$ as $\varepsilon \to 0$, for any time $t \in \mathbb{R}$?

To answer the question, we shall first clarify what the general behavior is of a quantum state Γ_ε, as $\varepsilon \to 0$. The intuition is that as the boson degrees of freedom become classical, the state—restricted to the boson subsystem—becomes classical as well (in the statistical mechanics sense, i.e. a probability measure); on the other hand, the spin subsystem retains its quantum nature, and thus its description shall still be given by a density matrix.

This picture is satisfactorily described mathematically in terms of a so-called *state-valued measure*, introduced in [15, 21]. A state-valued measure is a couple $\mathfrak{m} = (\mu, \gamma)$ consisting of a (Borel Radon) measure μ on the classical configuration space $\mathfrak{h}$ for the Boson subsystem, and a μ-almost-everywhere defined function $\mathfrak{h} \ni z \mapsto \gamma(z) \in \mathfrak{L}^1_{+,1}(\mathscr{H})$ with values in the density matrices of the Spin subsystem. The function $\gamma(z)$ acts as a vector-valued Radon-Nikodým derivative (it is in fact one), and thus the measure element $\mathrm{d}\mathfrak{m}(z)$ can be written as

$$\mathrm{d}\mathfrak{m}(z) = \gamma(z)\mathrm{d}\mu(z) .$$

Integrating a scalar measurable bounded function F with respect to $\mathfrak{m}$ gives an element in $\mathfrak{L}^1(\mathscr{H})$, that we denote by

$$\int_{\mathfrak{h}} F(z)\mathrm{d}\mathfrak{m}(z) = \int_{\mathfrak{h}} F(z)\gamma(z)\mathrm{d}\mu(z) .$$

It is also possible to integrate suitable functions $\mathfrak{F}$ with values in the bounded operators on $\mathscr{H}$, however in this case the relative order between the function and the measure matters: in general,

$$\int_{\mathfrak{h}} \mathfrak{F}(z)\mathrm{d}\mathfrak{m}(z) = \int_{\mathfrak{h}} \mathfrak{F}(z)\gamma(z)\mathrm{d}\mu(z) \neq \int_{\mathfrak{h}} \gamma(z)\mathfrak{F}(z)\mathrm{d}\mu(z) = \int_{\mathfrak{h}} \mathrm{d}\mathfrak{m}(z)\mathfrak{F}(z) .$$

A detailed study of state-valued measures and their properties is given in the aforementioned references [14, 15, 21]; we will make extensive use of the results proved in those papers, so the interested reader shall refer to them.

The last concept needed to understand the main results is that of the (non-commutative) Fourier transform of a quantum state, and of the Fourier transform of a state-valued measure. These tools allow to put quantum states and state-valued measures on the same grounds, to set up the quasi-classical convergence of the former to the latter. The *Fourier transform of a quantum state* Γ_ε is the function $\hat{\Gamma}_\varepsilon : \mathfrak{h} \to \mathfrak{L}^1(\mathscr{H})$ given by

$$\hat{\Gamma}_\varepsilon(\eta) = \mathrm{tr}_{\mathcal{G}^s_\varepsilon(\mathfrak{h})}\big(\Gamma_\varepsilon(W_\varepsilon(\eta))\big) ,$$

with $W_\varepsilon(\eta)$ being the bosonic Weyl operator

$$W_\varepsilon(\eta) = e^{i\varphi_\varepsilon(\eta)} = e^{i\left(a^*_\varepsilon(\eta)+a_\varepsilon(\eta)\right)} ,$$

and $\mathrm{tr}_{\mathcal{G}^s_\varepsilon(\mathfrak{h})}$ denoting the partial trace w.r.t. the bosonic degrees of freedom. The *Fourier transform of a state-valued measure* $\mathfrak{m}$ is the function $\hat{\mathfrak{m}} : \mathfrak{h} \to \mathfrak{L}^1(\mathscr{H})$ given by

$$\hat{\mathfrak{m}}(\eta) = \int_{\mathfrak{h}} e^{2i\mathrm{Re}\langle \eta, z\rangle_{\mathfrak{h}}} \mathrm{d}\mathfrak{m}(z) = \int_{\mathfrak{h}} e^{2i\mathrm{Re}\langle \eta, z\rangle_{\mathfrak{h}}} \gamma(z)\mathrm{d}\mu(z) .$$

We say that a state Γ_ε converges quasi-classically to a state-valued measure $\mathfrak{m}$, denoted by $\Gamma_\varepsilon \underset{\varepsilon\to 0}{\longrightarrow} \mathfrak{m}$, if and only if for all $\eta \in \mathfrak{h}$,

$$\text{w*-}\lim_{\varepsilon\to 0} \hat{\Gamma}_\varepsilon(\eta) = \hat{\mathfrak{m}}(\eta) ,$$

where w*-lim stands for the limit in the weak-* topology of $\mathfrak{L}^1(\mathscr{H})$, i.e. when tested with compact operators $\mathfrak{k} \in \mathfrak{L}^\infty(\mathscr{H})$:

$$\Gamma_\varepsilon \underset{\varepsilon\to 0}{\longrightarrow} \mathfrak{m} \quad \overset{\text{def}}{\Longleftrightarrow} \quad \mathrm{tr}_{\mathscr{H}}\big(\hat{\Gamma}_\varepsilon(\eta)\mathfrak{k}\big) \underset{\varepsilon\to 0}{\longrightarrow} \mathrm{tr}_{\mathscr{H}}\big(\hat{\mathfrak{m}}(\eta)\mathfrak{k}\big), \ \ \forall \eta \in \mathfrak{h}, \mathfrak{k} \in \mathfrak{L}^\infty(\mathscr{H}).$$

Proposition 2 (Quasi-Classical Convergence [15, Prop. 2.3]) *Let* Γ_ε *be a state such that there exist* $\delta, C > 0$ *with*

$$\mathrm{tr}\Big((\mathrm{d}\mathcal{G}_\varepsilon(1) + 1)^\delta \Gamma_\varepsilon\Big) \leqslant C . \tag{1}$$

Then there exists a sequence $\varepsilon_n \underset{n\to\infty}{\longrightarrow} 0$*, and a state-valued measure* $\mathfrak{m}$ *(in general depending on the sequence) such that*

$$\Gamma_{\varepsilon_n} \underset{n\to\infty}{\longrightarrow} \mathfrak{m} .$$

Proof of Sketch The proof of this proposition adapts to the quasi-classical setting the semiclassical analysis for quantum fields developed by Ammari and Nier in [2–5], with some crucial differences. A complete proof is given in [15], however the key ideas could be summarized as follows.

The Fourier transform of a quantum state enjoys some special properties [see 38, 39], namely:

- $\mathrm{tr}_{\mathscr{H}}\big(\hat{\Gamma}_\varepsilon(0)\big) = 1$;
- $\hat{\Gamma}_\varepsilon$ *is weak-* continuous when restricted to any finite-dimensional subspace of* $\mathfrak{h}$;
- $\hat{\Gamma}_\varepsilon$ is *"quantum-" completely positive definite*: for any finite collection $\{\eta_j\}_{j=1}^J \subset \mathfrak{h}$, and $\{\mathrm{t}_j\}_{j=1}^J \subset \mathscr{B}(\mathscr{H})$,

$$\sum_{j,k=1}^{J} \mathrm{t}_j \hat{\Gamma}_\varepsilon(\eta_j - \eta_k)\mathrm{t}_k^* e^{i\varepsilon \mathrm{Im}\langle \eta_j, \eta_k\rangle} \geqslant 0$$

as an operator on $\mathscr{H}$.

Intuitively, by taking the ($\mathfrak{h}$-pointwise) weak-* limit (using a compactness argument), one tries to prove that there exists a sequence $\varepsilon_n \to 0$ such that $\hat{\Gamma}_0 = \lim_{n\to\infty} \hat{\Gamma}_{\varepsilon_n}$ satisfies

- $\mathrm{tr}_{\mathscr{H}}\big(\hat{\Gamma}_0(0)\big) = 1$;
- $\hat{\Gamma}_0$ is *weak-* continuous* when restricted to any finite-dimensional subspace of $\mathfrak{h}$;
- $\hat{\Gamma}_0$ is *completely positive definite*: for any finite collection $\{\eta_j\}_{j=1}^J \subset \mathfrak{h}$, and $\{\mathrm{t}_j\}_{j=1}^J \subset \mathscr{B}(\mathscr{H})$,

$$\sum_{j,k=1}^{J} \mathrm{t}_j \hat{\Gamma}_0(\eta_j - \eta_k)\mathrm{t}_k^* \geqslant 0 .$$

It turns out that, under the assumption (1) above, the second and third properties are indeed satisfied, thus by the infinite dimensional version of Bochner's theorem [21], $\hat{\Gamma}_0$ identifies uniquely a *cylindrical state-valued measure*[3] $\mathfrak{m}$. In addition, (1) also implies that $\mathfrak{m}$ is tight, and thus a Borel Radon measure. The first property, namely that the mass is preserved in the limit, *does not hold* in general in the quasi-classical case, contrarily to the semiclassical case where it is again ensured by (1). This is due to the fact that some mass may be lost "at infinity" if the spin subsystem has infinitely many degrees of freedom, see Sect. 1.2 for a detailed discussion. □

We are now in a position to state the main result of this note, in an informal but intuitive manner.

[3] A cylindrical measure is a finitely additive measure that is σ-additive on any subalgebra of cylinders generated by a finite number of vectors.

Theorem 1 (Quasi-Classical Dynamics) *Let $\Gamma_\varepsilon \in \mathfrak{L}^1_{+,1}\big(\mathscr{H} \otimes \mathcal{G}^{\mathrm{s}}_\varepsilon(\mathfrak{h})\big)$ be such that there exists $\delta, C > 0$ such that, uniformly w.r.t. $\varepsilon \in (0, 1)$,*

$$\mathrm{tr}\Big(\big(\mathrm{d}\mathcal{G}_\varepsilon(1) + 1\big)^\delta \Gamma_\varepsilon\Big) \leqslant C \; .$$

Then there is a sequence $\varepsilon_n \to 0$ such that, with $\nu := \lim_{\varepsilon\to 0} \varepsilon\nu(\varepsilon)$, the following diagram is commutative, for any $t \in \mathbb{R}$:

$$\begin{array}{ccc}
\Gamma_{\varepsilon_n} & \xmapsto{\text{quantum evolution}} & \Gamma_{\varepsilon_n}(t) \\
\Big\downarrow {\scriptstyle \varepsilon_n \underset{n\to\infty}{\longrightarrow} 0} & & \Big\downarrow {\scriptstyle \varepsilon_n \underset{n\to\infty}{\longrightarrow} 0} \\
\gamma(z)\,\mathrm{d}\mu(z) & \xmapsto[\text{quasi-classical evolution}]{} & \mathfrak{U}_{t,0}(z)\gamma(z)\mathfrak{U}^*_{t,0}(z)\,\mathrm{d}\left(e^{-it\nu\omega}{}_\star\,\mu\right)(z).
\end{array}$$

In the above theorem, the symbol $(\,\cdot\,)_\star(\,\cdot\,)$ stands for the pushforward of the measure on the right by means of the map on the left, and $\mathfrak{U}_{t,s}(z)$ is the two-parameter unitary group on $\mathscr{H}$ generated by the self-adjoint, generally *time-dependent* effective Hamiltonian[4]

$$\mathfrak{H}(z) = \mathfrak{S} + 2\mathrm{Re}\langle e^{-it\nu\omega} z, g\rangle_{\mathfrak{h}}\, \mathfrak{s} \; .$$

The operator $\mathfrak{H}(z)$ is a time-dependent generator if $\nu = 1$, and it is time-independent if $\nu = 0$.[5] More precisely, for $\nu = 1$ the classical bosonic field described by $e^{-it\omega}{}_\star\,\mu$ evolves freely, while for $\nu = 0$ it does not evolve at all and is described by μ at all times; in both cases it drives the spin state through $\mathfrak{U}_{t,0}(z)$, mediated over all possible configurations z in the support of μ. Let us also remark that $\mathfrak{U}_{t,0}(z)\gamma(z)\mathfrak{U}^*_{t,0}(z)$ shall be seen as a Radon-Nikodým derivative, and as such the pushforward does not act on it.

Theorem 1 therefore explains how the semiclassical bosonic subsystem becomes an environment driving the spin system, unaffected by the latter, if the quasi-classical parameter ε is small enough. This also motivates the terminology used so far, i.e., the identification of the spin component as the 'small' system, while the bosonic field is the 'large' environment or reservoir. In addition, the effective dynamics of the spin system can be characterized explicitly, being unitary and described by $\mathfrak{U}_{t,0}(z)$ for any fixed configuration z of the classical field, but not

---

[4] $\mathfrak{U}_{t,s}(z)$ is the unique solution of $i\partial_t \mathfrak{U}_{t,s}(z) = \mathfrak{H}(z)\mathfrak{U}_{t,s}(z)$ and $\mathfrak{U}_{t,t}(z) = \mathbf{1}$.

[5] We restrict our attention only to the limits $\nu = 1$ and $\nu = 0$, since they encode all different and well-defined outcomes that one could obtain for the effective dynamics. In fact, every choice of $\nu(\varepsilon)$ such that $\lim_{\varepsilon\to 0} \varepsilon\nu(\varepsilon) = \lambda > 0$ would amount in a rescaling of the field dispersion relation, while any choice such that either $\lim_{\varepsilon\to 0} \varepsilon\nu(\varepsilon) = \lambda = \infty$ or such that the limit does not exist would prevent an explicit definition of the effective dynamics in the limit $\varepsilon \to 0$.

unitary (and not even Markovian[6]) in general, due to the integration over all configurations reached by the classical bosonic state $e^{-it\nu\omega}{}_\star\,\mu$. Both a stationary and a freely evolving bosonic environment can be obtained, tuning the microscopic initial state accordingly in a way that makes $\nu(\varepsilon)$ either 1 (stationary) or $\frac{1}{\varepsilon}$ (freely evolving). Let us stress that even if $\nu(\varepsilon)$ appears in the Hamiltonian, it should be thought as a feature of the chosen initial state, fixing the scale of energy for the bosonic subsystem.

1.2 Loss of Mass in the Quasi-Classical Limit

An interesting feature of quasi-classical systems is that some mass can be lost in the limit $\varepsilon \to 0$, due to the entanglement between the two subsystems, when the spin part is infinite dimensional. By loss of mass we mean that the measure in the quasi-classical limit satisfies $\mu(\mathfrak{h}) < 1$. There are many well-known examples of loss of mass (also called loss of compactness) in semiclassical analysis, both in finite and infinite dimensions [see, e.g., 2, 28]. In those cases, however, conditions like (1) are enough to guarantee that no mass is lost.

Here, on the contrary, mass can be lost "through the spin system", provided the systems are entangled, and the spin system components could escape to infinity. In fact, if the microscopic state is unentangled (in a natural quasi-classical way), i.e. it is of the form

$$\Gamma_\varepsilon = \gamma_0 \otimes \xi_\varepsilon \,,$$
$$\text{with } \gamma_0 \in \mathfrak{L}^1_{+,1}(\mathscr{H}) \,,\; \xi_\varepsilon \in \mathfrak{L}^1_{+,1}\big(G^{\mathrm{s}}_\varepsilon(\mathfrak{h})\big) \,,$$

the quasi-classical convergence in Proposition 2 "decouples" and no mass can be lost: for this class of states (1) implies $\mathfrak{m} = (\mu, \gamma_0)$, with $\mu(\mathfrak{h}) = 1$. Similarly, if the spin subsystem is finite dimensional or its particles are confined, again no mass can be lost. More precisely, if either $\dim(\mathscr{H}) < +\infty$ or there exists an operator $\mathfrak{A} > 0$ on $\mathscr{H}$ with compact resolvent[7] such that there exists $C > 0$ with

$$\operatorname{tr}\big(\Gamma_\varepsilon(\mathfrak{A} \otimes 1)\big) \leqslant C \,, \tag{2}$$

then the measure $\mathfrak{m} = (\mu, \gamma)$ in Proposition 2 is such that $\mu(\mathfrak{h}) = 1$.

In general however, part or all of the mass can be lost in the limit $\varepsilon \to 0$ of a generic quantum state Γ_ε. Theorem 1 is also interesting if (some) mass is lost. In

---

[6] We plan to investigate the non-Markovian character of the quasi-classical effective dynamics in an upcoming paper.

[7] If $\mathfrak{S}$ has compact resolvent (and it is bounded from below), $\mathfrak{A} = \mathfrak{S} + |\inf \sigma(\mathfrak{S})| + 1$ would be a natural choice, and the associated condition (2) would mean that mass is not lost if one restricts to states with ε-uniformly-bounded Spin kinetic energy.

fact, *the mass is preserved by the quasi-classical dynamics*: this means that the same amount of mass is lost at any time, and therefore that one should check if any mass is lost only at the initial time. We think that this loss of mass phenomenon peculiar to the quasi-classical entanglement is worth pointing out, and could be explored further in concrete applications.

2 Heuristic Derivation

If the initial state is quasi-classically unentangled, i.e.

$$\Gamma_\varepsilon = \gamma_0 \otimes \xi_\varepsilon$$

(see Sect. 1.2 above), it is possible to use the factorized nature of the spin-boson interaction to formally obtain a result akin to Theorem 1 in a very intuitive way, that hopefully helps to illustrate the main ideas behind the general proof.

In order to discuss the strategy, let us set some useful notation. Define the free Hamiltonian

$$H_\varepsilon^{\mathrm{f}} := H_\varepsilon\big|_{g=0} = \mathfrak{S} \otimes 1 + \nu(\varepsilon)\, 1 \otimes \mathrm{d}\mathcal{G}_\varepsilon(\omega) =: H^{\mathrm{fs}} + H_\varepsilon^{\mathrm{fb}} ,$$

and define the interaction

$$H_\varepsilon^{\mathrm{i}} = H_\varepsilon - H_\varepsilon^{\mathrm{f}} .$$

The Dyson expansion for the evolution in the interaction picture is

$$e^{itH_\varepsilon^{\mathrm{f}}} e^{-itH_\varepsilon} = 1 + \sum_{n\in\mathbb{N}_*} (-i)^n \int_0^t \mathrm{d}s_1 \int_0^{s_1} \mathrm{d}s_2 \cdots \int_0^{s_{n-1}} \mathrm{d}s_n \; \mathfrak{s}_{s_1} \cdots \mathfrak{s}_{s_n} \otimes \varphi_{\varepsilon,s_1} \cdots \varphi_{\varepsilon,s_n} ,$$

where

$$\mathfrak{s}_s = e^{isH^{\mathrm{fs}}} \mathfrak{s} e^{-isH^{\mathrm{fs}}} ,$$

$$\varphi_{\varepsilon,s} = e^{isH_\varepsilon^{\mathrm{fb}}} \varphi_\varepsilon(g) e^{-isH_\varepsilon^{\mathrm{fb}}} = \varphi_\varepsilon(e^{i\nu(\varepsilon)s\omega} g) ;$$

and in addition

$$e^{itH_\varepsilon} e^{-itH_\varepsilon^{\mathrm{f}}} = 1 + \sum_{m\in\mathbb{N}_*} i^m \int_0^t \mathrm{d}u_1 \int_0^{u_1} \mathrm{d}u_2 \cdots \int_0^{u_{m-1}} \mathrm{d}u_m\, \mathfrak{s}_{u_m} \cdots \mathfrak{s}_{u_1} \otimes \varphi_{\varepsilon,u_m} \cdots \varphi_{\varepsilon,u_1} .$$

It follows that

$$\begin{aligned}\tilde{\gamma}_\varepsilon(t) &:= \mathrm{tr}_{\mathcal{G}_\varepsilon^{\mathrm{s}}(\mathfrak{h})}\left(e^{itH^{\mathrm{fs}}} e^{-itH_\varepsilon}(\gamma_0 \otimes \xi_\varepsilon) e^{itH_\varepsilon} e^{-itH^{\mathrm{fs}}}\right) \\ &= \sum_{m,n\in\mathbb{N}} i^{m-n} \int_0^t \mathrm{d}s_1 \cdots \int_0^{s_{n-1}} \mathrm{d}s_n \int_0^t \mathrm{d}u_1 \cdots \int_0^{u_{m-1}} \mathrm{d}u_m\, \mathfrak{s}_{s_1} \cdots \mathfrak{s}_{s_n}\, \gamma_0\, \mathfrak{s}_{u_m} \cdots \mathfrak{s}_{u_1} \\ &\qquad \mathrm{tr}_{\mathcal{G}_\varepsilon^{\mathrm{s}}(\mathfrak{h})}\left(\xi_\varepsilon \varphi_{\varepsilon,u_m} \cdots \varphi_{\varepsilon,u_1} \varphi_{\varepsilon,s_1} \cdots \varphi_{\varepsilon,s_n}\right) .\end{aligned}$$

Now, in order to take the limit $\varepsilon \to 0$, one should focus on the expectation with respect to ξ_ε:

$$\langle \varphi_{\varepsilon,u_m} \cdots \varphi_{\varepsilon,u_1} \varphi_{\varepsilon,s_1} \cdots \varphi_{\varepsilon,s_n} \rangle_{\xi_\varepsilon} := \mathrm{tr}_{\mathcal{G}_\varepsilon^{\mathrm{s}}(\mathfrak{h})}\left(\xi_\varepsilon \varphi_{\varepsilon,u_m} \cdots \varphi_{\varepsilon,u_1} \varphi_{\varepsilon,s_1} \cdots \varphi_{\varepsilon,s_n}\right) .$$

It is possible to write such an expectation as follows, where $f_1, \ldots, f_k \in \mathfrak{h}$,

$$\begin{aligned}&\langle \varphi_\varepsilon(f_1) \cdots \varphi_\varepsilon(f_k) \rangle_{\xi_\varepsilon} \\ &\quad = (-i)^k \partial_{\lambda_1} \cdots \partial_{\lambda_k} \left(\langle W_\varepsilon(\lambda_1 f_1) \cdots W_\varepsilon(\lambda_k f_k) \rangle_{\xi_\varepsilon} \right)\Big|_{\lambda_1=\cdots=\lambda_k=0} \\ &\quad =: D_k \langle W_\varepsilon(\lambda_1 f_1) \cdots W_\varepsilon(\lambda_k f_k) \rangle_{\xi_\varepsilon} \Big|_{\underline{\lambda}=0} .\end{aligned}$$

It then follows from the Weyl CCR

$$W_\varepsilon(\lambda_1 f_1) W_\varepsilon(\lambda_2 f_2) = e^{-i\varepsilon \mathrm{Im}\langle \lambda_1 f_1, \lambda_2 f_2 \rangle} W_\varepsilon(\lambda_1 f_1 + \lambda_2 f_2)$$

that

$$\lim_{\varepsilon\to 0} D_k \langle W_\varepsilon(\lambda_1 f_1) \cdots W_\varepsilon(\lambda_k f_k) \rangle_{\xi_\varepsilon} \Big|_{\underline{\lambda}=0} = \lim_{\varepsilon\to 0} D_k \langle W_\varepsilon(\lambda_1 f_1 + \cdots + \lambda_k f_k) \rangle_{\xi_\varepsilon} \Big|_{\underline{\lambda}=0} .$$

Now, in this formal reasoning we feel free to exchange $\lim_{\varepsilon\to 0}$ with D_k; thus we obtain, provided that[8] $\xi_\varepsilon \underset{\varepsilon\to 0}{\longrightarrow} \mu$,

$$D_k \lim_{\varepsilon\to 0}\langle W_\varepsilon(\lambda_1 f_1 + \cdots + \lambda_k f_k)\rangle_{\xi_\varepsilon}\bigg|_{\underline{\lambda}=0} = D_k\hat{\mu}(\lambda_1 f_1 + \cdots \lambda_k f_k)\bigg|_{\underline{\lambda}=0}$$
$$= \int_{\mathfrak{h}} \alpha_{f_1}(z)\cdots\alpha_{f_k}(z)\mathrm{d}\mu(z) ;$$

where

$$\alpha_f(z) := 2\mathrm{Re}\langle z, f\rangle_{\mathfrak{h}} .$$

Applying these results to $\lim_{\varepsilon\to 0}\tilde{\gamma}_\varepsilon(t)$ yields:

$$\lim_{\varepsilon\to 0}\tilde{\gamma}_\varepsilon(t) = \sum_{m,n\in\mathbb{N}} i^{m-n}\int_0^t \mathrm{d}s_1\cdots\int_0^{s_{n-1}}\mathrm{d}s_n\int_0^t \mathrm{d}u_1\cdots\int_0^{u_{m-1}}\mathrm{d}u_m\; \mathfrak{s}_{s_1}\cdots$$
$$\mathfrak{s}_{s_n}\,\gamma\,\mathfrak{s}_{u_m}\cdots\mathfrak{s}_{u_1}\int_{\mathfrak{h}}\alpha_{s_1}(z)\cdots\alpha_{s_n}(z)\alpha_{u_1}(z)\cdots\alpha_{u_m}(z)\mathrm{d}\mu(z) ,$$

with

$$\alpha_s(z) = 2\mathrm{Re}\langle z, e^{i\nu s\omega}g\rangle_{\mathfrak{h}} . \tag{3}$$

We have thus

$$\gamma(t) := \lim_{\varepsilon\to 0}\gamma_\varepsilon(t) := \lim_{\varepsilon\to 0}\mathrm{tr}_{\mathcal{G}^{\mathrm{s}}_\varepsilon(\mathfrak{h})}\Big(e^{-itH_\varepsilon}(\gamma_0\otimes\xi_\varepsilon)e^{itH_\varepsilon}\Big)$$
$$= \int_{\mathfrak{h}}\mathfrak{U}_{t,0}(z)\gamma(z)\mathfrak{U}^*_{t,0}(z)\mathrm{d}\big(e^{-it\nu\omega}{}_\star\,\mu\big)(z) , \tag{4}$$

where $\mathfrak{U}_{t,0}(z)$ is defined in Theorem 1, that can also formally be seen as

$$\mathfrak{U}_{t,0}(z) = e^{-itH^{\mathrm{fs}}}\sum_{n\in\mathbb{N}}(-i)^n\int_0^t \mathrm{d}s_1\cdots\int_0^{s_{n-1}}\mathrm{d}s_n\;\mathfrak{s}_{s_1}\alpha_{s_1}(z)\cdots\mathfrak{s}_{s_n}\alpha_{s_n}(z) .$$

One can see last equality in (4) as a 'resummation of the Dyson series'.

[8] The scalar convergence $\xi_\varepsilon \underset{\varepsilon\to 0}{\longrightarrow} \mu$ is perfectly analogous to the quasi-classical one, and could be seen as a particular case of it where the additional degrees of freedom are trivial. Let us remark again that for the scalar case—and thus also for the unentangled quasi-classical states considered here – (1) is sufficient to guarantee that $\mu(\mathfrak{h}) = 1$.

3 Proof of Theorem 1

As we have seen in Sect. 2, the factorized structure of the spin-boson interaction can be used to simplify the study of the quasi-classical limit, compared to, say, the Nelson, polaron, or Pauli-Fierz models, where such a factorization is not present [see 14, 15, for their quasi-classical analysis]. The proof of Theorem 1 reflects this as well, as illustrated below. Since the proof follows closely [15]—and directly utilizes some of its results—we will mostly focus on highlighting the features specific to the spin-boson model.

The proof is organized in a few steps, namely:

- write a Duhamel-type formula for the Fourier transform of evolved quantum states in the interaction representation;
- extract a subsequence ε_{n_k} of common quasi-classical convergence for regular enough evolved states at any given time;
- take the limit $\varepsilon_{n_k} \to 0$ along the aforementioned subsequence of the Duhamel formula;
- study the resulting transport equation to identify the evolved measure, and uniqueness of the limit;
- relax the regularity assumption needed at step two to the assumption in the theorem.

We will review these steps below separately.

3.1 *The Duhamel Formula*

For technical reasons, related mostly to the possible unboundedness of ω, it is convenient to pass to the so-called interaction representation. Let us define the evolution in the interaction representation as

$$\Upsilon_\varepsilon(t) := e^{itH_\varepsilon^{\mathrm{f}}}\Gamma_\varepsilon(t)e^{-itH_\varepsilon^{\mathrm{f}}} .$$

The Schrödinger differential equation of quantum evolution requires too much regularity for its solutions; it is more convenient to use its integral (or Duhamel) form. To write it, it is sufficient to suppose that for all $t \in \mathbb{R}$, $\mathrm{tr}\Big(\Upsilon_\varepsilon(t)(\mathrm{d}\mathcal{G}_\varepsilon(1) + 1)^{1/2}\Big) < +\infty$. Under this assumption the Fourier transform $\hat{\Upsilon}_\varepsilon(t)$ satisfies the following integral equation, weakly on $\mathfrak{L}^1(\mathscr{H})$, for any $t, s \in \mathbb{R}$ and $\eta \in \mathfrak{h}$:

$$\begin{aligned}[\hat{\Upsilon}_\varepsilon(t)](\eta) - [\hat{\Upsilon}_\varepsilon(s)](\eta) &= -i \int_s^t \mathrm{tr}_{\mathcal{G}^{\mathrm{s}}_\varepsilon(\mathfrak{h})}\Big(\Big[\mathfrak{s}(\tau) \otimes \varphi_\varepsilon(\tau), \Upsilon_\varepsilon(\tau)\Big] W_\varepsilon(\eta)\Big)\mathrm{d}\tau \\ &= i \int_s^t \Big(\mathrm{tr}_{\mathcal{G}^{\mathrm{s}}_\varepsilon(\mathfrak{h})}\Big(\Upsilon_\varepsilon(\tau)\varphi_\varepsilon(\tau) W_\varepsilon(\eta)\Big)\mathfrak{s}(\tau) \\ &\qquad - \mathfrak{s}(\tau)\mathrm{tr}_{\mathcal{G}^{\mathrm{s}}_\varepsilon(\mathfrak{h})}\Big(\varphi_\varepsilon(\tau)\Upsilon_\varepsilon(\tau) W_\varepsilon(\eta)\Big)\Big)\mathrm{d}\tau \,. \end{aligned} \tag{5}$$

Here, we write

$$\mathfrak{s}(\tau) = e^{i\tau H^{\mathrm{fs}}}\mathfrak{s} e^{-i\tau H^{\mathrm{fs}}} \quad \text{and} \quad \varphi_\varepsilon(\tau) = e^{i\tau H^{\mathrm{fb}}_\varepsilon}\varphi_\varepsilon(g) e^{-i\tau H^{\mathrm{fb}}_\varepsilon} .$$

Once the required regularity is taken care of, this equation follows directly from the algebraic properties of the quantum evolution (in interaction representation) $e^{itH^{\mathrm{f}}_\varepsilon}e^{-itH_\varepsilon}$, as already outlined in Sect. 2. The Duhamel formula is the starting point for our study of the dynamical quasi-classical limit.

The regularity bound concerning the average of the number operator at all times that we used above—especially in its form that is uniform w.r.t. $\varepsilon \in (0, 1)$—will be crucial also in what follows, so let us formulate it as an auxiliary "black box" result. Such propagation results are typically heavily dependent on the model under consideration; for the Spin-Boson system one could adapt very easily the results available for the Nelson model with ultraviolet cutoff [19, Proposition 4.2], obtaining the lemma below.

Lemma 1 *For any $\delta, C > 0$ and for all $t \in \mathbb{R}$, there exists $K(\delta, C, t) > 0$ such that*

$$\begin{aligned}\mathrm{tr}\Big(\Gamma_\varepsilon\big(\mathrm{d}\mathcal{G}_\varepsilon(1)+1\big)^\delta\Big) \leqslant C \implies \Big(&\mathrm{tr}\Big(\Gamma_\varepsilon(t)\big(\mathrm{d}\mathcal{G}_\varepsilon(1)+1\big)^\delta\Big) \leqslant K(\delta, C, t) \\ &\wedge\ \mathrm{tr}\Big(\Upsilon_\varepsilon(t)\big(\mathrm{d}\mathcal{G}_\varepsilon(1)+1\big)^\delta\Big) \leqslant K(\delta, C, t)\Big) .\end{aligned}$$

3.2 Common Subsequence Extraction at All Times

Thanks to the propagation lemma, Lemma 1, it is possible to prove that $t \mapsto \hat{\Upsilon}_\varepsilon(t)$ is *uniformly equicontinuous w.r.t.* $\varepsilon \in (0, 1)$. This in turn implies, by a diagonal extraction argument, that starting from any sequence $\varepsilon_n \to 0$, it is possible to extract a subsequence $\varepsilon_{n_k} \to 0$ that guarantees convergence of $\Upsilon_{\varepsilon_{n_k}}(\tau)$ to some state-valued measure $\mathfrak{n}_\tau$ for any τ in a given compact interval $[s, t]$ (actually for any given time). This is the crucial ingredient allowing to study the limit $\varepsilon \to 0$ of the Duhamel formula (5), for the latter involves the integral over all evolved states between s and t. The result reads as follows, and it has been proved in [15, Propositions 4.2 and 4.3], with a general argument that does not depend on the nature of $\mathscr{H}$ or on the Hamiltonian (one only requires that a form of Lemma 1 is available).

Proposition 3 *Let Γ_ε be such that*

$$\mathrm{tr}\Big(\Gamma_\varepsilon\big(\mathrm{d}\mathcal{G}_\varepsilon(1)+1\big)^{1/2}\Big) \leqslant C .$$

Then $\mathbb{R}\times\mathfrak{h}\ni(t,\eta)\mapsto[\hat{\Upsilon}_\varepsilon(t)](\eta)\in\mathfrak{L}^1(\mathscr{H})$ is uniformly equicontinuous w.r.t. $\varepsilon\in(0,1)$ on bounded subsets of $\mathbb{R}\times\mathfrak{h}$, having endowed $\mathfrak{L}^1(\mathscr{H})$ with the weak- topology.*

In addition, for any sequence $\varepsilon_n\to 0$, there exists a subsequence $\varepsilon_{n_k}\to 0$ and a family $\{\mathfrak{n}_t\}_{t\in\mathbb{R}}$ of state-valued measures such that for all $t\in\mathbb{R}$,

$$\Upsilon_{\varepsilon_{n_k}}(t)\underset{k\to\infty}{\longrightarrow}\mathfrak{n}_t .$$

As a byproduct (again this is a general result concerning unitary evolutions generated by operators of the type $\nu(\varepsilon)\mathrm{d}\mathcal{G}_\varepsilon(\cdot)$), we also get the following information on the limit of the "true" evolution $\Gamma_\varepsilon(t)$. Remember that we defined $\nu = \lim_{\varepsilon\to 0}\varepsilon\nu(\varepsilon)$.

Corollary 1 *Under the same assumptions as in Proposition 3, and given the subsequence $\varepsilon_{n_k}\to 0$ and measures $\{\mathfrak{n}_t\}_{t\in\mathbb{R}}$, we have that for any $t\in\mathbb{R}$,*

$$\Gamma_{\varepsilon_{n_k}}(t)\underset{k\to\infty}{\longrightarrow}\mathfrak{m}_t = e^{-itH^{\mathrm{fs}}}\big(e^{-it\nu\omega}{}_\star\,\mathfrak{n}_t\big)e^{itH^{\mathrm{fs}}} .$$

In other words, we are able to relate the quasi-classical evolution in the interaction picture to the one not in interaction picture "as it should be", i.e. by acting with the expected free evolution on both the Spin and classical Boson subsystems. It follows that once we have characterized the map $t\to\mathfrak{n}_t$, we have also a characterization for the map $t\to\mathfrak{m}_t$.

3.3 The Limit of the Duhamel Formula

We are now in a position to take the limit $\varepsilon\to 0$ of the Duhamel formula (5). In taking this step, the factorized nature of the spin-boson interaction helps greatly, essentially allowing to transform the problem from quasi-classical to semiclassical, allowing us to avoid completely the use the heavy machinery of quasi-classical calculus developed in [15, §2] (that is however *necessary* whenever the interaction is not factorized as for the spin boson). Let $\mathfrak{k} \in \mathfrak{L}^{\infty}(\mathscr{H})$ be a compact operator on the Spin subsystem, then Duhamel's formula (5) becomes

$$\begin{aligned}
&\mathrm{tr}_{\mathscr{H}}\Big([\hat{\Upsilon}_{\varepsilon}(t)](\eta)\mathfrak{k}\Big) - \mathrm{tr}_{\mathscr{H}}\Big([\hat{\Upsilon}_{\varepsilon}(s)](\eta)\mathfrak{k}\Big) \\
&\quad = i\int_s^t \Big(\mathrm{tr}\Big(\Upsilon_{\varepsilon}(\tau)\varphi_{\varepsilon}(\tau)W_{\varepsilon}(\eta)\mathfrak{s}(\tau)\mathfrak{k}\Big) - \mathrm{tr}\Big(\mathfrak{k}\mathfrak{s}(\tau)\varphi_{\varepsilon}(\tau)\Upsilon_{\varepsilon}(\tau)W_{\varepsilon}(\eta)\Big)\Big)\mathrm{d}\tau\,.
\end{aligned}$$

It is possible to exchange the trace w.r.t. $\mathscr{H}$ and the integral by dominated convergence, using the bound for Γ_{ε} assumed in Proposition 3, and its time propagation given by Lemma 1. By Proposition 3, and the definition of quasi-classical convergence, it follows immediately that, along the common subsequence $\varepsilon_{n_k} \to 0$,

$$\begin{aligned}
\lim_{k\to\infty} \mathrm{tr}_{\mathscr{H}}\Big([\hat{\Upsilon}_{\varepsilon_{n_k}}(t)](\eta)\mathfrak{k}\Big) &= \mathrm{tr}_{\mathscr{H}}\Big(\hat{\mathfrak{n}}_t(\eta)\mathfrak{k}\Big)\,, \\
\lim_{k\to\infty} \mathrm{tr}_{\mathscr{H}}\Big([\hat{\Upsilon}_{\varepsilon_{n_k}}(s)](\eta)\mathfrak{k}\Big) &= \mathrm{tr}_{\mathscr{H}}\Big(\hat{\mathfrak{n}}_s(\eta)\mathfrak{k}\Big)\,.
\end{aligned}$$

Let us now focus on the interaction term, and in particular on the expression

$$\mathrm{tr}\Big(\Upsilon_{\varepsilon}(\tau)\varphi_{\varepsilon}(\tau)W_{\varepsilon}(\eta)\mathfrak{s}(\tau)\mathfrak{k}\Big)\,,$$

the other one being analogous. Let us now decompose the operator $\mathfrak{s}(\tau)\mathfrak{k}$ in its real positive, negative, and imaginary positive, negative parts:

$$\mathfrak{s}(\tau)\mathfrak{k} = \mathfrak{sk}_{r+} - \mathfrak{sk}_{r-} + i(\mathfrak{sk}_{i+} - \mathfrak{sk}_{i-})\,,$$

with $\mathfrak{sk}_{r+}, \mathfrak{sk}_{r-}, \mathfrak{sk}_{i+}, \mathfrak{sk}_{i-} \geqslant 0$. Therefore, we have that

$$\begin{aligned}
\mathrm{tr}\Big(\Upsilon_{\varepsilon}(\tau)\varphi_{\varepsilon}(\tau)W_{\varepsilon}(\eta)\mathfrak{s}(\tau)\mathfrak{k}\Big) &= \mathrm{tr}\Big(\Upsilon_{\varepsilon}(\tau)\varphi_{\varepsilon}(\tau)W_{\varepsilon}(\eta)\mathfrak{sk}_{r+}\Big) \\
&\quad - \mathrm{tr}\Big(\Upsilon_{\varepsilon}(\tau)\varphi_{\varepsilon}(\tau)W_{\varepsilon}(\eta)\mathfrak{sk}_{r-}\Big) + i\Big(\mathrm{tr}\Big(\Upsilon_{\varepsilon}(\tau)\varphi_{\varepsilon}(\tau)W_{\varepsilon}(\eta)\mathfrak{sk}_{i+}\Big) \\
&\qquad - \mathrm{tr}\Big(\Upsilon_{\varepsilon}(\tau)\varphi_{\varepsilon}(\tau)W_{\varepsilon}(\eta)\mathfrak{sk}_{i-}\Big)\Big)\,.
\end{aligned}$$

Now, we would like to treat all these terms in the same fashion, so let us focus on the first one. We can split the total trace in the two partial traces, but we do it *in reverse order w.r.t. before*:

$$\mathrm{tr}\Big(\Upsilon_{\varepsilon}(\tau)\varphi_{\varepsilon}(\tau)W_{\varepsilon}(\eta)\mathfrak{sk}_{r+}\Big) = \mathrm{tr}_{\mathcal{G}_{\varepsilon}^{s}(\mathfrak{h})}\Big(\mathrm{tr}_{\mathscr{H}}\Big(\Upsilon_{\varepsilon}(\tau)\mathfrak{sk}_{r+}\Big)\varphi_{\varepsilon}(\tau)W_{\varepsilon}(\eta)\Big)\,.$$

The partial trace w.r.t. to $\mathscr{H}$ is the expectation over a state of a positive operator, so

$$\zeta_\varepsilon(\tau, \mathfrak{s}\mathfrak{k}_{r+}) := \mathrm{tr}_{\mathscr{H}}\Big(\Upsilon_\varepsilon(\tau)\mathfrak{s}\mathfrak{k}_{r+}\Big) \in \mathfrak{L}^1_+(\mathfrak{h}) ,$$

and we finally obtain

$$\mathrm{tr}\Big(\Upsilon_\varepsilon(\tau)\varphi_\varepsilon(\tau)W_\varepsilon(\eta)\mathfrak{s}\mathfrak{k}_{r+}\Big) = \mathrm{tr}_{\mathcal{G}^s_\varepsilon(\mathfrak{h})}\Big(\zeta_\varepsilon(\tau, \mathfrak{s}\mathfrak{k}_{r+})\varphi_\varepsilon(\tau)W_\varepsilon(\eta)\Big) .$$

The state $\zeta_\varepsilon(\tau, \mathfrak{s}\mathfrak{k}_{r+})$ is a semiclassical (scalar) state, living on the Fock space. On one hand, by Proposition 3 and the definition of quasi-classical convergence[9] we know that

$$\zeta_{\varepsilon_{n_k}}(\tau, \mathfrak{s}\mathfrak{k}_{r+}) \underset{k\to\infty}{\longrightarrow} \mathrm{d}\mu_{\tau,\mathfrak{s}\mathfrak{k}_{r+}}(z) = \mathrm{tr}_{\mathscr{H}}\big(\mathrm{d}\mathfrak{n}_\tau(z)\mathfrak{s}\mathfrak{k}_{r+}\big) .$$

On the other hand, by semiclassical calculus in infinite dimensions [see 2], we also know that

$$\lim_{k\to\infty} \mathrm{tr}_{\mathcal{G}^s_{\varepsilon_{n_k}}(\mathfrak{h})}\Big(\zeta_{\varepsilon_{n_k}}(\tau, \mathfrak{s}\mathfrak{k}_{r+})\varphi_{\varepsilon_{n_k}}(\tau)W_{\varepsilon_{n_k}}(\eta)\Big) = \int_{\mathfrak{h}} \alpha_\tau(z)e^{2i\mathrm{Re}\langle\eta,z\rangle_{\mathfrak{h}}}\mathrm{d}\mu_{\tau,\mathfrak{s}\mathfrak{k}_{r+}}(z) ,$$

where the shorthand $\alpha_\tau(z)$ has been defined in (3). Combining the two things, and repeating the same reasoning for all the other remaining terms, we end up obtaining the following integral equation for the map $t \to \mathfrak{n}_t$ (another dominated convergence argument allows to pass the limit $\varepsilon_{n_k} \to 0$ inside the time integral, this time exploiting the uniformity w.r.t. $\varepsilon \in (0, 1)$ of the number operator bounds at any time).

Proposition 4 *The family of state-valued measures $\{\mathfrak{n}_t\}_{t\in\mathbb{R}}$ of Proposition 3 satisfies the following transport equation for the Fourier transform, in the weak sense on $\mathfrak{L}^1(\mathscr{H})$:*

$$\hat{\mathfrak{n}}_t(\eta) - \hat{\mathfrak{n}}_s(\eta) = i\int_s^t \int_{\mathfrak{h}} [\gamma_{\mathfrak{n}_\tau}(z), \mathfrak{s}(\tau)]\alpha_\tau(z)e^{2i\mathrm{Re}\langle\eta,z\rangle_{\mathfrak{h}}}\mathrm{d}\mu_{\mathfrak{n}_\tau}(z)\mathrm{d}\tau .$$

[9] Quasi-classical convergence is the pointwise convergence of Fourier transforms in weak-* topology, i.e. when tested with compact operators. Since $\mathfrak{s}\mathfrak{k}_{r+}$ is compact, we have pointwise convergence of $\hat{\Upsilon}_\varepsilon(\tau)$ traced together with $\mathfrak{s}\mathfrak{k}_{r+}$.

3.4 Uniqueness of the Solution to the Transport Equation, Uniqueness of the Limit

The transport equation for the Fourier transform of $\mathfrak{n}_t$ can be easily translated in an equation for the measure:

$$\gamma_{\mathfrak{n}_t}(z)\mathrm{d}\mu_{\mathfrak{n}_t}(z) - \gamma_{\mathfrak{n}_s}(z)\mathrm{d}\mu_{\mathfrak{n}_s}(z) = i\int_s^t [\gamma_{\mathfrak{n}_\tau}(z), \mathfrak{s}(\tau)]\alpha_\tau(z)\mathrm{d}\mu_{\mathfrak{n}_\tau}(z)\mathrm{d}\tau \ .$$

Now, let us fix $s = 0$, and suppose that we have the quasi-classical convergence at initial time

$$\Gamma_{\varepsilon_n} \underset{n\to\infty}{\longrightarrow} \mathfrak{m} \ .$$

It then follows that

$$\mathfrak{n}_0 = \mathfrak{m} \ ,$$

and the transport equation reads

$$\gamma_{\mathfrak{n}_t}(z)\mathrm{d}\mu_{\mathfrak{n}_t}(z) - \gamma_{\mathfrak{m}}(z)\mathrm{d}\mu_{\mathfrak{m}}(z) = i\int_0^t [\gamma_{\mathfrak{n}_\tau}(z), \mathfrak{s}(\tau)]\alpha_\tau(z)\mathrm{d}\mu_{\mathfrak{n}_\tau}(z)\mathrm{d}\tau \ .$$

The family of state-valued measures $\{\mathfrak{n}_t\}_{t\in\mathbb{R}}$ given by

$$\mathrm{d}\mathfrak{n}_t(z) = \tilde{\mathfrak{U}}_{t,0}(z)\gamma_{\mathfrak{m}}(z)\tilde{\mathfrak{U}}^*_{t,0}(z) \ ,$$

with $\tilde{\mathfrak{U}}_{t,0}(z)$ the two-parameter unitary group on $\mathscr{H}$ generated by

$$\alpha_\tau(z)\mathfrak{s}(\tau)$$

is easily checked to be a solution to the transport equation. Such solution is actually *unique*, as is proved in a general fashion in [15, Proposition 5.3]. Therefore, we have proved that given

$$\Gamma_{\varepsilon_n} \underset{n\to\infty}{\longrightarrow} \mathfrak{m} \quad \text{and} \quad \operatorname{tr}\Bigl(\Gamma_\varepsilon\bigl(\mathrm{d}\mathcal{G}_\varepsilon(1) + 1\bigr)^{1/2}\Bigr) \leqslant C \ ,$$

there exists a subsequence ε_{n_k} along which for any $t \in \mathbb{R}$ we have the convergence

$$\Upsilon_{\varepsilon_{n_k}}(t) \underset{k\to\infty}{\longrightarrow} \mathfrak{n}_t \ ,$$

with

$$\mathrm{d}\mathfrak{n}_t(z) = \tilde{\mathfrak{U}}_{t,0}(z)\gamma_{\mathfrak{m}}(z)\tilde{\mathfrak{U}}^*_{t,0}(z) \ .$$

By Corollary 1, it also follows that for any $t \in \mathbb{R}$,

$$\Gamma_{\varepsilon_{n_k}}(t) \underset{k\to\infty}{\longrightarrow} \mathfrak{m}_t \ ,$$

with

$$\mathrm{d}\mathfrak{m}_t(z) = \mathfrak{U}_{t,0}(z)\gamma(z)\mathfrak{U}^*_{t,0}(z)\,\mathrm{d}\left(e^{-it\nu\omega} {}_\star \mu\right)(z) \ ,$$

as stated in Theorem 1. However, a couple of steps are still missing to complete the proof of the latter.

First of all, one shall prove convergence along the original sequence of convergence at initial time $\varepsilon_n \to 0$, rather than on some existing subsequence $\varepsilon_{n_k} \to 0$. This is readily established exploiting once more the uniqueness of the solution to the transport equation. Suppose in fact that we have another subsequence $\varepsilon_{n_j} \to 0$ of convergence for $\Upsilon_{\varepsilon_{n_j}}(t)$ at all times $t \in \mathbb{R}$, with possibly different limit measure $\{\mathfrak{n}'_t\}_{t\in\mathbb{R}}$. Then, by the same argument as in Sect. 3.3, $\mathfrak{n}'_t$ would satisfy the very same transport equation given in Proposition 4 for $\mathfrak{n}_t$. Since the solution to that transport equation is unique, this would imply $\mathfrak{n}'_t = \mathfrak{n}_t$. In other words, there is a unique possible cluster point for the sequence $\Upsilon_{\varepsilon_n}(t)$, thus it converges itself to the very same limit $\mathfrak{n}_t$. We can thus conclude that, that if

$$\Gamma_{\varepsilon_n} \underset{n\to\infty}{\longrightarrow} \mathfrak{m} \quad \text{and} \quad \mathrm{tr}\left(\Gamma_\varepsilon\left(\mathrm{d}\mathcal{G}_\varepsilon(1)+1\right)^{1/2}\right) \leqslant C \ ,$$

then for any $t \in \mathbb{R}$,

$$\Gamma_{\varepsilon_n}(t) \underset{k\to\infty}{\longrightarrow} \mathfrak{m}_t \ ,$$

with

$$\mathrm{d}\mathfrak{m}_t(z) = \mathfrak{U}_{t,0}(z)\gamma(z)\mathfrak{U}^*_{t,0}(z)\,\mathrm{d}\left(e^{-it\nu\omega} {}_\star \mu\right)(z) \ .$$

3.5 *Relaxing the Regularity Assumption on the Expectation of the Number Operator*

The final step for the proof is to relax the initial time assumption

$$\mathrm{tr}\left(\Gamma_\varepsilon\left(\mathrm{d}\mathcal{G}_\varepsilon(1)+1\right)^{1/2}\right) \leqslant C$$

used in the above, to

$$\mathrm{tr}\Big(\Gamma_\varepsilon\big(\mathrm{d}\mathcal{G}_\varepsilon(1)+1\big)^\delta\Big) \leqslant C$$

for some $\delta > 0$. This is done using standard approximation techniques and density arguments, as detailed in [4, §2]. This concludes the proof of Theorem 1.

References

1. Arai, A., Hirokawa, M.: On the existence and uniqueness of ground states of a generalized spin-boson model. J. Funct. Anal. **151**(2), 455–503 (1997). http://dx.doi.org/10.1006/jfan.1997.3140
2. Ammari, Z., Nier, F.: Mean field limit for bosons and infinite dimensional phase-space analysis. Ann. Henri Poincaré **9**(8), 1503–1574 (2008). http://dx.doi.org/10.1007/s00023-008-0393-5. arXiv:0711.4128
3. Ammari, Z., Nier, F.: Mean field limit for bosons and propagation of Wigner measures. J. Math. Phys. **50**(4), 042107 (2009). http://dx.doi.org/10.1063/1.3115046. arXiv:0807.3108
4. Ammari, Z., Nier, F.: Mean field propagation of Wigner measures and BBGKY hierarchies for general bosonic states. J. Math. Pures Appl. **95**(6), 585–626 (2011). http://dx.doi.org/10.1016/j.matpur.2010.12.004. arXiv:1003.2054
5. Ammari, Z., Nier, F.: Mean field propagation of infinite-dimensional Wigner measures with a singular two-body interaction potential. Ann. Sc. Norm. Super. Pisa Cl. Sci. **XIV**(1), 155–220 (2015). http://dx.doi.org/10.2422/2036-2145.201112_004. arXiv:1111.5918
6. Amour, L., Lascar, R., Nourrigat, J.: Weyl calculus in Wiener spaces and in QED (2016). arXiv:1610.06379
7. Amour, L., Jager, L., Nourrigat, J.: Infinite dimensional semiclassical analysis and applications to a model in NMR (2017). arXiv:1705.07097
8. Amour, L., Lascar, R., Nourrigat, J.: Weyl calculus in QED I. The unitary group. J. Math. Phys. **58**(1), 013501 (2017). http://dx.doi.org/10.1063/1.4973742. arXiv:1510.05293
9. Arai, A.: An asymptotic analysis and its application to the nonrelativistic limit of the Pauli-Fierz and a spin-boson model. J. Math. Phys. **31**(11), 2653–2663 (1990). http://dx.doi.org/10.1063/1.528966
10. Arai, A.: A theorem on essential selfadjointness with application to Hamiltonians in nonrelativistic quantum field theory. J. Math. Phys. **32**(8), 2082–2088 (1991)
11. Carlone, R., Correggi, M., Falconi, M., Olivieri, M.: Emergence of time-dependent point interactions in polaron models. SIAM J. Math. Anal. **53**(4), 4657–4691 (2021). http://dx.doi.org/10.1137/20M1381344. arXiv:1904.11012
12. Correggi, M., Falconi, M.: Effective potentials generated by field interaction in the quasi-classical limit. Ann. Henri Poincaré **19**(1), 189–235 (2018). . arXiv:1701.01317
13. Correggi, M., Falconi, M., Olivieri, M.: Magnetic Schrödinger operators as the quasi-classical limit of Pauli-Fierz-type models. J. Spectr. Theory **9**(4), 1287–1325 (2019). . arXiv:1711.07413
14. Correggi, M., Falconi, M., Olivieri, M.: Ground state properties in the quasi-classical regime. Anal. PDE (2022, in press). arXiv:2007.09442
15. Correggi, M., Falconi, M., Olivieri, M.: Quasi-classical dynamics. J. Eur. Math. Soc. **25**, 731–783 (2023)
16. Cook, J.M.: The mathematics of second quantization. Proc. Nat. Acad. Sci. U. S. A. **37**, 417–420 (1951)

17. Dereziński, J.: Van Hove Hamiltonians—exactly solvable models of the infrared and ultraviolet problem. Ann. Henri Poincaré **4**(4), 713–738 (2003). http://dx.doi.org/10.1007/s00023-003-0145-5
18. Dereziński, J., Gérard, C.: Mathematics of Quantization and Quantum Fields. Cambridge Monographs on Mathematical Physics. Cambridge University Press, Cambridge (2013). ISBN 978-1-107-01111-3, xii+674pp.
19. Falconi, M.: Classical limit of the Nelson model with cutoff. J. Math. Phys. **54**(1), 012303 (2013). http://dx.doi.org/10.1063/1.4775716. arXiv:1205.4367
20. Falconi, M.: Self-adjointness criterion for operators in Fock spaces. Math. Phys. Anal. Geom. **18**(1), Art. 2 (2015). arXiv:1405.6570
21. Falconi, M.: Cylindrical Wigner measures. Doc. Math. **23**, 1677–1756 (2018). http://dx.doi.org/10.25537/dm.2018v23.1677-1756. arXiv:1605.04778
22. Hasler, D., Herbst, I.: Ground states in the spin boson model. Ann. Henri Poincaré **12**(4), 621–677 (2011). http://dx.doi.org/10.1007/s00023-011-0091-6
23. Joye, A., Merkli, M., Spehner, D.: Adiabatic transitions in a two-level system coupled to a free boson reservoir. Ann. Henri Poincaré **21**(10), 3157–3199 (2020). http://dx.doi.org/10.1007/s00023-020-00946-w
24. Könenberg, M., Merkli, M.: On the irreversible dynamics emerging from quantum resonances. J. Math. Phys. **57**(3), 033,302 (2016). http://dx.doi.org/10.1063/1.4944614
25. Könenberg, M., Merkli, M., Song, H.: Ergodicity of the spin-boson model for arbitrary coupling strength. Commun. Math. Phys. **336**(1), 261–285 (2015). http://dx.doi.org/10.1007/s00220-014-2242-3
26. Leggett, A.J., Chakravarty, S., Dorsey, A.T., Fisher, M.P.A., Garg, A., Zwerger, W.: Dynamics of the dissipative two-state system. Rev. Mod. Phys. **59**, 1–85 (1987). http://dx.doi.org/10.1103/RevModPhys.59.1. Erratum: Rev. Mod. Phys. **67**(1), 725 (1995)
27. Lonigro, D.: Generalized spin-boson models with non-normalizable form factors. J. Math. Phys. **63**(7), 072,105 (2022). http://dx.doi.org/10.1063/5.0085576
28. Lions, P.-L., Paul, T.: Sur les mesures de Wigner. Rev. Mat. Iberoamericana **9**(3), 553–618 (1993)
29. Merkli, M.: Quantum markovian master equations: resonance theory shows validity for all time scales. Ann. Phys. **412**, 167,996 (2020). http://dx.doi.org/https://doi.org/10.1016/j.aop.2019.167996
30. Merkli, M.: Dynamics of open quantum systems i, oscillation and decay. Quantum **6**, 615 (2022)
31. Merkli, M.: Dynamics of open quantum systems ii, markovian approximation. Quantum **6**, 616 (2022)
32. Merkli, M., Sigal, I.M., Berman, G.P.: Decoherence and thermalization. Phys. Rev. Lett. **98**(13), 130401 (2007). http://dx.doi.org/10.1103/PhysRevLett.98.130401
33. Merkli, M., Berman, G.P., Borgonovi, F., Gebresellasie, K.: Evolution of entanglement of two qubits interacting through local and collective environments. Quant. Inf. Comput. **11**(5–6), 390–419 (2011)
34. Merkli, M., Berman, G.P., Sayre, R.: Electron transfer reactions: generalized spin-boson approach. J. Math. Chem. **51**(3), 890–913 (2013). http://dx.doi.org/10.1007/s10910-012-0124-5
35. Merkli, M., Berman, G.P., Sayre, R.T., Gnanakaran, S., Könenberg, M., Nesterov, A.I., Song, H.: Dynamics of a chlorophyll dimer in collective and local thermal environments. J. Math. Chem. **54**(4), 866–917 (2016). http://dx.doi.org/10.1007/s10910-016-0593-z
36. Mohseni, M., Omar, Y., Engel, G.S., Plenio, M.B. (eds.): Quantum Effects in Biology. Cambridge University Press, Cambridge (2014)
37. Palma, G.M., Suominen, K.A., Ekert, A.K.: Quantum computers and dissipation. Proc. Roy. Soc. Lond. Ser. A **452**(1946), 567–584 (1996). http://dx.doi.org/10.1098/rspa.1996.0029
38. Segal, I.E.: Foundations of the theory of dynamical systems of infinitely many degrees of freedom. I. Mat.-Fys. Medd. Danske Vid. Selsk. **31**(12), 39pp. (1959)

39. Segal, I.E.: Foundations of the theory of dyamical systems of infinitely many degrees of freedom. II. Can. J. Math. **13**, 1–18 (1961)
40. Xu, D., Schulten, K.: Coupling of protein motion to electron transfer in a photosynthetic reaction center: investigating the low temperature behavior in the framework of the spin—boson model. Chem. Phys. **182**(2), 91–117 (1994). https://doi.org/10.1016/0301-0104(94)00016-6

Some Remarks on Semi-classical Analysis on Two-Step Nilmanifolds

Clotilde Fermanian Kammerer, Véronique Fischer, and Steven Flynn

1 Introduction

1.1 Subelliptic Operators and Subelliptic Estimates

Sub-elliptic operators are an important class of operators containing sub-Laplacians—also known as Hörmander's sums of squares of vector fields [25] that generate the tangent space by iterated commutation. These operators also appear naturally in stochastic analysis as the Kolmogorov equations of stochastic ordinary differential equations are described in terms of second order differential operators which are often sub-Laplacians. In complex geometry, Kohn Laplacian (acting on functions) on Cauchy-Riemann manifolds also gives an example of sub-elliptic operators. More generally, sub-elliptic operators appear in contact geometry, thereby having significant place.

One of their specific properties relies on the sub-elliptic estimates proved independently by Rothschild and Stein [28] on the one hand, and Fefferman and Phong [12], on the other one. While, in the elliptic case, if $\Delta u \in H^s(\mathbb{R}^d)$, then $u \in H^{s+2}(\mathbb{R}^d)$, the gain of regularity is smaller for a sub-elliptic operator $\mathbb{L} = X_1^2 + \cdots + X_p^2$. Indeed, one then has

$$\mathbb{L}u \in H^s(\mathbb{R}^d) \implies u \in H^{s+2/r}(\mathbb{R}^d)$$

C. Fermanian Kammerer (✉)
Univ Paris Est Creteil, Univ Gustave Eiffel, CNRS, LAMA, UMR8050, Creteil, France
e-mail: clotilde.fermanian@u-pec.fr

V. Fischer · S. Flynn
University of Bath, Department of Mathematical Sciences, Bath, UK
e-mail: v.c.m.fischer@bath.ac.uk; spf34@bath.ac.uk

M. Correggi, M. Falconi (eds.), *Quantum Mathematics I*,
Springer INdAM Series 57, https://doi.org/10.1007/978-981-99-5894-8_4

where r is the mean length to obtain spanning commutators. The Rothschild and Stein proof in [28] is based on Harmonic analysis on Lie groups, as developed in [20, 28], via a lifting procedure consisting in the construction of a nilpotent stratified Lie group for which the sub-elliptic operator is a sub-Laplacian. It is in that spirit that we work here and we are interested in sublaplacians associated with a special type of manifolds called nilmanifolds, that are naturally attached to a nilpotent Lie group.

1.2 Analysis on Nilmanifolds

In this paper, as is often the case in harmonic analysis, we restrict our attention to nilpotent Lie groups that are stratified. We will further assume that their step is two later on.

1.2.1 Stratified Lie Groups

A stratified Lie group G is a connected simply connected Lie group whose (finite dimensional, real) Lie algebra $\mathfrak{g}$ admits an $\mathbb{N}$-stratification into linear subspaces, i.e.

$$\mathfrak{g} = \mathfrak{g}_1 \oplus \mathfrak{g}_2 \oplus \dots \quad \text{with} \quad [\mathfrak{g}_1, \mathfrak{g}_j] = \mathfrak{g}_{1+j}, \quad 1 \le i \le j.$$

In this case, the group G and its Lie algebra are nilpotent. The step of nilpotency is the largest number $s \in \mathbb{N}$ such that $\mathfrak{g}_s$ is not trivial. In this paper, all the nilpotent Lie groups are assumed connected and simply connected.

Once a basis $X_1, \dots, X_n$ for $\mathfrak{g}$ has been chosen, we may identify the points $(x_1, \dots, x_n) \in \mathbb{R}^n$ with the points $x = \exp_G(x_1 X_1 + \dots + x_n X_n)$ in G via the exponential mapping $\exp_G : \mathfrak{g} \to G$. By choosing a basis adapted to the stratification, we derive the product law from the Baker-Campbell-Hausdorff formula. We can also define the (topological vector) spaces $C^\infty(G)$ and $\mathcal{S}(G)$ of smooth and Schwartz functions on G identified with $\mathbb{R}^n$. This induces a Haar measure dx on G which is invariant under left and right translations and defines Lebesgue spaces on G, together with a (non-commutative) convolution for functions $f_1, f_2 \in \mathcal{S}(G)$ or in $L^2(G)$,

$$(f_1 * f_2)(x) := \int_G f_1(y) f_2(y^{-1}x) dy, \quad x \in G.$$

The Lie algebra $\mathfrak{g}$ is naturally equipped with the family of dilations $\{\delta_r, r > 0\}$, $\delta_r : \mathfrak{g} \to \mathfrak{g}$, defined by $\delta_r X = r^\ell X$ for every $X \in \mathfrak{g}_\ell$, $\ell \in \mathbb{N}$ [20]. The associated group dilations derive from

$$\delta_r(\exp_G X) = \exp_G(\delta_r X), \quad r > 0, \ X \in \mathfrak{g}.$$

In a canonical way, this leads to a notion of homogeneity for functions (measurable functions as well as distributions) and operators. For instance, the Haar measure is Q-homogeneous where

$$Q := \sum_{\ell \in \mathbb{N}} \ell \dim \mathfrak{g}_\ell$$

is called the *homogeneous dimension* of G. Another example is obtained by identifying the elements of the Lie algebra $\mathfrak{g}$ with the left-invariant vector fields on G: we check readily that the elements of $\mathfrak{g}_j$ are homogeneous differential operators of degree j.

When a scalar product is fixed on the first stratum $\mathfrak{g}_1$ of the Lie algebra $\mathfrak{g}$, the group G is said to be Carnot. The intrinsic sub-Laplacian on G is then the differential operator given by

$$\mathbb{L}_G := V_1^2 + \cdots + V_q^2,$$

for any orthonormal basis $V_1, \ldots, V_q$ of $\mathfrak{g}_1$. We fix such a basis that will be used in different places of the paper.

1.2.2 Nilmanifolds

A nilmanifold is the one-sided quotient of a nilpotent Lie group G by a discrete subgroup Γ of G. In this paper, we will choose the left quotient of G and denote it by $M = \Gamma\backslash G$. We will consider compact nilmanifolds, or equivalently cocompact subgroups Γ. We denote by $x \mapsto \dot{x}$ the canonical projection which associates to $x \in G$ its class modulo Γ in M.

Recall that the Haar measure dx on G is unique up to a constant and, once it is fixed, $d\dot{x}$ is the only G-invariant measure on M satisfying for any function $f : G \to \mathbb{C}$, for instance continuous with compact support,

$$\int_G f(x)dx = \int_M \sum_{\gamma \in \Gamma} f(\gamma x)\, d\dot{x}. \tag{1}$$

We may allow ourselves to write dx for the measure on M when the variable of integration is $x \in M$ and no confusion with the Haar measure is possible.

The canonical projection $G \to M$ induces a one-to-one correspondence between the set of functions on M with the set of Γ-left periodic functions on G, that is, the set of functions f on G satisfying

$$\forall x \in G, \quad \forall \gamma \in \Gamma, \quad f(\gamma x) = f(x).$$

With a function f defined on M, we associate the Γ-left periodic function $f_G : x \mapsto f(\dot{x})$ defined on G. Conversely, a Γ-left periodic function f on G naturally defines a function $f_M : \dot{x} \mapsto f(x)$ on M.

Consider a linear continuous mapping $T : \mathcal{S}(G) \to \mathcal{S}'(G)$ which is invariant under Γ in the sense that

$$\forall F \in \mathcal{S}(G), \quad \forall \gamma \in \Gamma, \quad T(F(\gamma \cdot)) = (TF)(\gamma \cdot).$$

Then it naturally induces [18] an operator T_M on M via

$$T_M f = (T f_G)_M.$$

Furthermore, $T_M : \mathcal{D}(M) \to \mathcal{D}'(M)$ is a linear continuous mapping. Note that if T is invariant under G, then it is invariant under Γ. For instance, any left-invariant differential operator T on G induces a corresponding differential operator T_M on M.

Let us now assume that G is a Carnot group. The intrinsic sub-Laplacian on M is the operator $\mathbb{L}_M$ induced by $\mathbb{L}_G$ on M. It is a differential operator that is essentially self-adjoint on $L^2(M)$; we will keep the same notation for its self-adjoint extension. The spectrum of $-\mathbb{L}_M$ is a discrete and unbounded subset of $[0, +\infty)$. Each eigenspace of $\mathbb{L}_M$ has finite dimension. The constant functions on M form the 0-eigenspace of $\mathbb{L}_M$, see e.g. [18].

1.2.3 Objectives

In this paper, we consider nilpotent Lie groups G of step $s = 2$ whose Lie algebra is equipped with a scalar product. They are naturally stratified, (see Sect. 1.3.1) and so they will also be Carnot. We will focus our attention on sequences of eigenfuctions $(\psi_k)_{k\in\mathbb{N}}$ and eigenvalues $(E_k)_{k\in\mathbb{N}}$ of $-\mathbb{L}_M$, ordered in increasing order and repeated according to multiplicity:

$$-\mathbb{L}_M \psi_k = E_k \psi_k, \qquad E_1 \le E_2 \le \cdots \le E_k \le \cdots, \qquad E_k \underset{k\to\infty}{\longrightarrow} +\infty. \tag{2}$$

We are interested in the measures on M that are limit points of the densities $|\psi_k(x)|^2 dx$ as k tends to $+\infty$. Our result extends to operators

$$-\mathbb{L}_M^{\mathbb{U}} = -\mathbb{L}_M + \mathbb{U}(x)$$

where $x \mapsto \mathbb{U}(x)$ is a smooth potential on M. Our analysis will be using a semi-classical approach based on the harmonic analysis on the group G in order to derive invariance properties of these measures.

1.3 *Fourier Analysis of Step-Two Groups*

Our semi-classical approach is based on the Fourier theory of the group, as developed in Harmonic analysis (see for example [19, 20]). In the rest of this paper, we will consider only a nilpotent Lie group G of step two and its associated compact nilmanifolds $M = \Gamma\backslash G$.

1.3.1 Step-Two Groups

As G is step two, the derived algebra $\mathfrak{z} := [\mathfrak{g}, \mathfrak{g}]$ lies in the centre of $\mathfrak{g}$. Moreover, denoting by $\mathfrak{v}$ a complement of $\mathfrak{z}$, we have the decomposition:

$$\mathfrak{g} = \mathfrak{v} \oplus \mathfrak{z}.$$

Note that $\mathfrak{z} = [\mathfrak{v}, \mathfrak{v}]$ and that this decomposition yields a stratification of $\mathfrak{g}$ with $\mathfrak{g}_1 = \mathfrak{v}$, $\mathfrak{g}_2 = \mathfrak{z}$. Hence G is naturally stratified with dilations given by $\delta_\varepsilon(V + Z) = \varepsilon V + \varepsilon^2 Z$ where $\varepsilon > 0$, $V \in \mathfrak{v}$, $Z \in \mathfrak{z}$. Its topological dimension is $n = \dim \mathfrak{v} + \dim \mathfrak{z}$ while the homogeneous dimension is $Q = \dim \mathfrak{v} + 2 \dim \mathfrak{z}$. We also assume that a scalar product has been fixed on $\mathfrak{g}$, and that $\mathfrak{v}$ is an orthogonal complement of $\mathfrak{z}$.

1.3.2 The Dual Set

The dual set $\widehat{G}$ of G is the set of the equivalence classes of the irreducible unitary representations of G. We will often allow ourselves to identify a class of such representations with one of its representatives. Since G is a nilpotent Lie group, its dual is the disjoint union of the (classes of unitary irreducible) representations of dimension one and of infinite dimension:

$$\widehat{G} = \widehat{G}_1 \sqcup \widehat{G}_\infty, \quad \widehat{G}_1 := \{\text{class of } \pi, \ \dim \pi = 1\}, \quad \widehat{G}_\infty := \{\text{class of } \pi, \ \dim \pi = \infty\}.$$

As G is step two, $\widehat{G}_1$ and $\widehat{G}_\infty$ can be described in a relatively simple manner.

i. The (classes of unitary irreducible) one-dimensional representations are parametrized by the elements $\omega \in \mathfrak{v}^*$ of the dual of $\mathfrak{v}$ and consists of the characters

$$\pi^\omega(x) = \mathrm{e}^{i\omega(V)}, \quad x = \exp_G(V + Z), \quad V \in \mathfrak{v}, \quad Z \in \mathfrak{z}.$$

ii. The (classes of unitary irreducible) infinite dimensional representations are parametrised by a non-zero element $\lambda \in \mathfrak{z}^* \setminus \{0\}$ of the dual of $\mathfrak{z}$ and another parameter $\nu \in \mathfrak{v}^*$ which we now describe. For any $\lambda \in \mathfrak{z}^\star$, we consider the skew-symmetric bilinear form on $\mathfrak{v}$ defined by

$$\forall U, V \in \mathfrak{v}, \quad B(\lambda)(U, V) := \lambda([U, V]) . \tag{3}$$

We denote by $\mathfrak{r}_\lambda$ the radical of $B(\lambda)$. The other parameter ν will be in the dual $\mathfrak{r}_\lambda^*$ of this radical.

Using the scalar product on $\mathfrak{g}$, we can construct the representation $\pi^{\lambda,\nu}$ for each $\lambda \in \mathfrak{z}^* \setminus \{0\}$ and $\nu \in \mathfrak{r}_\lambda^*$ as follows. First, we will allow ourselves to keep the same notation for the skew-symmetric form $B(\lambda)$ and the corresponding skew-symmetric linear map on $\mathfrak{v}$. Hence $\mathfrak{r}_\lambda = \ker B(\lambda)$. As $B(\lambda)$ is skew symmetric, we find an orthonormal basis of $\mathfrak{v}$

$$\left(P_1^\lambda, \dots, P_d^\lambda, Q_1^\lambda, \dots, Q_d^\lambda, R_1^\lambda, \dots, R_k^\lambda\right)$$

with

$$k = k_\lambda := \dim \mathfrak{r}_\lambda, \qquad d = d_\lambda := \frac{\dim \mathfrak{v} - k}{2},$$

where the matrix of $B(\lambda)$ takes the block form

$$\begin{pmatrix} 0_{d,d} & D(\lambda) & 0_{d,k} \\ -D(\lambda) & 0_{d,d} & 0_{d,k} \\ 0_{k,d} & 0_{k,d} & 0_{k,k} \end{pmatrix} . \tag{4}$$

Here $D(\lambda)$ is a diagonal matrix with positive diagonal entries depending on λ. Note that $\mathfrak{r}_\lambda = \mathrm{Span}\left(R_1^\lambda, \dots, R_k^\lambda\right)$ and we decompose $\mathfrak{v}$ as

$$\mathfrak{v} = \mathfrak{p}_\lambda + \mathfrak{q}_\lambda + \mathfrak{r}_\lambda \quad \text{where} \quad \mathfrak{p}_\lambda := \mathrm{Span}\left(P_1^\lambda, \dots, P_d^\lambda\right), \quad \mathfrak{q}_\lambda := \mathrm{Span}\left(Q_1^\lambda, \dots, Q_d^\lambda\right).$$

The representation $\pi^{\lambda,\nu}$ acts on $L^2(\mathfrak{p}_\lambda)$ via

$$\pi^{\lambda,\nu}(x)\phi(\xi) = e^{i\lambda\left(Z + \left[D(\lambda)^{\frac{1}{2}}\xi + \frac{1}{2}P, Q\right]\right)} e^{i\nu(R)} \phi\left(D(\lambda)^{\frac{1}{2}}\xi + P\right), \tag{5}$$

for $\phi \in L^2(\mathfrak{p}_\lambda)$, $\xi \in \mathfrak{p}_\lambda$, where x is written as $x = \exp_G(P + Q + R + Z)$ with $P \in \mathfrak{p}_\lambda$, $Q \in \mathfrak{q}_\lambda$, $R \in \mathfrak{r}_\lambda$, $Z \in \mathfrak{z}$. If $\nu = 0$, we will use the shorthand $\pi^{\lambda,0} = \pi^\lambda$.

With the representations described in (i) and (ii) above, the dual set of G is: $\widehat{G} = \widehat{G}_1 \sqcup \widehat{G}_\infty$ with

$$\widehat{G}_1 = \{\text{class of } \pi^\omega,\ \omega \in \mathfrak{v}^*\} \quad \text{and} \quad \widehat{G}_\infty = \{\text{class of } \pi^{\lambda,\nu},\ \lambda \in \mathfrak{z}^* \setminus \{0\},\ \nu \in \mathfrak{r}_\lambda^*\}.$$

This can be justified in this case with the von Neumann theorem characterising the representations of the Heisenberg groups. Equivalently, we can also use the orbit method which states that there is a one-to-one correspondence between $\pi \in \widehat{G}$ and the co-adjoint orbits $\mathfrak{g}^*/G$. The advantage of the orbit method is that the Kirillov map $\mathfrak{g}^*/G \to \widehat{G}$ is a homeomorphism [7], giving us easy information on the topology of subsets of $\widehat{G}$. Furthermore, one can check that the co-adjoint action of G on $\mathfrak{g}^* = \mathfrak{v}^* \oplus \mathfrak{z}^*$ leaves the $\mathfrak{z}^*$-component invariant. Hence, we can describe the co-adjoint orbit of any $\nu + \lambda \in \mathfrak{g}^* = \mathfrak{v}^* \oplus \mathfrak{z}^*$ by choosing the unique representative as the linear form $\omega = \nu$ if $\lambda = 0$, and $\lambda + \nu$ with $\nu \in \mathfrak{r}_\lambda^*$ if $\lambda \neq 0$. Via the Kirillov map, they correspond respectively to π^ω and $\pi^{\lambda,\nu}$.

1.3.3 The Subsets Ω_k and Λ_0

As a set, $\mathfrak{z}^* \setminus \{0\}$ decomposes as the disjoint union of

$$\Omega_k := \{\lambda \in \mathfrak{z}^* \setminus \{0\} \ : \ \dim \mathfrak{r}_\lambda = k\}, \quad k \in \mathbb{N}.$$

Observe that $\Omega_k = \emptyset$ when $k > \dim \mathfrak{v}$ and also when $k = \dim \mathfrak{v}$ because if $k_\lambda = \dim \mathfrak{v}$ then $\mathfrak{r}_\lambda = \mathfrak{v}^*$, thus $B_\lambda = 0$ and $\lambda = 0$. We denote by k_0 the smallest $k \in \mathbb{N}$ such that $\Omega_k \neq \emptyset$; roughly speaking, this is the set of $\lambda \in \mathfrak{z}^*$ for which $B(\lambda)$ is of smallest kernel. We have

$$\mathfrak{z}^* \setminus \{0\} = \sqcup_{k_0 \le k < \dim \mathfrak{v}} \Omega_k.$$

We can describe $\cup_{k' \ge k} \Omega_{k'}$ as the set of $\lambda \in \mathfrak{z}^* \setminus \{0\}$ such that all the minors of $B(\lambda)$ (viewed as a matrix in the basis that we have fixed) of order $\le \dim \mathfrak{v} - k$ cancel, and Ω_k as the subset of $\cup_{k' \ge k} \Omega_{k'}$ formed by the $\lambda' s$ such that at least one minor of order $= \dim \mathfrak{v} - k$ does not vanish. Since $B(\lambda)$ is linear in λ, $\cup_{k' \ge k} \Omega_{k'}$ is an algebraic variety, and Ω_k is an open subset of it. Moreover, if $\Omega_k \neq \emptyset$ then $\cup_{k' > k} \Omega_{k'}$ is an algebraic subvariety with $\dim \cup_{k' > k} \Omega_{k'} < \dim \cup_{k' \ge k} \Omega_{k'}$. Consequently, Ω_k is an open subset of $\cup_{k' \ge k} \Omega_{k'}$ and it is either empty or dense in $\cup_{k' \ge k} \Omega_{k'}$.

We can decompose each Ω_k into further subsets, according to the multiplicity of the eigenvalues of $B(\lambda)$ viewed as a matrix in a canonical basis. Here, we will be only considering the case $k = k_0$ and denote by Λ_0 the set of $\lambda \in \Omega_{k_0}$ for which $B(\lambda)$ has the maximal number of distinct eigenvalues. Recall that, by the Cauchy residue formula, the multiplicity of a zero z_0 of a polynomial $p(z)$ is equal to $\oint_{|z-z_0|=r} \frac{p'(z)}{p(z)} dz$ for r small enough. Applying this to $\det(B(\lambda)^2 - z)$ in the case of maximal multiplicities implies that the multiplicities of the eigenvalues of $B(\lambda)^2$ for $\lambda \in \Lambda_0$ are locally constant and that the subset Λ_0 is open in Ω_{k_0}. Moreover, by the implicit function theorem, the eigenvalues of $B(\lambda)^2$ can be written locally as smooth functions (even algebraic expressions) of $\lambda \in \Lambda_0$. Similar properties hold for each subset of Ω_{k_0} with fewer constraints on the multiplicities, implying that Λ_0 is dense in Ω_{k_0}.

The Heisenberg groups correspond to the case when $\dim \mathfrak{z} = 1$ while the Heisenberg-type groups are exactly the step-two nilpotent groups G for which

$B(\lambda)^2 = -|\lambda|^2 \mathrm{I}_{\mathfrak{v}}$. Heisenberg-type groups and their nilmanifolds have an H-type foliation as in [4], and so do the groups G and their nilmanifolds when, more generally, every $B(\lambda)$, $\lambda \in \mathfrak{z}^* \setminus \{0\}$, has a trivial radical $\mathfrak{r}_\lambda = \{0\}$. Geometrically, these nilmanifolds are contact manifolds when the radicals are all trivial and $\dim \mathfrak{z} = 1$, and they are quasi-contact manifolds when the radicals may not be trivial. The analysis of the properties of weak limits of densities of eigenvalues of the sub-Laplacian for contact manifolds was studied in [10] and for quasi-contact manifold of dimension four with radical generically of dimension one in [29].

As the co-adjoint action is trivial on the $\mathfrak{z}^*$-component, the sets Ω_k may be viewed as the unions of the co-adjoint orbits of $\nu + \lambda \in \mathfrak{g}^* = \mathfrak{v}^* \oplus \mathfrak{z}^*$ with $\lambda \in \Omega_k$, or our chosen representatives for those co-orbits:

$$\Omega_k \sim \{(\lambda, \nu) \in \mathfrak{z}^* \times \mathfrak{v}^*, \ \lambda \in \Omega_k, \ \nu \in \mathfrak{r}_\lambda^*\}, \tag{6}$$

and therefore identified via Kirillov's map with the following subset of $\widehat{G}$

$$\Omega_k \sim \{\pi = \pi^{\lambda,\nu} \in \widehat{G}_\infty, \ \lambda \in \Omega_k, \ \nu \in \mathfrak{r}_\lambda^*\}.$$

We also proceed similarly for Λ_0. As subsets of $\widehat{G}_\infty$, they enjoy the same topological properties; for instance, Ω_{k_0} which is an open dense subset of $\widehat{G}_\infty$.

1.3.4 The Fourier Transform

Let $f \in L^1(G)$, the Fourier transform of f is the field of operators

$$\mathcal{F}(f) := \{\widehat{f}(\pi) : \mathcal{H}_\pi \to \mathcal{H}_\pi, \pi \in \widehat{G}\} \quad \text{given by } \widehat{f}(\pi) = \int_G f(x)\pi(x)^* dx,$$

for any (continuous unitary) representation π of G.

The unitary dual $\widehat{G}$ is a standard Borel space, and there exists a unique positive Borel measure μ on $\widehat{G}$ such that for any continuous function $f : G \to \mathbb{C}$ with compact support we have

$$\int_G |f(x)|^2 dx = \int_{\widehat{G}} \|\widehat{f}(\pi)\|^2_{HS(\mathcal{H}_\pi)} d\mu(\pi).$$

The measure μ is called the Plancherel measure and the formula above the Plancherel formula. For instance, in the case of step-two groups, the Plancherel measure is given by $d\mu(\pi^{\lambda,\nu}) = c_0 \det(D(\lambda))\, d\lambda d\nu$, for a known constant $c_0 > 0$ [11, 27]; note that it is supported on the subsets Ω_{k_0} or even Λ_0 of $\widehat{G}_\infty$ defined in Sect. 1.3.3.

The Plancherel formula extends the group Fourier transform unitarily to functions $f \in L^2(G)$: their Fourier transforms are then a Hilbert-Schmidt fields of operators satisfying the Plancherel formula. The group Fourier transform also extends readily to classes of distributions, for instance the distributions with compact support and the distributions whose associated right convolution operators are bounded on $L^2(G)$ or act continuously on the Sobolev spaces adapted to it. If T is the associated operator, we denote by $\widehat{T}$ or $\pi(T) = \widehat{T}(\pi)$ the associated field of operators with

$$\mathcal{F}(Tf)(\pi) = \pi(T) \circ \mathcal{F}f(\pi), \quad \forall f \in \mathcal{S}(G).$$

In particular, the group Fourier transform extends to left-invariant differential operators.

The considerations above are known for any nilpotent Lie group, and let us consider the case of step-two groups. The group Fourier transform of $f \in L^1(G)$ gives a scalar number at $\pi = \pi^\omega$ and a bounded operator on $\mathcal{H}_{\pi^{\lambda,\nu}} = L^2(\mathfrak{p}^\lambda)$ for $\pi = \pi^{\lambda,\nu}$. It is easy to compute that for the 1-dimensional representation, we have $\pi^\omega(-\mathbb{L}_G) = |\omega|^2$. In the remainder of the paper, we will use the notation $\pi(\mathbb{L})$ and $\widehat{\mathbb{L}} = \{\pi(\mathbb{L}), \pi \in \widehat{G}\}$ and omit the index G in this context. The case of representations of infinite dimension is more involved. The following is known in great generality [19]:

1. $\mathbb{L}_G$ and $\pi(\mathbb{L})$ for $\pi \in \widehat{G}$ are essentially self-adjoint on $L^2(G)$ and $\mathcal{H}_\pi$; we keep the same notation for their self-adjoint extensions. Hence they both admit spectral decompositions.
2. For each $\pi \in \widehat{G} \setminus \{1_{\widehat{G}}\}$, the spectrum $\mathrm{sp}(\pi(-\mathbb{L}))$ of $\pi(-\mathbb{L})$ is discrete and lies in $(0, \infty)$ and each eigenspace is finite dimensional, while for $\pi = 1_{\widehat{G}}$, $\pi(\mathbb{L}) = 0$.
3. Consider the spectral decomposition $\mathbb{P}_\zeta$, $\zeta \geq 0$, of $-\mathbb{L}_G$, i.e. $-\mathbb{L}_G = \int_0^\infty \zeta \, d\mathbb{P}_\zeta$. For each $\pi \in \widehat{G} \setminus \{1_{\widehat{G}}\}$, the group Fourier transform $\pi(\mathbb{P}_\zeta)$ of the projections $\mathbb{P}_\zeta$ are orthogonal projections of $\mathcal{H}_\pi$. Furthermore, they yield a spectral decomposition of $-\widehat{\mathbb{L}}$: $\pi(-\mathbb{L}) = \sum_{\zeta \in \mathrm{sp}(\pi(-\mathbb{L}))} \zeta \pi(\mathbb{P}_\zeta)$.

In the step-two case, some of the properties above are easy to see. Indeed, denoting by

$$\eta_j = \eta_j(\lambda), 1 \leq j \leq d, \quad \text{with the convention } 0 < \eta_1(\lambda) \leq \ldots \leq \eta_d(\lambda),$$

the positive entries of $D(\lambda) = \mathrm{diag}(\eta_1, \ldots, \eta_d)$, we readily compute

$$\pi^{\lambda,\nu}(P_j^\lambda) = \sqrt{\eta_j(\lambda)} \partial_{\xi_j} \quad \text{and} \quad \pi^{\lambda,\nu}(Q_j^\lambda) = i\sqrt{\eta_j(\lambda)} \xi_j \tag{7}$$

and deduce from the additional observation $\pi^{\lambda,\nu}(R_\ell^\lambda) = i\nu_\ell$, $1 \leq l \leq k$.

$$\pi^{\lambda,\nu}(-\mathbb{L}) = H(\lambda) + |\nu|^2,$$

where $H(\lambda)$ is the operator on $\mathcal{H}_\lambda$ given by

$$H(\lambda) = \sum_{1\leq j\leq d} \eta_j(\lambda)(-\partial_{\xi_j}^2 + \xi_j^2).$$

which is up to multiplicative factors the harmonic oscillator of $L^2(\mathbb{R}^d)$. Recall that Hermite functions give an orthonormal basis of eigenfunctions of $H(\lambda)$ with eigenvalues

$$\zeta(\alpha,\lambda) := \sum_{1\leq j\leq d} (2\alpha_j + 1)\eta_j(\lambda), \qquad \alpha \in \mathbb{N}^d, \tag{8}$$

see Sect. 4.2.2. Hence, the spectrum of $\pi^{\lambda,\nu}(-\mathbb{L})$ is

$$\mathrm{sp}(\pi^{\lambda,\nu}(-\mathbb{L})) = \left\{\zeta(\alpha,\lambda) + |\nu|^2,\ \alpha \in \mathbb{N}^d\right\},$$

giving in this special case Property (2) above. Furthermore, the spectral projections $\pi^{\lambda,\nu}(\mathbb{P}_\zeta)$ onto the eigenspaces of $H(\lambda)$ are either zero or orthogonal projections onto subspaces generated by Hermite functions.

The properties above hold for any $\lambda \in \mathfrak{z}^* \setminus \{0\}$. Restricting to Λ_0, each $\eta_j(\lambda)$ is a smooth function of $\lambda \in \Lambda_0$ since the η_j^2's are the eigenvalues of $B(\lambda)^2$ which are diagonalisable linear morphisms with eigenvalues of constant multiplicities depending smoothly on λ. Therefore, $\zeta(\alpha,\lambda)$ in (8) also depends smoothly on λ in Λ_0.

1.4 Main Result

Let $x \mapsto \mathbb{U}(x)$ be a smooth potential on M. Let $(\psi_k^{\mathbb{U}})_{k\in\mathbb{N}}$ be a sequence of eigenfunctions of $-\mathbb{L}_M^{\mathbb{U}} = -\mathbb{L}_M + \mathbb{U}$ according to

$$-\mathbb{L}_M^{\mathbb{U}}\psi_k^{\mathbb{U}} = E_k^{\mathbb{U}}\psi_k^{\mathbb{U}}, \quad k \in \mathbb{N}. \tag{9}$$

Without loss of generality, we may assume $E_k^{\mathbb{U}} \geq 0$ for all $k \in \mathbb{N}$ (if not, we modify $\mathbb{U}$ by a constant). Let ϱ be a weak limit of the density $|\psi_k^{\mathbb{U}}(x)|^2 dx$, then ϱ decompose according to the structure of $\widehat{G}$ and each of the elements of this decomposition enjoys its own invariances. These invariances are expressed in terms of the elements ω, λ and ν characterizing the points of $\widehat{G}$. We will need the following notation to state the result.

(a) For each $\lambda \in \mathfrak{z}^*$ and $\nu \in \mathfrak{r}_\lambda^*$, we associate

$$\nu \cdot R^\lambda := \nu_1 R_1^\lambda + \cdots + \nu_k R_k^\lambda \ \in \mathfrak{r}_\lambda,$$

where the ν_j's are the coordinates of ν in the dual of the orthonormal basis $(R_1^\lambda, \cdots, R_k^\lambda)$, i.e. $\nu = \nu_1(R_1^\lambda)^* + \ldots + \nu_k(R_k^\lambda)^*$. This definition is independent of the choice of the orthonormal basis $(R_1^\lambda, \cdots, R_k^\lambda)$ for $\mathfrak{r}_\lambda$.

(b) In the same spirit, for any $\omega \in \mathfrak{v}^*$, we associate

$$\omega \cdot V := \omega_1 V_1 + \cdots \omega_q V_q \ \in \mathfrak{v},$$

where the ω_j's are the coordinates of ω in the dual of an orthonormal basis $(V_1, \cdots, V_q)$: $\omega = \omega_1 V_1^* + \ldots + \omega_q V_q^*$. Here, $q = \dim \mathfrak{v}$. This definition is independent of the choice of the orthonormal basis $(V_1, \cdots, V_q)$ for $\mathfrak{v}$.

(c) If $k_0 = 0$, each eigenvalue $\zeta = \zeta(\alpha, \lambda)$ in (8) of $\pi^\lambda(\mathbb{L})$ depends smoothly on λ in Λ_0. The vector in $\mathfrak{z}$ corresponding to the gradient at λ is denoted by

$$\nabla_\lambda \zeta(\alpha, \lambda) = \nabla_\lambda \zeta \in \mathfrak{z}.$$

Theorem 1 ([15–17]) *Let $(\psi_k^{\mathbb{U}})_{k\in\mathbb{N}}$ be a sequence of eigenfunctions of $-\mathbb{L}_M^{\mathbb{U}} = -\mathbb{L}_M + \mathbb{U}$ according to (9). Then a weak limit ϱ of the density $|\psi_k^{\mathbb{U}}(x)|^2 dx$ decomposes as*

$$\varrho = \varrho^{\mathfrak{v}} + \varrho^{\mathfrak{z}} \tag{10}$$

with

1. $\varrho^{\mathfrak{v}}(x) = \int_{\mathfrak{v}^*} \varsigma(x, d\omega)$ *where the measure ς is invariant under the flow*

$$(x, \omega) \mapsto (\mathrm{Exp}(s\,\omega \cdot V)x, \omega), \quad s \in \mathbb{R}$$

2. $\varrho^{\mathfrak{z}}(x) = \sum_{k=0}^{\dim \mathfrak{v}-1} \int_{(\lambda,\nu)\in\Omega_k} \gamma_k(x, d\lambda, d\nu)$ *with the identification (6) for Ω_k, with each measure $\gamma_k(x, \lambda, \nu)$ being supported in $M \times \Omega_k$ where it is invariant under the flow given by*

$$(x, (\lambda, \nu)) \longmapsto (\mathrm{Exp}(s\,\nu \cdot R^\lambda)x, (\lambda, \nu)), \quad s \in \mathbb{R}.$$

3. *Furthermore, in the case when $\Omega_0 \neq \emptyset$, omitting $\nu = 0$,*

$$\gamma_0(x, \lambda) = \sum_{\alpha\in\mathbb{N}} \gamma_0^{(\alpha)}(x, \lambda),$$

with each measure $1_{\lambda \in \Lambda_0} \gamma_0^{(\alpha)}$ being supported on $M \times \Lambda_0$ where it is invariant under the flow given by

$$(x, \lambda) \longmapsto (\text{Exp}(s\, \nabla_\lambda \zeta(\alpha, \lambda))x, \lambda), \qquad s \in \mathbb{R}.$$

In the case of the groups of Heisenberg type, $\eta_j(\lambda) = |\lambda|$ for all j, so $\Lambda_0 = \Omega_0 = \mathfrak{z}^* \setminus \{0\}$ and

$$\nabla_\lambda \zeta(\alpha, \lambda) = \mathcal{Z}^\lambda \sum_{j=1}^{\dim \mathfrak{v}/2} (2\alpha_j + 1), \qquad \text{where} \quad \mathcal{Z}^\lambda := |\lambda|^{-1} \lambda^*, \tag{11}$$

and $\lambda^* \in \mathfrak{z}$ corresponds to λ by duality via the scalar product. We therefore recover with Theorem 1 the results of the first two authors in [15].

Theorem 1 is a consequence of Theorem 2 below. It is based on a microlocal approach and the measures γ that appear in the statement above are microlocal objects that can be compared with the semi-classical measures introduced in the 90s in the Euclidean context in [21–24]. The difference here is that the semi-classical calculus we use is based on the Harmonic analysis of the group G and on the Fourier transform introduced via representation theory as presented above. This setting has been introduced in [19] in a microlocal context where no specific semi-classical scale ε is specified. It uses a pseudo-differential calculus with operator-valued symbols that can be composed with the Fourier transform of the functions (that are also operator-valued).

The construction of a pseudodifferential calculus on groups is an old question from the 1980s [5, 6, 9, 30] that have known recent developments with an abstract point of view from the theory of algebra of operators in [31–33], and with a PDEs approach in [2, 14, 19] with applications in control theory and observability [16].

We conclude this section with some comments about Theorem 1. It is noticeable that there is coexistence of two kinds of behaviour, with a splitting of the measure γ corresponding to the different types of elements of $\widehat{G}$. In the context of the Heisenberg group, $\Lambda_0 = \Omega_0 \neq \emptyset$ and $\nabla_\lambda \zeta$ is colinear to $\mathcal{Z}^\lambda$ (see (11)) and this is linked to the wave aspect of the sub-Laplacian in this group pointed out in [1, 3, 8, 10]. On other nilpotent Lie groups where $\Omega_0 = \emptyset$, the other vector fields involved, $\nu \cdot R^\lambda$, are more of Schrödinger's type.

2 Noncommutative Semi-classical Setting

2.1 Semi-classical Pseudodifferential Operators

We consider the set $\mathcal{A}_0$ of fields of operators $\{\sigma(x,\pi) \in \mathcal{L}(\mathcal{H}_\pi),\ (x,\pi) \in M \times \widehat{G}\}$ such that

$$\sigma(x,\pi) = \mathcal{F}\kappa_x(\pi) = \int_G \kappa_x(z)\pi(z)^* dz,$$

where $x \mapsto \kappa_x(\cdot)$ is in $C^\infty(M, \mathcal{S}(G))$. We call the function κ_x the *convolution kernel* associated with the symbol σ. In the spirit of the works [3, 19], and when $\varepsilon \ll 1$ is a semi-classical parameter, the ε-quantization of the symbols $\sigma \in \mathcal{A}_0$ is given by

$$\mathrm{Op}_\varepsilon(\sigma) f(x) = \int_{\widehat{G}} \mathrm{tr}\left(\pi(x)\sigma(x, \varepsilon\cdot\pi)\widehat{f}(\pi)\right) d\mu(\pi), \quad f \in \mathcal{S}(M), \quad x \in M.$$

Here, $\varepsilon \cdot \pi$ denotes the class in $\widehat{G}$ of the irreducible representation $x \mapsto \pi(\delta_\varepsilon x)$. Setting

$$\kappa_x^\varepsilon(z) = \varepsilon^{-Q}\kappa_x(\delta_{\varepsilon^{-1}} z),$$

the ε-quantization then obeys to

$$\mathrm{Op}_\varepsilon(\sigma) f(x) = \int_G \kappa_x^\varepsilon(y^{-1}x) f(y) dy = \sum_{\gamma\in\Gamma} \int_{y\in M} \kappa_x^\varepsilon(\gamma y^{-1}x) dy, \tag{12}$$

for $f \in \mathcal{S}(M)$, $x \in M$. As in the case of groups (see [13]), the family $\left(\mathrm{Op}_\varepsilon(\sigma)\right)_{\varepsilon>0}$ is a bounded family in $\mathcal{L}(L^2(M))$:

Proposition 1 *There exists $C > 0$ such that for all $\sigma \in \mathcal{A}_0$ and $\varepsilon > 0$,*

$$\|\mathrm{Op}_\varepsilon(\sigma)\|_{\mathcal{L}(L^2(M))} \leq \int_G \sup_{x\in M} |\kappa_x(z)| dz.$$

Proof By Young's convolution inequality

$$\|f * \kappa_x^\varepsilon(\gamma\cdot)\|_{L^2(M)} \leq \|\sup_{x\in M} |\kappa_x^\varepsilon(\gamma\cdot)|\|_{L^1(M)} \|f\|_{L^2(M)},$$

with

$$\|\sup_{x\in M} |\kappa_x^\varepsilon(\gamma\cdot)|\|_{L^1(M)} = \varepsilon^{-Q} \int_M \sup_{x\in M} |\kappa_x(\varepsilon^{-1}\cdot\gamma y)| dy = \int_{\gamma^{-1}M} \sup_{x\in M} |\kappa_x(y)| dy.$$

Therefore, using (12), we deduce

$$\begin{aligned}\|\mathrm{Op}_\varepsilon(\sigma) f\|_{L^2(M)} &\leq \sum_{\gamma\in\Gamma} \|f * \kappa_x^\varepsilon(\gamma\cdot)\|_{L^2(M)} \\ &\leq \|f\|_{L^2(M)} \sum_{\gamma\in\Gamma} \int_{\gamma^{-1}M} \sup_{x\in M} |\kappa_x(y)| dy = \|f\|_{L^2(M)} \int_G \sup_{x\in M} |\kappa_x(y)| dy.\end{aligned}$$

□

Besides, this semi-classical pseudodifferential calculus enjoys symbolic calculus (see Proposition 3.6 in [13] in the case of groups and Proposition 2.2 in [16] for the extension to nilmanifolds).

2.2 Semi-classical Measures

Let us first introduce our notion of operator-valued measures introduced in the earlier papers of the first two authors. We will use the same notation as in those paper, even if it means using the Greek letter Γ for the trace-class operators $\Gamma(x, \pi)$. We think that there is no possible confusion with our current notation for the co-compact discrete subgroup Γ of G, and thus will allow this small conflict of notation

We consider pairs (Γ, γ) consisting in a positive Radon measure γ on $M \times \widehat{G}$ and a measurable field over $(x, \pi) \in M \times \widehat{G}$ of trace-class operators $\Gamma(x, \pi)$ on $\mathcal{H}_\pi$ satisfying

$$\int_{M\times\widehat{G}} \mathrm{Tr}\, |\Gamma(x, \pi)|\, d\gamma(x, \pi) < \infty.$$

We equip the set of such pairs with the equivalence relation $(\Gamma, \gamma) \sim (\Gamma', \gamma')$ given by the existence of a measurable function $f : M \times \widehat{G} \to \mathbb{C}$ such that

$$\gamma' = f\gamma \quad \text{and} \quad \Gamma' = f^{-1}\Gamma, \quad \gamma - a.e.$$

We denote by $\mathcal{M}_{ov}(M \times \widehat{G})$ the set of equivalence classes for this relation and by $\Gamma d\gamma$ the class of the pair (Γ, γ). If $\Gamma \geq 0$, then we say that the operator valued measure $\Gamma d\gamma$ is positive, and we denote by $\mathcal{M}_{ov}^+(M \times \widehat{G})$ the set of the positive operator-valued measures on $M \times \widehat{G}$. They characterize bounded families in $L^2(M)$ according to the following theorem.

Proposition 2 ([13, 15]) *Let $(\psi^\varepsilon)_{\varepsilon>0}$ be a bounded family in $L^2(M)$. There exist a subsequence $\varepsilon_k \to 0$ as $k \to \infty$, and an operator-valued measure $\Gamma d\gamma \in \mathcal{M}_{ov}^+(M \times \widehat{G})$ satisfying*

$$\forall \sigma \in \mathcal{A}_0, \qquad (\mathrm{Op}_{\varepsilon_k}(\sigma)\psi^{\varepsilon_k}, \psi^{\varepsilon_k}) \underset{k\to\infty}{\longrightarrow} \int_{M\times\widehat{G}} \mathrm{Tr}\,(\sigma(x, \pi)\Gamma(x, \pi))\, d\gamma(x, \pi).$$

Continuing with the setting of the statement above, we say then that the operator-valued measure $\Gamma d\gamma$ is a *semi-classical measure* of $(\psi^\varepsilon)_{\varepsilon>0}$ at the scale ε. A given family $(\psi^\varepsilon)_{\varepsilon>0}$ may have several semi-classical measures, depending on different subsequences $(\varepsilon_k)_{k\in\mathbb{N}}$. The knowledge of all these families indicates the obstruction to strong convergence in $L^2(M)$ of the family $(\psi^\varepsilon)_{\varepsilon>0}$.

The scale ε is particularly interesting for analyzing the oscillations of a family $(\psi^\varepsilon)_{\varepsilon>0}$ that satisfies weighted Sobolev estimates such as

$$\exists s, C > 0, \quad \forall \varepsilon > 0, \quad \|(-\varepsilon^2 \mathbb{L}_M)^{\frac{s}{2}} \psi^\varepsilon\|_{L^2(M)} \leq C. \tag{13}$$

Indeed, one can then link the weak limits of the energy densities with the semi-classical measures:

Proposition 3 ([15]) *Assume $(\psi^\varepsilon)_{\varepsilon>0}$ satisfies (13) and that $\Gamma d\gamma$ is a semi-classical measure of $(\psi^\varepsilon)_{\varepsilon>0}$ for the subsequence $(\varepsilon_k)_{k\in\mathbb{N}}$. Then for all $\phi \in C^\infty(M)$,*

$$\limsup_{k\to+\infty} \int_M \phi(x)|\psi^{\varepsilon_k}(t,x)|^2 dx = \int_{M\times\widehat{G}} \phi(x) \mathrm{Tr}\,(\Gamma(x,\pi))\, d\gamma(x,\pi). \tag{14}$$

2.3 *Application to Quantum Limits*

Let us now come back to the sequence (2) of eigenfunctions $(\psi_k^{\mathbb{U}})_{k\in\mathbb{N}}$ of the sub-Laplacian operator $-\mathbb{L}_M^{\mathbb{U}} = -\mathbb{L}_M + \mathbb{U}(x)$ for a compact nilmanifod $M = \Gamma\backslash G$ whose underlying group G is step two. Denoting by $E_k^{\mathbb{U}}$ the associated sequence of eigenvalues; we set

$$\varepsilon_k = (E_k^{\mathbb{U}})^{-1/2},$$

we obtain a semi-classical scale such that the sequence $(\psi_k^{\mathbb{U}})_{k\in\mathbb{N}}$ is ε_k-oscillating. Thus any weak limit ϱ of the energy density $|\psi_k^{\mathbb{U}}(x)|^2 dx$ is the marginal of a semi-classical measure $\Gamma d\gamma$ of the family $(\psi_k^{\mathbb{U}})_{k\in\mathbb{N}}$ according to (14). Therefore, the properties of the semi-classical measures of the sequence $(\psi_k^{\mathbb{U}})_{k\in\mathbb{N}}$ will reflect on any weak limit of the energy density.

We now omit the index $k \in \mathbb{N}$ and focus on the semi-classical measures of a family of normalized functions $(\psi^\varepsilon)_{\varepsilon>0}$ that satisfy

$$-\varepsilon^2 \mathbb{L}_M^{\mathbb{U}} \psi^\varepsilon = \psi^\varepsilon, \tag{15}$$

where $\mathbb{U} \in C^\infty(M)$ is a potential on M.

As G is a nilpotent Lie group, the elements of $\mathcal{M}_{ov}^{+}(M \times \widehat{G})$ split into two parts

$$\Gamma d\gamma \ = \ \mathbf{1}_{M\times\widehat{G}_1}\Gamma d\gamma \ + \ \mathbf{1}_{M\times\widehat{G}_\infty}\Gamma d\gamma$$

In particular, on $M \times \widehat{G}_1$, we may assume $\Gamma = 1$, while on $M \times \widehat{G}_\infty$, the trace-class operator $\Gamma(x, \pi^{\lambda,\nu})$ acts on $\mathcal{H}_{\pi^{\lambda,\nu}} = L^2(\mathfrak{p}_\lambda)$ in the case of a step-two group G.

We can already observe that the decomposition (10) in Theorem 1 is due to the split above: the measure $\varrho^{\mathfrak{v}}$ is the restriction of $\Gamma d\gamma$ to $M \times \widehat{G}_1$, while the restriction to $M \times \widehat{G}_\infty$ yields a more involved measure $\varrho^{\mathfrak{z}}$. The invariance then comes from the theorem below. In this statement, we will allow ourselves to use the identifications (see Sects. 1.3.2 and 1.3.3):

$$\widehat{G}_1 \sim \mathfrak{v}^* \qquad \text{and} \qquad \widehat{G}_\infty \sim \sqcup_{k=k_0}^{\dim \mathfrak{v}-1} \Omega_k.$$

Theorem 2 *Let $(\psi^\varepsilon)_{\varepsilon>0}$ be a family of normalized functions satisfying (15) and $\Gamma d\gamma$ one of its semi-classical measures. Then we have the following properties:*

(i) Localization:

$$\pi(\mathbb{L})\Gamma(x,\pi) = \Gamma(x,\pi)\pi(\mathbb{L}) = -\Gamma(x,\pi), \qquad \gamma(x,\pi)\, a.e.$$

which implies

a. The scalar measure $\mathbf{1}_{M\times\widehat{G}_1}\gamma$ on $M \times \widehat{G}_1$ is supported in the set

$$\{(x, \pi^\omega) \in M \times \widehat{G}_1, |\omega| = 1\}.$$

b. Setting $\Gamma_\zeta := \mathbf{1}_{M\times\widehat{G}_\infty}\widehat{\mathbb{P}}_\zeta\Gamma$ for each $\zeta > 0$, we have

$$\Gamma(x,\pi) = \sum_{\zeta\in \mathrm{sp}(\pi(-\mathbb{L}))} \Gamma_\zeta(x,\pi)$$

for γ-almost every $(x,\pi) \in M\times\widehat{G}_\infty$. Moreover, it satisfies $\zeta\Gamma_\zeta d\gamma = \Gamma_\zeta d\gamma$ in $\mathcal{M}_{ov}^{+}(M \times \widehat{G})$. In other words, $\zeta = 1$ on the support of the measure $\mathrm{Tr}(\Gamma_\zeta(x,\pi))\gamma(x,\pi)$.

(ii) Invariance:

a. The scalar measure $\mathbf{1}_{M\times\widehat{G}_1}\gamma$ is invariant under the flow

$$(x,\pi^\omega) \longmapsto (\mathrm{Exp}(s\omega\cdot V)x, \pi^\omega), \quad s \in \mathbb{R}.$$

b.

i. *For each* $\zeta > 0$, *the operator valued measure* $\Gamma_\zeta d\gamma = \mathbf{1}_{M\times\widehat{G}_\infty}\widehat{\mathbb{P}}_\zeta \Gamma d\gamma$ *is supported in* $M \times \widehat{G}_\infty$ *where it is invariant under the flow*

$$(x, \pi^{\lambda,\nu}) \longmapsto (\mathrm{Exp}(s\nu \cdot R^\lambda)x, \pi^{\lambda,\nu}), \quad s \in \mathbb{R}.$$

ii. *Assume* $\Omega_0 \neq \emptyset$. *For each* $\zeta > 0$ *parametrized smoothly by* λ, *the operator valued measure* $\mathbf{1}_{M\times\Lambda_0}\Gamma_\zeta d\gamma$ *is supported on* $M \times \Lambda_0$ *where it is invariant under the flow*

$$(x, \pi^\lambda) \longmapsto (\mathrm{Exp}(s\nabla_\lambda \zeta)x, \pi^\lambda), \quad s \in \mathbb{R}.$$

Note that the flow invariances may be different for various ζ in Part (2) (b). This was already observed on the groups of Heisenberg type where $\Omega_0 = \Lambda_0 = \mathfrak{z}^*\setminus\{0\} \sim \widehat{G}_\infty$ (see [15, 16]). The invariance of Point (2)(a) is empty in that case since the flow map of (2)(a) reduces to identity on Ω_0.

Theorem 2 implies Theorem 1 through the identification that has been mentioned above:

$$\varrho^{\mathfrak{v}}(x) = \int_{\omega\in\mathfrak{v}^*} d\gamma(x, \pi^\omega) \qquad \text{and} \qquad \varrho^{\mathfrak{z}}(x) = \int_{\pi\in\widehat{G}_\infty} \mathrm{Tr}(\Gamma(x, \pi))d\gamma(x, \pi).$$

2.4 Main Ideas of the Proof

Theorem 2 is inspired by the results [13, 16] where the group G was assumed to be of Heisenberg type. We follow here the ideas developed in these papers and extend them to general two-step groups. We explain below the main elements of the proof that rely on technical lemmata that are discussed in Sect. 4.

One can notice that, formally,

$$-\varepsilon^2\mathbb{L}_M^{\mathbb{U}} = -\mathrm{Op}_\varepsilon(\pi(\mathbb{L})) + \varepsilon^2\mathrm{Op}_\varepsilon(\mathbb{U}), \tag{16}$$

which implies that the term involving the potential $\mathbb{U}$ is of lower order than the operator $\varepsilon^2\mathbb{L}_M$ itself. For both the proof of the localisation results and the invariance ones, we start from some relations coming from the $(\psi^\varepsilon)_{\varepsilon>0}$ being eigenfunctions of the subLaplacian. We then use symbolic calculus as developed in [13, 16] to analyse these algebraic relations and compute precisely the symbols involved in the calculus. Finally, passing to the limit $\varepsilon \to 0$, we investigate what the resulting equations mean for the semi-classical measure. We restrict ourselves to the zone $\widehat{G}_1$ or $\widehat{G}_\infty$ by using symbol belonging to the von Neumann algebra generated by $\mathcal{A}_0$. Another important ingredient of the proof consists in analyzing the different behavior of symbols that commute with $\widehat{\mathbb{L}}$ and those who don't. These technical points are developed in Sect. 4.

(i) *Localization*. Let $\sigma \in \mathcal{A}_0$. By the definition of the family $(\psi^\varepsilon)_{\varepsilon>0}$, we have (by Eq. (15))

$$\left(\mathrm{Op}_\varepsilon(\sigma)(-\varepsilon^2 \mathbb{L}_M^{\mathbb{U}} \psi^\varepsilon), \psi^\varepsilon\right)_{L^2(M)} = \left(\mathrm{Op}_\varepsilon(\sigma)\psi^\varepsilon, -\varepsilon^2 \mathbb{L}_M^{\mathbb{U}} \psi^\varepsilon\right)_{L^2(M)}$$
$$= \left(\mathrm{Op}_\varepsilon(\sigma)\psi^\varepsilon, \psi^\varepsilon\right)_{L^2(M)}.$$

By passing to the limit and using (16), the definition of the semi-classical measures as in Proposition 2 and the properties of the calculus [13, 16], give that any semi-classical measure $\Gamma d\gamma$ of $(\psi^\varepsilon)_{\varepsilon>0}$ satisfies

$$\int_{M\times\widehat{G}} \mathrm{Tr}\,(\sigma(x,\pi)\pi(\mathbb{L})\Gamma(x,\pi))\, d\gamma(x,\pi) \tag{17}$$
$$= \int_{M\times\widehat{G}} \mathrm{Tr}\,(\pi(\mathbb{L})\sigma(x,\pi)\Gamma(x,\pi))\, d\gamma(x,\pi)$$
$$= -\int_{M\times\widehat{G}} \mathrm{Tr}\,(\sigma(x,\pi)\Gamma(x,\pi))\, d\gamma(x,\pi).$$

This readily implies the first localization property in (i). The rest of (i) follows as we can now apply (17) not only to symbols σ in $\mathcal{A}_0$, but also in the von Neumann algebra generated by $\mathcal{A}_0$, in particular to $\mathbf{1}_{M\times\widehat{G}_1}\sigma$ and to $\mathbf{1}_{M\times\widehat{G}_\infty}\sigma$, see Lemma 3. Furthermore, (17) shows the commutation of Γ with $\widehat{\mathbb{L}}$ so also with the spectral projectors $\widehat{\mathbb{P}}_\zeta$ for $\zeta > 0$. Therefore, with the notation of Sect. 4.1, our semi-classical measure $\Gamma d\gamma$ is in $\mathcal{M}_{ov}(M \times \widehat{G})^{(\widehat{\mathbb{L}})}$, the subspace of semi-classical measures that commute with $\widehat{\mathbb{L}}$. Hence, by the analysis in Sect. 4.1, we only need to consider symbols σ in $\mathcal{B}_0$ which is the space of the symbols in $\mathcal{A}_0$ that commutes with $\widehat{\mathbb{L}}$.

(ii) *Invariance*. We now take advantage of the fact that for all $\sigma \in \mathcal{A}_0$,

$$\left([\mathrm{Op}_\varepsilon(\sigma), -\varepsilon^2 \mathbb{L}_M^{\mathbb{U}}]\psi^\varepsilon, \psi^\varepsilon\right)_{L^2(M)} = 0. \tag{18}$$

Setting $\pi(V)\cdot V := \sum_{j=1}^q \pi(V_j)V_j$ for any orthonormal basis of $V_1, \ldots, V_q$ of $\mathfrak{v}$, a computation gives for $\sigma \in \mathcal{A}_0$,

$$\frac{1}{\varepsilon}[\mathrm{Op}_\varepsilon(\sigma), -\varepsilon^2 \mathbb{L}_M^{\mathbb{U}}] = -\frac{1}{\varepsilon}\mathrm{Op}_\varepsilon([\sigma, \pi(\mathbb{L})]) + 2\,\mathrm{Op}_\varepsilon(\pi(V)\cdot V\sigma)$$
$$+ \varepsilon\,\mathrm{Op}_\varepsilon(\mathbb{L}\sigma) + \varepsilon[\mathrm{Op}_\varepsilon(\sigma), \mathbb{U}(x)]. \tag{19}$$

For symbols $\sigma \in \mathcal{B}_0$ (which then commute with $\widehat{\mathbb{L}}$), the term in $\frac{1}{\varepsilon}$ in the right-hand side vanishes and we deduce by passing to the limit that any semi-classical measure $\Gamma d\gamma$ of $(\psi^\varepsilon)_{\varepsilon>0}$ satisfies

$$\forall \sigma \in \mathcal{B}_0, \qquad \int_{M\times\widehat{G}} \mathrm{Tr}\,(\pi(V)\cdot V\sigma(x,\pi)\Gamma(x,\pi))\, d\gamma(x,\pi) = 0. \tag{20}$$

Let us prove Part (2)(a). As for Part (1), we can apply this to the elements $\mathbf{1}_{M\times\widehat{G}_1}\sigma$ and to $\mathbf{1}_{M\times\widehat{G}_\infty}\sigma$ of the von Neumann algebra generated by $\mathcal{B}_0$, see Lemma 3. We obtain first that (20) holds with integration over $M \times \widehat{G}_1$; Part (ii)(1) then follows from this and Corollary 1. Then, we obtain that (20) holds with integration on $M \times \widehat{G}_\infty$. This yields

$$\begin{aligned} 0 &= \int_{M\times\widehat{G}_\infty} \mathrm{Tr}\,(\pi(V)\cdot V\sigma(x,\pi)\Gamma(x,\pi))\,d\gamma(x,\pi) \qquad (21) \\ &= \int_{M\times\widehat{G}_\infty} \sum_{\zeta\in\mathrm{sp}(\pi(\mathbb{L}))} \mathrm{Tr}\left(\pi(\mathbb{P}_\zeta)(\pi(V)\cdot V)\pi(\mathbb{P}_\zeta)\,\sigma(x,\pi)\Gamma(x,\pi)\right)d\gamma(x,\pi), \end{aligned}$$

since $\sum_{\zeta\in\mathrm{sp}(\pi(\mathbb{L})}\pi(\mathbb{P}_\zeta)$ is the identity operator on $\mathcal{H}_\pi$ and $\pi(\mathbb{P}_\zeta) = \pi(\mathbb{P}_\zeta)^2$ commutes with $\sigma(x,\pi)$ and $\Gamma(x,\pi)$. Furthermore, for $\pi = \pi^{\lambda,\nu}$, it follows from Sect. 4 (see (31))

$$\forall\zeta>0, \qquad \pi(\mathbb{P}_\zeta)(\pi(P^\lambda)\cdot P^\lambda)\pi(\mathbb{P}_\zeta)=0, \qquad \pi(\mathbb{P}_\zeta)(\pi(Q^\lambda)\cdot Q^\lambda)\pi(\mathbb{P}_\zeta)=0.$$

Hence (21) becomes

$$\forall\sigma\in\mathcal{B}_0, \qquad \int_{M\times\widehat{G}_\infty} \mathrm{Tr}\left(\nu\cdot R^\lambda\sigma(x,\pi^{\lambda,\nu})\Gamma(x,\pi^{\lambda,\nu})\right)d\gamma(x,\pi^{\lambda,\nu})=0.$$

This implies Part (ii)(2)(a) by Proposition 4 as $\Gamma d\gamma$ is in the set $\mathcal{M}_{ov}(M\times\widehat{G})^{(\widehat{\mathbb{L}})}$ of operator-valued measures that commute with $\widehat{\mathbb{L}}$.

Let us prove Part (2)(b). We now assume $\Omega_0 \neq \emptyset$. Indeed, on Ω_0, the analysis above does not yield anything since $\nu\cdot R^\lambda = 0$ on Ω_0. We will need the following observation:

Lemma 1 *If $\sigma\in\mathcal{A}_0$ and $\eta\in\mathcal{S}(\mathfrak{z}^*)$, then the symbol $\sigma\eta$ given by $(\sigma\eta)(x,\pi^{\lambda,\nu}) = \sigma(x,\pi^{\lambda,\nu})\eta(\lambda)$ is in $\mathcal{A}_0$. If $\sigma\in\mathcal{B}_0$ then $\sigma\eta\in\mathcal{B}_0$.*

Proof If $\kappa_x(y)$ is the kernel of σ, then we check readily that $(y_\mathfrak{v}, y_\mathfrak{z}) \mapsto (\kappa_x(y_\mathfrak{v},\cdot) *_\mathfrak{z} \mathcal{F}_\mathfrak{z}^{-1}\eta)(y_\mathfrak{z})$ is the kernel of $\sigma\eta$. The rest follows. □

By Lemma 1, if $\sigma_1\in\mathcal{B}_0$ and if $\eta\in\mathcal{S}(\mathfrak{z}^*)$ is supported in the dense open subset Ω_0 of $\mathfrak{z}^*\setminus\{0\}$, then $\sigma := \sigma_1\eta$ is supported in $M\times\Omega_0$. Moreover, by Lemma 4, there exists a symbol $T\sigma\in\mathcal{A}_0$ such that

$$\pi(V)\cdot V\sigma = [T\sigma, \pi(-\mathbb{L})].$$

Therefore, using the additional fact

$$[\mathrm{Op}_\varepsilon(T\sigma), \mathbb{U}(x)] = O(\varepsilon) \text{ in } \mathcal{L}(L^2(M)),$$

the Eq. (19) gives

$$\frac{1}{\varepsilon}\left(\mathrm{Op}_\varepsilon(\pi(V)\cdot V\sigma)\psi^\varepsilon,\psi^\varepsilon\right)_{L^2(M)}=\frac{1}{\varepsilon}\left(\mathrm{Op}_\varepsilon([T\sigma,\pi(-\mathbb{L})])\psi^\varepsilon,\psi^\varepsilon\right)_{L^2(M)}$$
$$=\frac{1}{\varepsilon}\left([\mathrm{Op}_\varepsilon(T\sigma),-\varepsilon^2\mathbb{L}_M^{\mathbb{U}}]\psi^\varepsilon,\psi^\varepsilon\right)_{L^2(M)}$$
$$-2\left(\mathrm{Op}_\varepsilon((\pi(V)\cdot V)\circ T\sigma)\psi^\varepsilon,\psi^\varepsilon\right)_{L^2(M)}+O(\varepsilon).$$

By (18), the first term of the right-hand side is 0 and we have

$$\frac{1}{\varepsilon}\left(\mathrm{Op}_\varepsilon(\pi(V)\cdot V\sigma)\psi^\varepsilon,\psi^\varepsilon\right)_{L^2(M)}$$
$$=-2\left(\mathrm{Op}_\varepsilon((\pi(V)\cdot V)\circ T\sigma)\psi^\varepsilon,\psi^\varepsilon\right)_{L^2(M)}+O(\varepsilon).$$

Plugging this expression of $\left(\mathrm{Op}_\varepsilon(\pi(V)\cdot V\sigma)\psi^\varepsilon,\psi^\varepsilon\right)_{L^2(M)}$ in (19) and using one more time (18), we finally get

$$O(\varepsilon)=\frac{2}{\varepsilon}\left(\mathrm{Op}_\varepsilon(\pi(V)\cdot V\sigma)\psi^\varepsilon,\psi^\varepsilon\right)_{L^2(M)}+\left(\mathrm{Op}_\varepsilon(\mathbb{L}\sigma)\psi^\varepsilon,\psi^\varepsilon\right)_{L^2(M)}$$
$$=\left(\mathrm{Op}_\varepsilon(-4\pi(V)\cdot V\circ T\sigma+\mathbb{L}\sigma)\psi^\varepsilon,\psi^\varepsilon\right)_{L^2(M)}\cdot$$

We now pass to the limit $\varepsilon\to 0$ and transform the latter equation according to the equality

$$-4\pi(V)\cdot V\circ T\sigma+\mathbb{L}\sigma=i\sum_{\zeta\in\mathrm{Sp}\widehat{\mathbb{L}}}\nabla_\lambda\zeta\,\sigma\,\widehat{\mathbb{P}}_\zeta,$$

induced by Corollary 2 and the fact that $\sigma\in\mathcal{B}_0$. We are left with

$$\sum_{\zeta\in\mathrm{Sp}\widehat{\mathbb{L}}}\int_{M\times\Omega_0}\mathrm{Tr}\left(\nabla_\lambda\zeta\,\sigma(x,\pi^\lambda)\Gamma_\zeta(x,\pi^\lambda)\right)d\gamma(x,\pi^\lambda)=0,$$

and the relation holds for all $\sigma=\sigma_1\eta$ with $\sigma_1\in\mathcal{B}_0$ and $\eta\in\mathcal{S}(\mathfrak{z}^*)$ supported in the dense open set Ω_0. This concludes the proof.

3 Geometric Invariance

In this section, we address the geometric invariance of the objects that we have introduced above.

3.1 Nilmanifolds as Filtered Manifolds

A stratified Lie group G carries a natural filtration on its Lie algebra given by

$$\mathfrak{h}_1 \subset \mathfrak{h}_2 \subset \cdots \subset \mathfrak{h}_k = \mathfrak{g} = T_e G, \qquad \text{with} \qquad \mathfrak{h}_j = \mathfrak{g}_1 \oplus \cdots \oplus \mathfrak{g}_j.$$

One can view the nilmanifold M as a *filtered manifold* with associated filtration of subbundles

$$H^1_{\dot{x}} \subset H^2_{\dot{x}} \subset \cdots \subset H^r_{\dot{x}} = T_{\dot{x}} M, \quad x \in G, \quad [H^i, H^j] \subset H^{i+j}, \ 1 \leq i + j \leq r, \tag{22}$$

given by $H^i_{\dot{x}} = d\pi_\Gamma \circ dL_x(\mathfrak{h}_i)$. Here $\pi_\Gamma : G \to M = \Gamma \backslash G$ is the quotient map and $L_x : G \to G$ is the left-translation. In fact, G induces a left-invariant stratification by the subbundles $d\pi_\Gamma \circ d_{L_x}(\mathfrak{g}_i)$ of TM in the obvious way, but such a stratification will not respect the Lie bracket of vector fields on M unless one restricts to left-invariant vector fields. What's more, we will see that the semi-classical calculus only depends on the filtration, and not on the stratification or the metric.

When G is step 2, we have $\mathfrak{h}_1 = \mathfrak{v}$ and $\mathfrak{h}_2 = \mathfrak{g}$. In this case, the data of the filtration on G is almost the same as a stratification except that one forgets the second stratum $\mathfrak{g}_2 = \mathfrak{z}$. On M, the filtration is given by a single step 2 bracket generating subbundle $H^1 \subset TM$ without a preferred complement.

3.2 Filtration Preserving Maps

Let U be an open subset of M and $\Phi : U \to M$ a smooth map on M. We introduce two definitions.

Definition 1

1. The smooth map Φ is said to preserve the filtration at $\dot{x} \in U$ when

$$d_{\dot{x}} \Phi \left(H^i_{\dot{x}} \right) \subseteq H^i_{\Phi(\dot{x})}, \quad i = 1, \ldots, r.$$

2. The map Φ is Pansu differentiable at the point $\dot{x}$ when for any $z \in G$,

$$\lim_{\varepsilon \to 0} \delta_{\varepsilon^{-1}} \left(\Phi(\dot{x})^{-1} \Phi(\dot{x} \delta_\varepsilon z) \right) = \lim_{\varepsilon \to 0} \delta_{\varepsilon^{-1}} \left(\Phi_{\dot{x}}(\delta_\varepsilon z) \right) =: \mathrm{PD}_{\dot{x}} \Phi(z). \tag{23}$$

3. The map Φ is uniformly Pansu differentiable on U if it is Pansu differentiable at every point in U, and the limit (23) holds locally uniformly on $U \times G$.

Remark 1 Taking $U \subset M$ to be a sufficiently small neighborhood of $\dot{x} \in M$, we may consider U as a neighborhood of $x \in G$ and lift Φ to a smooth map Φ_G :

$U \subset G \to G$. Then the above definition is equivalent to saying Φ_G is Pansu differentiable (resp. uniformly Pansu differentiable) at $x \in G$ (resp. on $U \times G$).

On a neighborhood $U \subset M$ sufficiently small to identify with a neighborhood in G, the notions of Pansu differentiability and filtration preservation are related via the following result [17]:

Theorem 3 ([17]) *The map Φ is uniformly Pansu differentiable on U if and only if Φ preserves the filtration at every point $x \in U$.*

This result relates a morally algebraic property, Pansu differentiability, to a geometric property of being filtration-preserving. Consequently, the diffeomorphisms Φ we consider in the sequel are uniformly Pansu differentiable, and the transformation of pseudodifferential operators by the pull-back associated with Φ will involve the Pansu derivative of Φ. This leads us to employ the osculating Lie group and Lie algebra bundles in the next section.

3.3 Schwartz Vertical Densities

For a filtered manifold M, the *osculating Lie algebra bundle* $\mathfrak{G}M$ (and the *osculatig Lie group bundle* $\mathbb{G}M := \exp(\mathbb{G}M)$), defined in [17], play the role of the tangent bundle. When $M = \Gamma\backslash G$, with the filtration (22), the fibers of $\mathfrak{G}M$ and $\mathbb{G}M$, are all isomorphic to $\mathfrak{g}$ and G respectively. In particular, we have canonical identifications

$$\mathfrak{G}M \cong M \times \mathfrak{g} \quad \text{and} \quad \mathbb{G}M \cong M \times G.$$

The Haar measure on each fibers $\mathbb{G}_{\dot{x}}M$ is given by $d_{\dot{x}}z = dz$ and the dual sets by $\widehat{\mathbb{G}}_{\dot{x}}M = \widehat{G}$.

To any semi-classical pseudodifferential operator on a compact nilmanifold $M = \Gamma\backslash G$, its convolution kernel κ may be viewed as an element of $C^\infty(M, \mathcal{S}(G))$. However, this is not the right space for the general case of filtered manifolds. Indeed, Theorem 4 together with Eq. (26) below will imply that as a pseudodifferential operator transforms under diffeomorphisms on M preserving the filtration, its associated convolution kernel transforms like a density on the osculating group bundle. We show that it is natural to view the convolution kernels κ as elements of the bundle of *Schwartz vertical densities* on $\mathbb{G}M$, rather than functions in $C^\infty(M, \mathcal{S}(G))$. We briefly elaborate below.

Let $\mathcal{V}(\mathbb{G}M)$ be the vertical bundle of $\mathbb{G}M$, that is, the kernel of the map $\mathbb{G}M \to M$. Let $|\Lambda|\mathcal{V}(\mathbb{G}M)$ be the bundle of vertical densities, that is, the bundle over $\mathbb{G}M$ whose fibers are densities in the vertical spaces. Let $\mathcal{S}(\mathbb{G}M) = \coprod_{\dot{x}\in M} \mathcal{S}(\mathbb{G}_{\dot{x}}M)$ be the Fréchet vector bundle over M whose fibers are Schwartz class functions. Furthermore, let $\mathcal{S}(\mathbb{G}M, |\Lambda|\mathcal{V})$ be the Fréchet bundle over M whose fibers are Schwartz class densities on the vertical space. As in [17], denote by $\Gamma_c(\mathcal{S}(\mathbb{G}M, |\Lambda|\mathcal{V}))$, the space of its smooth compactly supported sections, which we call the Schwartz vertical densities. After making a choice of Haar measure on G, this space is identified with $C^\infty(M, \mathcal{S}(G))$.

Indeed, by left-invariance, we identify the fibers of $\mathcal{V}(\mathbb{G}M)$ with the Lie algebra $\mathfrak{g}$ and fibers of $|\Lambda|\mathcal{V}(\mathbb{G}M)$ with $|\Lambda|\mathfrak{g}$, the set of densities on the Lie algebra:

$$\mathcal{V}(\mathbb{G}M) \cong (M \times G) \times \mathfrak{g}, \quad \text{whence} \quad |\Lambda|\mathcal{V}(\mathbb{G}M) \cong (M \times G) \times |\Lambda|\mathfrak{g}. \tag{24}$$

The above trivializations give the identification

$$\Gamma_c\left(\mathcal{S}\left(\mathbb{G}M, |\Lambda|\mathcal{V}\right)\right) \cong C^\infty(M, \mathcal{S}(G, |\Lambda|\mathfrak{g})).$$

And a choice of Haar measure on G gives $\mathcal{S}(G, |\Lambda|\mathfrak{g}) \cong \mathcal{S}(G)$.

For a choice of Haar measure dz on G, which in turn gives a Haar system $\{d_{\dot{x}}z\}$ on $\mathbb{G}M$ through (24), the identifications of vertical Schwartz densities with functions is given explicitly by

$$\kappa \in \Gamma_c\left(\mathcal{S}\left(\mathbb{G}M, |\Lambda|\mathcal{V}\right)\right) : \dot{x} \mapsto \kappa_{\dot{x}} = \widetilde{\kappa}_{\dot{x}} d_{\dot{x}}z, \quad \widetilde{\kappa}_{\dot{x}} \in \mathcal{S}(\mathbb{G}_{\dot{x}}M).$$

The symbols $\sigma \in \mathcal{A}_0$ are defined as the images of the elements κ of the set $\Gamma_c\left(\mathcal{S}\left(\mathbb{G}M, |\Lambda|\mathcal{V}\right)\right)$ by the fiberwise Fourier transform:

$$\dot{x} \mapsto \sigma(\dot{x}, \pi) = \int_{z \in \mathbb{G}_{\dot{x}}M} \widetilde{\kappa}_{\dot{x}}(z)\pi(z)^* d_{\dot{x}}z, \quad \pi \in \widehat{\mathbb{G}}_{\dot{x}}M. \tag{25}$$

Since our convolution kernels are densities, the integral (25) is independent of the choice of Haar measure.

3.4 *Semi-classical Pseudodifferential Calculus and Filtration Diffeomorphisms*

We keep the notations of the preceding section except we suppose $\Phi : U \subset M \to M$ is a diffeomorphism onto its image. Let J_Φ be the Jacobian of Φ. We associate with Φ

(i) a unitary transformation $\mathcal{U}_\Phi$ of $L^2(U)$ induced by Φ

$$\mathcal{U}_\Phi(f) := J_\Phi^{1/2} f \circ \Phi, \quad f \in L^2(U),$$

(ii) a map $\mathcal{I}_\Phi$ on the space of Schwartz vertical densities that extends to an isometry of $L^1(|\Lambda|\mathcal{V}(\mathbb{G}M))$

$$(\mathcal{I}_\Phi\kappa)_x(z) := J_\Phi(x)\, \kappa_{\Phi(x)}(\mathrm{PD}_x\Phi(z)), \quad \forall (x, z) \in U \times G.$$

We are interested in the properties of the operator $\mathcal{U}_\Phi \circ \mathrm{Op}_\varepsilon(\sigma) \circ \mathcal{U}_\Phi^{-1}$, in particular in the asymptotics in ε of its semi-classical pseudodifferential symbol. The structure of the latter and the way it can be deduced from σ will give information of the geometric nature of the objects. Indeed, Φ induces several geometric transformations:

(i) Φ induces a map on representations

$$\widehat{\mathbb{G}}\Phi : \begin{cases} U \times \widehat{G} \longrightarrow \Phi(U) \times \widehat{G} \\ (x, \pi) \longmapsto \left(\Phi(x), \pi \circ (\mathrm{PD}_x \Phi)^{-1}\right) \end{cases}$$

(ii) The *generalized canonical transformation* $\widehat{\mathbb{G}}\Phi$ induces a pull-back on symbols

$$(\widehat{\mathbb{G}}\Phi)^* \sigma(x, \pi) := \sigma(\widehat{\mathbb{G}}\Phi(x, \pi)).$$

The maps $\widehat{\mathbb{G}}\Phi$ and I_Φ are intertwined by the group Fourier Transform: If $\sigma(x, \pi) = \widehat{\kappa_x}(\pi)$ for all $x \in U \subset M$ and $\pi \in \widehat{\mathbb{G}}_x M$, then for any filtration preserving differmorphism $\Phi : U \to M$

$$(\widehat{\mathbb{G}}\Phi)^* \sigma(x, \pi) = \widehat{I_\Phi \kappa_x}(\pi), \quad x \in U \subset M, \ \pi \in \widehat{G} = \widehat{\mathbb{G}}_x M. \tag{26}$$

These two maps are involved in the description of the first term of the expansion of the semi-classical symbol of the operator $\mathcal{U}_\Phi \circ \mathrm{Op}_\varepsilon(\sigma) \circ \mathcal{U}_\Phi^{-1}$:

Theorem 4 **([17])** *Assume that Φ is filtration preserving on U. Then in $\mathcal{L}(L^2(U))$,*

$$\mathcal{U}_\Phi \circ \mathrm{Op}_\varepsilon(\sigma) \circ \mathcal{U}_\Phi^{-1} = \mathrm{Op}_\varepsilon\left((\widehat{\mathbb{G}}\Phi)^* \sigma\right) + O(\varepsilon).$$

Remark 2 Theorem 4 establishes the geometric invariance of the semi-classical calculus by filtration preserving differmorphisms Φ. In particular, Φ does not need to preserve the action of G on M, or even preserve the gradation.

The results of this section suggest that the semi-classical symbols we defined in Sect. 2.1 ought to be the natural generalization of symbols for arbitrary filtered manifolds. So defined, the semi-classical symbols are invariant under generalized canonical transformations of $\widehat{\mathbb{G}}M$ associated to differmorphisms preserving the filtration on M.

4 Technical Tools

This section is devoted to several technical results used in the proof of Theorem 2.

4.1 Some C^*-Algebras and Their Properties

4.1.1 The von Neumann Algebra $L^\infty(M \times \widehat{G})$

A measurable symbol $\sigma = \{\sigma(x, \pi) : (x, \pi) \in M \times \widehat{G}\}$ is said to be bounded when there exists a constant $C > 0$ such that for $dxd\mu(\pi)$-almost all $(x, \pi) \in M \times \widehat{G}$, we have $\|\sigma(x, \pi)\|_{\mathcal{H}_\pi} \leq C$. We denote by $\|\sigma\|_{L^\infty(M\times\widehat{G})}$ the smallest of such constant $C > 0$ and by $L^\infty(M \times \widehat{G})$ the space of bounded measurable symbols. We check readily that $\|\cdot\|_{L^\infty(M\times\widehat{G})}$ is a norm on $L^\infty(M \times \widehat{G})$ which is a C^*-algebra. We will later use the fact that it is a von Neumann algebra.

4.1.2 The C^*-Algebra $\mathcal{A}$ and Its Topological Dual

Clearly, $\mathcal{A}_0$ is a subspace of $L^\infty(M \times \widehat{G})$. Its closure denoted by $\mathcal{A}$ for the norm $\|\cdot\|_{L^\infty(M\times\widehat{G})}$ is a sub-C^*-algebra of $L^\infty(M \times \widehat{G})$. Its topological dual $\mathcal{A}^*$ is isomorphic to the Banach space of operator-valued measures $\mathcal{M}_{ov}(M \times \widehat{G})$ via

$$\mathcal{M}_{ov}(M \times \widehat{G}) \ni \Gamma d\gamma \mapsto \ell_{\Gamma d\gamma} \in \mathcal{A}^*,$$

$$\ell_{\Gamma d\gamma}(\sigma) := \int_{M\times\widehat{G}} \operatorname{Tr}\left(\sigma(x, \pi)\Gamma(x, \pi)\right) d\gamma(x, \pi).$$

Moreover, the isomorphism is isometric:

$$\|\ell_{\Gamma d\gamma}\|_{\mathcal{A}^*} = \|\Gamma d\gamma\|_{\mathcal{M}_{ov}(M\times\widehat{G})}$$

where

$$\|\Gamma d\gamma\|_{\mathcal{M}_{ov}} := \int_{M\times\widehat{G}} \operatorname{Tr}|\Gamma(x, \pi)|\, d\gamma(x, \pi),$$

and the positive linear functionals on $\mathcal{A}$ are the $\ell = \ell_{\Gamma d\gamma}$'s with $\Gamma d\gamma \geq 0$.

4.1.3 The C^*-Algebra $\mathcal{B}$ and Its Topological Dual

Let $\mathcal{B}_0$ be the subspace of $\mathcal{A}_0$ of symbols commuting with $\widehat{\mathbb{L}}$. Clearly $\mathcal{B}_0$ contains all the symbols of the form $a(x)\psi(\widehat{\mathbb{L}})$, $a \in C^\infty(M)$, $\psi \in \mathcal{S}(\mathbb{R})$, by Hulanicki's theorem (see [26]):

Theorem 5 (Hulanicki) *The convolution kernel of a spectral multiplier $\psi(\mathbb{L}_G)$ of $\mathbb{L}_G$ for a Schwartz function $\psi \in \mathcal{S}(\mathbb{R})$ is Schwartz on G.*

We denote by $\mathcal{B}$ the closure of $\mathcal{B}_0$ for the norm $\|\cdot\|_{L^\infty(M\times\widehat{G})}$. Property (2) of $\widehat{\mathbb{L}}$ recalled in Sect. 1.3.4 implies that $\mathcal{B}$ is the subspace of $\mathcal{A}$ of symbols commuting with every $\widehat{\mathbb{P}}_\zeta$, $\zeta > 0$. We check readily that $\mathcal{B}$ is a sub-C^*-algebra of $\mathcal{A}$ and that $\mathcal{B}_0 = \mathcal{A}_0 \cap \mathcal{B}$. The next statement identifies the topological dual of $\mathcal{B}$:

Proposition 4 *Via $\Gamma d\gamma \mapsto \ell_{\Gamma d\gamma}|_{\mathcal{B}}$, the topological dual $\mathcal{B}^*$ of $\mathcal{B}$ is isomorphic with the closed subspace $\mathcal{M}_{ov}(M \times \widehat{G})^{(\widehat{\mathbb{L}})}$ of operator valued measures $\Gamma d\gamma \in \mathcal{M}_{ov}(M \times \widehat{G})$ such that the operator Γ commutes with $\widehat{\mathbb{P}}_\zeta$ for all $\zeta > 0$, in the sense that*

$$\forall \zeta > 0 \qquad \pi(\mathbb{P}_\zeta)\Gamma(x,\pi) = \Gamma(x,\pi)\pi(\mathbb{P}_\zeta) \quad \textit{for } \gamma-\textit{almost all } (x,\pi) \in M \times \widehat{G}.$$

Proof

Step 0. We observe that if two pairs (Γ, γ) and (Γ_1, γ_1) are equivalent and one of them satisfies the commutative condition with $\widehat{\mathbb{P}}_\zeta$ for all $\zeta > 0$, then so does the other. Hence, $\mathcal{M}_{ov}(M \times \widehat{G})^{(\widehat{\mathbb{L}})}$ is a well defined subset of $\mathcal{M}_{ov}(M \times \widehat{G})$. One checks that it is a closed subspace of $\mathcal{M}_{ov}(M \times \widehat{G})$.

Step 1. Let $\ell \in \mathcal{B}^*$. By the Hahn-Banach theorem, this functional extends to $\tilde{\ell} \in \mathcal{A}^*$, i.e. $\tilde{\ell}_{|\mathcal{B}} = \ell$. Denote by $\Gamma d\gamma \in \mathcal{M}_{ov}(M \times \widehat{G})$ the corresponding operator-valued measure: $\tilde{\ell} = \ell_{\Gamma d\gamma}$. Now set

$$\Gamma_1(x,\pi) := \sum_{\zeta \in \mathrm{sp}(\pi(\mathbb{L}))} \pi(\mathbb{P}_\zeta)\Gamma(x,\pi)\pi(\mathbb{P}_\zeta).$$

The operator-valued measure $\Gamma_1 d\gamma$ is a well defined element of $\mathcal{M}_{ov}(M \times \widehat{G})$ satisfying the condition of commutativity with $\widehat{\mathbb{L}}$ so $\Gamma_1 d\gamma \in \mathcal{M}_{ov}(M \times \widehat{G})^{(\widehat{\mathbb{L}})}$. Let us show that it coincides with $\ell_{\Gamma d\gamma}$ on $\mathcal{B}$. Let $\sigma \in \mathcal{B}$. Since $\sum_{\zeta \in \mathrm{sp}(\pi(\mathbb{L}))} \pi(\mathbb{P}_\zeta)$ is the identity operator on $\mathcal{H}_\pi$, we have

$$\begin{aligned} \ell_{\Gamma d\gamma}(\sigma) &= \int_{M\times\widehat{G}} \sum_{\zeta \in \mathrm{sp}(\pi(\mathbb{L}))} \mathrm{Tr}\left(\sigma(x,\pi)\Gamma(x,\pi)\pi(\mathbb{P}_\zeta)\right) d\gamma(x,\pi) \\ &= \int_{M\times\widehat{G}} \sum_{\zeta \in \mathrm{sp}(\pi(\mathbb{L}))} \mathrm{Tr}\left(\sigma(x,\pi)\pi(\mathbb{P}_\zeta)\Gamma(x,\pi)\pi(\mathbb{P}_\zeta)\right) d\gamma(x,\pi), \end{aligned}$$

since $\pi(\mathbb{P}_\zeta) = \pi(\mathbb{P}_\zeta)^2$ commutes with $\sigma(x,\pi)$. We recognise $\ell_{\Gamma_1 d\gamma}(\sigma)$ on the right-hand side. We have obtained that any $\ell \in \mathcal{B}^*$ may be written as the restriction to $\mathcal{B}$ of $\ell_{\Gamma_1 d\gamma}$, for some $\Gamma_1 d\gamma \in \mathcal{M}_{ov}(M \times \widehat{G})^{(\widehat{\mathbb{L}})}$.

In other words, we have proved that $\Gamma d\gamma \mapsto \ell_{\Gamma d\gamma}|_{\mathcal{B}}$ maps $\mathcal{M}_{ov}(M \times \widehat{G})^{(\widehat{\mathbb{L}})}$ onto $\mathcal{B}^*$. This map is continuous and linear. It remains to show that it is injective.

Side step. Let us open a parenthesis. The von Neumann algebra $L^\infty(M \times \widehat{G})$ is a C^* algebra containing $\mathcal{A}$ and we denote by $vN\mathcal{A}$ the von Neumann algebra generated by $\mathcal{A}$. This means that $vN\mathcal{A}$ is the closure of $\mathcal{A}$ for the strong operator topology in $L^\infty(M \times \widehat{G})$. We are going to use this von Neumann algebra by considering the natural unique extension of $\ell = \ell_{\Gamma d\gamma} \in \mathcal{A}^*$ to a continuous linear functional on the von Neumann algebra $vN\mathcal{A}$ of $\mathcal{A}$.

Since $\mathcal{B} \subset \mathcal{A}$, we also have $vN\mathcal{B} \subset vN\mathcal{A}$ where $vN\mathcal{B}$ denotes the von Neumann algebra generated by $\mathcal{B}$. Moreover, $vN\mathcal{B}$ is the subspace of the symbols $\sigma \in vN\mathcal{A}$ commuting with $\widehat{\mathbb{P}}_\zeta \ dxd\mu(\pi)$-almost everywhere for every , $\zeta > 0$.

Finally, we observe that for $\zeta > 0$ and $\sigma \in \mathcal{A}$, the symbol $\pi(\widehat{\mathbb{P}}_\zeta)\sigma\pi(\widehat{\mathbb{P}}_\zeta)$ is in $vN\mathcal{A}$. Indeed, using Hulanicki's theorem (Theorem 5) together with $\mathcal{S}(G)*\mathcal{S}(G) \subset \mathcal{S}(G)$, we obtain that if $\sigma \in \mathcal{A}_0$ then for any $\psi_1, \psi_2 \in \mathcal{S}(\mathbb{R})$, the symbol $\psi_1(\widehat{\mathbb{L}})\sigma\psi_2(\widehat{\mathbb{L}})$ is in $\mathcal{A}_0$. Taking limits for suitable sequences of σ, ψ_1, ψ_2 implies that the symbol $\pi(\widehat{\mathbb{P}}_\zeta)\sigma\pi(\widehat{\mathbb{P}}_\zeta)$ is in $vN\mathcal{A}$ for any $\sigma \in \mathcal{A}$.

Step 2. Let us now consider $\Gamma d\gamma \in \mathcal{M}_{ov}(M \times \widehat{G})^{(\widehat{\mathbb{L}})}$ such that $\ell := \ell_{\Gamma d\gamma}$ vanishes on $\mathcal{B}$. We want to show $\ell = 0$. We extend ℓ to a functional L on $vN\mathcal{A}$. This functional vanishes on $vN\mathcal{B}$. We set

$$L_\zeta(\sigma) := \int_{M\times\widehat{G}} \mathrm{Tr}\left(\sigma(x,\pi)\pi(\mathbb{P}_\zeta)\Gamma(x,\pi)\right) d\gamma(x,\pi), \qquad \zeta > 0.$$

We check readily that $\zeta \mapsto L_\zeta(\sigma)$ defines a complex measure on $[0,\infty)$ with total mass that is smaller or equal to $\|\sigma\|_{L^\infty(M\times\widehat{G})}\|\Gamma d\gamma\|_{\mathcal{M}_{ov}}$. Moreover, $\ell(\sigma) = \int_0^{+\infty} L_\zeta(\sigma)$ since $\sum_{\zeta\in\mathrm{sp}(\pi(\mathbb{L})} \pi(\mathbb{P}_\zeta)$ is the identity operator on $\mathcal{H}_\pi$.

Using $\mathbb{P}_\zeta^2 = \mathbb{P}_\zeta$ and the commutation of Γ with $\pi(\mathbb{P}_\zeta)\ d\gamma$-a.e., together with trace property, we obtain

$$\begin{aligned} L_\zeta(\sigma) &= \int_{M\times\widehat{G}} \mathrm{Tr}\left(\pi(\mathbb{P}_\zeta)\sigma(x,\pi)\pi(\mathbb{P}_\zeta)\Gamma(x,\pi)\right) d\gamma(x,\pi) \\ &= L_\zeta(\pi(\mathbb{P}_\zeta)\sigma(x,\pi)\pi(\mathbb{P}_\zeta)) \end{aligned}$$

with $\pi(\mathbb{P}_\zeta)\sigma(x,\pi)\pi(\mathbb{P}_\zeta) \in vN\mathcal{B}$. Arguing as above (in the side step), we deduce $L_\zeta = 0$, whence $L = 0$ and $\ell = 0$. This implies the injectivity of $\Gamma d\gamma \mapsto \ell_{\Gamma d\gamma}|_{\mathcal{B}}$ on $\mathcal{M}_{ov}(M \times \widehat{G})^{(\widehat{\mathbb{L}})}$. □

The proof above has an important consequence regarding the restriction of symbols to $M \times \widehat{G}_1$, a notion we now explain.

4.1.4 Restriction of Symbols to $M \times \widehat{G}_1$

The restriction $\sigma|_{M\times\widehat{G}_1}$ of $\sigma \in \mathcal{A}$ with kernel $\kappa_x(y)$, to $M \times \widehat{G}_1$ is given by

$$\sigma|_{M\times\widehat{G}_1}(x,\omega) = \sigma(x,\pi^\omega) = \mathcal{F}_{\mathfrak{v}} \int_{\mathfrak{z}} \kappa_x(\cdot,z)dz(\omega), \qquad (x,\omega) \in M \times \mathfrak{v}^*,$$

having identified $\widehat{G}_1$ with $\mathfrak{v}^*$. Moreover, we can therefore identify

$$\mathcal{A}|_{M\times\widehat{G}_1} := \{\sigma|_{M\times\widehat{G}_1}, \sigma \in \mathcal{A}\}$$

with a sub-space of $C_0(M \times \mathfrak{v}^*)$. In fact, we can show

Lemma 2 *We have* $C_0(M \times \mathfrak{v}^*) = \mathcal{A}|_{M\times\widehat{G}_1}$.

Proof Any element of $C_0(M\times\mathfrak{v}^*)$ may be viewed as a limit for the supremum norm on $M \times \mathfrak{v}^*$ of $\mathcal{F}_{\mathfrak{v}}\kappa_x^{(j)}(\omega)$ for a sequence of kernels $\kappa^{(j)} \in C^\infty(M, \mathcal{S}(\mathfrak{v}))$. We then consider the sequence of symbols $\sigma_j(x,\pi) = \pi(\kappa_x^{(j)}\eta)$ with $\eta \in \mathcal{S}(\mathfrak{z})$ satisfying $\mathcal{F}_{\mathfrak{z}}\eta(0) = \int_{\mathfrak{z}} \eta(Z)dz = 1$. We check readily that $\sigma_j|_{M\times\widehat{G}_1}(x,\omega) = \mathcal{F}_{\mathfrak{v}}\kappa_x^{(j)}(\omega)$. □

For a subspace S of $\mathcal{A}$, we denote by

$$S|_{M\times\widehat{G}_1} := \{\sigma|_{M\times\widehat{G}_1}, \sigma \in S\}$$

the resulting subspace in $\mathcal{A}|_{M\times\widehat{G}_1}$. The proof of Lemma 2 shows that if $\bar{S}$ denotes the closure of S in the C^*-algebra $\mathcal{A}$, then $\bar{S}|_{M\times\widehat{G}_1}$ is the closure of $S|_{M\times\widehat{G}_1}$ in $C_0(M \times \mathfrak{v}^*)$, that is, given by the supremum norm on $M \times \mathfrak{v}^*$. Hence $\bar{S}|_{M\times\widehat{G}_1} \subset \mathcal{A}|_{M\times\widehat{G}_1}$.

We will need the following property regarding the restriction of the symbols in $\mathcal{A}_0$ and $\mathcal{B}_0$ to $M \times \widehat{G}_1$; its proof relies on the proof of Proposition 4:

Corollary 1 *The following commutative C^* algebras coincide:*

$$\bar{\mathcal{B}}_0|_{M\times\widehat{G}_1} = \mathcal{B}|_{M\times\widehat{G}_1} = \bar{\mathcal{A}}_0|_{M\times\widehat{G}_1} = \mathcal{A}|_{M\times\widehat{G}_1} = C_0(M \times \mathfrak{v}^*).$$

Proof Clearly, $\bar{\mathcal{B}}_0|_{M\times\widehat{G}_1} = \mathcal{B}|_{M\times\widehat{G}_1} \subset \bar{\mathcal{A}}_0|_{M\times\widehat{G}_1} = \mathcal{A}|_{M\times\widehat{G}_1} = C_0(M \times \mathfrak{v}^*)$. It remains to show the converse inequality.

Let ℓ be a continuous linear functional on $C_0(M \times \mathfrak{v}^*)$. This is given by integration against a complex Radon measure γ_1. Consider the operator-valued measure $\Gamma d\gamma \in \mathcal{M}_{ov}(M \times \widehat{G})$ defined by $1_{M\times\widehat{G}_\infty}\Gamma d\gamma = 0$ and $1_{M\times\widehat{G}_1}\Gamma d\gamma = \gamma_1$, that is,

$$\ell_{\Gamma d\gamma}(\sigma) = \int_{M\times\mathfrak{v}^*} \sigma|_{M\times\widehat{G}_1}(x,\pi^\omega)\, d\gamma_1(\omega), \quad \sigma \in \mathcal{A}.$$

We observe that Γ commutes with $\widehat{\mathbb{P}}_\zeta$, $\zeta > 0$. Hence, if $\ell = 0$ on $\mathcal{B}|_{M\times\widehat{G}_1}$ then $\ell_{\Gamma d\gamma} \equiv 0$ on $\mathcal{B}$ and therefore also on $\mathcal{A}$ by Proposition 4, or rather Step 2 of its proof; this implies $\Gamma d\gamma = 0$ thus $\gamma_1 = 0$ and $\ell = 0$. By the Hahn-Banach theorem, this shows that $\mathcal{B}|_{M\times\widehat{G}_1} = C_0(M \times \mathfrak{v}^*)$. □

4.1.5 Some Elements of $vN\mathcal{A}$ and $vN\mathcal{B}$

We will need the following properties:

Lemma 3 *If $\sigma \in \mathcal{A}_0$ then $\mathbf{1}_{M\times\widehat{G}_1}\sigma$ and $\mathbf{1}_{M\times\widehat{G}_\infty}\sigma$ are in $vN\mathcal{A}$. Similarly, if $\sigma \in \mathcal{B}_0$ then $\mathbf{1}_{M\times\widehat{G}_1}\sigma$ and $\mathbf{1}_{M\times\widehat{G}_\infty}\sigma$ are in $vN\mathcal{B}$.*

Proof We consider $\sigma\eta$ as in Lemma 1 with a sequence of functions $\eta \in \mathcal{S}(\mathfrak{z}^*)$ satisfying $\eta(0) = 1$ and with support shrinking to $\{0\}$. We check readily that if $\sigma \in \mathcal{A}_0$ then the limit of these $\sigma\eta$ for the strong operator topology will be $\mathbf{1}_{M\times\widehat{G}_1}\sigma$ which is therefore in $vN\mathcal{A}$. It will also be the case for $\mathbf{1}_{M\times\widehat{G}_\infty}\sigma = \sigma - \mathbf{1}_{M\times\widehat{G}_1}\sigma$. The case of $\mathcal{B}_0$ follows. □

4.2 *The Lowering and Raising Operators Associated with $H(\lambda)$*

4.2.1 Preliminaries

Before proving several useful identities, we introduce some notations. If π_1, π_2 are two representations of $\mathfrak{g}$, and $A : \mathfrak{v} \to \mathfrak{v}$ is a linear morphism, then we set

$$(A\pi_1(V)) \cdot \pi_2(V) = \sum_{j,k} A_{j,k}\pi_1(V_k) \otimes \pi_2(V_j) \in \mathcal{H}_{\pi_1} \otimes \mathcal{H}_{\pi_2}$$

where $(A_{j,k})$ is the matrix representing A in the orthonormal basis (V_j). We can check that this is independent of the orthonormal basis (V_j). If the context is clear, we may allow ourselves to omit the notation for the tensor product $\otimes$ and may swap the order in the tensor product.

With $A = \mathrm{id}_\mathfrak{v}$, π_1 being the regular representation of $\mathfrak{g}$ on $L^2(M)$ and $\pi_2 = \pi \in \widehat{G}$, this yields the super-operator $V \cdot \pi(V)$ acting on $\mathcal{A}_0$. If we restrict this to $M \times \widehat{G}_1$, i.e. $\pi_2 = \pi^\omega \in \widehat{G}_1$, this defines $\omega \cdot V$ acting on $C^\infty(M, \mathcal{S}(\mathfrak{v}^*)) \sim \mathcal{A}_0|_{M\times\widehat{G}_1}$.

4.2.2 Technical Computations

Here, we assume that $k = 0$ and consider $\lambda \in \Omega_0$. Following Appendix B in [15], instead of the basis P_j^λ, Q_j^λ, $1 \le j \le d$, we will use the fields

$$W_j^\lambda := \frac{1}{2}(P_j^\lambda - iQ_j^\lambda) \quad \text{and} \quad \overline{W}_j^\lambda := \frac{1}{2}(P_j^\lambda + iQ_j^\lambda). \tag{27}$$

Direct computations show using Eq. (7),

$$\pi(P_j^\lambda) = \sqrt{\eta_j(\lambda)}\partial_{\xi_j} \text{ and } \pi(Q_j^\lambda) = i\sqrt{\eta_j(\lambda)}\xi_j,$$

so we obtain

$$\pi^\lambda(W_j^\lambda) = \frac{\sqrt{\eta_j(\lambda)}}{2}(\partial_{\xi_j} + \xi_j) \qquad \text{and} \qquad \pi^\lambda(\overline{W}_j^\lambda) = \frac{\sqrt{\eta_j(\lambda)}}{2}(\partial_{\xi_j} - \xi_j).$$

In particular, these new fields coincide up to normalisation with the lowering and raising operators of the harmonic oscillators $(-\partial_{\xi_j}^2 + \xi_j^2)$. Consequently, the family of Hermite functions $(h_\alpha)_{\alpha\in\mathbb{N}^d}$ given by

$$h_\alpha(\xi_1, \ldots, \xi_d) = h_{\alpha_1}(\xi_1) \ldots h_{\alpha_d}(\xi_d),$$

where

$$h_n(\xi) = \frac{(-1)^n}{\sqrt{2^n n! \sqrt{\pi}}} e^{\frac{\xi^2}{2}} \frac{d}{d\xi}(e^{-\xi^2}), \quad n \in \mathbb{N},$$

is an orthonormal basis of $L^2(\mathbb{R}^d)$ that satisfies:

$$\pi^\lambda(W_j^\lambda)h_\alpha = \sqrt{\frac{\eta_j(\lambda)}{2}}\sqrt{\alpha_j}h_{\alpha-\mathbf{1}_j} \quad \pi^\lambda(\overline{W}_j^\lambda)h_\alpha = -\sqrt{\frac{\eta_j(\lambda)}{2}}\sqrt{\alpha_j+1}h_{\alpha+\mathbf{1}_j}. \tag{28}$$

Here, $\mathbf{1}_j$ denotes the multi-index with j-th coordinate 1 and 0 elsewhere. We also have extended the notation h_α to $\alpha \in \mathbb{Z}^d$ with $h_\alpha = 0$ if $\alpha \notin \mathbb{N}^d$. We then deduce easily

$$\left[\pi^\lambda(W_j^\lambda), \pi^\lambda(-\mathbb{L})\right] = 2\eta_j(\lambda)\pi^\lambda(W_j^\lambda), \tag{29}$$

$$\left[\pi^\lambda(\overline{W}_j^\lambda), \pi^\lambda(-\mathbb{L})\right] = -2\eta_j(\lambda)\pi^\lambda(\overline{W}_j^\lambda), \tag{30}$$

and that both the operators $\widehat{\mathbb{P}}_\zeta\pi(W_j^\lambda)\widehat{\mathbb{P}}_\zeta$ and $\widehat{\mathbb{P}}_\zeta\pi(\overline{W}_j^\lambda)\widehat{\mathbb{P}}_\zeta$ are zero. Consequently, we also have

$$\widehat{\mathbb{P}}_\zeta\pi(P_j^\lambda)\widehat{\mathbb{P}}_\zeta = 0 \text{ and } \widehat{\mathbb{P}}_\zeta\pi(Q_j^\lambda)\widehat{\mathbb{P}}_\zeta = 0 \tag{31}$$

for all $\lambda \in \mathfrak{z}^* \setminus \{0\}$, $j \in \{1, \cdots, d\}$, $\zeta \in \mathbb{R}$.

Following the ideas and notation from Sect. 4.2.1, we define the operator

$$T = \frac{i}{2}(B(\lambda)^{-1}V \cdot \pi^\lambda(V)), \quad \lambda \in \Omega_0,$$

acting on the space of symbols in $\mathcal{A}_0$ restricted to $M \times \Omega_0$. This may also be viewed as acting on the space of symbols in $\mathcal{A}_0$ which are supported in $M \times \Omega_0$. The properties above imply:

Lemma 4

1. *For any* $\sigma \in \mathcal{A}_0$, *we have on* $M \times \Omega_0$*:*

$$[T\sigma, \pi(-\mathbb{L})] = \pi(V) \cdot V\sigma$$

2. *For any* $\lambda \in \Omega_0$ *and* $\zeta > 0$, *using the shorthand* $\pi(\mathbb{P}_\zeta)$ *for* $\mathrm{id}_{L^2(G)} \otimes \pi(\mathbb{P}_\zeta)$, *we have*

$$\pi^\lambda(\mathbb{P}_\zeta)\Big(V \cdot \pi^\lambda(V)) \circ T\Big)\pi^\lambda(\mathbb{P}_\zeta) = \frac{1}{4}\mathbb{L} - \frac{i}{4}\sum_{j=1}^{d}(2\alpha_j + 1)[P_j^\lambda, Q_j^\lambda]\pi^\lambda(\mathbb{P}_\zeta).$$

Proof Since $P_j^\lambda = \overline{W}_j^\lambda + W_j^\lambda$, and $Q_j^\lambda = \frac{1}{i}(\overline{W}_j^\lambda - W_j^\lambda)$, we deduce for $\pi = \pi^\lambda$, $\lambda \in \Omega_0$,

$$V \cdot \pi(V) = 2\sum_{j=1}^{d}\left(W_j^\lambda \pi(\overline{W}_j^\lambda) + \overline{W}_j^\lambda \pi(W_j^\lambda)\right).$$

As $B(\lambda)Q_j^\lambda = \eta_j(\lambda)P_j^\lambda$ and $B(\lambda)P_j^\lambda = -\eta_j(\lambda)Q_j^\lambda$, we obtain

$$\begin{aligned}(B(\lambda)^{-1}V) \cdot \pi(V) &= \sum_{j}^{d}\frac{1}{\eta_j}\left(-P_j^\lambda \pi(Q_j^\lambda) + Q_j^\lambda \pi(P_j^\lambda)\right)\\ &= \frac{2}{i}\sum_{j=1}^{d}\frac{1}{\eta_j}\left(\overline{W}_j^\lambda \pi(W_j^\lambda) - W_j^\lambda \pi(\overline{W}_j^\lambda)\right).\end{aligned}$$

By (29) and (30), we check readily Part (1).

For Part (2), we may assume that $\pi^\lambda(\mathbb{P}_\zeta) \neq 0$, that is, ζ is in the spectrum of the harmonic oscillator $\pi^\lambda(\mathbb{L})$, or in other words $\zeta = \sum_j \eta_j(\lambda)(2\alpha_j + 1)$ for some $\alpha \in \mathbb{N}^d$. For any such index α and for an arbitrary vector $w_1 \in \mathcal{S}(G)$, by the computations above and (28), we see with $\pi = \pi^\lambda$:

$$\begin{aligned}&\pi(\mathbb{P}_\zeta)\big(V \cdot \pi(V)\big) \circ \big(B(\lambda)^{-1}V \cdot \pi(V)\big)w_1 \otimes h_\alpha\\ &= \frac{4}{i}\sum_{j_1, j_2}\eta_{j_2}^{-1}\pi(\mathbb{P}_\zeta)\big(W_{j_1}^\lambda \pi(\overline{W}_{j_1}^\lambda) + \overline{W}_{j_1}^\lambda \pi(W_{j_1}^\lambda)\big)\\ &\quad\big(\overline{W}_{j_2}^\lambda \pi(W_{j_2}^\lambda) - W_{j_2}^\lambda \pi(\overline{W}_{j_2}^\lambda)\big)w_1 \otimes h_\alpha\\ &= \frac{2}{i}\sum_{j}\left(\overline{W}_j^\lambda W_j^\lambda(\alpha_j + 1) - W_j^\lambda \overline{W}_j^\lambda \alpha_j\right) w_1 \otimes h_\alpha.\end{aligned}$$

We can simplify each term in the sum above with:

$$\overline{W}_j^\lambda W_j^\lambda(\alpha_j+1) - W_j^\lambda \overline{W}_j^\lambda \alpha_j = \frac{1}{4}((P_j^\lambda)^2 + (Q_j^\lambda)^2) - \frac{i}{4}(2\alpha_j+1)[P_j^\lambda, Q_j^\lambda].$$

Part (2) follows □

We recall that the maps $\lambda \mapsto \eta_j(\lambda)$, $j = 1, \dots, d$, are smooth in Λ_0. Moreover, if $\lambda_0 \in \Lambda_0$, one can choose the vectors P_j^λ, Q_j^λ, $j = 1, \dots, d$, so that they depend smoothly on λ in a neighborhood of λ_0. We then have the following result.

Lemma 5 *Let P_j^λ, Q_j^λ, $j = 1, \dots, d$, be smooth eigenvectors in an open subset U of U. Then we have*

$$[P_j^\lambda, Q_j^\lambda] = \nabla_\lambda \eta_j(\lambda) \in \mathfrak{z}, \qquad j = 1, \dots, d, \quad \lambda \in \Lambda_0.$$

Proof The differentiation of the equality $B(\lambda) Q_j^\lambda = \eta_j(\lambda) P_j^\lambda$ with respect to λ gives

$$\forall \lambda' \in \mathfrak{z}^* \qquad B(\lambda') Q_j^\lambda + B(\lambda)\, \lambda' \cdot \nabla_\lambda Q_j^\lambda = \lambda' \cdot \nabla_\lambda \eta_j(\lambda) P_j^\lambda + \eta_j(\lambda)\, \lambda' \cdot \nabla_\lambda P_j^\lambda.$$

Taking the scalar product with P_j^λ and using $B(\lambda)^t = -B(\lambda)$ with $-B(\lambda) P_j^\lambda = \eta_j(\lambda) Q_j^\lambda$, we obtain:

$$\begin{aligned}(B(\lambda') Q_j^\lambda, P_j^\lambda) + \eta_j(\lambda)(\lambda' \cdot \nabla_\lambda Q_j^\lambda, Q_j^\lambda) &= \lambda' \cdot \nabla_\lambda \eta_j(\lambda)(P_j^\lambda, P_j^\lambda) \\ &\quad + \eta_j(\lambda)(\lambda' \cdot \nabla_\lambda P_j^\lambda, P_j^\lambda).\end{aligned}$$

Now, $(Q_j^\lambda, Q_j^\lambda) = 1 = (P_j^\lambda, P_j^\lambda)$. Differentiating this with respect to λ yields

$$(\lambda' \cdot \nabla_\lambda Q_j^\lambda, Q_j^\lambda) = 0 = (\lambda' \cdot \nabla_\lambda P_j^\lambda, P_j^\lambda),$$

and we have for all $\lambda' \in \mathfrak{z}^*$

$$(B(\lambda') Q_j^\lambda, P_j^\lambda) = \lambda' \cdot \nabla_\lambda \eta_j(\lambda).$$

Since the left-hand side is equal to $\lambda'([Q_j^\lambda, P_j^\lambda])$ by definition of $B(\lambda')$, the conclusion follows. □

The two lemmata above imply readily:

Corollary 2 *Using $\zeta = \zeta(\alpha, \lambda) = \sum_{j=1}^d \eta_j(\lambda)(2\alpha_j + 1)$, we deduce that for the choice of orthonormal basis of Lemma 5, we have*

$$\pi^\lambda(\mathbb{P}_\zeta)\Big(V \cdot \pi^\lambda(V)) \circ T\Big)\pi^\lambda(\mathbb{P}_\zeta) = \frac{1}{4}\mathbb{L} - \frac{i}{4}\nabla_\lambda \zeta.$$

Acknowledgments The authors acknowledges the support of the Leverhulme Trust via Research Project Grant 2020-037. The first author thanks Cyril Letrouit for inspiring discussions.

References

1. Bahouri, H., Gérard, P., Xu, C.-J.: Espaces de Besov et estimations de Strichartz généralisées sur le groupe de Heisenberg. J. d'Analyse Math. **82**, 93–118 (2000)
2. Bahouri, H., Fermanian Kammerer, C., Gallagher, I.: Phase space analysis on the Heisenberg group. Astérisque **345** 134 (2012)
3. Bahouri, H., Fermanian Kammerer, C., Gallagher, I.: Dispersive estimates for the Schrödinger operator on step 2 stratified Lie Groups, Anal. PDEs **9**(3), 545–574 (2016)
4. Baudoin, F., Grong, E., Molino, G.: H-type foliations (2018). arXiv:1812.02563
5. Beals, R.: Weighted distribution spaces and pseudodifferential operators. J. d'Analyse Math. **39**, 130–187 (1981)
6. Beals, R., Greiner, P.: Calculus on Heisenberg Manifolds. Annals of Mathematics Studies, vol. 119. Princeton University Press, Princeton (1988)
7. Brown, I.: Dual topology of a nilpotent Lie group. Ann. Sci. École Norm. Sup. **6**, 407–411 (1973)
8. Burq, N., Sun, C.: Time optimal observability for Grushin Schrödinger equation (2019). arXiv:1910.03691
9. Christ, M., Geller, D., Głowacki, P., Polin, L.: Pseudodifferential operators on groups with dilations. Duke Math. J. **68**(1), 31–65 (1992)
10. Colin de Verdière, Y., Hillairet, L., Trélat, E.: Spectral asymptotics for sub-Riemannian Laplacians. I: quantum ergodicity and quantum limits in the 3D contact case. Duke Math. J. **167**(1), 109–174 (2018)
11. Corwin, L.J., Greenleaf, F.P.: Representations of Nilpotent Lie Groups and Their Applications, Part 1: Basic Theory and Examples. Cambridge Studies in Advanced Mathematics, vol. 18. Cambridge University Press, Cambridge (1990)
12. Fefferman, C., Phong, D.: The uncertainty principle and sharp Gårding inequalities. Commun. Pure Appl. Math. **34**(3), 285–331 (1981)
13. Fermanian Kammerer, C., Fischer, V.: Semi-classical analysis on H-type groups. Sci. China Math. **62**(6), 1057–1086 (2019)
14. Fermanian Kammerer, C., Fischer, V.: Defect measures on graded Lie groups. Ann. Sc. Norm. Super. Pisa Cl. Sci. **21**(5), 207–291 (2020)
15. Fermanian Kammerer, C., Fischer, V.: Quantum evolution and sub-laplacian operators on groups of Heisenberg type. J. Spectr. Theory **11**(3), 1313–1367 (2021)
16. Fermanian Kammerer, C., Letrouit, C.: Observability and controllability for the Schroedinger equation on quotients of groups of Heisenberg type. J. l'École Polytechn. Math. **8**, 1459–1513 (2021)
17. Fermanian Kammerer, C., Fischer, V., Flynn, S.: Geometric invariance of the semi-classical calculus on nilpotent graded Lie groups (2021). arXiv:2112.11509
18. Fischer, V.: Asymptotics and zeta functions on compact nilmanifolds. J. Math. Pures Appl. **160**, 1–28 (2022)
19. Fischer, V., Ruzhansky, M.: Quantization on Nilpotent Lie Groups. Progress in Mathematics, vol. 314. Birkhäuser, Basel (2016)
20. Folland, G.B., Stein, E.: Hardy Spaces on Homogeneous Groups. Mathematical Notes, vol. 28. Princeton University Press, Princeton (1982)
21. Gérard, P.: Mesures semi-classiques et ondes de Bloch. In: Séminaire sur les Équations aux Dérivées Partielles, 1990–1991. Exp. No. XVI, 19. École Polytech. Palaiseau, France (1991)
22. Gérard, P., Leichtnam, E.: Ergodic properties of eigenfunctions for the Dirichlet problem. Duke Math. J. **71**(2), 559–607 (1993)

23. Gérard, P., Markowich, P.A., Mauser, N.J., Poupaud, F.: Homogenization limits and Wigner transforms. Commun. Pure Appl. Math. **50**(4), 323–379 (1997)
24. Helffer, B., Martinez, A., Robert, D.: Ergodicité et limite semi-classique. Commun. Math. Phys. **109**(2), 313–326 (1987)
25. Hörmander, L.: Hypoelliptic second order differential equations. Acta Math. **119**, 147–171 (1967)
26. Hulanicki, A.: A functional calculus for Rockland operators on nilpotent Lie groups. Stud. Math. **78**(3), 253–266 (1984)
27. Müller, D., Ricci, F.: Solvability for a class of doubly characteristic differential operators on two-step nilpotent groups. Ann. Math. **142**, 1–49 (1995)
28. Rothschild, L.P., Stein, E.M.: Hypoelliptic differential operators and nilpotent groups. Acta Math. **137**(3–4), 247–320 (1976)
29. Savale, N.: Spectrum and abnormals in sub-Riemannian geometry: the 4D quasi-contact case (2019). arXiv:1909.00409
30. Taylor, M.E.: Noncommutative microlocal analysis. I. Mem. Amer. Math. Soc. **52**(313), iv+182 (1984). Revised version accessible at http://math.unc.edu/Faculty/met/ncmlms.pdf
31. Van Erp, E.: The Atiyah-Singer index formula for subelliptic operators on contact manifolds. Part I. Ann. Math. **171**, 1647–1681 (2010)
32. Van Erp, E., Yuncken, R.: On the tangent groupoid of a filtered manifold. Bull. L.M.S. **49**(6), 1000–1012 (2017)
33. Van Erp, E., Yuncken, R.: A groupoid approach to pseudodifferential operators. Crelle's J. **759**, 151–182 (2019)

Part II
Operator Theory

Spectral Asymptotics for Two-Dimensional Dirac Operators in Thin Waveguides

William Borrelli, Nour Kerraoui, and Thomas Ourmières-Bonafos

1 Introduction and Main Result

1.1 Introduction

In this article, we continue the study of spectral properties of relativistic quantum waveguides, initiated in [5]. In particular, as explained below, we focus on the existence of discrete eigenvalues in the spectral gap in the thin-width regime.

The study of non-relativistic quantum waveguides started with the pioneering paper [11] (see also [8, 13, 15] for further improvements), where it was demonstrated that the quantum free Hamiltonian on a waveguide given by the Dirichlet Laplacian possesses discrete eigenvalues when the base curve is not a straight line. Roughly speaking, the corresponding particle gets trapped in any non-trivially curved quantum waveguide. Notice that this is in sharp contrast with the classical case, considering particles following Newton's law with regular reflection at the boundary. Indeed, except for a set of initial conditions of zero measure in the phase space, particles will eventually leave any bounded region in finite time. The existence and properties of the geometrically induced bound states have attracted a lot of attention in the last decades, and this research field is still very active. We refer the reader to the monograph [10] for a comprehensive discussion of the subject.

The study of the relativistic counterpart of this Hamiltonian started very recently in the two-dimensional case, in [5], considering the Dirac operator on a tubular

W. Borrelli (✉)
Dipartimento di Matematica, Politecnico di Milano, Milano, Italy
e-mail: william.borrelli@polimi.it

N. Kerraoui · T. Ourmières-Bonafos
Aix-Marseille Université, CNRS, Centrale Marseille, Marseille, France
e-mail: nour-el-houda.kerraoui@etu.univ-amu.fr; thomas.ourmieres-bonafos@univ-amu.fr

M. Correggi, M. Falconi (eds.), *Quantum Mathematics I*,
Springer INdAM Series 57, https://doi.org/10.1007/978-981-99-5894-8_5

neighborhood of a curve with *infinite mass* boundary conditions at the boundary. Generally speaking, the mathematical study of such operator on domains started recently [2–4, 16], motivated by models of hadrons confinement from high-energy physics [6] or by the description of graphene samples [1]. We also mention the work [9], where spectral properties of Dirac operators on tubes with *ziz-zag* type boundary conditions are considered.

Notice, however, that boundary conditions for Dirac operators on manifolds with boundary had been already considered previously in the geometry literature, see e.g. [12, 17] and references therein.

In [5, Thm.2], under suitable assumptions, it has been proved that the Dirac operator with infinite mass boundary conditions (see (1)), posed in the tubular neighborhood of a planar curve, is self-adjoint and its essential spectrum has been identified. Thus, a natural question is to understand the interplay between the geometry and relativistic setting. In particular, we focus on the existence of geometrically induced bound states in the thin-waveguide regime. For the Dirichlet Laplacian, it is known that in this regime, up to a renormalization factor, the splitting of the eigenvalues is given by an effective operator and this operator is the one-dimensional Schrödinger operator with the attractive potential given by

$$-\frac{d^2}{ds^2} - \frac{\kappa^2}{4},$$

where κ is the curvature of the underlying curve Γ, and $s \in \mathbb{R}$ is the arc-length parameter. For this reason, one speaks of geometrically induced bound states related to the non-trivial geometry of the curve/waveguide (see, e.g. [8, 10]).

On the other hand, in [5, Thm.4] it is proved that the Dirac operator with infinite mass boundary conditions (see (1)), after a suitable choice of renormalization, converges in the norm-resolvent sense to a one-dimensional *free* effective Dirac operator

$$-i\sigma_1\partial_s + \frac{2}{\pi}m\sigma_3$$

whose spectrum is purely absolutely continuous. Here σ_1 and σ_3 are the first and third Pauli matrices, respectively (as in (2)). Then in this case, the effective operator does not bear any geometrical information, and geometric effects are expected to appear at the next order in the asymptotic expansion in the thin-waveguide regime. The purpose of this paper is precisely to investigate this problem and in our main result Theorem 1 we provide an asymptotic expansion (5) for the eigenvalues, which provides both the splitting and exhibits an effective operator involving the geometry of the underlying curve. This is achieved using min-max techniques, working on the square of the operator (1) and relating its eigenvalues to those of a reference operator, defined using its quadratic form (6).

1.2 Main Result

Let $\gamma : \mathbb{R} \to \mathbb{R}^2$ be an arc-length parametrization of a C^4-planar curve Γ. For $s \in \mathbb{R}$, we define the normal $\nu(s)$ at the point $\gamma(s) \in \Gamma$ such that $(\gamma'(s), \nu(s))$ is an orthonormal basis of $\mathbb{R}^2$. We define the curvature $\kappa(s)$ of the curve Γ at the point $\gamma(s)$ by

$$\kappa(s) := \gamma''(s) \cdot \nu(s).$$

Remark that under the smoothness assumption on γ, $\kappa \in C^2(\mathbb{R})$, and all along this paper, we assume the following hypothesis:

(A) $\lim_{s\to\pm\infty} \kappa(s) = 0$,
(B) $\kappa', \kappa'' \in L^\infty(\mathbb{R})$.

Define the strip $\mathsf{Str} := \mathbb{R} \times (-1, 1)$ and $\varepsilon_0 := \|\kappa\|^{-1}_{L^\infty(\mathbb{R})}$. For $\varepsilon > 0$, we consider the map

$$\Phi_\varepsilon : \begin{cases} \mathsf{Str} & \to \mathbb{R}^2 \\ (s,t) & \mapsto \gamma(s) + \varepsilon t \nu(s). \end{cases}$$

and define the tubular neighborhood of Γ

$$\Omega_\varepsilon := \Phi_\varepsilon(\mathsf{Str}).$$

Thus $s \in \mathbb{R}$ and $t \in (-1, 1)$ are the arc-length parameter of the curve and the transverse coordinate with respect to the curve, respectively.

In order to guarantee that Φ_ε is a C^3-diffeomorphism from Str to Ω_ε we will always assume that

(C) the map Φ_ε is injective.

We are interested in the spectrum of the Dirac operator with infinite mass boundary conditions posed in the domain Ω_ε. Let $\mathcal{D}_\Gamma(\varepsilon)$ denote this operator it writes

$$\mathcal{D}_\Gamma(\varepsilon) := -i\sigma_1\partial_1 - i\sigma_2\partial_2 + m\sigma_3$$
$$\mathrm{dom}\mathcal{D}_\Gamma(\varepsilon) := \{u \in H^1(\Omega_\varepsilon, \mathbb{C}^2) : -i\sigma_3\sigma \cdot \nu_\varepsilon = u \text{ on } \partial\Omega_\varepsilon\}, \quad (1)$$

where ν_ε is the outward pointing normal vector field on $\partial\Omega_\varepsilon$ and $m \geq 0$ is a fixed parameter. Here $\sigma = (\sigma_1, \sigma_2)$ and $\sigma_1, \sigma_2, \sigma_3$ are the Pauli matrices

$$\sigma_1 := \begin{pmatrix} 0 & 1 \\ 1 & 0 \end{pmatrix}, \quad \sigma_2 := \begin{pmatrix} 0 & -i \\ i & 0 \end{pmatrix}, \quad \sigma_3 := \begin{pmatrix} 1 & 0 \\ 0 & -1 \end{pmatrix}. \quad (2)$$

Notice that in (1), $\sigma_3\sigma \cdot \nu_\varepsilon = \sigma_3(\sigma_1\nu_\varepsilon^1 + \sigma_2\nu_\varepsilon^2)$, where $\nu_\varepsilon = (\nu_\varepsilon^1, \nu_\varepsilon^2)$.

Thanks to [5, Thm. 2], we know that for ε small enough, $\mathcal{D}_\Gamma(\varepsilon)$ is self-adjoint, that its spectrum is symmetric with respect to 0 and that is essential spectrum is given by

$$\mathrm{Sp}_{ess}(\mathcal{D}_\Gamma(\varepsilon)) = \Big(-\infty, -\frac{E_1(m\varepsilon)}{\varepsilon}\Big] \cup \Big[\frac{E_1(m\varepsilon)}{\varepsilon}, +\infty\Big)$$

where for $\rho \geq 0$, $E_1(\rho) := \sqrt{\rho^2 + k_1(\rho)^2}$ and where $k_1(\rho)$ is defined as the unique root of

$$\rho \sin(2k) + k\cos(2k) = 0$$

lying in $[\frac{\pi}{4}, \frac{\pi}{2})$.

Remark 1 Notice that there is a slight change in notation compared to [5, Thm. 2]. Indeed, there $k_1(\cdot)$ is denoted by $E_1(\cdot)$ and then the thresholds of the essential spectrum are $\pm\sqrt{\varepsilon^{-2}E_1(m\varepsilon) + m^2}$.

Our aim is to investigate the possible existence of a discrete spectrum of $\mathcal{D}_\Gamma(\varepsilon)$ in the thin waveguide regime $\varepsilon \to 0$. To do so, we use the min-max principle for the operator $\mathcal{D}_\Gamma(\varepsilon)^2$, recalled below (see [7, Thm. 4.5.1 & 4.5.2]).

Definition 1 Let Q be a closed, lower semi-bounded, and densely defined quadratic form with domain $\mathrm{dom}Q$ in a Hilbert space $\mathcal{H}$. For $n \in \mathbb{N}$, the n-th min-max value of Q is defined as

$$\mu_n(Q) := \inf_{\substack{F \subset \mathrm{dom}Q \\ \dim F = n}} \sup_{u \in F \setminus \{0\}} \frac{Q[u]}{\|u\|^2_{\mathcal{H}}}. \tag{3}$$

If A is the unique self-adjoint operator associated with the sesquilinear form derived from Q *via* Kato's first representation theorem (see [14, Ch. VI, Thm. 2.1]) we shall refer to (3) as the n-th min-max level of A and we note $\mu_n(A) = \mu_n(Q)$.

Now, we can recall the min-max principle.

Proposition 1 *Let Q be a closed, lower semi-bounded densely defined quadratic form with domain* $\mathrm{dom}Q$ *in a Hilbert space $\mathcal{H}$. Let A be the unique self-adjoint operator associated with Q. Then, for $n \in \mathbb{N}$, the following alternative holds true:*

1. *either $\mu_n(A) < \inf \mathrm{Sp}_{ess}(A)$ and $\mu_n(A)$ is the n-th eigenvalue of A (counted with multiplicities),*
2. *or $\mu_n(A) = \inf \mathrm{Sp}_{ess}(A)$ and for all $k \geq n$ there holds $\mu_k(A) = \inf \mathrm{Sp}_{ess}(A)$.*

In order to state the main result of this paper, we introduce the one dimensional Schrödinger operator defined through its quadratic form as

$$q_e[f] = \int_{\mathbb{R}} (|f'|^2 - \frac{\kappa^2}{\pi^2}|f|^2)ds, \quad \mathrm{dom} q_e = H^1(\mathbb{R}) \tag{4}$$

and set $J := \sharp\{\mu_n(q_e) < 0\}$.

Theorem 1 *If $J \geq 1$ then there exists $\varepsilon_1 > 0$ such that for all $\varepsilon \in (0, \varepsilon_1)$ there holds*

$$\mathrm{Sp}_{dis}(\mathcal{D}_\Gamma(\varepsilon)) \neq \emptyset.$$

Moreover, if $\lambda_j(\mathcal{D}_\Gamma(\varepsilon))$ denotes the j-th positive discrete eigenvalue of $\mathcal{D}_\Gamma(\varepsilon)$ counted with multiplicity then for all $j \in \{1, \dots, J\}$, there holds

$$\lambda_j(\mathcal{D}_\Gamma(\varepsilon)) = \frac{E_1(m\varepsilon)}{\varepsilon} + \frac{2}{\pi}\mu_j(q_e)\varepsilon + O(\varepsilon^2). \tag{5}$$

Remark 2 A situation in which $J \geq 1$ is when $\kappa \in L^2(\mathbb{R}) \setminus \{0\}$. Indeed, let $\theta > 0$ and consider the map defined for $s \in \mathbb{R}$ by

$$\psi_\theta(s) := \begin{cases} \theta^{-1}(s + 2\theta) & \text{if} \quad s \in [-2\theta, -\theta), \\ 1 & \text{if} \quad s \in [-\theta, \theta], \\ -\theta^{-1}(s - 2\theta) & \text{if} \quad s \in (\theta, 2\theta], \\ 0 & \text{otherwise.} \end{cases}$$

One remarks that $\psi_\theta \in H^1(\mathbb{R})$, verifies $\|\psi_\theta\|_{L^\infty(\mathbb{R})} \leq 1$ and that

$$q_e[\psi_\theta] = \frac{2}{\theta} - \frac{1}{\pi^2}\int_{\mathbb{R}} \kappa^2|\psi_\theta|^2 ds \leq \frac{2}{\theta} - \frac{1}{\pi^2}\int_{-\theta}^{\theta} \kappa^2\, ds\,.$$

Hence, since $\kappa \in L^2(\mathbb{R})$, choosing θ sufficiently large, we get $q_e[\psi_\theta] < 0$ and the min-max principle (Prop. 1) gives $\mu_1(q_e) < 0$.

As already remarked, Theorem 1 proves that as long as the curvature creates bound states for the effective operator given by the quadratic form q_e, it also creates bound states for the operator $\mathcal{D}_\Gamma(\varepsilon)$. Note that in Theorem 1 there is no term of order 0 and the bound states are at a distance of order ε from the essential spectrum. This differs from the non-relativistic counter part of this problem studied in [8, Thm. 5.1.], where the splitting of eigenvalues appears at constant order. However, the result is consistent with [5, Thm. 4] where the authors prove that up to a unitary map, the operator (1), suitably renormalized, behaves at constant order as a massive (free) Dirac operator on the real line with effective mass $\frac{2}{\pi}m$. The spectrum of this operator being purely absolutely continuous, this is consistent with (5), where the constant term arising in the expansion of $\frac{E_1(m\varepsilon)}{\varepsilon^2}$ is precisely given by such effective mass $\frac{2}{\pi}m$.

2 Preliminaries

The purpose of Sects. 2.1 and 2.2 is to gather several results on one dimensional operators, which play an important role in the proof of Theorem 1. This proof is based on the min-max principle applied to the quadratic form of the square of $\mathcal{D}_\Gamma(\varepsilon)$, exploiting suitable lower and upper bounds for it, given in Sect. 2.3.

2.1 The Effective Operator

In what follows, we deal with the following one dimensional operator, defined through its quadratic form by

$$\tilde{q}_e[f] = \int_{\mathbb{R}} \Big(|f' - i\frac{\kappa}{\pi}\sigma_3 f|^2 - \frac{\kappa^2}{\pi^2}|f|^2 \Big) ds, \quad \mathrm{dom}\tilde{q}_e = H^1(\mathbb{R}, \mathbb{C}^2). \tag{6}$$

It turns out that the spectral properties of the operator associated with $\tilde{q}_e$ are related to the one of the operator associated with q_e defined in (4). Notice that the former is defined for vector valued functions, while the latter is defined for scalar ones.

Proposition 2 *There exists a unitary map $U : L^2(\mathbb{R}, \mathbb{C}^2) \to L^2(\mathbb{R}, \mathbb{C}^2)$ such that for all $f = (f^+, f^-)^\top \in \mathrm{dom}(q_e \oplus q_e)$*

$$(q_e \oplus q_e)[f] = q_e[f^+] + q_e[f^-] = \tilde{q}_e[Uf].$$

Proof Let us consider the following gauge transform

$$U : L^2(\mathbb{R}^2, \mathbb{C}^2) \to L^2(\mathbb{R}^2, \mathbb{C}^2), \quad (Uf) = e^{i\frac{\rho}{2}\sigma_3} f,$$

where for all $s \in \mathbb{R}$ we have set $\rho(s) = \int_0^s \kappa(\eta)d\eta$. Remark that there holds

$$|(Uf)' - i\frac{\kappa}{2}\sigma_3(Uf)| = |i\frac{\kappa}{2}\sigma_3 e^{i\frac{\rho}{2}\sigma_3} f + e^{i\frac{\rho}{2}\sigma_3} f' - i\frac{\kappa}{2}\sigma_3 e^{i\frac{\rho}{2}\sigma_3} f| = |e^{i\frac{\rho}{2}\sigma_3} f'| = |f'|$$

because for all $s \in \mathbb{R}$, $e^{i\frac{\rho(s)}{2}\sigma_3}$ is a unitary matrix. For the same reason, there holds $|(Uf)| = |f|$ and this yields

$$(q_e \oplus q_e)[f] = \tilde{q}_e[f].$$

2.2 The Transverse Dirac Operator

When proving Theorem 1, we need to use some spectral properties of a one dimensional operator. It is defined for $m \geq 0$ by

$$\mathcal{T}(m) := -i\sigma_2 \frac{d}{dt} + m\sigma_3,$$
$$\mathrm{dom}\mathcal{T}(m) := \{u = (u_1, u_2)^\top \in H^1((-1,1), \mathbb{C}^2) : u_2(\pm 1) = \mp u_1(\pm 1)\}.$$

The following proposition holds.

Proposition 3 *Let $m \geq 0$. The operator $\mathcal{T}(m)$ is self-adjoint and has compact resolvent. Moreover, the following holds:*

1. *for all $u \in \mathrm{dom}\mathcal{T}(m)$ there holds*

$$\|\mathcal{T}(m)u\|^2_{L^2(-1,1)} = \|u'\|^2_{L^2(-1,1)} + m^2\|u\|^2_{L^2(-1,1)} + m(|u(1)|^2 + |u(-1)|^2), \tag{7}$$

2. $Sp\big(\mathcal{T}(m)\big) \cap [-m, m] = \emptyset$,
3. *for all $p \geq 1$, define $k_p(m)$ as the only root lying in $[(2p-1)\frac{\pi}{4}, p\frac{\pi}{2}]$ of*

$$m \sin(2k) + k \cos(2k) = 0,$$

now if one sets $E_p(m) = \sqrt{m^2 + k_p(m)^2}$ there holds $Sp\big(\mathcal{T}(m)\big) = \bigcup_{p\geq 1} \{\pm E_p(m)\}$,

4. *there holds*

$$k_1(m) = \frac{\pi}{4} + \frac{2}{\pi}m - \frac{16}{\pi^3}m^2 + O(m^3),$$

5. *for $p \geq 1$, a normalized eigenfunction associated with $E_p(m)$ is given by*

$$\varphi_p^{m,+}(t) := N_{m,p}\Big(k_p \cos(k_p(t+1)) \begin{pmatrix}1\\1\end{pmatrix} + \sin(k_p(t+1)) \begin{pmatrix}E_p+m\\-(E_p-m)\end{pmatrix}\Big)$$

where $N_{m,p}$ is a normalization constant. We consider $\varphi_p^{m,-} := \sigma_1\varphi_p^{m,+}$; a normalized eigenfunction associated with $-E_p(m)$ and if one sets $\varphi_p^{\pm} := \varphi_p^{0,\pm}$ there holds

$$\varphi_1^{m,\pm} = \varphi_1^{\pm} + O(m),$$

where the remainder is understood in the L^∞-norm on $(-1,1)$.

Proof The proof of Points (1)–(3) can be found, *e.g.*, in [5, Proposition 10]. Point (4) relies on the fact that

$$m\sin(2k_1(m)) + k_1(m)\cos(2k_1(m)) = 0. \tag{8}$$

Hence, as k_1 is defined near $m = 0$ by this smooth implicit equation, k_1 is smooth near $m = 0$ and there holds

$$k_1(m) = k_1(0) + k_1'(0)m + \frac{1}{2}k_1''(0)m^2 + O(m^3), \quad m \to 0.$$

One can compute thanks to (8) that

$$k_1(0) = \frac{\pi}{4}, \quad k_1'(0) = \frac{2}{\pi}, \quad k_1''(0) = -\frac{32}{\pi^3},$$

which yields Point (4). To prove Point (5), again by Borrelli et al. [5, Proposition 10], any eigenfunction associated with $E_p(m)$ is of the form

$$u_p^m(t) = \cos(k_p(m)(t+1)) \begin{pmatrix}\alpha\\ \frac{k_p(m)}{E_p(m)+m}\beta\end{pmatrix} + \sin(k_p(m)(t+1)) \begin{pmatrix}\beta\\ -\frac{k_p(m)}{E_p(m)+m}\alpha\end{pmatrix},$$

for some constants $\alpha, \beta \in \mathbb{C}$. The boundary condition at $t = -1$ gives $\alpha = \frac{k_p(m)}{E_p(m)+m}\beta$ so that, choosing $\beta = (E_p(m)+m)$:

$$u_p^m(t) = k_p(m)\cos(k_p(m)(t+1)) \begin{pmatrix}1\\1\end{pmatrix} + \sin(k_p(m)(t+1)) \begin{pmatrix}E_p(m)+m\\-(E_p(m)-m)\end{pmatrix}.$$

Hence, we take $N_{m,p} := \|u_p^m\|^{-1}_{L^2((-1,1),\mathbb{C}^2)}$ and remark that $\varphi_p^{m,+} := N_{m,p}u_p^m$. Note that

$$\|u_1^m\|^2_{L^2((-1,1),\mathbb{C}^2)} = 2k_1(m)^2\big(1 + \frac{\sin(4k_1(m))}{4k_1(m)}\big) + 2(E_1(m)^2 + m^2)\big(1 - \frac{\sin(4k_1(m))}{4k_1(m)}\big) + 2mk_1(m)\big(1 - \frac{\cos(4k_1(m))}{2k_1(m)}\big)$$

which gives using Point (4)

$$N_{1,m} = \frac{1}{2k_1(m)} + O(m), \quad m \to 0. \tag{9}$$

Now, remark that for all $t \in (-1, 1)$ there holds

$$|\varphi_1^{m,+}(t) - \varphi_1^+(t)| \le 2\Big|k_1(m)N_{m,p} - \frac{1}{2}\Big| + 2\Big|E_1(m)N_{m,p} - \frac{1}{2}\Big| + 2m.$$

which gives Point (5) for $\varphi_1^{m,+}$ thanks to (9). For $\varphi_1^{m,-}$ one only has to note that for all $t \in (-1, 1)$ there holds $|\varphi_1^{m,-}(t) - \varphi_1^-(t)| = |\sigma_1(\varphi_1^{m,-}(t) - \varphi_1^-(t))| = |\varphi_1^{m,+}(t) - \varphi_1^+(t)|$.

Remark 3 The explicit expression of the functions $\varphi_1^\pm$ is of crucial importance in what follows. They are defined, for all $t \in (-1, 1)$, as

$$\varphi_1^\pm(t) = \frac{1}{2}\cos(\frac{\pi}{4}(t+1))\begin{pmatrix}1\\1\end{pmatrix} \pm \frac{1}{2}\sin(\frac{\pi}{4}(t+1))\begin{pmatrix}1\\-1\end{pmatrix}. \tag{10}$$

2.3 *The Quadratic Form of the Square*

By Borrelli et al. [5, Prop. 3] we know that the operator $\mathcal{D}_\Gamma(\varepsilon)$ is unitarily equivalent to

$$\mathcal{E}_\Gamma(\varepsilon) := \frac{1}{1-\varepsilon t\kappa}(-i\sigma)\partial_s + \frac{1}{\varepsilon}(-i\sigma_2)\partial_t + \frac{\varepsilon t\kappa'}{2(1-\varepsilon t\kappa)^2}(-i\sigma_1) + m\sigma_3,$$

$$\mathrm{dom}\mathcal{E}_\Gamma(\varepsilon) := \{u = (u_1, u_2)^\top \in H^1(\mathsf{Str}, \mathbb{C}^2) : u_2(\cdot, \pm 1) = \mp u_1(\cdot, \pm 1)\}$$

and that the quadratic form of its square is given, for every $u \in \mathrm{dom}\mathcal{E}_\Gamma(\varepsilon)$, by

$$\begin{aligned}\|\mathcal{E}_\Gamma(\varepsilon)u\|^2_{L^2(\mathsf{Str},\mathbb{C}^2)} = &\int_{\mathsf{Str}} \frac{1}{(1-\varepsilon t\kappa)^2}|\partial_s u - i\frac{\kappa}{2}\sigma_3 u|^2 ds dt + \frac{1}{\varepsilon^2}\int_{\mathsf{Str}} |\partial_t u|^2 ds dt \\ &+ \frac{m}{\varepsilon}\int_{\mathbb{R}}\left(|u(s,1)|^2 + |u(s,-1)|^2\right) ds + m^2\|u\|^2_{L^2(\mathsf{Str},\mathbb{C}^2)} \\ &- \int_{\mathsf{Str}} \frac{\kappa^2}{4(1-\varepsilon t\kappa)^2}|u|^2 ds dt - \frac{5}{4}\int_{\mathsf{Str}} \frac{(\varepsilon t\kappa')^2}{(1-\varepsilon t\kappa)^4}|u|^2 ds dt \\ &- \frac{1}{2}\int_{\mathsf{Str}} \frac{\varepsilon t\kappa''}{(1-\varepsilon t\kappa)^3}|u|^2 ds dt\,. \end{aligned} \tag{11}$$

The main result of this section reads as follows.

Proposition 4 *There exists $\varepsilon' > 0$ and $c > 0$ such that for all $\varepsilon \in (0, \varepsilon')$ and all $u \in \mathrm{dom}\mathcal{E}_\Gamma(\varepsilon, m)$ there holds*

$$a_-[u] \le \|\mathcal{E}_\Gamma(\varepsilon)u\|^2_{L^2(\mathsf{Str},\mathbb{C}^2)} \le a_+[u], \tag{12}$$

where we have introduced the quadratic forms $a_\pm$ defined by

$$\begin{aligned} a_\pm[u] := (1 \pm c\varepsilon)&\int_{\mathsf{Str}}\left(|\partial_s u - i\frac{\kappa}{2}\sigma_3 u|^2 - \frac{\kappa^2}{4}|u|^2\right) ds dt \\ &+ \frac{1}{\varepsilon^2}\int_{\mathbb{R}}\left(\|\mathcal{T}(m\varepsilon)u\|^2_{L^2((-1,1),\mathbb{C}^2)}\right) ds \pm c\varepsilon\|u\|^2, \\ &\qquad \mathrm{dom}\, a_\pm := \mathrm{dom}\mathcal{E}_\Gamma(\varepsilon). \end{aligned}$$

The proof of Proposition 4 is straightforward, taking into account (11) and the fact that $\kappa, \kappa', \kappa'' \in L^\infty(\mathbb{R})$. To this aim, observe that $(1 - \varepsilon t\kappa(s))^{-1} = 1 + \mathcal{O}(\varepsilon)$, uniformly in $(s, t) \in \mathbb{R} \times [-1, 1]$, and recall (7). By comparison, the bounds (12) will be used to get the desired spectral asymptotics.

3 Proof of the Main Result

In Sect. 3.1 we give an upper bound on the j-th min-max level of $\mathcal{D}_\Gamma(\varepsilon)^2$, while a lower bound is obtained in Sect. 3.2. Combining these results, Theorem 1 is proved in Sect. 3.3.

3.1 An Upper Bound

The goal of this paragraph is to prove the following Proposition.

Proposition 5 *Let $j \in \mathbb{N}$. There exists $\varepsilon_1 > 0$ and $c > 0$ such that for all $\varepsilon \in (0, \varepsilon_1)$ there holds*

$$\mu_j(\mathcal{D}_\Gamma(\varepsilon)^2) \le \frac{E_1(m\varepsilon)^2}{\varepsilon^2} + \mu_j(q_e \oplus q_e) + c\varepsilon. \tag{13}$$

Proof Let $f = (f^+, f^-) \in H^1(\mathbb{R}, \mathbb{C}^2)$ and set $u = f^+\varphi_1^{m\varepsilon,+} + f^-\varphi_1^{m\varepsilon,-}$. By construction $u \in \mathrm{dom}\mathcal{E}_\Gamma(\varepsilon)$ and for ε small enough there holds

$$a_+[u] = (1+c\varepsilon)\int_{\mathsf{Str}}\Big(|\partial_s u - i\frac{\kappa}{2}\sigma_3 u|^2\Big)dsdt - (1+c\varepsilon)\int_{\mathbb{R}}\frac{\kappa^2}{4}|f|^2 ds$$
$$+\frac{E_1(m\varepsilon)^2}{\varepsilon}\|f\|^2_{L^2(\mathbb{R})} + c\varepsilon\|f\|^2_{L^2(\mathbb{R})}.$$

Now, one remarks that

$$\int_{\mathsf{Str}}\Big(|\partial_s u - i\frac{\kappa}{2}\sigma_3 u|^2\Big)dsdt = \int_{\mathbb{R}}|f'|^2 ds + \int_{\mathbb{R}}\frac{\kappa^2}{4}|f|^2 ds$$
$$+\int_{\mathbb{R}}\kappa\Re\Big(\int_{-1}^{1}\langle\partial_s u, -i\sigma_3 u\rangle\, dt\Big)ds$$

and there holds

$$\langle\partial_s u, -i\sigma_3 u\rangle = (f^+)'\overline{f^+}\langle\varphi_1^{m\varepsilon,+}, -i\sigma_3\varphi_1^{m\varepsilon,+}\rangle + (f^+)'\overline{f^-}\langle\varphi_1^{m\varepsilon,+}, -i\sigma_3\varphi_1^{m\varepsilon,-}\rangle$$
$$+(f^-)'\overline{f^+}\langle\varphi_1^{m\varepsilon,-}, -i\sigma_3\varphi_1^{m\varepsilon,+}\rangle + (f^-)'\overline{f^-}\langle\varphi_1^{m\varepsilon,-}, -i\sigma_3\varphi_1^{m\varepsilon,-}\rangle.$$

Now by Point (5) in Proposition 3, there holds

$$\int_{-1}^{1}\langle\varphi_1^{m\varepsilon,+}, -i\sigma_3\varphi_1^{m\varepsilon,+}\rangle dt = i\int_{-1}^{1}\langle\varphi_1^{+}, \sigma_3\varphi_1^{+}\rangle dt + O(\varepsilon) = i\frac{2}{\pi} + O(\varepsilon),$$

where we have used the explicit expression of φ_1^+ given in (10). Similarly, one gets

$$\int_{-1}^{1}\langle\varphi_1^{m\varepsilon,-}, -i\sigma_3\varphi_1^{m\varepsilon,-}\rangle dt = -i\frac{2}{\pi} + O(\varepsilon)$$

as well as

$$\int_{-1}^{1}\langle\varphi_1^{m\varepsilon,+}, -i\sigma_3\varphi_1^{m\varepsilon,-}\rangle\,dt = O(\varepsilon), \quad \int_{-1}^{1}\langle\varphi_1^{m\varepsilon,-}, -i\sigma_3\varphi_1^{m\varepsilon,+}\rangle\,dt = O(\varepsilon).$$

Hence, we get

$$\begin{aligned}\int_{-1}^{1}\langle\partial_s u, -i\sigma_3 u\rangle\,dt &= i\frac{2}{\pi}\left((f^+)'\overline{f^+} - (f^-)'\overline{f^-}\right) + \langle f', f + \sigma_1 f\rangle O(\varepsilon)\\ &= \langle\partial_s f, -i\frac{2}{\pi}\sigma_3 f\rangle + \langle f', f + \sigma_1 f\rangle O(\varepsilon).\end{aligned}$$

Thus, there exists $\varepsilon_1 > 0$ small enough and $k > 0$ such that for all $\varepsilon \in (0, \varepsilon_1)$ there holds

$$\begin{aligned}a_+[u] \le (1 + c\varepsilon)\Big(\int_{\mathbb{R}}|f'|^2 ds + \int_{\mathbb{R}} 2\Re(\langle f', -i\frac{\kappa}{\pi}\sigma_3 f\rangle)ds\Big)\\ + \frac{E_1(m\varepsilon)^2}{\varepsilon^2}\|f\|^2_{L^2(\mathbb{R})} + (1 + c\varepsilon)k\varepsilon(\|f\|^2_{L^2(\mathbb{R})} + \|f'\|^2_{L^2(\mathbb{R})}).\end{aligned}$$

Now, remark that

$$\int_{\mathbb{R}}|f'|^2 ds + \int_{\mathbb{R}} 2\Re(\langle f', -i\frac{\kappa}{\pi}\sigma_3 f\rangle)ds = \int_{\mathbb{R}}\Big(|f' - i\frac{\kappa}{\pi}\sigma_3 f|^2 - \frac{\kappa^2}{\pi^2}|f|^2\Big)ds.$$

Thus, for a $c_1 > 0$ there holds

$$\begin{aligned}a_+[u] &\le (1 + c_1\varepsilon)\int_{\mathbb{R}}\Big(|f' - i\frac{\kappa}{\pi}\sigma_3 f|^2 - \frac{\kappa^2}{\pi^2}|f|^2\Big)ds\\ &\quad + \frac{E_1(m\varepsilon)^2}{\varepsilon^2}\|f\|^2_{L^2(\mathbb{R})} + c_1\varepsilon\|f\|^2_{L^2(\mathbb{R})}\\ &= (1 + c_1\varepsilon)\quad \tilde{q}_e[f] + \frac{E_1(m\varepsilon)^2}{\varepsilon^2}\|f\|^2_{L^2(\mathbb{R})} + c_1\varepsilon\|f\|^2_{L^2(\mathbb{R})}.\end{aligned}$$

Now, the min-max principle of Propositions 1, 4 and 2 give for all $j \in \mathbb{N}$:

$$\mu_j(\mathcal{D}_\Gamma(\varepsilon)^2) \le (1 + c_1\varepsilon)\mu_j(q_e \oplus q_e) + \frac{E_1(m\varepsilon)^2}{\varepsilon^2} + c_1\varepsilon$$

so that (13) follows.

3.2 A Lower Bound

The aim of this paragraph is to prove the following lower bound.

Proposition 6 *Let $j \in \mathbb{N}$. There exists $\varepsilon_1 > 0$ and $c > 0$ such that for all $\varepsilon \in (0, \varepsilon_1)$ there holds*

$$\mu_j(\mathcal{D}_\Gamma(\varepsilon)^2) \geq \frac{E_1(m\varepsilon)^2}{\varepsilon^2} + \mu_j(q_e \oplus q_e) - c\varepsilon\,. \tag{14}$$

To prove Proposition 6, we need to introduce the projector in $L^2(\mathsf{Str}, \mathbb{C}^2)$ defined for all $\delta > 0$ and $u \in L^2(\mathsf{Str}, \mathbb{C}^2)$ by

$$\Pi^\delta u := \langle u, \varphi_1^{\delta,+}\rangle_{L^2(-1,1)}\varphi_1^{\delta,+} + \langle u, \varphi_1^{\delta,-}\rangle_{L^2(-1,1)}\varphi_1^{\delta,-}$$

and set $\Pi := \Pi^0$. Thanks to Point 5 in Proposition 3, there holds

$$\Pi^\delta = \Pi + \mathcal{O}(\delta), \qquad \text{as } \delta \to 0,$$

where the remainder is estimated in the operator norm. We also set $(\Pi^\delta)^\perp := Id - \Pi^\delta$ and $\Pi^\perp = Id - \Pi$.

Proof Let $u \in \mathrm{dom}\mathcal{E}_\Gamma(\varepsilon)$ and remark that there holds

$$\begin{aligned} a_-[u] - \frac{E_1(m\varepsilon)^2}{\varepsilon^2}\|u\|^2 \geq (1 - c\varepsilon) \int_{\mathsf{Str}} \Big(&|(\partial_s - i\frac{\kappa}{2}\sigma_3)(\Pi^{m\varepsilon} + (\Pi^{m\varepsilon})^\perp)u|^2 \\ &- \frac{\kappa^2}{4}(|\Pi^{m\varepsilon}u|^2 + |(\Pi^{m\varepsilon})^\perp u|^2)\Big)dsdt \\ &+ \frac{E_2(m\varepsilon)^2 - E_1(m\varepsilon)^2}{\varepsilon^2}\|(\Pi^{m\varepsilon})^\perp u\|^2 \\ &- c\varepsilon(\|\Pi^{m\varepsilon}u\|^2 + \|(\Pi^{m\varepsilon})^\perp u\|^2). \end{aligned} \tag{15}$$

We focus on the first term on the right-hand side of the last equation which gives

$$\begin{aligned} \int_{\mathsf{Str}} \big(|(\partial_s - i\frac{\kappa}{2}\sigma_3)(\Pi^{m\varepsilon} + (\Pi^{m\varepsilon})^\perp)u|^2\big)dsdt &= \int_{\mathsf{Str}} |(\partial_s - i\frac{\kappa}{2}\sigma_3)\Pi^{m\varepsilon}u|^2 dsdt \\ &+ \int_{\mathsf{Str}} |(\partial_s - i\frac{\kappa}{2}\sigma_3)(\Pi^{m\varepsilon})^\perp u|^2 dsdt \\ + 2\Re\Big(\int_{\mathsf{Str}} \langle(\partial_s - i\frac{\kappa}{2}\sigma_3)\Pi^{m\varepsilon}u,\, &(\partial_s - i\frac{\kappa}{2}\sigma_3)(\Pi^{m\varepsilon})^\perp u\rangle dsdt\Big) \end{aligned}$$

$$\geq \int_{\mathsf{Str}} |(\partial_s - i\frac{\kappa}{2}\sigma_3)\Pi^{m\varepsilon}u|^2 dsdt$$
$$+ 2\Re\Big(\int_{\mathsf{Str}} \langle(\partial_s - i\frac{\kappa}{2}\sigma_3)\Pi^{m\varepsilon}u, (\partial_s - i\frac{\kappa}{2}\sigma_3)(\Pi^{m\varepsilon})^{\perp}u\rangle dsdt\Big).$$

Let us deal with the last term. Remark that there holds

$$\langle(\partial_s - i\frac{\kappa}{2}\sigma_3)\Pi^{m\varepsilon}u, (\partial_s - i\frac{\kappa}{2}\sigma_3)(\Pi^{m\varepsilon})^{\perp}u\rangle$$
$$= \langle\big([(\partial_s - i\frac{\kappa}{2}\sigma_3), \Pi^{m\varepsilon}] + \Pi^{m\varepsilon}(\partial_s - i\frac{\kappa}{2}\sigma_3)\big)\Pi^{m\varepsilon}u, \big([(\partial_s - i\frac{\kappa}{2}\sigma_3), (\Pi^{m\varepsilon})^{\perp}]$$
$$+ (\Pi^{m\varepsilon})^{\perp}(\partial_s - i\frac{\kappa}{2}\sigma_3)\big)(\Pi^{m\varepsilon})^{\perp}u\rangle,$$

where the scalar product is taken in $L^2(\mathsf{Str}, \mathbb{C}^2)$. Taking into account that $\Pi^{m\varepsilon}$ commutes with ∂_s, we obtain

$$[(\partial_s - i\frac{\kappa}{2}\sigma_3), \Pi^{m\varepsilon}] = -[(\partial_s - i\frac{\kappa}{2}\sigma_3), (\Pi^{m\varepsilon})^{\perp}] = -i\frac{\kappa}{2}[\sigma_3, \Pi^{m\varepsilon}],$$

which gives

$$\langle(\partial_s - i\frac{\kappa}{2}\sigma_3)\Pi^{m\varepsilon}u, (\partial_s - i\frac{\kappa}{2}\sigma_3)(\Pi^{m\varepsilon})^{\perp}u\rangle$$
$$= \langle\big(-i\frac{\kappa}{2}[\sigma_3, \Pi^{m\varepsilon}] + \Pi^{m\varepsilon}(\partial_s - i\frac{\kappa}{2}\sigma_3)\big)\Pi^{m\varepsilon}u,$$
$$\times \big(i\frac{\kappa}{2}[\sigma_3, \Pi^{m\varepsilon}] + (\Pi^{m\varepsilon})^{\perp}(\partial_s - i\frac{\kappa}{2}\sigma_3)\big)(\Pi^{m\varepsilon})^{\perp}u\rangle$$
$$= \langle -i\frac{\kappa}{2}[\sigma_3, \Pi^{m\varepsilon}]\Pi^{m\varepsilon}u, i\frac{\kappa}{2}[\sigma_3, \Pi^{m\varepsilon}](\Pi^{m\varepsilon})^{\perp}u\rangle$$
$$+ \langle -i\frac{\kappa}{2}[\sigma_3, \Pi^{m\varepsilon}]\Pi^{m\varepsilon}u, (\Pi^{m\varepsilon})^{\perp}(\partial_s - i\frac{\kappa}{2}\sigma_3)(\Pi^{m\varepsilon})^{\perp}u\rangle$$
$$+ \langle \Pi^{m\varepsilon}(\partial_s - i\frac{\kappa}{2}\sigma_3)\Pi^{m\varepsilon}u, i\frac{\kappa}{2}[\sigma_3, \Pi^{m\varepsilon}](\Pi^{m\varepsilon})^{\perp}u\rangle$$
$$:= J_1 + J_2 + J_3.$$

One notices that there exists $c_1 > 0$ such that

$$|J_1| \leq c_1\|\Pi^{m\varepsilon}u\|\|(\Pi^{m\varepsilon})^{\perp}u\| \leq \frac{c_1}{2}\varepsilon\|\Pi^{m\varepsilon}u\|^2 + \frac{c_1}{2\varepsilon}\|(\Pi^{m\varepsilon})^{\perp}u\|^2. \tag{16}$$

Similarly, there exists $c_2 > 0$ such that

$$|J_3| \leq c_2 \|(\partial_s - i\frac{\kappa}{2})\Pi^{m\varepsilon} u\| \|(\Pi^{m\varepsilon})^{\perp} u\| \\ \leq \frac{c_2}{2}\varepsilon \|(\partial_s - i\frac{\kappa}{2})\Pi^{m\varepsilon} u\|^2 + \frac{c_2}{2\varepsilon}\|(\Pi^{m\varepsilon})^{\perp} u\|^2. \tag{17}$$

Concerning the term J_2, there holds

$$J_2 = \langle -i\frac{\kappa}{2}(\Pi^{m\varepsilon})^{\perp}\sigma_3 \Pi^{m\varepsilon} u, \partial_s (\Pi^{m\varepsilon})^{\perp} u\rangle \\ + \langle -i\frac{\kappa}{2}(\Pi^{m\varepsilon})^{\perp}\sigma_3 \Pi^{m\varepsilon} u, -i\frac{\kappa}{2}\sigma_3 (\Pi^{m\varepsilon})^{\perp} u\rangle$$

and integration by parts in the s-variable gives

$$J_2 = -\langle -i\frac{\kappa}{2}(\Pi^{m\varepsilon})^{\perp}\sigma_3 \Pi^{m\varepsilon} \partial_s u, (\Pi^{m\varepsilon})^{\perp} u\rangle - \langle -i\frac{\kappa'}{2}(\Pi^{m\varepsilon})^{\perp}\sigma_3 \Pi^{m\varepsilon} u, (\Pi^{m\varepsilon})^{\perp} u\rangle \\ + \langle -i\frac{\kappa}{2}(\Pi^{m\varepsilon})^{\perp}\sigma_3 \Pi^{m\varepsilon} u, -i\frac{\kappa}{2}\sigma_3 (\Pi^{m\varepsilon})^{\perp} u\rangle.$$

Hence, there exists $c_3 > 0$ such that

$$|J_2| \leq c_3\big(\|\partial_s \Pi^{m\varepsilon} u\| \|(\Pi^{m\varepsilon})^{\perp} u\| + \|\Pi^{m\varepsilon} u\| \|(\Pi^{m\varepsilon})^{\perp} u\|\big) \\ \leq c_3\big(\frac{\varepsilon}{2}(\|\partial_s \Pi^{m\varepsilon} u\|^2 + \|\Pi^{m\varepsilon} u\|^2) + \frac{1}{2\varepsilon}\|(\Pi^{m\varepsilon})^{\perp} u\|^2\big).$$

Noting that

$$\|\partial_s \Pi^{m\varepsilon} u\| \leq \|(\partial_s - i\frac{\kappa}{2}\sigma_3)u\| + \|\frac{\kappa}{2}\Pi^{m\varepsilon} u\|$$

there exists $c_4 > 0$ such that

$$|J_2| \leq c_4\big(\frac{\varepsilon}{2}(\|(\partial_s - i\frac{\kappa}{2}\sigma_3)\Pi^{m\varepsilon} u\|^2 + \|\Pi^{m\varepsilon} u\|^2) + \frac{1}{2\varepsilon}\|(\Pi^{m\varepsilon})^{\perp} u\|^2\big). \tag{18}$$

In estimates (16), (17) and (18) we have used that $\kappa \in L^{\infty}(\mathbb{R})$, and the following bounds on the operator norms

$$\|\Pi^{m\varepsilon}\| \leq 1\,, \qquad \|(\Pi^{m\varepsilon})^{\perp}\| \leq 1$$

as $\Pi^{m\varepsilon}$ and $(\Pi^{m\varepsilon})^{\perp}$ are projectors, as well as the elementary identity $2ab \leq \varepsilon a^2 + \frac{1}{\varepsilon}b^2$, with $a, b, \varepsilon > 0$.

Combining the above observations, coming back to (15), we get, for some $c_5 \geq 0$:

$$a_-[u] - \frac{E_1(m\varepsilon)^2}{\varepsilon^2} \geq (1 - c_5\varepsilon) \int_{\mathsf{Str}} \Big(|(\partial_s - i\frac{\kappa}{2}\sigma_3)\Pi^{m\varepsilon}u|^2 - \frac{\kappa^2}{4}|\Pi^{m\varepsilon}u|^2 \Big) dt ds$$
$$+ \Big(\frac{E_2(m\varepsilon)^2 - E_1(m\varepsilon)^2}{\varepsilon^2} - \frac{c_5}{\varepsilon} - c_5 \Big) \|(\Pi^{m\varepsilon})^{\perp}u\|^2 - c_5\|\Pi^{m\varepsilon}u\|^2 .$$

Notice that the first term on the right-hand-side of (18) has been absorbed in the integral above, taking a new constant c_5, and the terms involving $\|\Pi^{m\varepsilon}u\|^2$ and $\|(\Pi^{m\varepsilon})^{\perp}u\|^2$ contribute to the last two terms in the above formula.

Now, set $f^{\pm} := \langle u, \varphi_1^{m\varepsilon,\pm} \rangle_{L^2(-1,1)}$ and remark that the computation of the term

$$\int_{\mathsf{Str}} \Big(|(\partial_s - i\frac{\kappa}{2}\sigma_3)\Pi^{m\varepsilon}u|^2 - \frac{\kappa^2}{4}|\Pi^{m\varepsilon}u|^2 ds dt.$$

is similar to the one performed in the proof of the upper bound (see the proof of Proposition 5) and it yields

$$\int_{\mathsf{Str}} \Big(|(\partial_s - i\frac{\kappa}{2}\sigma_3)\Pi^{m\varepsilon}u|^2 - \frac{\kappa^2}{4}|\Pi^{m\varepsilon}u|^2 ds dt \geq (q_e \oplus q_e)[f] - c_6\varepsilon\|f\|^2_{L^2(\mathbb{R})},$$

for some constant c_6. All in all, we have obtained that there exists $k > 0$ such that provided ε is small enough there holds

$$a_-[u] - \frac{E_1(m\varepsilon)^2}{\varepsilon^2} \geq (1 - k\varepsilon)(q_e \oplus q_e)[f] - k\varepsilon\|f\|^2_{L^2(\mathbb{R})}$$
$$+ \Big(\frac{E_2(m\varepsilon)^2 - E_1(m\varepsilon)^2}{\varepsilon^2} - \frac{k}{\varepsilon} - k \Big) \|(\Pi^{m\varepsilon})^{\perp}u\|^2$$
$$\geq (1 - k\varepsilon)(q_e \oplus q_e)[f] - k\varepsilon\|f\|^2_{L^2(\mathbb{R})} + \Big(\frac{5\pi^2}{16\varepsilon^2} - \frac{k}{\varepsilon} - k \Big) \|(\Pi^{m\varepsilon})^{\perp}u\|^2,$$

where for the last inequality we have used Point (3) Proposition 3. As the quadratic form on the right-hand side is the quadratic form of the direct sum of two operators, if one fixes $j \in \mathbb{N}$ the min-max principle of Proposition 1 yields

$$\mu_j(\mathcal{D}_\Gamma(\varepsilon)^2) - \frac{E_1(m\varepsilon)^2}{\varepsilon^2}$$
$$\geq j - \text{th element of the set} \Big(\{(1 - k\varepsilon)\mu_j(q_e \oplus q_e) - k\varepsilon\} \cup \{ \frac{5\pi^2}{16\varepsilon^2} - \frac{k}{\varepsilon} - k \} \Big).$$

Hence, for ε small enough (depending on j), this reads

$$\mu_j(\mathcal{D}_\Gamma(\varepsilon)^2) - \frac{E_1(m\varepsilon)^2}{\varepsilon^2} \geq (1 - k\varepsilon)\mu_j(q_e \oplus q_e) - k\varepsilon.$$

which is precisely Proposition 6.

3.3 *Proof of Theorem 1*

Let $J \geq 1$ and remark that due to the symmetry of the spectrum of $\mathcal{D}_\Gamma(\varepsilon)$ with respect to zero, for all $j \in \{1, \ldots, J\}$ there holds

$$\lambda_j(\mathcal{D}_\Gamma(\varepsilon)) = \sqrt{\mu_{2j}(\mathcal{D}_\Gamma(\varepsilon)^2)}. \tag{19}$$

Combining Propositions 5 and 6, we have for all $j \in \mathbb{N}$ that when $\varepsilon \to 0$

$$\begin{aligned}\mu_j(\mathcal{D}_\Gamma(\varepsilon)^2) &= \frac{E_1(m\varepsilon)^2}{\varepsilon^2} + \mu_j(q_e \oplus q_e) + O(\varepsilon)\\ &= \frac{E_1(m\varepsilon)^2}{\varepsilon^2}\Big(1 + \frac{\varepsilon^2}{E_1(m\varepsilon)^2}\mu_j(q_e \oplus q_e) + O(\varepsilon^3)\Big),\end{aligned}$$

where we have used that $E_1(m\varepsilon) = O(1)$ when $\varepsilon \to 0$. Hence, there holds

$$\sqrt{\mu_j(\mathcal{D}_\Gamma(\varepsilon)^2)} = \frac{E_1(m\varepsilon)}{\varepsilon} + \frac{1}{2E_1(m\varepsilon)}\mu_j(q_e \oplus q_e)\varepsilon + O(\varepsilon^2)$$

and by Point 4 in Proposition 3, there holds

$$\sqrt{\mu_j(\mathcal{D}_\Gamma(\varepsilon)^2)} = \frac{E_1(m\varepsilon)}{\varepsilon} + \frac{2}{\pi}\mu_j(q_e \oplus q_e)\varepsilon + O(\varepsilon^2).$$

Thus, for $j \in \{1, \ldots, J\}$, (19) yields

$$\lambda_j(\mathcal{D}_\Gamma(\varepsilon)) = \frac{E_1(m\varepsilon)}{\varepsilon} + \frac{2}{\pi}\mu_{2j}(q_e \oplus q_e)\varepsilon + O(\varepsilon^2) = \frac{E_1(m\varepsilon)}{\varepsilon} + \frac{2}{\pi}\mu_j(q_e)\varepsilon + O(\varepsilon^2),$$

concluding the proof.

Acknowledgments We acknowledge the support of Istituto Nazionale di Alta Matematica "F. Severi", through the Intensive Period "INdAM Quantum Meetings (IQM22)". W.B. has been partially supported by Gruppo Nazionale per Analisi Matematica, la Probabilità e le loro Applicazioni (GNAMPA) of the Istituto Nazionale di Alta Matematica (INdAM), through the 2022 project "Modelli matematici con singolarità per fenomeni di interazione".

References

1. Akhmerov, A.R., Beenakker, C.W.J.: Boundary conditions for Dirac fermions on a terminated honeycomb lattice. Phys. Rev. B **77** (2008). https://doi.org/10.1103/PhysRevB.77.085423
2. Arrizabalaga, N., Le Treust, L., Raymond, N.: On the MIT bag model in the non-relativistic limit. Commun. Math. Phys. **354**(2), 641–669 (2017). https://doi.org/10.1007/s00220-017-2916-8
3. Barbaroux, J.-M., Cornean, H.D., Le Treust, L., Stockmeyer, E.: Resolvent convergence to Dirac operators on planar domains. Ann. Henri Poincaré **20**, 1877–1891 (2019). https://doi.org/10.1007/s00023-019-00787-2
4. Benguria, R.D., Fournais, S., Stockmeyer, E., Van Den Bosch, H.: Self-adjointness of two-dimensional Dirac operators on domains. Ann. Henri Poincaré **18**, 1371–1383 (2017). https://doi.org/10.1007/s00023-017-0554-5
5. Borrelli, W., Briet, P., Krejčiřík, D., Ourmieres-Bonafos, T.: Spectral properties of relativistic quantum waveguides. Ann. Henri Pincaré (2022). https://doi.org/10.1007/s00023-022-01179-9
6. Chodos, A., Jaffe, R.L., Johnson, K., Thorn, C.B., Weisskopf, V.F.: New extended model of hadrons. Phys. Rev. D (3) **9**(12), 3471–3495 (1974). https://doi.org/10.1103/PhysRevD.9.3471
7. Davies, E.B.: Spectral Theory and Differential Operators. Cambridge Studies in Advanced Mathematics, vol. 42. Cambridge University Press, Cambridge (1995). https://doi.org/10.1017/CBO9780511623721
8. Duclos, P., Exner, P.: Curvature-induced bound states in quantum waveguides in two and three dimensions. Rev. Math. Phys. **7**, 73–102 (1995). https://doi.org/10.1142/S0129055X95000062
9. Exner, P., Holzmann, M.: Dirac operator spectrum in tubes and layers with a zigzag type boundary. Lett. Math. Phys. **112**, 102 (2022). https://doi.org/10.1007/s11005-022-01594-3
10. Exner, P., Kovařík, H.: Quantum Waveguides. Springer, Berlin (2015). https://doi.org/10.1007/978-3-319-18576-7
11. Exner, P., Šeba, P.: Bound states in curved quantum waveguides. J. Math. Phys. **30**, 2574–2580 (1989). https://doi.org/10.1063/1.528538
12. Ginoux, N.: The Dirac Spectrum. Lecture notes in Mathematics. Springer-Verlag, Berlin/Heidelberg (2009). https://doi.org/10.1007/978-3-642-01570-0
13. Goldstone, J., Jaffe, R.L.: Bound states in twisting tubes. Phys. Rev. B **45**, 14100–14107 (1992). https://doi.org/10.1103/PhysRevB.45.14100
14. Kato, T.: Perturbation Theory for Linear Operators. Springer-Verlag, Berlin (1995). Reprint of the Corr. Print. of the 2nd ed. 1980. https://doi.org/10.1007/978-3-642-66282-9
15. Krejčiřík, D., Kříž, J.: On the spectrum of curved quantum waveguides. Publ. RIMS Kyoto Univ. **41**(3), 757–791 (2005). https://doi.org/10.2977/PRIMS/1145475229
16. Le Treust, L., Ourmières-Bonafos, T.: Self-adjointness of Dirac operators with infinite mass boundary conditions in sectors. Ann. H. Poincaré **19**, 1465–1487 (2018). https://doi.org/10.1007/s00023-018-0661-y
17. Hijazi, O., Montiel, S., Zhang, X.: Eigenvalues of the dirac operator on manifolds with boundary. Commun. Math. Phys. **221**, 255–265 (2001). https://doi.org/10.1007/s002200100475

Topological Polarization in Disordered Systems

Giuseppe De Nittis and Danilo Polo Ojito

1 Introduction

In nature, there are materials in which a *macroscopic polarization* at the edges of the sample appears when subjected to mechanical strains, *i.e.*, to the accumulation of charge whenever the materials are deformed. This phenomenon is known as *piezoelectric effect*, and its microscopically description was only understood in the last 50 years. In the 70's, Martin [11] noticed that the previous approach in terms of dipole momenta for the macroscopic polarization was incomplete and unsatisfactory. This fact is due to that the total polarization comes from two contributions: the *relative displacements* of the ionic cores in a unit cell (whose computation is straightforward), and electrical conduction which is called *orbital polarization*. Resta [17] and King-Smith and Vanderbilt [8] shifted the attention to the orbital polarization and derived a formula using linear response theory, which allows calculating the polarization in terms of the *Berry connection*. Namely, the change in polarization $\Delta\mathscr{P}$ accumulated during a (periodic) deformation in the interval $[0, T]$ is given by

$$\Delta\mathscr{P} := \frac{1}{(2\pi)^d}\sum_{m=0}^{M}\int_{\mathbb{B}} \mathrm{d}k\big(\mathscr{A}_m(k,T) - \mathscr{A}_m(k,0)\big). \tag{1}$$

Here $\mathbb{B} \simeq \mathbb{T}^d$ denotes the first Brillouin zone, d is the space dimension, $\mathscr{A}_m(k,t)$ is the Berry connection for the m-th Bloch band at time t, and the sum runs over all the occupied M Bloch bands. Panati et al. [14] generalized Eq. (1) for continuous and periodic systems, showing that in the adiabatic limit of slow deformations

G. De Nittis (✉) · D. Polo Ojito
Pontificia Universidad Catolica de Chile, Santiago, Chile
e-mail: gidenittis@mat.uc.cl; djpolo@mat.uc.cl

M. Correggi, M. Falconi (eds.), *Quantum Mathematics I*,
Springer INdAM Series 57, https://doi.org/10.1007/978-981-99-5894-8_6

the macroscopic piezoelectric current is determined by the geometry of the Bloch bundle. Using an adaption of Nenciu's super-adiabatic theory [13] to C^*-dynamical systems, Schulz-Baldes and Teufel [18] established formula (1) for discrete random systems. They obtained that in the adiabatic limit it holds true that

$$\Delta\mathscr{P}_k = \mathrm{i}\int_0^T \mathrm{d}t\,\mathscr{T}\big(P(t)[\,\partial_t P(t),\ \nabla_k P(t)\,]\big) + \mathcal{O}(\varepsilon^N), \qquad (2)$$

where $\mathscr{T}$ denotes the trace per unit volume, $P(t) = \chi_{(-\infty,\epsilon_F)}\big(H(t)\big)$ is the spectral projection onto all states below the *Fermi energy* ϵ_F, $H(t)$ is the instantaneous Hamiltonian of the system at time t, $N \in \mathbb{N}$ is related to the regularity of the map $t \mapsto H(t)$ and $k = 1, \dots, d$ indicates the direction of the polarization in the physical space. It is important to point out that the works [14] and [18] it is also explored the topological nature of orbital polarization. They proved that $\Delta\mathscr{P}$ is quantized up to a small error (in the *adiabatic parameter* ε) whenever the slow deformation is periodic. The latter fact is in agreement with the observation of Thouless [21] in a more restricted context. In [5] one of the authors and Lein carried out a topological study of the orbital polarization in discrete graphene-like systems, where they showed that the polarization depends only on the class of homotopy paths in the gapped parameter space. Therefore, a necessary condition for the existence of piezoelectric effects is that the fundamental group of the gapped parameter space is non-trivial.

In this work, we focus on deriving rigorously Eq. (2) for continuous and disordered systems of independent electrons in the regime of an adiabatic periodic deformation of the background potential. The main strategy is to use the mathematical framework introduced in [6], along with tools from (super)adiabatic theory [13, 20], for the derivation of the formula for $\Delta\mathscr{P}$ in a wide range of *covariant* (random) systems, which in principle are defined over a topological group $\mathcal{G}$ that can be chosen equal to $\mathbb{R}^d$ (continuous case) or $\mathbb{Z}^d$ (discrete case) in concrete applications. Our main result, Theorem 2, establishes the expression for the orbital polarization in this generalized setting, along with its main topological consequences when the deformation is periodic.

Organization of the Paper Section 2 is devoted to the construction of the semi-finite von Neumann algebra of observables and its trace per unit volume. In Sect. 3, we briefly review all the necessary mathematical notions and we state the main hypotheses needed for the derivation of Eq. (2). In Sect. 4 we present the main results. We start this section with a notion of differentiability for affiliated self-adjoint operators to the observable algebra, and after that, we prove an equivalence for the current expectation value (Theorem 1). We will use the later facts to derive the *King-Smith and Vanderbilt formula*. We finish this section with the topological quantization of the polarization for periodic deformations. Section 5 provides the physical models where our results apply. We will present in detail the case of continuous disordered systems and we will build the Landau Hamiltonian which fulfills all the required hypotheses. In order to maintain the clarity in the proof of the Theorem 2, in Appendix 5.2 we have included the technical proofs needed for the construction of the superadiabatic projections.

2 Description of the Physical Models

The background material contained in this section is based on [6, Chapter 4] where the relevant references are also provided.

Let $\mathfrak{h}$ be a (separable) Hilbert space and $\mathcal{B}(\mathfrak{h})$ the set of linear bounded operators on $\mathfrak{h}$. The physical relevant *observables* (like the Hamiltonians) will be modeled by strongly continuous[1] families $(H_\omega)_{\omega\in\Omega}$ of (self-adjoint) operators affiliated to a von Neumann algebra $\mathcal{A}$. Here Ω denotes a compact[2] space which describes the possible configurations of the *interacting potential* between particles and medium (e.g. random interaction). In order to construct a von Neumann algebra $\mathcal{A}$ which contains homogeneous models[3] we assume that there is an *ergodic topological dynamical system* $(\Omega, \mathcal{G}, \tau, \mathbb{P})$ consisting of a locally compact[4] abelian group $\mathcal{G}$ (with a given Haar measure $\mu_\mathcal{G}$), a probability space $(\Omega, \mathcal{F}, \mathbb{P})$, where $\mathcal{F}$ is the Borel σ-algebra and $\mathbb{P}$ is a probability measure, and a representation $\tau : \mathcal{G} \to \mathrm{Homeo}(\Omega)$. These structures are related by the following assumptions:

(i) The group action $\mathcal{G} \times \Omega \ni (g, \omega) \mapsto \tau_g(\omega) \in \Omega$ is jointly continuous;
(ii) $\mathbb{P}$ is a τ-invariant ergodic measure, namely $\mathbb{P}(\tau_g(B)) = \mathbb{P}(B)$ for all $B \in \mathcal{F}$, and if $\tau_g(B) = B$ for all $g \in \mathcal{G}$ then $\mathbb{P}(B) = 1$ or $\mathbb{P}(B) = 0$.

In the next we will consider the Hilbert space $\mathfrak{h} = L^2(\mathcal{G})\otimes\mathbb{C}^N$, where N depends on the spin-type degrees of freedom (e.g. the isospin) and we will introduce the direct integral [7, Part II, Chapters 1–5]

$$\mathfrak{H} := \int_\Omega^\oplus \mathrm{d}\,\mathbb{P}(\omega)\, \mathfrak{h}_\omega \;\simeq\; L^2(\Omega, \mathfrak{h})\,,$$

with the assumption that $\mathfrak{h}_\omega = \mathfrak{h}$ for (almost) all $\omega \in \Omega$. A random operator is a bounded-operator valued map $\Omega \ni \omega \mapsto A_\omega \in \mathcal{B}(\mathfrak{h})$ such that the map $\Omega \ni \omega \mapsto \langle \phi, A_\omega \psi\rangle_\mathfrak{h}$ is measurable for all $\phi, \psi \in \mathfrak{h}$, and $\mathrm{ess}-\sup \|A_\omega\|_{\mathcal{B}(\mathfrak{h})} < \infty$. We will denote the set of random operators by $\mathrm{Rand}(\mathfrak{H}) \subset \mathcal{B}(\mathfrak{H})$. Furthermore, any random operator $A := \{A_\omega\}_{\omega\in\Omega}$ fulfills

$$\|A\|_{\mathcal{B}(\mathfrak{H})} \;=\; \mathrm{ess} - \sup_{\omega\in\Omega} \|A_\omega\|_{\mathcal{B}(\mathfrak{h})}\,.$$

[1] In the sense of the resolvent.

[2] We will assume that Ω is also metrizable, and in turn separable. This assumption implies that $L^2(\Omega)$ is a separable Hilbert space.

[3] In the sense of Bellissard [2].

[4] In the interesting examples $\mathcal{G}$ is also separable and metrizable (e.g. $\mathcal{G} = \mathbb{R}^d, \mathbb{Z}^d, \mathbb{T}^d$) and this implies that $L^2(\mathcal{G})$ is a separable Hilbert space.

Let $\Theta \colon \mathcal{G} \times \mathcal{G} \to \mathbb{U}(1)$ be a *twisting group* 2*-cocycle* [6, Definition 4.1.2], and for every $g \in \mathcal{G}$ consider the operator $U_g \in \mathcal{B}(\mathfrak{H})$ defined by

$$\big(U_g\,\psi\big)_{\tau_g(\omega)}(h) \;:=\; \Theta(g, hg^{-1})\;\psi_\omega(hg^{-1})\;, \qquad \forall\, h \in \mathcal{G}$$

where $\psi := \{\psi_\omega\}_{\omega\in\Omega}$ is any element of $\mathfrak{H}$, and on the left-hand side the symbol $(\cdot)_{\tau_g(\omega)}$ means that the value of the vector $U_g\psi$ on the fiber of $\mathfrak{H}$ at $\tau_g(\omega)$. It is evident from the definition that U_g doesn't respect the fiber structure of the direct integral $\mathfrak{H}$. One can check that the mapping $\mathcal{G} \ni g \mapsto U_g \in \mathcal{B}(\mathfrak{H})$ forms a projective unitary representation of $\mathcal{G}$.

Definition 1 The von Neumann algebra of observables is the set

$$\mathcal{A} \;\equiv\; \mathcal{A}(\Omega, \mathbb{P}, \mathcal{G}, \Theta) \;=\; \mathrm{Span}_{\mathcal{G}}\{U_g\}' \;\cap\; \mathrm{Rand}(\mathfrak{H})$$

where $\mathrm{Span}_{\mathcal{G}}\{U_g\}$ denotes the linear space generated by the U_g and the symbol $'$ denotes the commutant.

For sake of notational simplicity, we write $\mathcal{A}$ instead of $\mathcal{A}(\Omega, \mathbb{P}, \mathcal{G}, \Theta)$. Said differently $\mathcal{A}$ consists of those random operators A which are covariant with respect to the projective unitary representation of $\mathcal{G}$ provided by the U_g, i.e.

$$U_{g,\tau_g(\omega)}\, A_\omega\, U_{g,\tau_g(\omega)}^{-1} \;=\; A_{\tau_g(\omega)}\;, \qquad \forall\, g \in \mathcal{G}\;,\;\; \forall\, \omega \in \Omega$$

where $U_{g,\tau_g(\omega)}$ denotes the action of U_g from the fiber at ω into the fiber at $\tau_g(\omega)$.

It is known that $\mathcal{A}$ is a semi-finite von Neumman algebra, hence $\mathcal{A}$ admits a faithful normal semi-finite (f.n.s.) trace [7, Part I, Chapter 6, Proposition 9]. On the domain of definition, such a trace can be constructed following the procedure described in [9, Proposition 2.1.6 and Theorem 2.2.2], i.e.

$$\mathscr{T}_{\mathbb{P}}(A) \;:=\; \int_\Omega \mathrm{d}\mathbb{P}(\omega)\, \mathrm{Tr}_{\mathfrak{h}}(M_\lambda A_\omega M_\lambda)\;, \qquad A \in \mathcal{A}^+\;,$$

where $\lambda \in L^\infty(\mathcal{G}) \cap L^2(\mathcal{G})$ is any positive function of unitary norm $\|\lambda\|_{L^2} = 1$, and M_λ is the operator which acts on $\mathfrak{h}$ as the multiplication by the diagonal matrix $\lambda \otimes \mathbf{1}_N$. It turns out that $\mathscr{T}_{\mathbb{P}}$ coincides with the *trace per unit volume*, namely

$$\mathscr{T}_{\mathbb{P}}(A) \;=\; \lim_{n\to\infty} \frac{1}{|\Lambda_n|}\, \mathrm{Tr}_{\mathfrak{H}}(P_{\Lambda_n} A_\omega P_{\Lambda_n})\;, \qquad \mathbb{P} - \text{a. e.}$$

where P_{Λ_m} is the multiplication operator by the characteristic function of the compact set $\Lambda_n \subset \mathcal{G}$, $|\Lambda_n|$ is its volume, and $\{\Lambda_n\}_{n\in\mathbb{N}}$ forms a *Følner exhausting sequence* for $\mathcal{G}$.

3 Main Hypotheses for a Linear Response Theory

In this section, we will briefly review all the necessary mathematical notions and we will state the main hypotheses needed for a rigorous derivation of the linear response theory as formulated in [6, Sect. 2]. These notions and hypotheses will be used in the following sections of this work.

Let $\mathtt{AFF}(\mathcal{A})$ be the set of *closed* and *densely defined* operators affiliated with $\mathcal{A}$ [6, Sect. 3.1.2], and $\mathfrak{L}^p(\mathcal{A})$ the *L^p-spaces* (or p-Schatten classes) associate to the semi-finite von Neumman algebra $\mathcal{A}$ with its f.n.s. trace $\mathscr{T}_{\mathbb{P}}$ [7, 12, 19, 22] or [6, Sect. 3.2]. The non-commutative Hölder inequalities allow defining the commutators

$$[A, B]_{(r)} := AB - BA \in \mathfrak{L}^r(\mathcal{A}) , \qquad A \in \mathfrak{L}^p(\mathcal{A}) , \quad B \in \mathfrak{L}^q(\mathcal{A}) ,$$

with $r^{-1} = p^{-1} + q^{-1}$.

Hypothesis 1 (Unperturbed Dynamics) Let $H \in \mathtt{AFF}(\mathcal{A})$ be a (possibly unbounded) self-adjoint operator (or *Hamiltonian*) which prescribes the *unperturbed dynamics* of the system. The affiliation of H to $\mathcal{A}$ implies that the unperturbed dynamics

$$\alpha_t(A) := \mathrm{e}^{-\mathrm{i}tH} A \, \mathrm{e}^{\mathrm{i}tH} , \qquad t \in \mathbb{R}, \ A \in \mathcal{A},$$

generated by H is a strongly continuous one-parameter group of isometries on each Banach space $\mathfrak{L}^p(\mathcal{A})$. The generator $\mathscr{L}_H^{(p)}$ of α_t on $\mathfrak{L}^p(\mathcal{A})$ has a core $\mathcal{D}_{H,p}$ where it acts as a generalized commutator [6, Proposition 5.1.3], i.e.

$$\mathscr{L}_H^{(p)}(A) = -\mathrm{i}\big(HA - (HA^*)^*\big) =: -\mathrm{i}\,[H, A]_\dagger , \qquad A \in \mathcal{D}_{H,p} .$$

we will refer to $\mathscr{L}_H^{(p)}$ as the *p-Liouvillian* of H. An (initial) *equilibrium configuration* for H is any positive element $\rho \in \mathcal{A}^+$ such that $\alpha_t(\rho) = \rho$ for every $t \in \mathbb{R}$. It will be called *equilibrium state* if in addition $\mathscr{T}_{\mathbb{P}}(\rho) = c < +\infty$ (and up to a multiplicative factor one can always impose the normalization condition $c = 1$). For instance $\rho = f(H)$, with $f \in L^\infty(\mathbb{R})$ and positive, is an equilibrium configuration.

Hypothesis 2 (Spatial Derivation) Let $\{X_1, X_2, ..., X_d\}$ be a set of (possibly unbounded) self-adjoint operators which are *$\mathscr{T}_{\mathbb{P}}$-compatible* in the sense that for all $k = 1, 2, ..., d$ and for all $s \in \mathbb{R}$ they satisfy

(i) $\mathrm{e}^{\mathrm{i}sX_k} A \, \mathrm{e}^{-\mathrm{i}sX_k} \in \mathcal{A}$ for all $A \in \mathcal{A}$;
(ii) $\mathscr{T}_{\mathbb{P}}(\mathrm{e}^{\mathrm{i}sX_k} A \, \mathrm{e}^{-\mathrm{i}sX_k}) = \mathscr{T}_{\mathbb{P}}(A)$ for all A in the domain of $\mathscr{T}_{\mathbb{P}}$;
(iii) $\mathrm{e}^{\mathrm{i}sX_j} \mathrm{e}^{\mathrm{i}sX_k} = \mathrm{e}^{\mathrm{i}sX_k} \mathrm{e}^{\mathrm{i}sX_j}$ for all $j, k = 1, 2, ..., d$.

This assumption allows to introduce the spatial derivations on $\mathfrak{L}^p(\mathcal{A})$ as generators of an $\mathbb{R}$-flow, i.e.

$$\nabla_k(A) := \lim_{s\to 0} \frac{\mathrm{e}^{\mathrm{i}sX_k}\, A\, \mathrm{e}^{-\mathrm{i}sX_k} - A}{s}.$$

The ∇_k are densely defined closed operators on each $\mathfrak{L}^p(\mathcal{A})$ with a common core where they act as commutators, i.e. $\nabla_k(A) = \mathrm{i}\,[X_k, A]$ [6, Sect. 3.4.1]. The domain of the associated gradient $\nabla := (\nabla_1, ..., \nabla_d)$ is the (non-commutative) *Sobolev space* [6, Sect. 3.4.2]

$$\mathfrak{M}^{1,p}(\mathcal{A}) := \{A \in \mathfrak{L}^p(\mathcal{A}) \mid \nabla_k(A) \in \mathfrak{L}^p(\mathcal{A}),\ k = 1, 2, ..., d\}.$$

Hypothesis 3 (Current Operator) The self-adjoint Hamiltonian $H \in \mathtt{AFF}(\mathcal{A})$ with dense domain $\mathcal{D}(H)$ and the set of $\mathscr{T}_{\mathbb{P}}$-compatible generators $\{X_1, X_2, ..., X_d\}$ with common localizing domain $\mathcal{D}_c \subset \mathfrak{H}$ [6, Remark 3.4.7] meet the following assumptions:

(i) The joint core $\mathcal{D}_c(H) := \mathcal{D}_c \cap \mathcal{D}(H)$ is a densely defined core for H, and $X_k[\mathcal{D}_c(H)] \subset \mathcal{D}_c(H)$ for all $k = 1, \dots, d$;
(ii) $H[\mathcal{D}_c(H)] \subset \mathcal{D}_c$ and the formal commutators

$$J_k := -\mathrm{i}\,(X_k H - H X_k)\,, \qquad k = 1, \dots, d \tag{3}$$

are essentially self-adjoint on $\mathcal{D}_c(H)$, and therefore uniquely extend to self-adjoint operators denoted (with abuse of notation) by $J_k = \nabla_k(H)$.
(iii) All the J_k are infinitesimally H-bounded, i.e., for any $\delta > 0$ there are constants $a > 0$ and $\delta > b > 0$ such that

$$\|J_k\varphi\|_{\mathfrak{H}} \leq a\|\varphi\|_{\mathfrak{H}} + b\|H\varphi\|_{\mathfrak{H}}, \qquad \varphi \in \mathcal{D}_c(H)$$

for all $k = 1, \dots, d$.
(iv) $J_k \in \mathtt{AFF}(\mathcal{A})$ for every $k = 1, \dots, d$.

The vector-valued operator

$$J := \nabla(H) = (\nabla_1(H), \dots, \nabla_d(H))$$

will be called *current operator*.

Hypothesis 4 (Perturbed Dynamics) Let $\mathbb{R} \supseteq I \ni t \mapsto H(t) \in \mathtt{AFF}(\mathcal{A})$ be a path such that:

(i) $H(0) = H$ and $\mathcal{D}(H(t)) = \mathcal{D}(H)$ for every $t \in \mathbb{R}$;
(ii) For every $t \in I$ the operator $H(t)$ meets the properties of Hypotheses 3 and therefore there exists the time-dependent current $J(t) = \nabla(H(t))$;

(iii) There exists a unique strongly jointly continuous map $I^2 \ni (s,t) \mapsto U(s,t) \in \mathcal{A}$, called *unitary propagator*, which leaves invariant the domain $\mathcal{D}(H)$ and solves the differential equation

$$\mathrm{i}\,\partial_t \psi_s(t) \;=\; H(t)\psi_s(t)\,, \qquad \psi_s(s) = \psi_0 \in \mathcal{D}(H)$$

in the sense that $\psi_s(t) = U(t,s)\psi_0$.

The unitary propagator verifies the properties $U(t,t) = \mathbf{1}$ and $U(t,s)U(s,r) = U(t,r)$ for every $t,s,r \in I$. Suitable conditions for the existence of the unitary propagator are given in [6, Theorem 5.2.4]. Since $U(t,s) \in \mathcal{A}$, it can be used to define dynamics on $\mathcal{A}$ and $\mathfrak{L}^p(\mathcal{A})$ by

$$\eta_{(t,s)}(A) \;:=\; U(t,s)AU(s,t)\,, \qquad (t,s) \in \mathbb{R}^2\,, \quad A \in \mathcal{A} \text{ or } \mathfrak{L}^p(\mathcal{A}). \tag{4}$$

These are isometries jointly strongly continuous in t and s on $\mathcal{A}$ and in each $\mathfrak{L}^p(\mathcal{A})$. Moreover, it turns out that the map $I \ni t \mapsto \eta_{(t,s)}(A) \in \mathfrak{L}^p(\mathcal{A})$ is differentiable for every fixed s, and

$$\mathrm{i}\,\partial_t \eta_{(t,s)}(A) \;=\; [H(t), \eta_{(t,s)}(A)]_\dagger$$

whenever HA and HA^* are in $\mathfrak{L}^p(\mathcal{A})$ [6, Proposition 5.2.6].

Hypothesis 5 (Gap Condition) Let $\sigma_*(t) \subset \sigma(H(t))$ be a subset of spectrum of $H(t)$ such that there exist continuous function $f_\pm : I \to \mathbb{R}$ defining intervals $G(t) = [f_-(t), f_+(t)]$ so that $\sigma_*(t) \subset G(t)$ and

$$g \;:=\; \inf_{t\in I} \mathrm{dist}\big(G(t), \sigma(H(t)) \setminus \sigma_*(t)\big)$$

is strictly positive. We will denote by $P_*(t) := \chi_{\sigma_*(t)}(H(t))$ the spectral projection of $H(t)$ on the gapped spectral patch $\sigma_*(t)$.

Hypothesis 6 (Regularity of the Equilibrium State) Let ρ be an equilibrium state for H. We assume that ρ is p-regular, i.e.

(i) $\rho \in \mathcal{A}^+ \cap \mathfrak{M}^{1,1}(\mathcal{A}) \cap \mathfrak{M}^{1,p}(\mathcal{A})$;
(ii) $H(t)\rho(t) \in \mathfrak{M}^{1,1}(\mathcal{A}) \cap \mathfrak{M}^{1,p}(\mathcal{A})$ for all $t \in I$.

The state ρ can be evolved also by the perturbed dynamics $\eta_{(t,s)}$ through the prescription

$$\rho(t) \;:=\; \eta_{(t,0)}(\rho) \;=\; U(t,0)\rho U(0,t)\,, \qquad t \in \mathbb{R}. \tag{5}$$

Since $\rho(t)^* = \rho(t)$ for every $t \in I$ it follows that the generalized commutator $[H(t), \rho(t)]_\dagger$ is well defined and from [6, Theorem 5.2.6] one gets that $\rho(t)$ is the unique solution of

$$\begin{cases} \mathrm{i}\,\partial_t \rho(t) = [H(t), \rho(t)]_\dagger \,, \\ \\ \rho(0) = \rho \,, \end{cases} \tag{6}$$

where the derivative is taken in $\mathfrak{L}^1(\mathcal{A})$ or $\mathfrak{L}^p(\mathcal{A})$.

4 The King-Smith and Vanderbilt Formula for the Orbital Polarization

In this section we present the main results of this paper, i.e., the derivation of the *King-Smith and Vanderbilt formula* for the orbital polarization.

Let us start by saying that a self-adjoint map $\mathbb{R} \supseteq I \ni t \mapsto H(t) \in \mathtt{AFF}(\mathcal{A})$ is N-differentiable in the uniform sense (in the interval I) if the map

$$I \ni t \longmapsto \big(\mathrm{i}\mathbf{1} - H(t)\big)^{-1} \in \mathcal{A}$$

is N-differentiable with respect to the norm topology of $\mathcal{A}$. We will denote with $C^N(I, \mathcal{A}) \subset \mathcal{A}$ the space of $\mathcal{A}$-valued maps which are N-differentiable.

Remark 4 Notice that if the map $I \ni t \mapsto H(t) \in \mathtt{AFF}(\mathcal{A})$ is N-differentiable in the uniform sense, then it is also true that

$$\big(z\mathbf{1} - H(\cdot)\big)^{-1} \in C^N(I, \mathcal{A})$$

for each $z \in \mathbb{C}$ which lies in the resolvent set of $H(t)$, for any $t \in I$. Indeed, one has that

$$\big(z\mathbf{1} - H(t)\big)^{-1} - \big(\mathrm{i}\mathbf{1} - H(t)\big)^{-1} = (\mathrm{i} - z)\big(z\mathbf{1} - H(t)\big)^{-1}\big(\mathrm{i}\mathbf{1} - H(t)\big)^{-1}.$$

Thus,

$$\big(z\mathbf{1} - H(t)\big)^{-1} = F(z, t)\big(\mathrm{i}\mathbf{1} - H(t)\big)^{-1}$$

where

$$F(z, \cdot) := \Big(\mathbf{1} - (\mathrm{i} - z)\big(\mathrm{i}\mathbf{1} - H(\cdot)\big)^{-1}\Big)^{-1} \in C^N(I, \mathcal{A}) \,.$$

Therefore

$$\partial_t^n \big(z\mathbf{1} - H(t)\big)^{-1} = \partial_t^n F(z,t)\big(\mathrm{i}\,\mathbf{1} - H(t)\big)^{-1} + F(z,t)\partial_t^n \big(\mathrm{i}\,\mathbf{1} - H(t)\big)^{-1}$$

for $0 < n \leq N$ in consequence of the fact that $(\mathrm{i} - z)^{-1}$ lies in the resolvent of $\big(\mathrm{i}\,\mathbf{1} - H(t)\big)^{-1}$ for every $t \in I$, and of the identity

$$\partial_t^n F(z,t) = -(\mathrm{i} - z)F(z,t)\partial_t^n \big(\mathrm{i}\,\mathbf{1} - H(t)\big)^{-1} F(z,t)\,, \qquad 0 < n \leq N\,.$$

Our first result is a generalization of [18, Proposition 4].

Theorem 1 *Let $\mathbb{R} \supseteq I \ni t \mapsto H(t) \in \mathrm{AFF}(\mathcal{A})$ be a path of self-adjoint operators which meets Hypothesis 1, 2, 3, 4. Let $P \in \mathcal{A}$ be an orthogonal projection which satisfies Hypothesis 6 with $p = 1, 2$. Let $P(t) := \eta_{(t,0)}(P)$ and $J_k(t)$ the k-th component of the current operator $J(t) = \nabla(H(t)) \in \mathrm{AFF}(\mathcal{A})$. Then, the current expectation value can be rewritten as*

$$\mathscr{T}_{\mathbb{P}}\big(J_k(t)P(t)\big) = \mathrm{i}\,\mathscr{T}_{\mathbb{P}}\big(P(t)[\partial_t P(t), \nabla_k(P(t))]_{(1)}\big) \tag{7}$$

for every $k = 1, \ldots, d$.

Proof From the hypothesis we have that $H(t)P(t) \in \mathfrak{L}^1(\mathcal{A})$ and $J_k(t) \in \mathrm{AFF}(\mathcal{A})$ for all $t \in I$. Then, using [6, Lemma 3.3.7] one obtains that

$$J_k(t)P(t) = \big(J_k(t)(H(t) - z\mathbf{1})^{-1}\big)\big((H(t) - z\mathbf{1})P(t)\big) \in \mathfrak{L}^1(\mathcal{A}),$$

where z (which can depend on t) lies in the resolvent set of $H(t)$. Therefore, the left-hand side of the expressions (7) is well defined. From the hypothesis, it also follows that $\nabla_k(P(t)) \in \mathfrak{L}^2(\mathcal{A})$ and $H(t)P(t) \in \mathfrak{L}^2(\mathcal{A})$ which implies

$$\partial_t P(t) = -\mathrm{i}\big(H(t)P(t) - (H(t)P(t))^*\big) \in \mathfrak{L}^2(\mathcal{A})\,.$$

Therefore, the commutator $[\partial_t P(t), \nabla_k(P(t))]_{(1)}$ is a well defined element in $\mathfrak{L}^1(\mathcal{A})$. For sake of notational simplicity, we suppress the t dependencies in the following computation. From [6, Lemma 3.2.14], one gets

$$\begin{aligned} \mathrm{i}\,\mathscr{T}_{\mathbb{P}}\big(P[\partial_t P, \nabla_k(P)]_{(1)}\big) &= \lim_{n\to\infty} \mathrm{i}\,\mathscr{T}_{\mathbb{P}}\big(P[\partial_t P P_n, \nabla_k(P)]_{(1)}\big) \\ &= \lim_{n\to\infty} \mathrm{i}\,\mathscr{T}_{\mathbb{P}}\big(P[\partial_t P P_n, \nabla_k(P)]_{(1)} P\big) \end{aligned}$$

where $P_n(t) := \chi_{[-n,n]}\big(H(t)\big)$ is the spectral projection of $H(t)$ on $[-n, n]$. Moreover, one has that

$$\begin{aligned} \mathrm{i}\,\partial_t P P_n &= [H, P]_{\dagger} P_n = H P P_n - (HP)^* P_n \\ &= H P P_n - P H P_n = (HP - PH)P_n, \end{aligned}$$

since $(HP)^* = PH$ when projected on P_n. Thus, beginning from the right-hand side of (7) and using the properties of the trace one finds

$$\begin{aligned}
&\mathrm{i}\,\mathcal{T}_{\mathbb{P}}\big(P[\partial_t P, \nabla_k(P)]_{(1)}\big)\\
&= \lim_{n\to\infty} \mathcal{T}_{\mathbb{P}}\big(P[(HP - PH)P_n, \nabla_k(P)]_{(1)}P\big)\\
&= \lim_{n\to\infty} \mathcal{T}_{\mathbb{P}}\big(P(HP - PH)P_n\nabla_k(P)P - P\nabla_k(P)(HP - PH)P_nP\big)\\
&= \lim_{n\to\infty} \mathcal{T}_{\mathbb{P}}\big(-PHP_n\nabla_k(P)P - P\nabla_k(P)HPP_nP\big)
\end{aligned}$$

where in the last equality we have used the identity $P\nabla_k(P)P = 0$ (which follows from $\nabla_k(P) = \nabla_k(P)^2$) to remove the term $PHPP_n\nabla_k(P)P$ which goes to 0 when $n \to \infty$, and the term $P\nabla_k(P)PHP_nP$. Since $PHP_n \in \mathcal{A}$ and using the "integration by part" between $\mathcal{T}_{\mathbb{P}}$ and ∇_k one gets

$$\begin{aligned}
&\mathrm{i}\,\mathcal{T}_{\mathbb{P}}\big(P[\partial_t P, \nabla_k(P)]_{(1)}\big)\\
&= \lim_{n\to\infty} \mathcal{T}_{\mathbb{P}}\big(\nabla_k(PHP_n)P\big) - \mathcal{T}_{\mathbb{P}}\big(P\nabla_k(P)HPP_n\big)\\
&= \lim_{n\to\infty} \mathcal{T}_{\mathbb{P}}\big(\nabla_k(P)HP_nP + P\nabla_k(HP_n)P\big) - \mathcal{T}_{\mathbb{P}}\big(P\nabla_k(P)HPP_n\big)\\
&= \lim_{n\to\infty} \mathcal{T}_{\mathbb{P}}\big(P\nabla_k(HP_n)P\big) + \lim_{n\to\infty} \mathcal{T}_{\mathbb{P}}\big(\nabla_k(P)HP_nP - P\nabla_k(P)HPP_n\big)\\
&= \lim_{n\to\infty} \mathcal{T}_{\mathbb{P}}\big(\nabla_k(HP_n)P\big) + \lim_{n\to\infty} \mathcal{T}_{\mathbb{P}}\big(\nabla_k(P)HP_nP - \nabla_k(P)HPP_nP\big)\\
&= \mathcal{T}_{\mathbb{P}}\big(\nabla_k(H)P\big) + 0\\
&= \mathcal{T}_{\mathbb{P}}\big(J_kP\big).
\end{aligned}$$

This concludes the proof.

Let $\rho_0 := \chi_{(-\infty,\epsilon_F]}(H)$ be the spectral projection of the Hamiltonian $H = H(0)$ with Fermi level $\epsilon_F \in \mathbb{R}$ in a gap of the spectrum of H. It is clear that ρ_0 is an initial equilibrium state for H. Let us assume that ρ_0 meets the regularity condition of Hypothesis 6 and let $\rho(t)$ be the solution of the Eq. (6). The variation of the *polarization* $\Delta\mathcal{P}_k$ between time $t = 0$ and $t = T$ due to the current J_k in the state ρ_0 is by definition

$$\Delta\mathcal{P}_k := \int_0^T \mathrm{d}t\; \mathcal{T}_{\mathbb{P}}\big(J_k(t)\rho(t)\big)\,, \qquad k = 1, ..., d\,. \tag{8}$$

By using Theorem 1 one can rewrite the quantity (8) as follows

$$\Delta\mathcal{P}_k := \mathrm{i}\int_0^T dt\; \mathcal{T}_{\mathbb{P}}\big(\rho(t)[\partial_t\rho(t), \nabla_k(\rho(t))]_{(1)}\big)\,, \qquad k = 1, 2, \ldots, d\,. \tag{9}$$

It is important to point out that Eq. (9) is not very useful in general, since it requires the knowledge of $\rho(t)$, which is not a function of $H(t)$. Thereby, we will use tools from *adiabatic perturbation theory* adapted from [18], in order to express the polarization in terms of the spectral projections of $H(t)$. For that, let us consider the Liouville equation

$$\varepsilon \partial_t \rho(t) = -\mathrm{i}\,[\, H(t),\ \rho(t)\,]_\dagger, \tag{10}$$

where $\varepsilon > 0$ is a small adiabatic parameter. With these ingredients, we present now the main Theorem of this work, which is based on some technical results described in Appendix 5.2.

Theorem 2 *Assume that the map* $[0, T] \ni t \mapsto H(t) \in AFF(\mathcal{A})$ *is* $N+2$-*differentiable in the uniform sense and meets Hypothesis 1, 2, 3, 4, 5 and 6. If* $\partial_t^n \big(\mathrm{i}\mathbf{1} - H(t)\big)^{-1}|_{t=0} = \partial_t^n \big(\mathrm{i}\mathbf{1} - H(t)\big)^{-1}|_{t=T} = 0$ *for all* $0 < n \leq N$, *then*

$$\Delta\mathscr{P}_k \;=\; \mathrm{i} \int_0^T \mathrm{d}t\ \mathscr{T}_{\mathbb{P}}\big(P_*(t)[\partial_t P_*(t),\ \nabla_k(P_*(t))]_{(1)}\big) + \mathcal{O}(\varepsilon^N) \tag{11}$$

where $P_*(t)$ *is the instantaneous spectral projection of* $H(t)$ *on the gapped spectral patch* $\sigma_*(t)$.

Proof From Theorem 3, there are projections P_N^ε such that

$$\|P_N^\varepsilon(t) - \rho(t)\| + \|\nabla_k\big(P_N^\varepsilon(t) - \rho(t)\big)\| + \|\partial_t\big(P_N^\varepsilon(t) - \rho(t)\big)\| = \mathcal{O}(\varepsilon^N)\,.$$

Since ρ is a p-regular initial equilibrium state, then for some $\varepsilon > 0$ small enough $\nabla_k H(t) P_N^\varepsilon(t) \in \mathfrak{L}^1(\mathcal{A})$. Furthermore, the Eq. (9), the norm bound property of the trace, and Corollary 2 yield

$$\Delta\mathscr{P}_k \;=\; \mathrm{i} \int_0^T \mathrm{d}t\ \mathscr{T}_{\mathbb{P}}\big(P_N^\varepsilon(t)[\,\partial_t P_N^\varepsilon(t),\ \nabla_k(P_N^\varepsilon(t))]_{(1)}\big) + \mathcal{O}(\varepsilon^N)\,.$$

Now let us show that the above integral is independent of ϵ. Indeed, since the first N derivates of $t \to \big(\mathrm{i}\mathbf{1} - H(t)\big)^{-1}$ vanish at the endpoints then by Theorem 3 one has that $P_N^\varepsilon(0) = P_*(0)$ and $P_N^\varepsilon(T) = P_*(T)$. As a consequence of the dominated convergence theorem [1, Corollary 5.8], and following the same algebraic steps used in the proof of [18, Theorem 1], one gets

$$\begin{aligned} &\partial_\epsilon \int_0^T \mathrm{d}t\ \mathscr{T}_{\mathbb{P}}\big(P_N^\varepsilon[\partial_t P_N^\varepsilon,\ \nabla_k(P_N^\varepsilon)]_{(1)}\big) \\ &= \int_0^T \mathrm{d}t\ \mathscr{T}_{\mathbb{P}}\big(P_N^\varepsilon[\partial_\varepsilon \partial_t P_N^\varepsilon,\ \nabla_k(P_N^\varepsilon)]_{(1)} + P_N^\varepsilon[\partial_t P_N^\varepsilon,\ \partial_\varepsilon \nabla_k(P_N^\varepsilon)]_{(1)}\big) \end{aligned}$$

$$= \mathcal{T}_{\mathbb{P}}\big(P_N^\varepsilon[\,\partial_\varepsilon P_N^\varepsilon,\, \nabla_k(P_N^\varepsilon)]_{(1)}\big)\Big|_0^T - \int_0^T \mathrm{d}t\; \mathcal{T}_{\mathbb{P}}(P_N^\varepsilon[\partial_\varepsilon P_N^\varepsilon,\, \nabla_k(\partial_t P_N^\varepsilon)]_{(1)})$$

$$+ \int_0^T \mathrm{d}t\; \mathcal{T}_{\mathbb{P}}\big(P_N^\varepsilon[\partial_t P_N^\varepsilon,\, \nabla_k(\partial_\varepsilon P_N^\varepsilon)]_{(1)}\big) = 0\,.$$

In both equalities above it was used that $\mathcal{T}_{\mathbb{P}}(\partial_\varepsilon P_N^\varepsilon \partial_t P_N^\varepsilon \nabla_k P_N^\varepsilon) = 0$, and the differentiability of the map $\varepsilon \mapsto P_N^\varepsilon$, which follows from Theorem 3, implies existence and equality of the mixed derivatives. Now, making $\varepsilon \to 0$ one obtains $P_N^\varepsilon(t) \to P_*(t)$ and in turn

$$\Delta\mathcal{P}_k \;=\; \mathrm{i}\int_0^T \mathrm{d}t\; \mathcal{T}_{\mathbb{P}}\big(P_*(t)[\partial_t P_*(t),\, \nabla_k(P_*(t))]_{(1)}\big) + \mathcal{O}(\varepsilon^N)\,.$$

This concludes the proof.

It is important to notice that the leading order term of (11) is invariant under diffeotopies. Consider a diffeotopy F between the projection-valued paths P_0 and P_1, i.e., a smooth function $F\colon [0, T]\times[0, 1] \to \mathfrak{M}^{1,1}(\mathcal{A})$ such that $F(t, 0) = P_0(t)$ and $F(t, 1) = P_1(T)$ for all $t \in [0, T]$. By replacing ε with the diffeotopy parameter $s \in [0, 1]$ in the proof of Theorem 2, one obtains immediately the equality

$$\Delta\mathcal{P}_k^0[P_0] \;=\; \Delta\mathcal{P}_k^0[P_1]\,, \qquad k = 1, 2, \ldots, d$$

where

$$\Delta\mathcal{P}_k^0[P_j] \;:=\; \int_0^T \mathrm{d}t\; \mathcal{T}_{\mathbb{P}}\big(P_j(t)[\,\partial_t P_j(t),\, \nabla_k(P_j(t))\,]_{(1)}\big)\,, \qquad j = 0, 1$$

stands for the leading order term of $\Delta\mathcal{P}_k$ with respect to the path P_j.

The last important step consists in proving that the leading order term $\Delta\mathcal{P}_k^0$ is topologically quantized. This can be shown following the argument of [18].

Corollary 1 *Under the assumptions of the Theorem 2, if the deformation is cyclic, that is $H(0) = H(T)$, it holds true that*

$$\Delta\mathcal{P}_k^0 \;=\; 2\pi\;\mathrm{Ch}(P_*)$$

where

$$\mathrm{Ch}(P_*) \;:=\; \frac{1}{2\pi\,\mathrm{i}}\int_0^T \mathrm{d}t\; \mathcal{T}_{\mathbb{P}}\big(P_*[\,\mathrm{i}\,\partial_t P_*,\, \nabla_k(P_*)]_{(1)}\big) \;\in\; \mathbb{Z}$$

is the Chern number of the differentiable map $[0, T] \ni t \mapsto P_*(t) \in \mathfrak{M}^{1,1}(\mathcal{A})$.

Proof If the deformation is cyclic, we can consider P_* as a projection-valued map in the C^*-algebra $C(\mathbb{S}^1)\otimes\mathcal{A}$, where $C(\mathbb{S}^1)$ are the continuous functions on the circle $\mathbb{S}^1 \cong [0, T)$. We can endow this C^*-algebra with the spatial derivation ∇_k, the time derivation $\mathrm{i}\,\partial_t$, and the trace given by

$$\widehat{\mathscr{T}_{\mathbb{P}}}(\widehat{A}) := \int_0^T \mathrm{d}t\ \mathscr{T}_{\mathbb{P}}\big(A(t)\big), \qquad \widehat{A} \in C(\mathbb{S}^1)\otimes\mathcal{A}\,.$$

Thus, it follows that $\Delta\mathscr{P}_k = 2\pi\,\mathrm{Ch}(P_*) + \mathcal{O}(\varepsilon^N)$, where $\mathrm{Ch}(P_*)$ is the Chern number of the element $\widehat{P}_* \in C(\mathbb{S}^1)\otimes\mathcal{A}$ defined by $t \mapsto P_*(t)$. It is well known that Chern numbers of projections take value in $\mathbb{Z}$ [4, 15].

The last result can be rephrased by saying that up to arbitrarily small corrections in the adiabatic parameter ε, the orbital polarization $\Delta\mathscr{P}_k$ is topologically quantized.

5 Applications

The mathematical framework described in the previous sections applies directly to the two most common cases, namely $\mathcal{G} = \mathbb{Z}^d$ and $\mathbb{R}^d$, which describe discrete (tight-binding) models and continuum systems, respectively. The case of discrete random models has been considered in detail in [18] and it will not be considered here. On the other hand, the treatment of the continuous random case is one of the main motivations for the writing of this work. In the following part of this section, we will present the formalism to describe the continuous random system and we will show that all the **Hypothesis 1–6** listed in Sect. 3 are satisfied for such models.

5.1 Continuous Models in Disordered Media

Let us focus on *ergodic magnetic media* [3, Section 4], i.e., non-interacting systems of charge particles submitted to a constant magnetic field $\mathtt{B}$, and to random potentials A_ω and V_ω (solids that can be either random, periodic or quasi-periodic), where ω runs in the ergodic probability space $(\Omega, \mathbb{P})$ of disorder with the ergodic $\mathbb{R}^n$-action τ. Let us consider the one-particle Hilbert space $\mathfrak{h} = L^2(\mathbb{R}^d)$, which describes the quantum states of the system. The constant magnetic field $\mathtt{B}$ can be represented by a $d\times d$ antisymmetric matrix with entries $\{\mathtt{B}_{j,k}\}$ and the associated vector potential $A : \mathbb{R}^d \to \mathbb{R}^d$ can be chosen as

$$A_j(x) = -\frac{1}{2}\sum_{k=1}^{d} \mathtt{B}_{j,k}x_k\,, \qquad j = 1,\ldots,d\,.$$

It turns out that

$$\frac{\partial A_k}{\partial x_j} - \frac{\partial A_j}{\partial x_k} = \mathrm{B}_{j,k}\,, \qquad \frac{\partial A_k}{\partial x_j} + \frac{\partial A_j}{\partial x_k} = 0\,, \qquad j,k = 1,\ldots,d\,.$$

On $\mathfrak{h}$ acts the *free Landau Hamiltonian*

$$H_0^A := (-\mathrm{i}\nabla - A)^2\,,$$

and the family of *random* magnetic Hamiltonians

$$H_\omega^A \equiv H_\omega^A(A_\omega, V_\omega) := (-\mathrm{i}\nabla - A - A_\omega)^2 + V_\omega\,, \qquad \omega \in \Omega\,.$$

In order to ensure the self-adjointness of the Hamiltonians H_0^A and H_ω^A, we assume the *Leinfelder–Simader conditions* on the potentials A, A_ω and V_ω (see [10] or [3, Section 2.1]). It turns out that H_ω^A is essentially self-adjoint on $C_0^\infty(\mathbb{R}^d)$. We will denote with $\mathcal{D}_\omega := \mathcal{D}(H_\omega^A)$ the domain of H_ω^A, i.e., the closure of $C_0^\infty(\mathbb{R}^d)$ with respect to the graph norm induced by H_ω^A. Observe that H_ω^A meets the *gauge covariance* property

$$H_\omega^{A+\nabla\chi} = \mathrm{e}^{-\mathrm{i}\chi}\, H_\omega^A\, \mathrm{e}^{\mathrm{i}\chi} \tag{12}$$

where $\chi : \mathbb{R}^d \to \mathbb{R}$ is considered as a multiplication operator on $\mathfrak{h}$.

Let us consider the direct integral

$$\mathfrak{H} := \int_\Omega^\oplus \mathrm{d}\,\mathbb{P}(\omega)\, \mathfrak{h}_\omega \simeq L^2(\Omega) \otimes L^2(\mathbb{R}^d) \simeq L^2(\Omega \times \mathbb{R}^d)\,,$$

and the subspaces $\mathcal{D} := L^2(\Omega) \otimes \mathcal{D}_\omega$ and $\mathcal{D}_c := L^2(\Omega) \otimes C_0^\infty(\mathbb{R}^d)$. The family of Hamiltonians $H^A := \{H_\omega^A\}_{\omega\in\Omega}$ defines an operator acting on the Hilbert space $\mathfrak{H}$. It turns out that H^A is essentially self-adjoint with core $\mathcal{D}_c$ and domain $\mathcal{D}$. Moreover, it follows that the maps $\omega \mapsto f(H_\omega^A)$ are measurable for every $f \in L^\infty(\mathbb{R})$ (see [3, Section 4.1] and references therein). In particular, the spectral projections of H^A define measurable maps and this is equivalent to say that H^A is affiliated to $\mathrm{Rand}(\mathfrak{H})$.

Let us introduce the vector-valued operators $G := -\mathrm{i}\nabla + A$. It turns out that every component of G commutes with H_0^A. For every y we consider the unitary operator $T_y := \mathrm{e}^{-\mathrm{i}y\cdot G}$ which acts on $\varphi \in \mathfrak{h}$ as

$$(T_y\varphi)(x) = \Theta^{\mathrm{B}}(y,x)\varphi(x-y) = \Theta^{\mathrm{B}}(y,x-y)\varphi(x-y) \tag{13}$$

where

$$\Theta^{\mathrm{B}}(y,x) := \mathrm{e}^{\frac{\mathrm{i}}{2}y\cdot\mathrm{B}\cdot x} = \mathrm{e}^{\frac{\mathrm{i}}{2}\sum_{j,k=1}^d \mathrm{B}_{j,k}y_j x_k}\,.$$

It follows that $T_y H_0^A T_y^* = H_0^A$ for every $y \in \mathbb{R}^d$. Furthermore, one can check that the map $\Theta^{\mathrm{B}} : \mathbb{R}^d \times \mathbb{R}^d \to \mathbb{U}(1)$ is a *twisting group 2-cocycle* according to [6, Definition 4.1.2]. We assume that the potentials A_ω and V_ω are covariant random variables, i.e., they meet

$$V_\omega(x - y) = V_{\tau_y(\omega)}(x) , \qquad A_\omega(x - y) = A_{\tau_y(\omega)}(x)$$

for $\mathbb{P}$-almost every $\omega \in \Omega$ and Lebesgue-almost every $x \in \mathbb{R}^d$. Then, one obtains the *covariance* relations

$$T_y H_\omega^A T_y^* = H_{\tau_y(\omega)}^A .$$

If one defines the unitary $U_y \in \mathcal{B}(\mathfrak{H})$ as

$$\big(U_y\, \psi\big)_{\tau_y(\omega)}(x) := \Theta^{\mathrm{B}}(y, x)\, \psi_\omega(x - y) , \tag{14}$$

where $\psi := \{\psi_\omega\}_{\omega \in \Omega}$ is any element of $\mathfrak{H}$ and on the left-hand side the symbol $(\cdot)_{\tau_y(\omega)}$ means the value of the vector $U_y \psi$ on the fiber of $\mathfrak{H}$ at $\tau_y(\omega)$, then one gets the *invariance* relations

$$U_y H^A U_y^* = H^A , \qquad \forall y \in \mathbb{R}^d .$$

Moreover, one has that the spectral projections of H^A commute with the U_y, and in turn H^A results affiliated with the von Neumann algebra

$$\mathcal{A} = \mathrm{Span}_{\mathbb{R}^d}\{U_y\}' \cap \mathrm{Rand}(\mathfrak{H}) .$$

Ultimately $H^A \in \mathtt{AFF}(\mathcal{A})$ according to **Hypothesis** 1.

Hypothesis 2 and 3 are verified if the $\{X_1, X_2, ..., X_d\}$ are the usual position operators which act constantly on the fibers of $\mathfrak{H}$, namely $(X_j \psi)_\omega(x) := x_j \psi_\omega(x)$ for every $\{\psi_\omega\}_{\omega \in \Omega} \in \mathfrak{H}$. Observe that the localization domain $\mathcal{D}_c := L^2(\Omega) \otimes C_0^\infty(\mathbb{R}^d)$ is also a core for H^A, therefore $\mathcal{D}_c(H^A) := \mathcal{D}_c \cap \mathcal{D}(H^A) = \mathcal{D}_c$. Finally, the components of the current are $J_k := \{J_{k,\omega}\}_{\omega \in \Omega}$ with

$$J_{k,\omega} := 2(-\mathrm{i}\,\partial_k - A_k - A_{\omega,k}) , \qquad k = 1, \dots, d .$$

The effects of the external deformation on the system are modeled by a sufficiently regular function $w : [0, T] \to \mathbb{R}$ with the boundary conditions $w(0) = 0 = w(T)$, which enters in the definition of the time-dependent Hamiltonian $H^A(t) := \{H_\omega^A(t)\}_{\omega \in \Omega}$ defined by

$$H_\omega^A(t) := H_\omega^A + w(t)\, W_\omega ,$$

where $W := \{W_\omega\}_{\omega\in\Omega} \in \mathrm{Rand}(\mathfrak{H})$ is a bounded random potential. In view of the Kato-Rellich theorem [16, Theorem X.12] one has that $\mathcal{D}(H^A(t)) = \mathcal{D}(H^A)$ for every $t \in [0, T]$. Moreover, it is straightforward to check $J_k(t) = J_k$ for every $k = 1, \ldots, d$, namely the time-dependent current equals the stationary current. Let us assume that there is a *Fermi energy* $\epsilon_F \in \mathbb{C} \setminus \sigma(H^A)$ inside the resolvent set of H^A. If $\|w\|_\infty \ll 1$ is sufficiently small the gap around ϵ_F doesn't closed during the time-dependent perturbation and one gets that $\epsilon_F \in \mathbb{C} \setminus \sigma(H^A(t))$ for every $t \in [0, T]$. This is in particular a gap condition stronger than that assumed in **Hypothesis 5**, which is therefore automatically satisfied. In particular the relevant spectral patch can be chosen as $\sigma_*(t) := (-\infty, \epsilon_F] \cap \sigma(H^A(t))$. In order to complete the check of the validity of **Hypothesis 4** we need to prove that there exists the *unitary propagator* $[0, T]^2 \ni (s, t) \mapsto U^A(s, t) \in \mathcal{A}$ associated to $H^A(t)$. For that, it is sufficient to show that the conditions listed in [16, Theorem X.70] (see also [23, Section XIV.4]) are satisfied. The main object is the operator

$$\begin{aligned} C(t, s) \; &:= \; \left(H^A(t) - H^A(s)\right) \frac{1}{H^A(s) - \xi \mathbf{1}} \\ &= \; (w(t) - w(s))\, W \frac{1}{H^A(s) - \xi \mathbf{1}} \, . \end{aligned}$$

If one assumes that $w \in C^1([0, T])$, then $C(t, s)$ automatically fulfills all the conditions for the construction of the unitary propagator.

Finally, the relevant initial equilibrium state for H^A is given by the spectral projection of H^A on the Fermi energy $\rho_0 := \chi_{(-\infty, \epsilon_F]}(H^A)$. Let us observe that, in view of the gap condition, the step function $\chi_{(-\infty, \epsilon_F]}$ can be replaced by a smooth and compactly supported function. Therefore, from [3, Proposition 4.2] one has that also **Hypothesis 6** is verified.

5.2 *Continuous Periodic Models*

The case of a continuous periodic model has been rigorously studied in [14]. However, it represents a special case of the model described in Sect. 5.1 when the ergodic topological dynamical system $(\mathbb{T}^d, \mathbb{R}^d, \tau, \mu)$ is given by a d-dimensional torus $\mathbb{T}^d := \mathbb{R}^d/\Gamma$, with $\Gamma \simeq \mathbb{Z}^d$ a lattice, and its normalized Haar measure μ. Evidently, the action of $\mathbb{R}^d$ on $\mathbb{T}^d$ is given by translations and the resulting dynamical system is minimal, which means that the orbit of any point $\omega \in \mathbb{T}^d$ under the action of $\mathbb{R}^d$ is dense.

Let us fix the reference point $\omega_0 = [0]$. In view of the covariance conditions one gets

$$V_{\omega_0}(x - \gamma) \; = \; V_{\tau_\gamma(\omega_0)}(x) \; = \; V_{\omega_0}(x) \, , \qquad \forall\, \gamma \in \Gamma$$

since $\tau_\gamma(\omega_0) = [0+\gamma] = [0]$. Moreover, this is independent of the election of the reference point ω_0. Therefore it turns out that V_{ω_0}, and similarly A_{ω_0} are Γ-periodic potentials which will be denoted simply by V_Γ and A_Γ, respectively. Note also that for any $\omega \in \mathbb{T}^d$ and $A \in \mathcal{A}$ it holds true that

$$T_y A_{\omega_0} T_y^* \;=\; A_{\tau_y(\omega_0)} \;=\; A_\omega\,,$$

where $\omega = [y]$. Therefore, if one factor the action of $\mathbb{R}^d$ as $\mathbb{R}^d = \mathbb{T}^d \times \Gamma$ one can decompose the algebra $\mathcal{A}$ as follows

$$\mathcal{A} \;=\; \int_{\mathbb{T}^d}^{\oplus} \mathrm{d}\,\mu(\omega)\,\mathfrak{A}_\Gamma$$

where

$$\begin{aligned}\mathfrak{A}_\Gamma \;&:=\; \{A \in \mathcal{B}(L^2(\mathbb{R}^d)) \mid [T_\gamma, A] = 0\,,\;\; \forall\,\gamma \in \Gamma\}\\ &=\; \mathrm{Span}_\Gamma\{T_\gamma\}'\end{aligned}$$

is the von Neumann algebra of the bounded operators on the Hilbert space $L^2(\mathbb{R}^d)$ which are invariant under the action of the translations T_γ defined by (13). Thus, there is a $*$-isomorphism of von Neumann algebras $\mathcal{A} \simeq \mathfrak{A}_\Gamma$ given by the identification $\mathcal{A} \ni A \mapsto A_{\omega_0} \in \mathfrak{A}_\Gamma$. Hence, in the case of continuous periodic models, it is sufficient to work with the algebra $\mathfrak{A}_\Gamma$ defined on the Hilbert space $L^2(\mathbb{R}^d)$.

In the case that the algebra $\mathrm{Span}_\Gamma\{T_\gamma\}$ contains a commutative C^*-subalgebra $\mathfrak{I}_\Gamma$ (rational magnetic flux), then the von Neumann's spectral Theorem [7, Part II, Chap.6, Theorem 1], provides a (new) direct integral decomposition

$$L^2(\mathbb{R}^d) \;:=\; \int_{\sigma(\mathfrak{I}_\Gamma)}^{\oplus} \mathrm{d}\nu(k)\,\mathfrak{H}_k \tag{15}$$

where ν is a basic spectral measure and $\sigma(\mathfrak{I}_\Gamma)$ is the Gelfand spectrum of $\mathfrak{I}_\Gamma$. Moreover, there is a unitary map $\mathcal{F}$, called the Bloch-Floquet transform, such that $\mathcal{F}\mathfrak{A}_\Gamma\mathcal{F}^{-1}$ is contained in the bounded decomposable operators over the direct integral, that is,

$$\mathcal{F}A\mathcal{F}^{-1} \;=\; \int_{\sigma(\mathfrak{I}_\Gamma)}^{\oplus} \mathrm{d}\nu(k)\,A(k)\,, \qquad A \in \mathfrak{A}_\Gamma,$$

where $A(k) \in \mathcal{B}(\mathfrak{H}_k)$. Note also that the trace per unite of volume $\mathscr{T}$ on $\mathfrak{A}_\Gamma$ is given by

$$\mathscr{T}(A) \;=\; \frac{1}{\mu\big(\sigma(\mathfrak{I}_\Gamma)\big)} \int_{\sigma(\mathfrak{I}_\Gamma)}^{\oplus} \mathrm{d}\nu(k)\,\mathrm{Tr}_{\mathfrak{H}_k}\big(A(k)\big)\,. \tag{16}$$

Appendix: Adiabatic Theorem

The aim of this section is to extend the *adiabatic Theorem* proved in [18, Appendix A] to our setting.

The first result concerns the regularity of the spectral projections on the gap of $H(t)$.

Lemma 1 *Suppose that the map* $[0, T] \ni t \mapsto H(t) \in \mathtt{AFF}(\mathcal{A})$ *is* N*-differentiable in the uniform sense and that the Hypothesis 1, 2 and 5 hold. Then, the spectral projection map* $P_*(t) = \chi_{\sigma_*(t)}(H(t))$ *fulfills* $P_* \in C^N([0, T], \mathcal{A})$.

Proof Let $\gamma(t) \subset \mathbb{C}$ be a closed curve in the resolvent set of $H(t)$ surrounding $\sigma_*(t)$ in the positive sense with

$$\mathrm{dist}\big(\gamma(t), \sigma(H(t)) \setminus \sigma_*(t)\big) \;\leqslant\; \frac{g}{2} \,,$$

where g is defined in Hypotheses 5. Then

$$P_*(t) \;=\; \frac{1}{\mathrm{i}2\pi} \oint_{\gamma(t)} \mathrm{d}z \; \big(z\mathbf{1} - H(t)\big)^{-1} \,.$$

Since $f_\pm(t)$ are continuous functions, one has that $\gamma(t+h)$ is homotopic to $\gamma(t)$ in the resolvent set of $H(t+h)$ for $|h|$ small enough, and hence

$$\begin{aligned} P_*(t+h) \;&=\; \frac{1}{\mathrm{i}2\pi} \oint_{\gamma(t+h)} \mathrm{d}z \; \big(z\mathbf{1} - H(t+h)\big)^{-1} \\ &=\; \frac{1}{\mathrm{i}2\pi} \oint_{\gamma(t)} \mathrm{d}z \; \big(z\mathbf{1} - H(t+h)\big)^{-1} \,. \end{aligned}$$

Since $\partial_t^n (z\mathbf{1} - H(t))^{-1} \in \mathcal{A}$ for all $n \leq N$, then one deduce with an induction on n that

$$\partial_t^n P_*(t) \;=\; \frac{1}{\mathrm{i}2\pi} \oint_{\gamma(t)} \mathrm{d}z \; \partial_t^n \big(z\mathbf{1} - H(t)\big)^{-1} \;\in\; \mathcal{A} \,.$$

This concludes the proof.

The next two results concern the existence of the *superadiabatic projections* and are adaptions of [18, Proposition 7 and Theorem 9].

Proposition 1 *Under the assumptions of the Lemma 1, there exist unique maps* $P_n \in C^{N+2-n}([0, T], \mathcal{A})$, *with* $1 \leqslant n \leqslant N$, *such that the functions*

$$\widetilde{P}^{\varepsilon}_m(t) \;=\; \sum_{n=0}^{m} \varepsilon^n P_n(t)$$

for $0 \leqslant m \leqslant N$ *and* $P_0(t) = P_*(t) = \chi_{\sigma_*(t)}(H(t))$*, satisfies*

$$\big(\widetilde{P}_m^\varepsilon\big)^2 \;=\; \widetilde{P}_m^\varepsilon + \varepsilon^{m+1} G_{m+1} + \mathcal{O}(\varepsilon^{m+2}) \tag{17}$$

with $G_{m+1} := \sum_{n=1}^m P_n P_{m+1-n}$ *and*

$$\mathrm{i}\varepsilon \partial_t \widetilde{P}_m^\varepsilon(t) \;-\; [H(t), \widetilde{P}_m^\varepsilon(t)]_\dagger \;=\; \mathrm{i}\,\varepsilon^{m+1} \partial_t P_m(t)\,. \tag{18}$$

Furthermore, if $\partial_t^n\big(\mathrm{i}\mathbf{1} - H(t)\big)^{-1}|_{t=t_0} = 0$ *for some* $t_0 \in [0, T]$ *and all* $n \leqslant N$*, then* $P_n(t_0) = 0$ *for all* $1 \leqslant n \leqslant N$.

Proof This result can be obtained by using induction in m. For $m = 0$, with $\widetilde{P}_0^\varepsilon(t) := P_*(t)$ the instantaneous spectral projection of $H(t)$, it follows that

$$\big(\widetilde{P}_0^\varepsilon\big)^2 \;=\; \widetilde{P}_0^\varepsilon\,, \qquad \mathrm{i}\varepsilon\partial_t \widetilde{P}_0^\varepsilon - [H(t), \widetilde{P}_0^\varepsilon]_\dagger \;=\; \mathrm{i}\,\varepsilon \partial_t P_*(t) \;=\; \mathcal{O}(\varepsilon)\,.$$

Assume now that (17) and (18) holds for P_j with $j = 0, \ldots, m$. Thus, if we define P_{m+1} as

$$\begin{aligned} P_{m+1} \;:=&\; P_*^\perp G_{m+1} P_*^\perp - P_* G_{m+1} P_* \\ &+ \frac{1}{\mathrm{i}2\pi} \oint_{\gamma(t)} \mathrm{d}z\ (z\mathbf{1} - H)^{-1} [\partial_t P_m, P_*]_\dagger (z\mathbf{1} - H)^{-1}\,, \end{aligned}$$

with $\gamma(t)$ any curve encircling $\sigma_*(t)$ once in the positive sense, one can show that (17) and (18) hold for P_{m+1} just following the same steps in [18, Proposition 7]. Moreover, since $\mathcal{A}$ is closed under holomorphic functional calculus, one gets $P_{m+1} \in \mathcal{A}$. Finally, if $\partial_t^n\big(\mathrm{i}\mathbf{1} - H(t)\big)^{-1}|_{t=t_0} = 0$ then it is also true that $\partial_t^n\big(z\mathbf{1} - H(t)\big)^{-1}|_{t=t_0} = 0$ for each z in the resolvent of $H(t)$ for every t in $[0, T]$ (see Remark 4). Thus, $\dot{P}_*(t_0) = 0$ and by the construction of P_{m+1}, it follows also that $P_1(t_0) = 0$. Using induction again we conclude the last statement.

In order to simplify the notation, we introduce the following norm

$$\|A(t)\|_{\mathcal{S},k} \;:=\; \|A(t)\| + \|\partial_t A(t)\| + \|\nabla_k A(t)\| \qquad k = 1, \ldots, d$$

for any differentiable path $t \mapsto A(t)$ in $C^1([0, T], \mathcal{A})$.

Theorem 3 *Let the map* $[0, T] \ni t \mapsto H(t) \in \mathrm{AFF}(\mathcal{A})$ *be* N*-differentiable in the uniform sense for some* $N \in \mathbb{N}$ *and assume the hypothesis of Lemma 1. Then, there are constants* $\varepsilon_N > 0$, $c_N < \infty$ *and orthogonal projections* $P_N^\varepsilon(t) \in \mathfrak{M}^{1,1}(\mathcal{A})$ *such that the map* $(0, \varepsilon_N) \ni \varepsilon \to P_N^\varepsilon(\cdot) \in C^2([0, T], \mathfrak{M}^{1,1}(\mathcal{A}))$*, and the following properties hold uniformly in* t*:*

$$\|P_N^\varepsilon(t) - P_*(t)\|_{\mathcal{S},k} \;<\; c_N \varepsilon\,, \tag{19}$$

$$\left\| \mathrm{i}\varepsilon\partial_t P_N^\varepsilon - \left[H(t), P_N^\varepsilon \right]_\dagger \right\|_{S,k} < c_N \varepsilon^{N+1} . \tag{20}$$

Moreover, if $\partial_t^n \left(\mathrm{i}\mathbf{1} - H(t)\right)^{-1}|_{t=t_0} = 0$ *for some* $t_0 \in [0, T]$*, then* $P_N^\varepsilon(t_0) = P(t_0)$.

Proof We know by (17) that there is a constant c_N such that

$$\|(\widetilde{P}_m^\varepsilon)^2 - \widetilde{P}_m^\varepsilon\| \leqslant c_N \varepsilon^{N+1} .$$

Therefore, the spectral mapping theorem provides

$$\begin{aligned}\sigma(\widetilde{P}_m^\varepsilon) &\subset [-c_N\varepsilon^{N+1}, c_N\varepsilon^{N+1}] \cup [1 - c_N\varepsilon^{N+1}, 1 + c_N\varepsilon^{N+1}] \\ &\subset \left[-\frac{1}{4}, \frac{1}{4}\right] \cup \left[\frac{3}{4}, \frac{5}{4}\right]\end{aligned}$$

where the latter holds for $\varepsilon < \varepsilon_N = (4c_N)^{-\frac{1}{N+1}}$. Thus, one can define for any $\varepsilon < \varepsilon_N$

$$P_N^\varepsilon := \frac{1}{\mathrm{i}2\pi} \oint_{|z-1|=\frac{1}{2}} \mathrm{d}z\ (z\mathbf{1} - \widetilde{P}_m^\varepsilon)^{-1},$$

where the integral is taken in the positive sense. It follows that $P_N^\varepsilon \in \mathcal{A}$. Moreover by adapting the arguments used in [3, Proposition 4.2] one can show that $P_N^\varepsilon(t) \in \mathfrak{M}^{1,1}(\mathcal{A})$ for every $t \in [0, T]$. Using the fact that $\widetilde{P}_m^\varepsilon$ is differentiable one obtains that $\varepsilon \mapsto P_N^\varepsilon(\cdot)$ is in $C^2([0, T], \mathfrak{M}^{1,1}(\mathcal{A}))$. Now one can obtain (19) by taking the norms of

$$P_N^\varepsilon - P_* = \frac{1}{12\pi} \oint_{|z-1|=\frac{1}{2}} \mathrm{d}z\ (z\mathbf{1} - \widetilde{P}_m^\varepsilon)^{-1}(\widetilde{P}_m^\varepsilon - P_*)(z\mathbf{1} - P_*)^{-1} ,$$

of its time derivate ∂_t and of its gradient ∇_k. In the same way, we can use

$$\begin{aligned}&\mathrm{i}\varepsilon\partial_t P_N^\varepsilon - [H, P_N^\varepsilon]_\dagger \\ &= \frac{1}{\mathrm{i}2\pi} \oint_{|z-1|=\frac{1}{2}} \mathrm{d}z\ \left(\mathrm{i}\varepsilon\partial_t (z\mathbf{1} - \widetilde{P}_m^\varepsilon)^{-1} - [H, (z\mathbf{1} - \widetilde{P}_m^\varepsilon)^{-1}]_\dagger\right) \\ &= \frac{1}{\mathrm{i}2\pi} \oint_{|z-1|=\frac{1}{2}} \mathrm{d}z\ (z\mathbf{1} - \widetilde{P}_m^\varepsilon)^{-1} \left(\mathrm{i}\varepsilon\partial_t \widetilde{P}_m^\varepsilon - [H, \widetilde{P}_m^\varepsilon]_\dagger\right) (z\mathbf{1} - \widetilde{P}_m^\varepsilon)^{-1} \\ &= \frac{\varepsilon^{N+1}}{\mathrm{i}2\pi} \oint_{|z-1|=\frac{1}{2}} \mathrm{d}z\ (z\mathbf{1} - \widetilde{P}_m^\varepsilon)^{-1} \partial_t P_N (z\mathbf{1} - \widetilde{P}_m^\varepsilon)^{-1}\end{aligned}$$

to show (20). The last claim follows directly from Proposition 1.

The proof of the following result is an adaption of [18, Corollary 5].

Corollary 2 *Let $\rho^{\varepsilon}_{sa}(t)$ be the unique solution of the equation*

$$\mathrm{i}\varepsilon\partial_t\rho^{\varepsilon}_{sa}(t) \;=\; \mathrm{i}\,[H(t),\rho^{\varepsilon}_{sa}(t)]_{\dagger}\;, \qquad \rho^{\varepsilon}_{sa}(0) \;:=\; P^{\varepsilon}_{N}(0).$$

Then under the hypothesis of Lemma 1 one gets that

$$\rho^{\varepsilon}_{sa}(t) \;=\; P^{\varepsilon}_{N}(t) + \Delta^{\varepsilon}(t)$$

with $\|\Delta^{\varepsilon}(t)\|_{\mathcal{S}} = \mathcal{O}(\varepsilon^{N}|t|)$.

Acknowledgments GD's research is supported by the grant Fondecyt Regular - 1230032. DP's research is supported by ANID-Subdirecci6n de Capital Humano/ Doctorado Nacional/ 2022-21220144. GD's would like to thank the Alexander von Humboldt Foundation for supporting his stay at the University of Erlangen-Nürnberg during July 2022 where the large part of this work was completed. He is also grateful to Camping due barche (Scanzano, Italy) where the peaceful atmosphere of this place provided an invaluable help for the preparation of the final version of this manuscript. The authors are indebted to M. Lein and S. Teufel for many stimulating discussions.

References

1. Bartle, G.: The Elements of Integration and Lebesgue Measure. John Wiley & Sons, New York (1995)
2. Bellissard, J.: K-theory of C^*-Algebras in solid state physics. In: Dorlas, T.C., Hugenholtz, N.M., Winnink, M. (eds.) Statistical Mechanics and Field Theory: Mathematical Aspects. Lecture Notes in Physics, vol. 257, pp. 99–156. Springer-Verlag, Berlin (1986)
3. Bouclet, J.M., Germinet, F., Klein, A., Schenker, J.H.: Linear response theory for magnetic Schrödinger operators in disordered media. J. Func. Anal. **226**, 301–372 (2005)
4. Connes, A.: Noncommutative Geometry. Academic Press, Cambridge (1994)
5. De Nittis, G., Lein, M.: Topological Polarization in Graphene-like Systems. J. Phys. A Math. Theor. **46**, 385001 (2013)
6. De Nittis, G., Lein, M.: Linear Response Theory: An Analytic-Algebraic Approach. Springer Briefs in Mathematical Physics, vol. 21. Springer, Berlin (2017)
7. Dixmier, J.: Von Neumann Algebras. North-Holland Publishing, Amsterdam (1981)
8. King-Smith, R., Vanderbilt, D.: Theory of polarization of crystalline solids. Phys. Rev. B **47**, 1651–1654 (1993)
9. Lenz, D.H.: Random operators and crossed products. Math. Phys. Anal. Geom. **2**(2), 197–220 (1999)
10. Leinfelder, H., Simader, C.: Schrödinger operators with singular magnetic vector potentials. Math. Z. **176**, 1–19 (1981)
11. Martin, R.: Comment on calculations of electric polarization in crystals. Phys. Rev. B **9**, 1998–1203 (1974)
12. Nelson, E.: Notes on non-commutative integration. J. Func. Anal. **15**, 103–116 (1974)
13. Nenciu, G.: Linear adiabatic theory: exponential estimates. Commun. Math. Phys. **152**, 479–496 (1993)
14. Panati, G., Sparber, C., Teufel, S.: Geometric currents in piezoelectricity. Arch. Rat. Mech. Anal. **191**, 387–422 (2009)
15. Prodan, E., Schulz-Baldes, H.: Bulk and Boundary Invariants for Complex Topological Insulators: From K-Theory to Physics. Mathematical Physics Studies. Springer, Berlin (2016)

16. Reed, M., Simon, B.: Fourier analysis. Self-Adjointness (Methods of Modern Mathematical Physics II). Academic Press, New York-London (1975)
17. Resta, R.: Macroscopic polarization in crystalline dielectrics: the geometric phase approach. Rev. Mod. Phys. **66**(3), 899–915 (1994)
18. Schulz-Baldes, H., Teufel, S.: Orbital Polarization and Magnetization for Independent Particles in Disordered Media. Commun. Math. Phys. **319**, 649–681 (2013)
19. Terp, M.: Lp-Spaces Associated with von Neumann Algebras. Copenhagen University, Copenhagen (1981)
20. Teufel, S.: Adiabatic Perturbation Theory in Quantum Dynamics. Lecture Notes in Mathematics, vol. 1821. Springer, Heidelberg (2003)
21. Thouless, D.J.: Quantization of particle transport. Phys. Rev. B **27**, 6083–6087 (1983)
22. Yeadon, F.J.: Convergence of measurable operators. Proc. Camb. Philos. Soc. **74**, 257–268 (1973)
23. Yosida, K.: Functional Analysis, 6th edn. Springer, Berlin (1980)

Quadratic Forms for Aharonov-Bohm Hamiltonians

Davide Fermi

***2020 Mathematics Subject Classification*:** 47A07, 49J45, 81Q10

1 Introduction

In a pioneering work dating back to 1949 [17], Ehrenberg and Siday foretold that charged quantum particles confined within regions where the electromagnetic field vanishes do still experience a phase shift, which can be expressed in terms of non-zero electromagnetic potentials. Their prediction did not attract great attention until Aharonov and Bohm re-discovered it in an independent research of 1959 [2], eventually reaching a much larger audience. Ever since then, people referred to the said phenomenon as the "Aharonov-Bohm effect" [5, 29]. Despite sound experimental evidence [33], this groundbreaking discovery generated conflicting views and a long-standing debate about the reality of electromagnetic potentials and the tenability of the locality principle in quantum mechanics [3]. This controversy somehow continues even nowadays [4, 20, 34], though there are strong indications that an explanation in terms of local field interactions can actually be attained within the framework of QED [21, 25].

The prototypical Aharonov-Bohm configuration consists of a single, non-relativistic, spinless and electrically charged quantum particle moving outside of a long thin solenoid. More precisely, attention is restricted to a low-energy regime in which the De Broglie wavelength of the particle is much larger than the solenoid section diameter and, at the same time, much smaller than the solenoid longitudinal length. The natural first order approximation considers an ideal solenoid of infinite

D. Fermi (✉)
Dipartimento di Matematica, Politecnico di Milano, Milano, Italy
e-mail: davide.fermi@polimi.it

M. Correggi, M. Falconi (eds.), *Quantum Mathematics I*,
Springer INdAM Series 57, https://doi.org/10.1007/978-981-99-5894-8_7

length, zero cross section and finite total magnetic flux. Against this background, the Schrödinger operator ruling the dynamics of the particle reads

$$H_{3D} = \frac{1}{2m}\big(-i\hbar\nabla - q\mathbf{A}_{3D}\big)^2, \qquad \mathbf{A}_{3D}(x,y,z) = \frac{\Phi}{2\pi}\left(-\frac{y}{x^2+y^2}, \frac{x}{x^2+y^2}, 0\right),$$

where $\hbar$ is the reduced Plank constant, m is the particle mass, q is the electric charge, Φ is the total magnetic flux across the solenoid, and (x, y, z) are coordinates in $\mathbb{R}^3$ such that the solenoid coincides with the z-axis. The corresponding singular magnetic field is given by $\mathbf{B} = \operatorname{curl} \mathbf{A}_{3D} = (0, 0, \Phi\, \delta_{(x,y)=(0,0)})$.

Upon factorizing the axial direction and passing to natural units of measure, the model is described by the reduced Schrödinger operator

$$H_\alpha := \big(-i\nabla + \mathbf{A}_\alpha\big)^2, \qquad \mathbf{A}_\alpha(\mathbf{x}) := \alpha\, \frac{\mathbf{x}^\perp}{|\mathbf{x}|^2}, \tag{1}$$

acting in $L^2(\mathbb{R}^2)$. Here and in the sequel, $\mathbf{x} = (x, y) \in \mathbb{R}^2$ and $\mathbf{x}^\perp \equiv (-y, x)$. Besides, α is a dimensionless parameter measuring the magnetic flux Φ in units of the flux quantum $2\pi\hbar/q$. It entails no loss of generality to assume[1]

$$\alpha \in (0, 1)\,.$$

Due to the singularity of $\mathbf{A}_\alpha$ at $\mathbf{x} = \mathbf{0}$, the self-adjointness of H_α is not granted a priori. Assessing this feature is in fact a crucial task. Decomposing in angular harmonics and exploiting the exact solvability of the radial problems, all admissible self-adjoint extensions of the symmetric restriction $H_\alpha \restriction C_c^\infty(\mathbb{R}^2 \setminus \{\mathbf{0}\})$ were originally characterized via Krein-von Neumann methods in [1] and [12] (see also [7, 15, 28]). These extensions include the Friedrichs one and finite rank perturbations thereof, forming a family labeled by 2×2 complex Hermitian matrices. It is worth noting that a complete analysis of the spectral and scattering properties of the resulting Hamiltonians can be derived by resolvent techniques. A complementary approach relies on considering first finite size solenoids, partially shielded by electrostatic potentials, and then examining suitable scaling regimes [14, 22, 24, 30, 31]. Different self-adjoint realizations of H_α are obtained as limits (in strong resolvent sense) of Hamiltonian operators comprising just regular potentials. In this connection, zero-energy resonances produced by the shielding electrostatic

[1] For any $\alpha \in \mathbb{R}$, consider the decomposition $\alpha = 2\ell + \tilde{\alpha}$ with $\ell \in \mathbb{Z}$, $\tilde{\alpha} \in (-1, 1)$. For any fixed determination of arctan, the map $(U\psi)(\mathbf{x}) = e^{-2i\ell \arctan(x/y)}\psi(\mathbf{x})$ defines a unitary operator in $L^2(\mathbb{R}^2)$. A direct computation gives $U\big(-i\nabla + \alpha\frac{\mathbf{x}^\perp}{|\mathbf{x}|^2}\big)^2 U^{-1} = \big(-i\nabla + (\alpha - 2\ell)\frac{\mathbf{x}^\perp}{|\mathbf{x}|^2}\big)^2$, showing that H_α is unitarily equivalent to $H_{\tilde{\alpha}}$ with $\tilde{\alpha} \in (-1, 1)$. The condition $\tilde{\alpha} \in [0, 1)$ can then be realized exploiting conjugation symmetry. The case $\alpha = 0$ is here discarded because of its triviality ($H_{\alpha=0}$ is just the free Laplacian in $\mathbb{R}^2$, with no magnetic flux).

potential play a key role. This also allows to gain some intuition about the physical meaning of different self-adjoint realizations.

The present work examines magnetic perturbations of the pure Aharonov-Bohm Hamiltonian H_α, continuing the analysis begun in [8, 11]. The main goal is to characterize self-adjoint realizations in $L^2(\mathbb{R}^2)$ of the Schrödinger operator

$$H_{\alpha,S} := (-i\nabla + \mathbf{A}_\alpha + \mathbf{S})^2 ,$$

where $\mathbf{A}_\alpha$ is like in (1) and $\mathbf{S} \in L^\infty_{\mathrm{loc}}(\mathbb{R}^2, \mathbb{R}^2)$ is the vector potential associated to a regular, external axial magnetic field, to be regarded as a perturbation of the Aharonov-Bohm singular flux. The said perturbation allows to account, in particular, for magnetic traps and leakages of magnetic field lines outside of the solenoid coils.

The case of a homogeneous magnetic field, corresponding to $\mathbf{S} = \frac{B}{2}\mathbf{x}^\perp$ ($B > 0$ being the magnetic field intensity), was previously analyzed in [18]. An exhaustive classification of all self-adjoint realizations of $H_{\alpha,S}$ (with the said choice of $\mathbf{S}$) and of their spectral properties was derived therein by means of Krein-von Neumann methods, exploiting again decomposition in angular harmonics.

For a generic perturbation $\mathbf{S}$, lacking rotational symmetry, the decomposition in angular harmonics appears to be somewhat artificial and the Krein-von Neumann construction cannot be implemented straightaway. On the contrary, a quadratic form approach is more natural and flexible. In the context under analysis the use of quadratic forms was first proposed in [11] for the pure Aharonov-Bohm setting with $\mathbf{S} = \mathbf{0}$. In [8] similar techniques were employed to characterize the Friedrichs realization of $H_{\alpha,S}$ and a one-parameter family of singular s-wave perturbations.

In this paper we extend the previous analysis, including p-wave and mixed singular perturbations of the Friedrichs Hamiltonian (see Theorem 1 and Corollary 1). In view of the results derived in [1, 12, 18], this is expected to encompass all admissible self-adjoint realizations of $H_{\alpha,S}$ in $L^2(\mathbb{R}^2)$. The focus is not on identifying minimal regularity hypotheses for $\mathbf{S}$, but rather on providing techniques which can be generalized to the case of multiple Aharonov-Bohm fluxes [9]. Building on the quadratic form construction, we further derive a natural convergence result showing that the Friedrichs realization of $H_{0,S} = H_{\alpha,S}\big|_{\alpha=0}$ is the Γ-limit for $\alpha \to 0^+$ of the analogous realizations of $H_{\alpha,S}$ for $\alpha \in (0, 1)$ (see Theorem 2 and Corollary 2). Let us finally mention that some of the results presented in this work might be of interest also for applications to anyonic particle models [10, 11, 23, 27, 35, 36], though this issue is not addressed directly here.

Throughout the paper we often refer to polar coordinates $(r, \theta) : \mathbb{R}^2 \setminus \{\mathbf{0}\} \to (0, +\infty) \times [0, 2\pi)$ centered at $\mathbf{x} = \mathbf{0}$, and to the related angular averages of functions $f : \mathbb{R}^2 \to \mathbb{C}$, namely,

$$\langle f \rangle(r) := \frac{1}{2\pi r} \int_{\partial B_r(\mathbf{0})} d\Sigma_r \, f = \frac{1}{2\pi} \int_0^{2\pi} d\theta \, f\big(\mathbf{x}(r,\theta)\big) .$$

2 Main Results

2.1 Self-Adjoint Realizations of $H_{\alpha,S}$

Self-adjoint realizations in $L^2(\mathbb{R}^2)$ of the Schrödinger operator $H_{\alpha,S}$ are generally characterized as suitable extensions of the densely defined, symmetric operator $H_{\alpha,S} \restriction C_c^\infty(\mathbb{R}^2 \setminus \{\mathbf{0}\})$. The simplest of such extensions is the Friedrichs one, to be denoted with $H_{\alpha,S}^{(F)}$ henceforth. This is obtained introducing the quadratic form

$$\mathcal{Q}_{\alpha,S}[\psi] = \int_{\mathbb{R}^2} d\mathbf{x}\, \left|(-i\nabla + \mathbf{A}_\alpha + \mathbf{S})\psi\right|^2, \qquad \text{for } \psi \in C_c^\infty(\mathbb{R}^2 \setminus \{\mathbf{0}\}),$$

and considering its Friedrichs realization

$$D\big[\mathcal{Q}_{\alpha,S}^{(F)}\big] := \overline{C_c^\infty(\mathbb{R}^2 \setminus \{\mathbf{0}\})}^{\,\|\cdot\|_{\alpha,S}}, \qquad \mathcal{Q}_{\alpha,S}^{(F)}[\psi] = \mathcal{Q}_{\alpha,S}[\psi],$$

where $\|\psi\|_{\alpha,S} := \|\psi\|_2 + \mathcal{Q}_{\alpha,S}[\psi]$. We recall the following Proposition from [8].

Proposition 1 (Friedrichs Realization) *Let $\alpha \in (0,1)$ and $\mathbf{S} \in L^\infty_{loc}(\mathbb{R}^2)$. Then:*

(i) The quadratic form $\mathcal{Q}_{\alpha,S}^{(F)}$ is closed and non-negative on its domain, and

$$D\big[\mathcal{Q}_{\alpha,S}^{(F)}\big] = \big\{\psi \in L^2(\mathbb{R}^2) \;\big|\; (-i\nabla + \mathbf{S})\psi \in L^2(\mathbb{R}^2),\ \mathbf{A}_\alpha \psi \in L^2(\mathbb{R}^2)\big\}. \tag{2}$$

Moreover, any $\psi \in D\big[\mathcal{Q}_{\alpha,S}^{(F)}\big]$ fulfills

$$\lim_{r \to 0^+} \left\langle |\psi|^2 \right\rangle(r) = 0, \qquad \lim_{r \to 0^+} r^2 \left\langle |\partial_r \psi|^2 \right\rangle(r) = 0. \tag{3}$$

(ii) The unique self-adjoint operator $H_{\alpha,S}^{(F)}$ associated to $\mathcal{Q}_{\alpha,S}^{(F)}$ is

$$D\big(H_{\alpha,S}^{(F)}\big) = \big\{\psi \in D\big[\mathcal{Q}_{\alpha,S}^{(F)}\big] \;\big|\; H_{\alpha,S}\,\psi \in L^2(\mathbb{R}^2)\big\}, \qquad H_{\alpha,S}^{(F)}\,\psi = H_{\alpha,S}\,\psi. \tag{4}$$

It is well known that the Friedrichs extension of a symmetric operator is the one with the smallest domain of self-adjointness. Accordingly, other self-adjoint realizations can only be obtained by suitably enlarging the domain. The standard approach to achieve this goal is to consider elements of the form $\psi = \phi_\lambda + q\,\mathcal{G}_\lambda$, where $\phi_\lambda \in D\big(H_{\alpha,S}^{(F)}\big)$, $q \in \mathbb{C}$ and $\mathcal{G}_\lambda \in L^2(\mathbb{R}^2)$, $\lambda > 0$, is a solution of the defect equation $(H_{\alpha,S} + \lambda^2)\mathcal{G}_\lambda = 0$ in $\mathbb{R}^2 \setminus \{\mathbf{0}\}$. A key obstacle in this construction is that, for generic choices of $\mathbf{S}$, an explicit expression of $\mathcal{G}_\lambda$ is not available. At the same time, we expect that the asymptotic behavior of $\mathcal{G}_\lambda$ near $\mathbf{x} = \mathbf{0}$ should not be affected by the perturbation $\mathbf{S}$, at least to leading order.

Trusting the latter surmise, we proceed to consider the solutions $G_\lambda^{(k)} \in L^2(\mathbb{R}^2)$, $k \in \{0, -1\}$, of the unperturbed defect equation

$$\big((-i\nabla + \mathbf{A}_\alpha)^2 + \lambda^2\big)G_\lambda^{(k)} = 0\,, \qquad \text{in} \quad \mathbb{R}^2 \setminus \{\mathbf{0}\}\,. \tag{5}$$

The parameter k coincides with the angular momentum number of the wavefunction $G_\lambda^{(k)}$. For this reason, we shall refer to $G_\lambda^{(0)}$ and $G_\lambda^{(-1)}$, respectively, as the *s-wave* and *p-wave Green functions*.[2]

Using angular coordinates and writing K_ν for the modified Bessel function of second kind (*a.k.a.* Macdonald function), their explicit expressions are given by

$$G_\lambda^{(k)}(r,\theta) \equiv G_\lambda^{(k)}\big(\mathbf{x}(r,\theta)\big) = \lambda^{|k+\alpha|}\, K_{|k+\alpha|}(\lambda r)\, e^{ik\theta}\,, \qquad \text{for} \quad k \in \{0, -1\}\,. \tag{6}$$

For later reference, let us mention that (see [19, Eq. 6.521.3])

$$\|G_\lambda^{(k)}\|_2^2 = \frac{\pi^2\,|k+\alpha|}{\sin(\pi\alpha)}\, \lambda^{2|k+\alpha|-2}\,, \qquad \text{for} \quad k \in \{0, -1\}\,. \tag{7}$$

To say more, for $r \to 0^+$ there holds (see [26, §10.31])

$$G_\lambda^{(k)}(r,\theta) = \left[\frac{\Gamma\big(|k+\alpha|\big)}{2^{1-|k+\alpha|}}\,\frac{1}{r^{|k+\alpha|}} + \frac{\Gamma\big(-|k+\alpha|\big)}{2^{1+|k+\alpha|}}\,\lambda^{2|k+\alpha|}\, r^{|k+\alpha|} + O\big(r^{2-|k+\alpha|}\big)\right] e^{ik\theta}\,, \tag{8}$$

So far, we made no assumption concerning the regularity of $\mathbf{S}$ near the origin, where the Aharonov-Bohm potential $\mathbf{A}_\alpha$ is singular. We henceforth require that

$$\mathbf{S} \in L^\infty_{\text{loc}}(\mathbb{R}^2, \mathbb{R}^2) \text{ is Lipschitz continuous at } \mathbf{x} = \mathbf{0}\,. \tag{9}$$

Without loss of generality we also fix the Coulomb gauge, which entails

$$\nabla \cdot \mathbf{S} = 0\,. \tag{10}$$

[2] By decomposition in angular harmonics, it appears that (5) admits non-trivial solutions only for $k = 0$ and $k = -1$.

Under the above hypotheses, for $\lambda > 0$ we consider trial functions of the form

$$\psi = \phi_\lambda + e^{-i\,\mathbf{S}(\mathbf{0})\cdot\mathbf{x}}\,\chi \sum_{k\in\{0,-1\}} q^{(k)}\, G_\lambda^{(k)}\,, \tag{11}$$

where $\phi_\lambda \in D\big[Q^{(F)}_{\alpha,S}\big]$, $q^{(k)} \in \mathbb{C}$ for $k \in \{0,-1\}$ and $\chi : \mathbb{R}^2 \to [0,1]$ is a smooth cut-off function fulfilling

$$\chi \in C^2_c(\mathbb{R}^2)\,, \qquad \chi(\mathbf{x}) \equiv 1 \quad \text{for any } \mathbf{x} \in B_a(\mathbf{0}), \text{ for some } a > 0\,. \tag{12}$$

The latter cut-off is necessary to include in the present analysis the case of perturbations $\mathbf{S}$ which are unbounded at infinity, comprising especially configurations with magnetic traps. For the sake of simplicity, we assume χ to be radial, *i.e.*,

$$\chi(\mathbf{x}) \equiv \chi\big(|\mathbf{x}|\big) \equiv \chi(r)\,. \tag{13}$$

Remark 1 One could fix $\mathbf{S}(\mathbf{0}) = \mathbf{0}$, on top of the Coulomb gauge (10). This would make the phase factor in (11) irrelevant and would even allow to abridge some of the expressions to be derived in the following. Yet, in this work we choose not to fix the value of $\mathbf{S}$ at $\mathbf{x} = \mathbf{0}$ in order to exhibit a construction which can be generalized to the case of multiple fluxes with a moderate effort [9]. With the same objective in mind, we stick to the Lipschitz condition in (9), though most of the results in this work are still valid requiring just some Hölder regularity of $\mathbf{S}$ at the origin.

A heuristic evaluation of the expectation value $\langle \psi | H_{\alpha,S}\, \psi \rangle$ for functions ψ like (11) suggests the educated guess

$$\begin{aligned}
Q^{(\beta)}_{\alpha,S}[\psi] &:= Q^{(F)}_{\alpha,S}[\phi_\lambda] - \lambda^2\, \|\psi\|_2^2 + \lambda^2\, \|\phi_\lambda\|_2^2 \\
&+ 2 \sum_{k\in\{0,-1\}} \mathrm{Re}\Big[q^{(k)} \Big(2 \Big\langle (-i\nabla + \mathbf{A}_\alpha)\phi_\lambda \,\Big|\, e^{-i\,\mathbf{S}(\mathbf{0})\cdot\mathbf{x}} \big((\mathbf{S} - \mathbf{S}(\mathbf{0}))\chi - i\nabla\chi \big) G_\lambda^{(k)} \Big\rangle \\
&+ \Big\langle \phi_\lambda \,\Big|\, e^{-i\,\mathbf{S}(\mathbf{0})\cdot\mathbf{x}} \big[(\mathbf{S} - \mathbf{S}(\mathbf{0}))^2 \chi + 2\mathbf{S}(\mathbf{0}) \cdot \big((\mathbf{S} - \mathbf{S}(\mathbf{0}))\chi - i\nabla\chi \big) + \Delta\chi \big] G_\lambda^{(k)} \Big\rangle \Big) \Big] \\
&+ \sum_{k,k'\in\{0,-1\}} \overline{q^{(k)}}\, q^{(k')} \left[\beta_{kk'} + \frac{\pi^2}{\sin(\pi\alpha)}\, \lambda^{2|k+\alpha|}\, \delta_{kk'} + \Xi_{kk'}(\lambda) \right],
\end{aligned} \tag{14}$$

where we have introduced the 2×2 complex Hermitian matrix $\beta = (\beta_{kk'})$, labeling the quadratic form, and we have set

$$\begin{aligned}
\Xi_{kk'}(\lambda) &:= \Big\langle \chi\, G_\lambda^{(k)} \,\Big|\, \big[(\mathbf{S} - \mathbf{S}(\mathbf{0}))^2 + 2 (\mathbf{S} - \mathbf{S}(\mathbf{0})) \cdot \mathbf{A}_\alpha \big] \chi\, G_\lambda^{(k')} \Big\rangle \\
&\quad + \|(\nabla\chi)\, G_\lambda^{(k)}\|_2^2\, \delta_{kk'} + 2 \Big\langle \chi\, G_\lambda^{(k)} \,\Big|\, (\mathbf{S} - \mathbf{S}(\mathbf{0})) \cdot (-i\nabla) \big(\chi\, G_\lambda^{(k')} \big) \Big\rangle .
\end{aligned} \tag{15}$$

The expression (14) was derived integrating by parts and deliberately discarding some contributions, an operation to be justified a posteriori in the proof of Theorem 1. We also used that $\mathbf{A}_\alpha \cdot \nabla\chi = 0$, since χ is radial (see (13)), and the identity

$$\big\langle G_\lambda^{(k)} \,\big|\, \eta\, G_\lambda^{(k')} \big\rangle = \big\langle G_\lambda^{(k)} \,\big|\, \eta\, G_\lambda^{(k)} \big\rangle\, \delta_{kk'}\,, \qquad \text{for any radial } \eta : \mathbb{R}^2 \to \mathbb{R}\,. \tag{16}$$

Remark 2 For any $\lambda > 0$, the matrix $\Xi_{kk'}(\lambda)$, $k, k' \in \{0, -1\}$, defined by (15) is itself Hermitian. This feature is evident for the first two addenda in (15), given that $\mathbf{S}$, $\mathbf{A}_\alpha$ and χ are real-valued. As regards the last addendum in (15), integrating by parts and checking that the boundary contribution vanishes (recall (8) and (9)), we have in fact

$$\begin{aligned}\Big\langle \chi\, G_\lambda^{(k)} \,\Big|\, (\mathbf{S} - \mathbf{S}(\mathbf{0}))\cdot(-i\nabla)\big(\chi\, G_\lambda^{(k')}\big) \Big\rangle &= \Big\langle (-i\nabla)\big(\chi\, G_\lambda^{(k)}\big) \,\Big|\, (\mathbf{S} - \mathbf{S}(\mathbf{0}))\,\chi\, G_\lambda^{(k')} \Big\rangle \\ &= \overline{\Big\langle \chi\, G_\lambda^{(k')} \,\Big|\, (\mathbf{S} - \mathbf{S}(\mathbf{0}))\cdot(-i\nabla)\big(\chi\, G_\lambda^{(k)}\big) \Big\rangle}\,.\end{aligned}$$

Besides, building on the fact that $G_\lambda^{(0)}$ is real-valued, c.f. (6), it can be inferred that

$$\Big\langle \chi\, G_\lambda^{(k)} \,\Big|\, (\mathbf{S} - \mathbf{S}(\mathbf{0}))\cdot(-i\nabla)\big(\chi\, G_\lambda^{(k')}\big) \Big\rangle = 0\,, \qquad \text{for} \quad k = k' = 0\,. \tag{17}$$

A family of admissible singular perturbations of the Friedrichs realization is described by the upcoming Theorem 1 and Corollary 1.

Theorem 1 (Quadratic Forms for Singular Perturbations) *Let* $\alpha \in (0, 1)$ *and* $\mathbf{S} \in L^\infty_{loc}(\mathbb{R}^2)$ *be Lipschitz continuous at* $\mathbf{x} = \mathbf{0}$, *with* $\nabla \cdot \mathbf{S} = 0$. *Then, for any Hermitian matrix* $\beta = (\beta_{kk'})$, $k, k' \in \{0, -1\}$, *the quadratic form* $Q^{(\beta)}_{\alpha,S}$ *defined in* (14) *satisfies the following:*

(i) It is well-posed on the domain

$$\begin{aligned} D\big[Q^{(\beta)}_{\alpha,S}\big] := \Big\{ \psi = \phi_\lambda + e^{-i\,\mathbf{S}(\mathbf{0})\cdot\mathbf{x}}\, \chi \sum_{k\in\{0,-1\}} q^{(k)}\, G_\lambda^{(k)} \in L^2(\mathbb{R}^2) \quad s.t. \\ \phi_\lambda \in D\big[Q^{(F)}_{\alpha,S}\big],\ \lambda > 0\,,\ \ \chi \ \ \textit{fulfills}\ (12)(13)\,,\ \ q^{(k)} \in \mathbb{C}\,, k \in \{0, -1\} \Big\}\,. \end{aligned} \tag{18}$$

(ii) It is independent of $\lambda > 0$ *and of the cut-off* χ, *provided that* (12)(13) *hold true.*
(iii) It is closed and bounded from below on the domain (18).

Corollary 1 (Self-Adjoint Realizations for Singular Perturbations) *Assume the hypotheses of Theorem 1 to hold. Then, for any* 2×2 *Hermitian matrix* β*, the self-adjoint operator* $H^{(\beta)}_{\alpha,S}$ *associated to the quadratic form* $Q^{(\beta)}_{\alpha,S}$ *is given by*

$$D\big(H^{(\beta)}_{\alpha,S}\big) = \Bigg\{ \psi = \phi_\lambda + e^{-i\,\mathbf{S}(\mathbf{0})\cdot\mathbf{x}} \chi \sum_{k\in\{0,-1\}} q^{(k)} G^{(k)}_\lambda \in D\big[Q^{(\beta)}_{\alpha,S}\big] \quad s.t.$$

$$\phi_\lambda \in D\big(H^{(F)}_{\alpha,S}\big) \quad and \tag{19}$$

$$\frac{2^{1-|k+\alpha|}}{\Gamma\big(|k+\alpha|\big)} \sum_{k'\in\{0,-1\}} q^{(k')} \bigg(\beta_{kk'} + \frac{\pi^2 \lambda^{2|k+\alpha|}}{\sin(\pi\alpha)}\, \delta_{kk'} \bigg)$$

$$= \lim_{r\to 0^+} \frac{|k+\alpha|\, \langle e^{-ik\theta}\phi_\lambda\rangle(r) + r\,\langle e^{-ik\theta}\partial_r \phi_\lambda\rangle(r)}{r^{|k+\alpha|}}\,, \quad for \quad k\in\{0,-1\} \Bigg\}\,;$$

$$\begin{aligned}\big(H^{(\beta)}_{\alpha,S} + \lambda^2\big)\psi = {} & \big(H^{(F)}_{\alpha,S} + \lambda^2\big)\phi_\lambda \\ & + \sum_{k\in\{0,-1\}} q^{(k)} e^{-i\,\mathbf{S}(\mathbf{0})\cdot\mathbf{x}} \Big[2\Big(\big(\mathbf{S} - \mathbf{S}(\mathbf{0})\big)\chi - i\nabla\chi \Big)\cdot\big(-i\nabla + \mathbf{A}_\alpha\big) G^{(k)}_\lambda \\ & + \Big(\big(\mathbf{S} - \mathbf{S}(\mathbf{0})\big)^2 \chi + 2\big(\mathbf{S} - \mathbf{S}(\mathbf{0})\big)\cdot(-i\nabla\chi) - \Delta\chi \Big) G^{(k)}_\lambda \Big]\,.\end{aligned} \tag{20}$$

The set of operators $H^{(\beta)}_{\alpha,S}$*,* β *Hermitian, identifies a family of self-adjoint extensions of* $H_{\alpha,S} \restriction C^\infty_c(\mathbb{R}^2 \setminus \{\mathbf{0}\})$ *in* $L^2(\mathbb{R}^2)$*, labeled by four real parameters.*[3] *If* $\mathbf{S} \in L^\infty(\mathbb{R}^2)$*, this family comprises all admissible self-adjoint realizations of* $H_{\alpha,S}$ *in* $L^2(\mathbb{R}^2)$*.*

Remark 3 The characterization (19) of the operator domain is quite standard. Especially, it incorporates boundary conditions relating the "charges" $q^{(k)}$ to the asymptotic behavior of the "regular part" ϕ_λ at $\mathbf{x} = \mathbf{0}$. Considering that the matrix $\beta_{kk'} + \frac{\pi^2\lambda^{2|k+\alpha|}}{\sin(\pi\alpha)}\,\delta_{kk'}$ is certainly invertible for λ large enough, it is always possible to derive an explicit expression for $q^{(k)}$ in terms of boundary values of ϕ_λ. Let us also stress that, in agreement with our expectations, only the leading order term of the asymptotic expansion (8) for $G^{(k)}_\lambda$ is relevant here (*cf.* the proof of Corollary 1).

Remark 4 The Hermitian matrix β labeling the self-adjoint operator $H^{(\beta)}_{\alpha,S}$ only appears in the boundary conditions for $D\big(H^{(\beta)}_{\alpha,S}\big)$. In this sense, it parametrizes a singular interaction affecting just the *s-wave* and *p-wave* modes of the wave-functions. In [8] attention was restricted to pure *s-wave* perturbations, corresponding to $\beta_{kk'} = b\,\delta_{k,0}\,\delta_{k',0}$ with $b \in \mathbb{R}$. Here we also include pure *p-wave* perturbations, as well as mixed interactions coupling *s-wave* and *p-wave* modes.

[3] Notice that 2×2 complex Hermitian matrices form a 4-dimensional real vector space.

Remark 5 The Friedrichs realization is formally recovered fixing "$\beta_{kk'} = \infty\, \delta_{kk'}$" and, accordingly, $q^{(0)} = q^{(-1)} = 0$. In this case, the boundary conditions appearing in the characterization (19) of $D\big(H^{(F)}_{\alpha,S}\big)$ become, for $k \in \{0, -1\}$,

$$|k+\alpha|\, \langle e^{-ik\theta}\phi_\lambda\rangle(r) + r\, \langle e^{-ik\theta}\partial_r\phi_\lambda\rangle(r) = O\big(r^{|k+\alpha|}\big)\,, \qquad \text{for} \quad r \to 0^+.$$

These conditions are otherwise concealed in the position $H^{(F)}_{\alpha,S}\, \phi_\lambda \in L^2(\mathbb{R}^2)$ of (4). In particular, notice that $D\big(H^{(F)}_{\alpha,S}\big)$ certainly contains elements $\psi \in H^2_{loc}(\mathbb{R}^2 \setminus \{\mathbf{0}\})$ such that $\langle\phi_\lambda\rangle = O\big(r^{\alpha}\big)$ and $\langle e^{i\theta}\phi_\lambda\rangle = O\big(r^{1-\alpha}\big)$ for $r \to 0^+$ (cf. [11, Prop. 2.1]).

Remark 6 The action of the operator described in (20) is somehow unorthodox. Making reference to the standard theory of self-adjoint extensions, one would rather expect the simpler relation $\big(H^{(\beta)}_{\alpha,S} + \lambda^2\big)\psi = \big(H^{(F)}_{\alpha,S} + \lambda^2\big)\phi_\lambda$. The expressions in the last two lines of (20) are in fact necessary corrections, produced by the use of surrogates in place of true defect functions for $H_{\alpha,S}$.

Remark 7 Electrostatic potentials regular enough at the Aharonov-Bohm singularity could be easily incorporated in the construction provided here. We omit the discussion of this further development for the sake of brevity.

2.2 *Γ-Convergence for the Friedrichs Hamiltonian*

Consider now a regime where the Aharonov-Bohm flux is negligible, in suitable units, compared to the external magnetic perturbation or to the angular momentum of the particle. In this context, the dynamics of the particle should be properly described by some self-adjoint realization in $L^2(\mathbb{R}^2)$ of the Schrödinger operator

$$H_{0,S} \equiv H_{\alpha,S}\big|_{\alpha=0} = \big(-i\nabla + \mathbf{S}\big)^2. \tag{21}$$

At the same time, due to the local singularity at $\mathbf{x} = \mathbf{0}$ of the Aharonov-Bohm potential $\mathbf{A}_\alpha$, establishing the convergence $H_{\alpha,S} \to H_{0,S}$ for $\alpha \to 0^+$ (in any reasonable operator topology) is not a plain task. Building on the quadratic form approach described in the previous subsection, we present hereafter a result based on the classical notion of Γ-convergence [6, 13].

For the sake of simplicity, let us assume that[4]

$$\mathbf{S} \in L^\infty(\mathbb{R}^2, \mathbb{R}^2) \ \text{ is Lipschitz continuous at}\, \mathbf{x} = \mathbf{0}\,. \tag{22}$$

[4] Notice the similarity with (9). Here we are making a stronger requirement: $\mathbf{S}$ must be uniformly bounded on the whole space $\mathbb{R}^2$, not just on compact subsets of it. This excludes magnetic traps.

Besides, we restrict the attention to the Friedrichs Hamiltonian $H^{(F)}_{\alpha,S}$ of Proposition 1, postponing the discussion of the singular perturbations $H^{(\beta)}_{\alpha,S}$ characterized in Theorem 1 and Corollary 1 to future investigations.

Let us consider the Friedrichs quadratic form $Q^{(F)}_{\alpha,S}$ (see Proposition 1) and extend it to the whole Hilbert space $L^2(\mathbb{R}^2)$ setting

$$Q^{(F)}_{\alpha,S}[\psi] := \begin{cases} \|(-i\nabla + \mathbf{A}_\alpha + \mathbf{S})\psi\|_2^2 & \text{if } \psi \in D\big[Q^{(F)}_{\alpha,S}\big]\,; \\ +\infty & \text{if } \psi \in L^2(\mathbb{R}^2)\setminus D\big[Q^{(F)}_{\alpha,S}\big]\,. \end{cases} \tag{23}$$

Notice that, under the hypothesis (22), the identity (2) in Proposition 1 reduces to

$$D\big[Q^{(F)}_{\alpha,S}\big] = \big\{\psi \in H^1(\mathbb{R}^2)\;\big|\; \mathbf{A}_\alpha \psi \in L^2(\mathbb{R}^2)\big\}\,.$$

In a similar fashion, for $\alpha = 0$ we put

$$Q^{(F)}_{0,S}[\psi] := \begin{cases} \|(-i\nabla + \mathbf{S})\psi\|_2^2 & \text{if } \psi \in D\big[Q^{(F)}_{0,S}\big] \equiv H^1(\mathbb{R}^2)\,; \\ +\infty & \text{if } \psi \in L^2(\mathbb{R}^2)\setminus H^1(\mathbb{R}^2)\,. \end{cases} \tag{24}$$

For later reference, let us mention that the self-adjoint operators associated to the above quadratic forms are respectively given by (*cf.* (4))

$$D\big(H^{(F)}_{\alpha,S}\big) = \big\{\psi \in H^1(\mathbb{R}^2)\;\big|\; \mathbf{A}_\alpha\psi,\, H_{\alpha,S}\,\psi \in L^2(\mathbb{R}^2)\big\}\,, \qquad H^{(F)}_{\alpha,S}\,\psi = H_{\alpha,S}\,\psi\,;$$
$$D\big(H^{(F)}_{0,S}\big) = H^2(\mathbb{R}^2)\,, \qquad H^{(F)}_{0,S}\,\psi = H_{0,S}\,\psi\,.$$

Theorem 2 *Let $\mathbf{S} \in L^\infty(\mathbb{R}^2)$ be Lipschitz continuous at $\mathbf{x} = \mathbf{0}$, with $\nabla \cdot \mathbf{S} = 0$, and $\{\alpha_n\}_{n\in\mathbb{N}} \subset (0,1)$ be any sequence such that $\alpha_n \to 0$ for $n \to +\infty$. Then, the family of quadratic forms $Q^{(F)}_{\alpha_n,S}$ Γ-converges to $Q^{(F)}_{0,S}$, that is:*

(i) Lower bound inequality. *For every sequence $\{\psi_{\alpha_n}\}_{n\in\mathbb{N}} \subset L^2(\mathbb{R}^2)$ such that $\psi_{\alpha_n} \to \psi_0 \in L^2(\mathbb{R}^2)$ as $n \to +\infty$, there holds*

$$Q^{(F)}_{0,S}[\psi_0] \leqslant \liminf_{n\to+\infty} Q^{(F)}_{\alpha_n,S}[\psi_{\alpha_n}]\,. \tag{25}$$

(ii) Upper bound inequality. *For every $\psi_0 \in L^2(\mathbb{R}^2)$ there exists a sequence $\{\psi_{\alpha_n}\}_{n\in\mathbb{N}} \subset L^2(\mathbb{R}^2)$ such that $\psi_{\alpha_n} \to \psi_0$ as $n \to +\infty$ and*

$$Q^{(F)}_{0,S}[\psi_0] \geqslant \limsup_{n\to+\infty} Q^{(F)}_{\alpha_n,S}[\psi_{\alpha_n}]\,. \tag{26}$$

From the previous theorem and classical results on Γ-convergence [13, §13], we readily deduce the following.

Corollary 2 *Under the same assumptions of Theorem 2, the family of operators* $H^{(F)}_{\alpha_n,S}$ *converges to* $H^{(F)}_{0,S}$ *in strong resolvent sense for* $n \to +\infty$. *More precisely, for any* $z \in \mathbb{C}\setminus[0,+\infty)$ *and any* $\psi \in L^2(\mathbb{R}^2)$, *there holds*

$$\left\| \left(H^{(F)}_{\alpha_n,S} - z\right)^{-1}\psi - \left(H^{(F)}_{0,S} - z\right)^{-1}\psi \right\|_2 \xrightarrow{n\to+\infty} 0\,. \tag{27}$$

Remark 8 The requirement $z \in \mathbb{C}\setminus[0,+\infty)$ in Corollary 2 matches the elementary inclusions $\sigma\left(H^{(F)}_{\alpha_n,S}\right) \subset [0,+\infty)$ and $\sigma\left(H^{(F)}_{0,S}\right) \subset [0,+\infty)$.

Remark 9 In the pure Aharonov-Bohm configuration, with $\mathbf{S} = \mathbf{0}$, it should be possible to infer strong resolvent convergence for $\alpha \to 0$ even by direct computations, starting from the explicit expression for the integral kernel of the resolvent operator derived in [1]. This alternative approach would however involve a rather complicate analysis, relying on non-elementary regularity features of the Bessel functions with respect to their order and further demanding non-trivial exchanges of limits and integrations. On top of that, the Γ- convergence method considered in this work appears to be more flexible. Especially, it should be possible to adapt it to multiple fluxes configurations with not too much effort.

Remark 10 Despite being quite natural, the results derived in Theorem 2 and Corollary 2 are not completely obvious, especially if one considers the topology of the underlying space domains. In fact, the Aharonov-Bohm configuration ($\alpha \neq 0$) refers to the domain $\mathbb{R}^2\setminus\{\mathbf{0}\}$, with first homotopy group given by $\mathbb{Z}$, while the setting with no singular flux ($\alpha = 0$) corresponds to the plane $\mathbb{R}^2$, with trivial topology.

3 Proofs

Let us recall that Theorem 1 and Corollary 1 rely on the hypothesis (9) for $\mathbf{S}$, demanding $\mathbf{S}$ to be locally uniformly bounded and Lipschitz continuous at $\mathbf{x} = \mathbf{0}$.

Proof *(Theorem 1)* Each of the statements *(i)–(iii)* can be derived adapting some related arguments from [8]. Throughout the proof, $\mathbf{1}_\chi$ is the indicator function of the support of χ and $c \equiv c(\alpha, \mathbf{S})$ is a suitable positive constant independent of λ, which may vary from line to line.

(i) Upon identifying $(-i\nabla + \mathbf{A}_\alpha)\phi_\lambda$ with $\mathbf{1}_\chi(-i\nabla + \mathbf{A}_\alpha)\phi_\lambda$ in (14), all parings in (14) (15) are well-defined inner products in $L^2(\mathbb{R}^2)$. To account for this claim, firstly note that $(-i\nabla+\mathbf{A}_\alpha)\phi_\lambda \in L^2_{\mathrm{loc}}(\mathbb{R}^2)$ for any $\phi_\lambda \in D\big[Q^{(F)}_{\alpha,S}\big]$, see (2). Secondly, recall that $G^{(k)}_\lambda \in L^2(\mathbb{R}^2)$ for $k \in \{0,-1\}$, see (7). Hypotheses (9) and (12) further grant the uniform boundedness of $\nabla\chi$, $\Delta\chi$, $\big(\mathbf{S} - \mathbf{S}(\mathbf{0})\big)\chi$ and

$\big(\mathbf{S}-\mathbf{S}(\mathbf{0})\big)\cdot \mathbf{A}_\alpha \chi$. In view of the basic relation $\big|\nabla\big(\chi G^{(k)}_{\lambda}\big)\big| \leqslant \frac{c}{|\mathbf{x}|}\,\chi G^{(k)}_{\lambda}$, $k \in \{0,-1\}$, the same hypotheses also ensure that $\big(\mathbf{S}-\mathbf{S}(\mathbf{0})\big)\cdot(-i\nabla)\big(\chi\, G^{(k)}_{\lambda}\big) \in L^2(\mathbb{R}^2)$.

(ii) Let us show that the form is independent of $\lambda > 0$. To this purpose, fix $\lambda_1 \neq \lambda_2$ and consider, for any $\psi \in D\big[Q^{(\beta)}_{\alpha,S}\big]$, the two alternative representations $\psi = \phi_{\lambda_1} + e^{-i\mathbf{S}(\mathbf{0})\cdot\mathbf{x}}\chi \sum_{k\in\{0,-1\}} q^{(k)} G^{(k)}_{\lambda_1}$ and $\psi = \phi_{\lambda_2} + e^{-i\mathbf{S}(\mathbf{0})\cdot\mathbf{x}}\chi \sum_{k\in\{0,-1\}} q^{(k)} G^{(k)}_{\lambda_2}$. It is easy to check that $\chi\big(G^{(k)}_{\lambda_2} - G^{(k)}_{\lambda_1}\big) \in D\big[Q^{(F)}_{\alpha,S}\big]$ for $k \in \{0,-1\}$ (see (2) and (8)). This ensures that the "charges" $q^{(k)}$ are independent of λ, and further entails $\phi_{\lambda_1} = \phi_{\lambda_2} + e^{-i\mathbf{S}(\mathbf{0})\cdot\mathbf{x}}\chi \sum_{k\in\{0,-1\}} q^{(k)} (G^{(k)}_{\lambda_2} - G^{(k)}_{\lambda_1})$. Taking these facts into account and exploiting the identity (16), with a number of integrations by parts we obtain

$$
\begin{aligned}
&Q^{(\beta)}_{\alpha,S}\left[\phi_{\lambda_1} + e^{-i\mathbf{S}(\mathbf{0})\cdot\mathbf{x}}\chi \sum_{k\in\{0,-1\}} q^{(k)} G^{(k)}_{\lambda_1}\right] \\
&= \langle \phi_{\lambda_2} | (-i\nabla + \mathbf{A}_\alpha + \mathbf{S})^2 \phi_{\lambda_2}\rangle - \lambda_2^2\, \|\psi\|_2^2 + \lambda_2^2\, \|\phi_\lambda\|_2^2 \\
&\quad + 2 \sum_{k\in\{0,-1\}} \mathrm{Re}\Big[q^{(k)} \Big(2 \Big\langle (-i\nabla + \mathbf{A}_\alpha)\phi_{\lambda_2} \Big| e^{-i\mathbf{S}(\mathbf{0})\cdot\mathbf{x}} \big[(\mathbf{S}-\mathbf{S}(\mathbf{0}))\chi - i\nabla\chi\big] G^{(k)}_{\lambda_2} \Big\rangle \\
&\quad + \Big\langle \phi_{\lambda_2} \Big| e^{-i\mathbf{S}(\mathbf{0})\cdot\mathbf{x}} \big[(\mathbf{S}-\mathbf{S}(\mathbf{0}))^2\chi + 2\,\mathbf{S}(\mathbf{0})\cdot\big((\mathbf{S}-\mathbf{S}(\mathbf{0}))\chi - i\nabla\chi\big) + \Delta\chi\big] G^{(k)}_{\lambda_2} \Big\rangle \Big) \Big] \\
&\quad + \sum_{k,k'\in\{0,-1\}} \overline{q^{(k)}}\, q^{(k')} \Big[\beta_{kk'} + \frac{\pi^2}{\sin(\pi\alpha)}\, \lambda_2^{2|k+\alpha|}\, \delta_{ij} \\
&\quad + 2 \Big\langle \chi\, G^{(k)}_{\lambda_2} \,\Big|\, (\mathbf{S}-\mathbf{S}(\mathbf{0}))\cdot(-i\nabla)\big(\chi\, G^{(k')}_{\lambda_2}\big) \Big\rangle \\
&\quad + \Big\langle \chi G^{(k)}_{\lambda_2} \,\Big|\, \big[(\mathbf{S}-\mathbf{S}(\mathbf{0}))^2 + 2(\mathbf{S}-\mathbf{S}(\mathbf{0}))\cdot\mathbf{A}_\alpha\big]\chi G^{(k')}_{\lambda_2} \Big\rangle + \|(\nabla\chi)\, G^{(k)}_{\lambda_2}\|_2^2\, \delta_{kk'} \Big] \\
&\quad + \sum_{k\in\{0,-1\}} |q^{(k)}|^2 \Big[\frac{\pi^2}{\sin(\pi\alpha)} \Big(\lambda_1^{2|k+\alpha|} - \lambda_2^{2|k+\alpha|} \Big) + (\lambda_2^2 - \lambda_1^2) \langle \chi G^{(k)}_{\lambda_1} \,|\, \chi G^{(k)}_{\lambda_2} \rangle \\
&\quad + \langle G^{(k)}_{\lambda_1} \,|\, (\nabla\chi^2)\cdot\nabla G^{(k)}_{\lambda_2} \rangle - \langle \nabla G^{(k)}_{\lambda_1} \,|\, (\nabla\chi^2) G^{(k)}_{\lambda_2} \rangle \Big] \\
&\quad + 2 \sum_{k\in\{0,-1\}} q^{(k)} \Big[\lim_{r\to 0^+} \int_{\partial B_r(\mathbf{0})} d\Sigma_r\; e^{-i\mathbf{S}(\mathbf{0})\cdot\mathbf{x}} \Big(i\chi\, (\mathbf{S}-\mathbf{S}(\mathbf{0}))\cdot\hat{\mathbf{r}} + \partial_r\chi \Big) \\
&\quad + \overline{\phi_{\lambda_2}}\, \big(G^{(k)}_{\lambda_2} - G^{(k)}_{\lambda_1}\big) \Big]
\end{aligned}
$$

$$+\sum_{k\in\{0,-1\}} q_k^*\Big[\lim_{r\to0^+}\int_{\partial B_r(\mathbf{0})} d\Sigma_r\, e^{i\mathbf{S}(\mathbf{0})\cdot\mathbf{x}}\Big(\big(i\chi\,\mathbf{S}(\mathbf{0})\cdot\hat{\mathbf{r}}-\partial_r\chi\big)\,\overline{\big(G^{(k)}_{\lambda_2}-G^{(k)}_{\lambda_1}\big)}\,\phi_{\lambda_2}$$

$$+\chi\,\overline{\big(G^{(k)}_{\lambda_2}-G^{(k)}_{\lambda_1}\big)}\,\partial_r\phi_{\lambda_2}-\chi\,\overline{\partial_r\big(G^{(k)}_{\lambda_2}-G^{(k)}_{\lambda_1}\big)}\,\phi_{\lambda_2}\Big)\Big]$$

$$+\sum_{k\in\{0,-1\}}|q^{(k)}|^2\Big[-2i\lim_{r\to0^+}\int_{\partial B_r(\mathbf{0})} d\Sigma_r\,\chi^2\big(\mathbf{S}-\mathbf{S}(\mathbf{0})\big)\cdot\hat{\mathbf{r}}\,\overline{\big(G^{(k)}_{\lambda_2}-G^{(k)}_{\lambda_1}\big)}\,G^{(k)}_{\lambda_1}$$

$$+\lim_{r\to0^+}\int_{\partial B_r(\mathbf{0})} d\Sigma_r\,\chi\,\partial_r\chi\left(|G^{(k)}_{\lambda_2}-G^{(k)}_{\lambda_1}|^2-2\,\mathrm{Re}\Big(\overline{G^{(k)}_{\lambda_1}}\big(G^{(k)}_{\lambda_2}-G^{(k)}_{\lambda_1}\big)\Big)\right)\Big]. \tag{28}$$

By comparison with (14)(15), it appears that the terms from the second to the sixth line of (28) exactly reproduce $Q^{(\beta)}_{\alpha,S}\left[\phi_{\lambda_2}+e^{-i\mathbf{S}(\mathbf{0})\cdot\mathbf{x}}\chi\sum_{k\in\{0,-1\}}q^{(k)}G^{(k)}_{\lambda_2}\right]$. We now proceed to show that all other contributions vanish. On one side, consider the terms in the seventh and eighth lines of (28). An additional integration by parts gives

$$\begin{aligned}&\langle G^{(k)}_{\lambda_1}|\,(\nabla\chi^2)\cdot\nabla G^{(k)}_{\lambda_2}\rangle-\langle\nabla G^{(k)}_{\lambda_1}|\,(\nabla\chi^2)G^{(k)}_{\lambda_2}\rangle\\ &=\lim_{r\to0^+}\Big[\int_{\partial B_r(\mathbf{0})} d\Sigma_r\,\chi^2\Big[\overline{\partial_r G^{(k)}_{\lambda_1}}G^{(k)}_{\lambda_2}-\overline{G^{(k)}_{\lambda_1}}\partial_r G^{(k)}_{\lambda_2}\Big]\\ &\quad+\int_{\mathbb{R}^2\setminus B_r(\mathbf{0})} d\mathbf{x}\,\chi^2\left(\overline{\Delta G^{(k)}_{\lambda_1}}G^{(k)}_{\lambda_2}-\overline{G^{(k)}_{\lambda_1}}\Delta G^{(k)}_{\lambda_2}\right)\Big].\end{aligned}$$

From (8) we deduce, for $r\to0^+$,

$$\left(\overline{\partial_r G^{(k)}_{\lambda_1}}G^{(k)}_{\lambda_2}-\overline{G^{(k)}_{\lambda_1}}\partial_r G^{(k)}_{\lambda_2}\right)(r)=\frac{\pi\,(\lambda_2^{2|k+\alpha|}-\lambda_1^{2|k+\alpha|})}{2\sin(\pi\alpha)\,r}+O\big(r^{1-2|k+\alpha|}\big).$$

Moreover, in view of (5) and (6), an explicit computation gives

$$\Delta G^{(k)}_{\lambda}=\big(\mathbf{A}^2+\lambda^2+2\mathbf{A}\cdot(-i\nabla)\big)G^{(k)}_{\lambda}=\left(\mathbf{A}^2+\lambda^2+\frac{2\alpha k}{r}\right)G^{(k)}_{\lambda},\quad \text{in } \mathbb{R}^2\setminus\{\mathbf{0}\}.$$

Recalling that $\chi=1$ in an open neighborhood of $\mathbf{x}=\mathbf{0}$, see (12), we obtain

$$\begin{aligned}&\frac{\pi^2}{\sin(\pi\alpha)}\left(\lambda_1^{2|k+\alpha|}-\lambda_2^{2|k+\alpha|}\right)+(\lambda_2^2-\lambda_1^2)\langle\chi G^{(k)}_{\lambda_1}|\,\chi G^{(k)}_{\lambda_2}\rangle\\ &\qquad+\langle G^{(k)}_{\lambda_1}|\,(\nabla\chi^2)\cdot\nabla G^{(k)}_{\lambda_2}\rangle-\langle\nabla G^{(k)}_{\lambda_1}|\,(\nabla\chi^2)G^{(k)}_{\lambda_2}\rangle=0.\end{aligned}$$

On the other side, consider the boundary contributions in the last five lines of (28). For r small enough, the following holds true: $\partial_r \chi = 0$ on $\partial B_r(\mathbf{0})$, see (12); $|\mathbf{S}-\mathbf{S}(\mathbf{0})| \leqslant c\, r$, see (9); $|G^{(k)}_{\lambda_2} - G^{(k)}_{\lambda_1}| \leqslant c\, r^{|\alpha+k|}$ and $\big|\partial_r (G^{(k)}_{\lambda_2} - G^{(k)}_{\lambda_1})\big| \leqslant c\, r^{|\alpha+k|-1}$, see (8). Recalling as well the condition (3), by Cauchy-Schwarz inequality we get:

$$\left| \int_{\partial B_r(\mathbf{0})} d\Sigma_r\; e^{-i\mathbf{S}(\mathbf{0})\cdot\mathbf{x}} \Big(i\chi\, (\mathbf{S} - \mathbf{S}(\mathbf{0}))\cdot\hat{\mathbf{r}} + \partial_r \chi \Big)\, \overline{\phi_{\lambda_2}}\, (G^{(k)}_{\lambda_2} - G^{(k)}_{\lambda_1}) \right| \leqslant c\, r^{2+|\alpha+k|} \sqrt{\langle |\phi_{\lambda_2}|^2 \rangle} \xrightarrow{r\to 0^+} 0\,;$$

$$\left| \int_{\partial B_r(\mathbf{0})} d\Sigma_r\; e^{i\mathbf{S}(\mathbf{0})\cdot\mathbf{x}} \Big(\big(i\chi\, \mathbf{S}(0)\cdot\hat{\mathbf{r}} - \partial_r\chi \big)\, \overline{(G^{(k)}_{\lambda_2} - G^{(k)}_{\lambda_1})}\, \phi_{\lambda_2} + \chi\, \overline{(G^{(k)}_{\lambda_2} - G^{(k)}_{\lambda_1})}\, \partial_r \phi_{\lambda_2} - \chi\, \overline{\partial_r (G^{(k)}_{\lambda_2} - G^{(k)}_{\lambda_1})}\, \phi_{\lambda_2} \Big) \right|$$
$$\leqslant c\, r^{|\alpha+k|} \left(r \sqrt{\langle |\phi_{\lambda_2}|^2 \rangle} + r \sqrt{\langle |\partial_r \phi_{\lambda_2}|^2 \rangle} + \sqrt{\langle |\phi_{\lambda_2}|^2 \rangle} \right) \xrightarrow{r\to 0^+} 0\,;$$

$$\left| \int_{\partial B_r(\mathbf{0})} d\Sigma_r\; \chi^2 (\mathbf{S} - \mathbf{S}(\mathbf{0}))\cdot\hat{\mathbf{r}}\; \overline{(G^{(k)}_{\lambda_2} - G^{(k)}_{\lambda_1})}\, G^{(k)}_{\lambda_1} \right| \leqslant C r^2 \xrightarrow{r\to 0^+} 0\,;$$

$$\int_{\partial B_r(\mathbf{0})} d\Sigma_r\; \chi\, \partial_r \chi \left(|G^{(k)}_{\lambda_2} - G^{(k)}_{\lambda_1}|^2 - 2\,\mathrm{Re}\Big(\overline{G^{(k)}_{\lambda_1}}\, (G^{(k)}_{\lambda_2} - G^{(k)}_{\lambda_1}) \Big) \right) = 0\,.$$

Summing up, the previous results entail

$$\mathcal{Q}^{(\beta)}_{\alpha,S}\Big[\phi_{\lambda_1} + e^{-i\mathbf{S}(\mathbf{0})\cdot\mathbf{x}} \chi \sum_{k\in\{0,-1\}} q^{(k)} G^{(k)}_{\lambda_1} \Big] = \mathcal{Q}^{(\beta)}_{\alpha,S}\Big[\phi_{\lambda_2} + e^{-i\mathbf{S}(\mathbf{0})\cdot\mathbf{x}} \chi \sum_{k\in\{0,-1\}} q^{(k)} G^{(k)}_{\lambda_2} \Big],$$

whence the thesis. By similar arguments it can be shown that the form does not depend on the choice of χ, as long as hypotheses (12) (13) are fulfilled.

(iii) Closedness can be deduced by classical arguments [11, 32], once lower boundedness has been proved. Therefore, the thesis follows as soon as we show that

$$\mathcal{Q}^{(\beta)}_{\alpha,S}[\psi] + \lambda^2\, \|\psi\|_2^2 \geqslant 0\,, \qquad \text{for } \lambda > 0 \text{ large enough}\,. \tag{29}$$

To this avail, by minor variations of the arguments described in [8] (also recall (7)), we obtain the following for any $\varepsilon_1, \varepsilon_2, \varepsilon_3 \in (0,1)$ and suitable $c_1, c_2, c_3 > 0$:

$$Q^{(F)}_{\alpha,S}[\phi_\lambda] \geqslant \frac{1}{2}\, Q^{(F)}_{\alpha,S}[\phi_\lambda] + \frac{1-\varepsilon_1}{2}\, \|\mathbf{1}_\chi(-i\nabla + \mathbf{A}_\alpha)\phi_\lambda\|_2^2 - \frac{1-\varepsilon_1}{2\varepsilon_1}\, c_1\, \|\phi_\lambda\|_2^2\,;$$

$$\sum_{k\in\{0,-1\}} \mathrm{Re}\Big[q^{(k)}\Big\langle(-i\nabla + \mathbf{A}_\alpha)\phi_\lambda\,\Big|\, e^{-i\,\mathbf{S}(\mathbf{0})\cdot\mathbf{x}}\big((\mathbf{S}-\mathbf{S}(\mathbf{0}))\chi - i\nabla\chi\big)G^{(k)}_\lambda\Big\rangle\Big]$$
$$\geqslant -\frac{\varepsilon_2}{8}\, \|\mathbf{1}_\chi(-i\nabla + \mathbf{A}_\alpha)\phi_\lambda\|_2^2 - \frac{2c_2}{\varepsilon_2} \sum_{k\in\{0,-1\}} |q^{(k)}|^2\, \lambda^{2|k+\alpha|-2}\,;$$

$$\sum_{k\in\{0,-1\}} \mathrm{Re}\Big[q^{(k)}\Big\langle\phi_\lambda\,\Big|\, e^{-i\,\mathbf{S}(\mathbf{0})\cdot\mathbf{x}}\big[(\mathbf{S}-\mathbf{S}(\mathbf{0}))^2\chi$$
$$+ 2\mathbf{S}(\mathbf{0})\cdot\big((\mathbf{S}-\mathbf{S}(\mathbf{0}))\chi - i\nabla\chi\big) + \Delta\chi\big]G^{(k)}_\lambda\Big\rangle\Big]$$
$$\geqslant -\frac{\varepsilon_3}{2}\, \|\phi_\lambda\|_2^2 - \frac{c_3}{\varepsilon_3} \sum_{k\in\{0,-1\}} |q^{(k)}|^2\, \lambda^{2|k+\alpha|-2}\,.$$

Building on the basic inequality $\big|\nabla\big(\chi\, G^{(k)}_\lambda\big)\big| \leqslant \frac{c}{|\mathbf{x}|}\big(\chi\, G^{(k)}_\lambda\big)$ and (7) and (9), we further deduce $|\Xi_{kk'}(\lambda)| \leqslant c\, \|G^{(k)}_\lambda\|_2\, \|G^{(k')}_\lambda\|_2 \leqslant c\, \lambda^{|k+\alpha|+|k'+\alpha|-2}$. This allows us to infer that, for some suitable $c_4 > 0$,

$$\sum_{k,k'\in\{0,-1\}} \overline{q^{(k)}}\, q^{(k')}\Big[\beta_{kk'} + \frac{\pi^2}{\sin(\pi\alpha)}\, \lambda^{2|k+\alpha|}\, \delta_{kk'} + \Xi_{kk'}(\lambda)\Big]$$
$$\geqslant \Big[\frac{\pi^2}{\sin(\pi\alpha)} \min_{k\in\{0,-1\}}\big(\lambda^{2|k+\alpha|}\big) - \max_{k,k'\in\{0,-1\}}\Big(|\beta_{kk'}| + |\Xi_{kk'}(\lambda)|\Big)\Big]$$
$$\times \sum_{k\in\{0,-1\}} |q^{(k)}|^2$$
$$\geqslant c_4\Big(\min\big\{\lambda^{2\alpha}, \lambda^{2(1-\alpha)}\big\} - 1 - \max\big\{\lambda^{-2\alpha}, \lambda^{-2(1-\alpha)}\big\}\Big) \sum_{k\in\{0,-1\}} |q^{(k)}|^2\,.$$

Summing up, we have

$$Q^{(\beta)}_{\alpha,S}[\psi] + \lambda^2\, \|\psi\|_2^2$$
$$\geqslant \frac{1}{2}\, Q^{(F)}_{\alpha,S}[\phi_\lambda] + \frac{1-\varepsilon_1-\varepsilon_2}{2}\, \|\mathbf{1}_\chi(-i\nabla + \mathbf{A}_\alpha)\phi_\lambda\|_2^2$$

$$+ \left(\lambda^2 - \frac{1-\varepsilon_1}{2\varepsilon_1} c_1 - \varepsilon_3\right) \|\phi_\lambda\|_2^2$$
$$+ \Big[c_4 \min\{\lambda^{2\alpha}, \lambda^{2(1-\alpha)}\} - c_4$$
$$- \left(\frac{8c_2}{\varepsilon_2} + \frac{2c_3}{\varepsilon_3} + c_4\right) \max\{\lambda^{-2\alpha}, \lambda^{-2(1-\alpha)}\}\Big] \sum_{k\in\{0,-1\}} |q^{(k)}|^2 .$$

Upon fixing $\varepsilon_1, \varepsilon_2, \varepsilon_3 \in (0, 1)$ appropriately and $\lambda > 0$ large enough, the above relation suffices to infer (29), whence the thesis.

□

Proof *(Corollary 1)* For any Hermitian matrix β, we derive the self-adjoint operator $H^{(\beta)}_{\alpha,S}$ associated to the quadratic form $Q^{(\beta)}_{\alpha,S}$ by standard methods. To begin with, for any pair $\psi_\ell = \phi_{\ell,\lambda} + e^{-i\,\mathbf{S(0)}\cdot\mathbf{x}}\,\chi \sum_{k\in\{0,-1\}} q^{(k)}_\ell G^{(k)}_\lambda$, $\ell \in \{1, 2\}$, belonging to the form domain $D[Q^{(\beta)}_{\alpha,S}]$, consider the sesquilinear form defined by polarization

$$Q^{(\beta)}_{\alpha,S}[\psi_1, \psi_2] = Q^{(F)}_{\alpha,S}[\phi_{1,\lambda}, \phi_{2,\lambda}] - \lambda^2 \langle\psi_1|\psi_2\rangle + \lambda^2 \langle\phi_{1,\lambda}|\phi_{2,\lambda}\rangle$$
$$+ \sum_{k\in\{0,-1\}} \overline{q^{(k)}_1} \Big[2\Big\langle e^{-i\,\mathbf{S(0)}\cdot\mathbf{x}}\big((\mathbf{S}-\mathbf{S(0)})\chi - i\nabla\chi\big)G^{(k)}_\lambda \Big| (-i\nabla + \mathbf{A}_\alpha)\phi_{2,\lambda}\Big\rangle$$
$$+ \Big\langle e^{-i\,\mathbf{S(0)}\cdot\mathbf{x}}\big[(\mathbf{S}-\mathbf{S(0)})^2\chi + 2\mathbf{S(0)}\cdot\big((\mathbf{S}-\mathbf{S(0)})\chi - i\nabla\chi\big) + \Delta\chi\big]G^{(k)}_\lambda \Big| \phi_{2,\lambda}\Big\rangle\Big]$$
$$+ \sum_{k\in\{0,-1\}} q^{(k)}_2 \Big[2\Big\langle(-i\nabla + \mathbf{A}_\alpha)\phi_{1,\lambda}\Big| e^{-i\,\mathbf{S(0)}\cdot\mathbf{x}}\big((\mathbf{S}-\mathbf{S(0)})\chi - i\nabla\chi\big)G^{(k)}_\lambda\Big\rangle$$
$$+ \Big\langle\phi_{1,\lambda}\Big| e^{-i\,\mathbf{S(0)}\cdot\mathbf{x}}\big[(\mathbf{S}-\mathbf{S(0)})^2\chi + 2\mathbf{S(0)}\cdot\big((\mathbf{S}-\mathbf{S(0)})\chi - i\nabla\chi\big) + \Delta\chi\big]G^{(k)}_\lambda\Big\rangle\Big]$$
$$+ \sum_{k,k'\in\{0,-1\}} \overline{q^{(k)}_1}\, q^{(k')}_2 \left[\beta_{kk'} + \frac{\pi^2}{\sin(\pi\alpha)}\, \lambda^{2|k+\alpha|}\, \delta_{kk'} + \Xi_{kk'}(\lambda)\right]. \tag{30}$$

Here $Q^{(F)}_{\alpha,S}[\phi_{1,\lambda}, \phi_{2,\lambda}]$ is the sesquilinear form associated to the Friedrichs quadratic form, namely, $Q^{(F)}_{\alpha,S}[\phi_{1,\lambda}, \phi_{2,\lambda}] := \int_{\mathbb{R}^2} d\mathbf{x}\, \overline{(-i\nabla + \mathbf{A}_\alpha + \mathbf{S})\phi_{1,\lambda}} \cdot (-i\nabla + \mathbf{A}_\alpha + \mathbf{S})\phi_{2,\lambda}$.

Now assume $q^{(0)}_1 = q^{(-1)}_1 = 0$, so that $\psi_1 = \phi_{1,\lambda}$. Integrating by parts and checking that boundary contributions vanish by means of arguments similar to those outlined in the proof of Theorem 1, item *(ii)*, the sesquilinear form (30) reduces to

$$Q^{(\beta)}_{\alpha,S}[\phi_1, \psi_2] = \langle\phi_{1,\lambda} \,|\, H_{\alpha,S}\phi_{2,\lambda}\rangle$$
$$+ \sum_{k\in\{0,-1\}} q^{(k)}_2 \Big[2\Big\langle\phi_{1,\lambda}\Big| e^{-i\,\mathbf{S(0)}\cdot\mathbf{x}}\big((\mathbf{S}-\mathbf{S(0)})\chi - i\nabla\chi\big)\cdot(-i\nabla + \mathbf{A}_\alpha)G^{(k)}_\lambda\Big\rangle$$
$$+ \Big\langle\phi_{1,\lambda}\Big| e^{-i\,\mathbf{S(0)}\cdot\mathbf{x}}\Big((\mathbf{S}-\mathbf{S(0)})^2\chi - \lambda^2\chi + 2(\mathbf{S}-\mathbf{S(0)})\cdot(-i\nabla\chi) - \Delta\chi\Big)G^{(k)}_\lambda\Big\rangle\Big].$$

Considerations analogous to those reported in the proof of Theorem 1, item *(i)*, ensure that all pairings in the second and third lines of the above identity are well-defined inner products in $L^2(\mathbb{R}^2)$. So, to fulfill the condition $Q^{(\beta)}_{\alpha,S}[\phi_1, \psi_2] = \langle \phi_1 | w \rangle$ for some $w = H^{(\beta)}_{\alpha,S}\, \psi_2 \in L^2(\mathbb{R}^2)$, we must require $H_{\alpha,S}\, \phi_{2,\lambda} \in L^2(\mathbb{R}^2)$ (*cf.* (4) and the condition in the second line of (19)), as well as (*cf.* (20))

$$w = H_{\alpha,S}\, \phi_{2,\lambda} + \sum_{k\in\{0,-1\}} q_2^{(k)} e^{-i\,\mathbf{S(0)}\cdot\mathbf{x}} \Big[2\Big((\mathbf{S}-\mathbf{S(0)})\chi - i\nabla\chi \Big) \cdot (-i\nabla + \mathbf{A}_\alpha) G^{(k)}_\lambda$$
$$+ \Big((\mathbf{S}-\mathbf{S(0)})^2 \chi - \lambda^2 \chi + 2(\mathbf{S}-\mathbf{S(0)}) \cdot (-i\nabla\chi) - \Delta\chi \Big) G^{(k)}_\lambda \Big]. \tag{31}$$

In view of the previous results, the sesquilinear form (30) can be re-written as

$$Q^{(\beta)}_{\alpha,S}[\psi_1, \psi_2] = Q^{(\beta)}_{\alpha,S}[\phi_{1,\lambda}, \psi_2]$$
$$+ \sum_{k\in\{0,-1\}} \overline{q_1^{(k)}} \Big[2 \Big\langle e^{-i\,\mathbf{S(0)}\cdot\mathbf{x}} \big((\mathbf{S}-\mathbf{S(0)})\chi - i\nabla\chi \big) G^{(k)}_\lambda \Big| \, (-i\nabla + \mathbf{A}_\alpha)\phi_{2,\lambda} \Big\rangle$$
$$+ \Big\langle e^{-i\,\mathbf{S(0)}\cdot\mathbf{x}} \big[(\mathbf{S}-\mathbf{S(0)})^2 \chi - \lambda^2 \chi + 2\mathbf{S(0)} \cdot \big((\mathbf{S}-\mathbf{S(0)})\chi - i\nabla\chi \big)$$
$$+ \Delta\chi \big] G^{(k)}_\lambda \Big| \, \phi_{2,\lambda} \Big\rangle \Big]$$
$$+ \sum_{k,k'\in\{0,-1\}} \overline{q_1^{(k)}} q_2^{(k')} \Big[\beta_{kk'} + \frac{\pi^2}{\sin(\pi\alpha)}\, \lambda^{2|k+\alpha|}\, \delta_{kk'} + \Xi_{kk'}(\lambda) - \lambda^2 \Big\langle \chi\, G^{(k)}_\lambda \Big| \chi\, G^{(k')}_\lambda \Big\rangle \Big].$$

Building on this and recalling the definition (15) of $\Xi_{kk'}(\lambda)$, by simple (though lengthy) computations we deduce that the position $Q^{(\beta)}_{\alpha,S}[\psi_1, \psi_2] = \langle \psi_1 | w \rangle$, with w as in (31), can be satisfied for generic $q_1^{(0)}, q_1^{(-1)}$ only if, for $k \in \{0, -1\}$,

$$\Big\langle G^{(k)}_\lambda \Big| \big[(-i\nabla + \mathbf{A}_\alpha)^2 + \lambda^2 \big] \big(e^{i\,\mathbf{S(0)}\cdot\mathbf{x}}\, \chi\, \phi_{2,\lambda} \big) \Big\rangle$$
$$= \sum_{k'\in\{0,-1\}} q_2^{(k')} \Big[\beta_{kk'} + \frac{\pi^2}{\sin(\pi\alpha)}\, \lambda^{2|k+\alpha|}\, \delta_{kk'} +$$
$$+ \Big(\Big\langle G^{(k)}_\lambda \Big| \nabla\cdot(\chi\nabla\chi)\, G^{(k)}_\lambda \Big\rangle + 2 \Big\langle G^{(k)}_\lambda \Big| (\chi\nabla\chi)\cdot\nabla G^{(k)}_\lambda \Big\rangle \Big) \delta_{kk'} \Big]. \tag{32}$$

To derive the above identity we used in particular the identity (16) and the fact that $\mathbf{A}_\alpha \cdot \nabla\chi = 0$, both descending from (13). On one side, recalling the explicit expression (6) for $G^{(k)}_\lambda$ and that χ is radial, we get $2\, \overline{G^{(k)}_\lambda} (\chi\nabla\chi) \cdot \nabla G^{(k)}_\lambda = (\chi\nabla\chi) \cdot$

$\nabla|G_\lambda^{(k)}|^2$; then, integrating by parts and keeping in mind that $\chi \equiv 1$ near the origin, we obtain

$$\Big\langle G_\lambda^{(k)} \Big| \nabla\cdot(\chi\nabla\chi)\, G_\lambda^{(k)} \Big\rangle + 2 \Big\langle G_\lambda^{(k)} \Big| (\chi\nabla\chi)\cdot \nabla G_\lambda^{(k)} \Big\rangle = - \lim_{r\to 0^+} \int_{\partial B_r(\mathbf{0})} d\Sigma_r\, (\chi\,\partial_r \chi)\, |G_\lambda^{(k)}|^2 = 0\,. \tag{33}$$

On the other side, integrating by parts twice and using the basic identity (5), we get

$$\begin{aligned}
&\Big\langle G_\lambda^{(k)} \Big| [(-i\nabla + \mathbf{A}_\alpha)^2 + \lambda^2](e^{i\,\mathbf{S}(\mathbf{0})\cdot\mathbf{x}}\, \chi\, \phi_{2,\lambda}) \Big\rangle \\
&= \lim_{r\to 0^+} \int_{\partial B_r(\mathbf{0})} d\Sigma_r \Big[\overline{G_\lambda^{(k)}}\, \partial_r (e^{i\,\mathbf{S}(\mathbf{0})\cdot\mathbf{x}}\, \chi\, \phi_{2,\lambda}) - \overline{\partial_r G_\lambda^{(k)}}\, (e^{i\,\mathbf{S}(\mathbf{0})\cdot\mathbf{x}}\, \chi\, \phi_{2,\lambda}) \Big] \\
&\qquad + \lim_{r\to 0^+} \int_{\mathbb{R}^2\setminus B_r(\mathbf{0})} d\Sigma_r\, \overline{[(-i\nabla + \mathbf{A}_\alpha)^2 + \lambda^2] G_\lambda^{(k)}}\, (e^{i\,\mathbf{S}(\mathbf{0})\cdot\mathbf{x}}\, \chi\, \phi_{2,\lambda}) \\
&= \lim_{r\to 0^+} \int_{\partial B_r(\mathbf{0})} d\Sigma_r \Big[\overline{G_\lambda^{(k)}}\, \partial_r (e^{i\,\mathbf{S}(\mathbf{0})\cdot\mathbf{x}}\, \chi\, \phi_{2,\lambda}) - \overline{\partial_r G_\lambda^{(k)}}\, (e^{i\,\mathbf{S}(\mathbf{0})\cdot\mathbf{x}}\, \chi\, \phi_{2,\lambda}) \Big] \\
&= \frac{\Gamma(|k+\alpha|)}{2^{1-|k+\alpha|}} \lim_{r\to 0^+} \frac{1}{r^{|k+\alpha|}} \int_{\partial B_r(\mathbf{0})} d\Sigma_r\, e^{-ik\theta} \Big[\partial_r \phi_{2,\lambda} + \frac{|k+\alpha|}{r}\, \phi_{2,\lambda} \Big],
\end{aligned} \tag{34}$$

where the last identity follows from the asymptotic relations (3), (8), by arguments analogous to those mentioned in the proof of Theorem 1. Notably, only the leading order term in (8) plays a role here. Summing up, from (32), (33) and (34) we infer

$$\sum_{k'\in\{0,-1\}} q_2^{(k')} \left(\beta_{kk'} + \frac{\pi^2}{\sin(\pi\alpha)}\, \lambda^{2|k+\alpha|}\, \delta_{kk'} \right) = \frac{\Gamma(|k+\alpha|)}{2^{1-|k+\alpha|}} \lim_{r\to 0^+} \frac{1}{r^{|k+\alpha|}} \int_{\partial B_r(\mathbf{0})} d\Sigma_r\, e^{-ik\theta} \Big[\partial_r \phi_{2,\lambda} + \frac{|k+\alpha|}{r}\, \phi_{2,\lambda} \Big],$$

which proves the boundary condition in (19), thus completing the characterization of $D\big(H_{\alpha,\mathbf{S}}^{(\beta)}\big)$.

The fact that the family $H_{\alpha,\mathbf{S}}^{(\beta)}$, β any 2×2 Hermitian matrix, exhausts all self-adjoint extensions of $H_{\alpha,\mathbf{S}} \restriction C_c^\infty(\mathbb{R}^2\setminus\{\mathbf{0}\})$ in $L^2(\mathbb{R}^2)$ if $\mathbf{S}\in L^\infty(\mathbb{R}^2)$ can be deduced by exactly the same arguments reported in [8, Proof of Corollary 1.10]. □

Let us finally proceed to present the proof of Theorem 2, keeping in mind that it relies on the hypothesis (22). The latter implies that $\mathbf{S}$ is uniformly bounded on the whole space $\mathbb{R}^2$ and Lipschitz continuous at the origin.

Proof *(Theorem 2)* The derivation of both the lower and upper bound inequalities relies on the following algebraic identity, which can be easily deduced using the gauge transformation $\psi \mapsto e^{-i\mathbf{S}(\mathbf{0})\cdot\mathbf{x}}\,\psi$ and an elementary telescopic argument:

$$\begin{aligned} & Q^{(F)}_{\alpha_n,S}[\psi_{\alpha_n}] - Q^{(F)}_{0,S}[\psi_0] \\ & = \big\|\mathbf{A}_{\alpha_n}\psi_{\alpha_n}\big\|_2^2 + 2\,\mathrm{Re}\big[\big\langle\big(-i\nabla + \mathbf{S} - \mathbf{S}(\mathbf{0})\big)\psi_{\alpha_n}\,\big|\,\mathbf{A}_{\alpha_n}\psi_{\alpha_n}\big\rangle\big] \\ & \quad + \big\|\big(-i\nabla + \mathbf{S} - \mathbf{S}(\mathbf{0})\big)\psi_{\alpha_n}\big\|_2^2 - \big\|\big(-i\nabla + \mathbf{S} - \mathbf{S}(\mathbf{0})\big)\psi_0\big\|_2^2 . \end{aligned} \tag{35}$$

(i) *Lower bound inequality*. First of all, on account of the hypothesis $\mathbf{S}\in L^\infty(\mathbb{R}^2)$, from [8, Eq. (2.7)] we deduce that

$$Q^{(F)}_{\alpha_n,S}[\psi_{\alpha_n}] + \gamma\,\|\psi_{\alpha_n}\|_2^2 \geqslant C_\gamma\Big(\,\|\nabla\psi_{\alpha_n}\|_2^2 + \|\mathbf{A}_{\alpha_n}\psi_{\alpha_n}\|_2^2\Big)\,, \tag{36}$$

for any $\gamma > 0$ large enough and some suitable $C_\gamma > 0$. With obvious understandings, the above inequality is in fact valid for all $\psi_{\alpha_n}\in L^2(\mathbb{R}^2)$.

For any convergent sequence $\psi_{\alpha_n}\to\psi_0\in L^2(\mathbb{R}^2)\setminus H^1(\mathbb{R}^2)$, the thesis can be derived by *reductio ad absurdum*. In this case, the condition (25) reads (*cf.* (23))

$$+\infty = Q^{(F)}_{0,S}[\psi_0] \leqslant \liminf_{n\to+\infty} Q^{(F)}_{\alpha_n,S}[\psi_{\alpha_n}] \leqslant \limsup_{n\to+\infty} Q^{(F)}_{\alpha_n,S}[\psi_{\alpha_n}]\,.$$

By contradiction, assume there exists some sequence $\psi_{\alpha_n}\to\psi_0\in L^2(\mathbb{R}^2)\setminus H^1(\mathbb{R}^2)$ such that $\lim_{n\to+\infty} Q^{(F)}_{\alpha_n,S}[\psi_{\alpha_n}] \leqslant C_S < +\infty$. Then, from (36) it follows that the said sequence is uniformly bounded in $H^1(\mathbb{R}^2)$. By Banach-Alaoglu theorem, this implies in turn that $\psi_{\alpha_n} \rightharpoonup \varphi \in H^1(\mathbb{R}^2)$ (weak convergence, up to extraction of a subsequence). This contradicts the hypothesis $\psi_{\alpha_n}\to\psi_0\in L^2(\mathbb{R}^2)\setminus H^1(\mathbb{R}^2)$, since uniqueness of the limit implies $\psi_0 = \varphi$.

Next, consider any convergent sequence in $L^2(\mathbb{R}^2)$ fulfilling $\psi_{\alpha_n}\to\psi_0\in H^1(\mathbb{R}^2)$. The thesis (25) follows trivially if $Q^{(F)}_{\alpha_n,S}[\psi_{\alpha_n}] > Q^{(F)}_{0,S}[\psi_0]$ for almost all $n\in\mathbb{N}$. On the contrary, let us assume that $Q^{(F)}_{\alpha_n,S}[\psi_{\alpha_n}] \leqslant Q^{(F)}_{0,S}[\psi_0]$ for almost all $n\in\mathbb{N}$. Since $Q^{(F)}_{0,S}[\psi_0] < +\infty$ for $\psi_0\in H^1(\mathbb{R}^2)$, by arguments similar to those described before we deduce the existence of a uniformly bounded subsequence $\{\psi_{\tilde{\alpha}_n}\}_{n\in\mathbb{N}}$ in $H^1(\mathbb{R}^2)$, converging weakly to ψ_0. Taking this into account, let us now refer to (35). On one side, notice that $\|(-i\nabla + \mathbf{S} - \mathbf{S}(\mathbf{0}))\psi\|_2^2 + \|\psi\|_2^2$ defines an equivalent norm in $H^1(\mathbb{R}^2)$ for $\mathbf{S}\in L^\infty(\mathbb{R}^2)$. Then, keeping in mind that $\psi_{\tilde{\alpha}_n}\to\psi_0$ in the strong L^2-topology, by lower semicontinuity of the norm in $H^1(\mathbb{R}^2)$ we infer

$$\begin{aligned} & \liminf_{n\to+\infty} \big\|\big(-i\nabla + \mathbf{S} - \mathbf{S}(\mathbf{0})\big)\psi_{\tilde{\alpha}_n}\big\|_2^2 \\ & = \liminf_{n\to+\infty}\Big(\big\|\big(-i\nabla + \mathbf{S} - \mathbf{S}(\mathbf{0})\big)\psi_{\tilde{\alpha}_n}\big\|_2^2 + \|\psi_{\tilde{\alpha}_n}\|_2^2\Big) - \lim_{n\to+\infty}\|\psi_{\tilde{\alpha}_n}\|_2^2 \end{aligned}$$

$$\geqslant \Big(\big\|(-i\nabla+\mathbf{S}-\mathbf{S}(\mathbf{0}))\psi_0\big\|_2^2+\|\psi_0\|_2^2\Big)-\|\psi_0\|_2^2$$
$$=\big\|(-i\nabla+\mathbf{S}-\mathbf{S}(\mathbf{0}))\psi_0\big\|_2^2\,.$$

On the other side, using the angular harmonics decomposition

$$\psi_{\tilde{\alpha}_n}(r,\theta)=\sum_{k\in\mathbb{Z}}\psi_{\tilde{\alpha}_n}^{(k)}(r)\,\frac{e^{ik\theta}}{\sqrt{2\pi}}\,,$$

by a direct computation we infer

$$\big|\langle(-i\nabla)\psi_{\tilde{\alpha}_n}\,|\,\mathbf{A}_{\tilde{\alpha}_n}\psi_{\tilde{\alpha}_n}\rangle\big|=\left|\sum_{k\in\mathbb{Z}}\int_0^{+\infty}dr\;\frac{\tilde{\alpha}_n\,k}{r}\,\big|\psi_{\tilde{\alpha}_n}^{(k)}(r)\big|^2\right|$$
$$\leqslant\tilde{\alpha}_n\sum_{k\in\mathbb{Z}}\int_0^{+\infty}dr\;\frac{k^2}{r}\,\big|\psi_{\tilde{\alpha}_n}^{(k)}(r)\big|^2\leqslant\tilde{\alpha}_n\,\big\|\psi_{\tilde{\alpha}_n}\big\|_{H^1}^2\,.$$

At the same time, exploiting the Lipschitz continuity of $\mathbf{S}$ at $\mathbf{x}=\mathbf{0}$, we get

$$\big|\langle(\mathbf{S}-\mathbf{S}(\mathbf{0}))\psi_{\tilde{\alpha}_n}\,|\,\mathbf{A}_{\tilde{\alpha}_n}\psi_{\tilde{\alpha}_n}\rangle\big|\leqslant\big\|(\mathbf{S}-\mathbf{S}(\mathbf{0}))\cdot\mathbf{A}_{\tilde{\alpha}_n}\big\|_\infty\,\big\|\psi_{\tilde{\alpha}_n}\big\|_2^2\leqslant\tilde{\alpha}_n\,c\,\big\|\psi_{\tilde{\alpha}_n}\big\|_2^2\,.$$

Discarding the positive term $\|\mathbf{A}_{\alpha_n}\psi_{\alpha_n}\|_2^2$ and recalling that $\{\psi_{\tilde{\alpha}_n}\}_{n\in\mathbb{N}}$ is uniformly bounded in $H^1(\mathbb{R}^2)$, from the above arguments and (35) we deduce

$$\liminf_{n\to+\infty}Q_{\alpha_n,S}^{(F)}[\psi_{\alpha_n}]-Q_{0,S}^{(F)}[\psi_0]$$
$$\geqslant-2\limsup_{n\to+\infty}\big|\langle(-i\nabla+\mathbf{S}-\mathbf{S}(\mathbf{0}))\psi_{\alpha_n}\,|\mathbf{A}_{\alpha_n}\psi_{\alpha_n}\rangle\big|$$
$$\geqslant-C\limsup_{n\to+\infty}\Big(\tilde{\alpha}_n\,\big\|\psi_{\tilde{\alpha}_n}\big\|_{H^1}^2\Big)=0\,,$$

which proves the lower bound inequality (25).

(ii) *Upper bound inequality.* For $\psi_0\in L^2(\mathbb{R}^2)\setminus H^1(\mathbb{R}^2)$ the thesis (26) is trivial, since $Q_{0,S}^{(F)}[\psi_0]=+\infty$ by (24). Let us henceforth assume $\psi_0\in H^1(\mathbb{R}^2)$. For any given family $\{\alpha_n\}_{n\in\mathbb{N}}\subset(0,1)$ such that $\alpha_n\to0$ as $n\to+\infty$, we consider the sequence of approximants

$$\psi_{\alpha_n}:=\eta_{\alpha_n}\,\psi_0\in L^2(\mathbb{R}^2)\,,$$

where $\eta_{\alpha_n}(\mathbf{x})\equiv\eta_{\alpha_n}(|\mathbf{x}|):[0,+\infty)\to[0,1]$ is a monotone increasing, smooth radial function with downward concavity fulfilling

$$\eta_{\alpha_n}(\mathbf{x})=\begin{cases}\big(|\mathbf{x}|/\sqrt{\alpha_n}\big)^{\alpha_n} & \text{for}\quad \mathbf{x}\in B_{\sqrt{\alpha_n}}(\mathbf{0})\,,\\ 1 & \text{for}\quad \mathbf{x}\in\mathbb{R}^2\setminus B_{2\sqrt{\alpha_n}}(\mathbf{0})\,.\end{cases}\tag{37}$$

By monotone convergence, we readily infer

$$\|\psi_{\alpha_n} - \psi_0\|_2^2 = \int_{\mathbb{R}^2} d\mathbf{x}\, |\eta_{\alpha_n} - 1|^2 \, |\psi_0|^2 \xrightarrow{n\to+\infty} 0\,,$$

proving the required strong convergence $\psi_{\alpha_n} \to \psi_0$ in $L^2(\mathbb{R}^2)$.

In the sequel we proceed to deduce the upper bound (26), using the telescopic identity (35) to derive the stronger condition

$$\lim_{n\to+\infty} \left| Q^{(F)}_{\alpha_n,S}[\psi_{\alpha_n}] - Q^{(F)}_{0,S}[\psi_0] \right| = 0\,. \tag{38}$$

To this purpose, let us first consider the expression $\|\mathbf{A}_{\alpha_n}\psi_{\alpha_n}\|_2^2$ in (35) and refer to the decomposition

$$\|\mathbf{A}_{\alpha_n}\psi_{\alpha_n}\|_2^2 = \int_{\mathbb{R}^2\setminus B_{\sqrt{\alpha_n}}(\mathbf{0})} d\mathbf{x}\, |\mathbf{A}_{\alpha_n}\eta_{\alpha_n}\psi_0|^2 + \int_{B_{\sqrt{\alpha_n}}(\mathbf{0})} d\mathbf{x}\, |\mathbf{A}_{\alpha_n}\eta_{\alpha_n}\psi_0|^2\,.$$

By elementary estimates we get

$$\int_{\mathbb{R}^2\setminus B_{\sqrt{\alpha_n}}(\mathbf{0})} d\mathbf{x}\, |\mathbf{A}_{\alpha_n}\eta_{\alpha_n}\psi_0|^2 \leqslant \alpha_n \|\psi_0\|_2^2\,.$$

On the other side, keeping in mind that $\psi_0 \in H^1(\mathbb{R}^2)$, we use a sharp result on Sobolev embeddings [16] and dominated convergence to infer

$$\int_{B_{\sqrt{\alpha_n}}(\mathbf{0})} d\mathbf{x}\, |\mathbf{A}_{\alpha_n}\eta_{\alpha_n}\psi_0|^2 = \int_{B_{\sqrt{\alpha_n}}(\mathbf{0})} d\mathbf{x}\, \frac{\alpha_n^2}{|\mathbf{x}|^2}\left(\frac{|\mathbf{x}|}{\sqrt{\alpha_n}}\right)^{2\alpha_n} |\psi_0|^2$$

$$\leqslant \alpha_n^{2-\alpha_n} \operatorname*{ess\,sup}_{\mathbf{x}\in B_{\sqrt{\alpha_n}}(\mathbf{0})} \left(|\mathbf{x}|^{2\alpha_n}\big(1 + |\log|\mathbf{x}||\big)^2 \right) \int_{B_{\sqrt{\alpha_n}}(\mathbf{0})} d\mathbf{x}\, \frac{|\psi_0|^2}{|\mathbf{x}|^2\big(1+|\log|\mathbf{x}||\big)^2}$$

$$\leqslant e^{-2-\alpha_n \log\alpha_n + 2\alpha_n} \int_{B_1(\mathbf{0})} d\mathbf{x}\, \frac{\mathbf{1}_{B_{\sqrt{\alpha_n}}(\mathbf{0})}\, |\psi_0|^2}{|\mathbf{x}|^2\big(1+|\log|\mathbf{x}||\big)^2} \xrightarrow{n\to+\infty} 0\,.$$

The above arguments show that

$$\|\mathbf{A}_{\alpha_n}\psi_{\alpha_n}\|_2^2 \xrightarrow{n\to+\infty} 0\,. \tag{39}$$

Next, let us examine the behavior in $H^1(\mathbb{R}^2)$ of the sequence $\{\psi_{\alpha_n}\}_{n\in\mathbb{N}}$, taking into account that we already established strong convergence $\psi_{\alpha_n} \to \psi_0$ in $L^2(\mathbb{R}^2)$. By triangular inequality, we get

$$\|\nabla\psi_{\alpha_n} - \nabla\psi_0\|_2 \leqslant \|(\eta_{\alpha_n} - 1)\nabla\psi_0\|_2 + \|(\nabla\eta_{\alpha_n})\,\psi_0\|_2\,.$$

Recalling once more that $\psi_0 \in H^1(\mathbb{R}^2)$, by dominated convergence we obtain

$$\left\|(\eta_{\alpha_n}-1)\nabla\psi_0\right\|_2^2 = \int_{\mathbb{R}^2} d\mathbf{x}\; \left|\eta_{\alpha_n}-1\right|^2 \left|\nabla\psi_0\right|^2 \xrightarrow{n\to+\infty} 0\,.$$

On the other hand, from (37) we deduce

$$\left|\nabla\eta_{\alpha_n}(\mathbf{x})\right| = \begin{cases} \sqrt{\alpha_n}\left(|\mathbf{x}|/\sqrt{\alpha_n}\right)^{\alpha_n-1} = \left|\mathbf{A}_{\alpha_n}(\mathbf{x})\right|\eta_{\alpha_n}(\mathbf{x}) & \text{for } \mathbf{x}\in B_{\sqrt{\alpha_n}}(\mathbf{0})\,, \\ 0 & \text{for } \mathbf{x}\in\mathbb{R}^2\setminus B_{2\sqrt{\alpha_n}}(\mathbf{0})\,; \end{cases}$$

the downward concavity of η_{α_n} further ensures

$$\left|\nabla\eta_{\alpha_n}(\mathbf{x})\right| \leqslant \sqrt{\alpha_n} \qquad \text{for} \quad \mathbf{x}\in B_{2\sqrt{\alpha_n}}(\mathbf{0})\setminus B_{\sqrt{\alpha_n}}(\mathbf{0})\,.$$

The above relations, together with (39), entail

$$\begin{aligned} \left\|(\nabla\eta_{\alpha_n})\,\psi_0\right\|_2^2 &\leqslant \int_{B_{\sqrt{\alpha_n}}(\mathbf{0})} d\mathbf{x}\; \left|\mathbf{A}_{\alpha_n}(\mathbf{x})\right|^2 \left|\eta_{\alpha_n}(\mathbf{x})\psi_0\right|^2 + \alpha_n \int_{B_{2\sqrt{\alpha_n}}(\mathbf{0})\setminus B_{\sqrt{\alpha_n}}(\mathbf{0})} d\mathbf{x}\; |\psi_0|^2 \\ &\leqslant \left\|\mathbf{A}_{\alpha_n}\psi_{\alpha_n}\right\|_2^2 + \alpha_n\,\|\psi_0\|_2^2 \xrightarrow{n\to+\infty} 0\,. \end{aligned}$$

Summing up, we have $\left\|\nabla\psi_{\alpha_n} - \nabla\psi_0\right\|_2 \to 0$ for $n\to+\infty$, which entails strong convergence $\psi_{\alpha_n}\to\psi_0$ in $H^1(\mathbb{R}^2)$. In particular, we have that $\{\psi_{\alpha_n}\}_{n\in\mathbb{N}}$ is a uniformly bounded sequence in $H^1(\mathbb{R}^2)$.

Returning to (35) and recalling that $\mathbf{S}\in L^\infty(\mathbb{R}^2)$, on account of the results derived above we finally obtain

$$\begin{aligned} &\left| Q^{(F)}_{\alpha_n,S}[\psi_{\alpha_n}] - Q^{(F)}_{0,S}[\psi_0]\right| \\ &\leqslant C\Big[\left\|\mathbf{A}_{\alpha_n}\psi_{\alpha_n}\right\|_2^2 + \left\|\psi_{\alpha_n}\right\|_{H^1}\left\|\mathbf{A}_{\alpha_n}\psi_{\alpha_n}\right\|_2 \\ &\quad + \left(\left\|\psi_{\alpha_n}\right\|_{H^1} + \|\psi_0\|_{H^1}\right)\left\|\psi_{\alpha_n}-\psi_0\right\|_{H^1}\Big] \xrightarrow{n\to+\infty} 0\,, \end{aligned}$$

which proves (38), whence the thesis (26).

□

Acknowledgments I wish to thank Michele Correggi for stimulating conversations and valuable comments on the content of this work.

This work has been supported by: the European Research Council (ERC) under the European Union's Horizon 2020 research and innovation programme (ERC CoG UniCoSM, grant agreement n. 724939); INdAM-GNFM Progetto Giovani 2020 *Emergent Features in Quantum Bosonic Theories and Semiclassical Analysis*; Istituto Nazionale di Alta Matematica "F. Severi", through the Intensive Period "INdAM Quantum Meetings (IQM22)".

References

1. Adami, R., Teta, A.: On the Aharonov–Bohm Hamiltonian. Lett. Math. Phys. **43**, 43–54 (1998)
2. Aharonov, Y., Bohm, D.: Significance of electromagnetic potentials in the quantum theory. Phys. Rev. **115**(3), 485–491 (1959)
3. Aharonov, Y., Bohm, D.: Further Considerations on Electromagnetic Potentials in the Quantum Theory. Phys. Rev. **123**, 1511–1524 (1961)
4. Aharonov, Y., Cohen, E., Rohrlich, D.: Nonlocality of the Aharonov-Bohm effect. Phys. Rev. A **93**, 042110 (2016)
5. Batelaan, H., Tonomura, A., The Aharonov-Bohm effects: variations on a subtle theme. Phys. Today **62**, 38–43 (2009)
6. Braides, A.: Γ-Convergence for Beginners. Oxford University Press, Oxford/New York (2002)
7. Bruneau, L., Dereziński, J., Georgescu, V.: Homogeneous Schrödinger Operators on Half-Line. Ann. Henri Poincaré **12**, 547–590 (2011)
8. Correggi, M., Fermi, D.: Magnetic perturbations of anyonic and Aharonov–Bohm Schrödinger operators. J. Math. Phys. **62**, 032101 (2021)
9. Correggi, M., Fermi, D.: Schrödinger operators with multiple Aharonov-Bohm fluxes (2023), arXiv:2306.08910 [math-ph]
10. Correggi, M., Lundholm, D., Rougerie, N.: Local density approximation for the almost-bosonic anyon gas. Anal. PDE **10**, 1169–1200 (2017)
11. Correggi, M., Oddis, L.: Hamiltonians for two-anyons systems. R. Math. Appl. **39**, 277–292 (2018)
12. Dabrowski, L., Šťovíček, P.: Aharonov-Bohm effect with δ-type interaction. J. Math. Phys. **39**, 47–62 (1998)
13. Dal Maso, G.: An Introduction to Γ-Convergence. Birkhäuser, Basel (1993)
14. de Oliveira, C.R., Pereira, M.: Mathematical Justification of the Aharonov-Bohm Hamiltonian. J. Stat. Phys. **133**, 1175–1184 (2008)
15. Dereziński, J., Faupin, J., Nguyen, Q.N., Richard, S.: On radial Schrödinger operators with a Coulomb potential: general boundary conditions. Adv. Oper. Theory **5**, 1132–1192 (2020)
16. Edmunds, D.E., Triebel, H.: Sharp sobolev embeddings and related hardy inequalities: the critical case. Math. Nachr. **207**, 79–92 (1999)
17. Ehrenberg, W., Siday, R.E.: The Refractive Index in Electron Optics and the Principles of Dynamics. Proc. Phys. Soc. Series B **62**, 821 (1949)
18. Exner, P., Šťovíček, P., Vytřas, P.: Generalized boundary conditions for the Aharonov-Bohm effect combined with a homogeneous magnetic field. J. Math. Phys. **43**, 2151–2168 (2002)
19. Gradshteyn, I.S., Ryzhik, I.M.: Table of Integrals, Series, and Products. Academic Press/Elsevier, Amsterdam (2007)
20. Kang, K.: Locality of the Aharonov-Bohm-Casher effect. Phys. Rev. A **91**, 052116 (2015)
21. Kang, K.: Gauge invariance of the local phase in the Aharonov-Bohm interference: Quantum electrodynamic approach. Europhysics Letters **140**(4), 46001 (2022)
22. Kretzschmar, M.: Aharonov-Bohm Scattering of a Wave Packet of Finite Extension. Zeitschrift für Physik **185**, 84–96 (1965)
23. Leinaas, J.M., Myrheim, J.: On the theory of identical particles. Nuovo Cimento B **37**, 1–23 (1977)
24. Magni, C., Valz-Gris, F.: Can elementary quantum mechanics explain the Aharonov–Bohm effect? J. Math. Phys. **36**, 177–186 (1995)
25. Marletto, C., Vedral, V.: Aharonov-Bohm phase is locally generated like all other quantum phases. Phys. Rev. Lett. **125**, 040401 (2020)
26. Olver, F.W.J., Lozier, D.W., Boisvert, R.F., Clark, C.W.: NIST Handbook of Mathematical Functions. Cambridge University Press, Cambridge (2010)
27. Ouvry, S.: δ perturbative interactions in the Aharonov-Bohm and anyon models. Phys. Rev. D **50**, 5296–5299 (1994)

28. Pankrashkin, K., Richard, S.: Spectral and scattering theory for the Aharonov-Bohm operators. Rev. Math. Phys. **23**(01), 53–81 (2011)
29. Peshkin, M., Tonomura, A.: The Aharonov-Bohm Effect. Springer Verlag, Heidelberg (1989)
30. Tamura, H.: Magnetic scattering at low energy in two dimensions. Nagoya Math. J. **155**, 95–151 (1999)
31. Tamura, H.: Norm resolvent convergence to magnetic Schrödinger operators with point interactions. Rev. Math. Phys. **13**(04), 465–511 (2001)
32. Teta, A.: Quadratic forms for singular perturbations of the Laplacian. Publ. Res. Inst. Math. Sci. **26**(5), 803–817 (1990)
33. Tonomura, A., Osakabe, N., Matsuda, T., Kawasaki, T., Endo, J., Yano, S., Yamada, H.: Evidence for Aharonov-Bohm effect with magnetic field completely shielded from electron wave. Phys. Rev. Lett. **56**(8), 792–795 (1986)
34. Vaidman, L.: Role of potentials in the Aharonov-Bohm effect. Phys. Rev. A **86**, 040101(R) (2012)
35. Wilczek, F.: Quantum mechanics of fractional-spin particles. Phys. Rev. Lett. **49**, 957–959 (1982)
36. Yakaboylu, E., Ghazaryan, A., Lundholm, D., Rougerie, N., Lemeshko, M., Seiringer, R.: Fermionic behavior of ideal anyons. Phys. Rev. B **102**(14), 144109 (2020)

Some Remarks on the Regularized Hamiltonian for Three Bosons with Contact Interactions

Daniele Ferretti and Alessandro Teta

1 Introduction

In this note we discuss some properties of a model Hamiltonian describing the dynamics of three identical bosons interacting via zero-range forces in dimension three. Since the seminal papers by Minlos and Faddeev [11, 12], it is known that a natural candidate for such Hamiltonian turns out to be unbounded from below, giving rise to the so-called Thomas effect. Here natural means that the boundary condition defining the Hamiltonian, known as Ter-Martirosyan Skornyakov (TMS) boundary condition, is the direct generalization to the three-body case of the boundary condition characterizing the Hamiltonian of the two-body problem. Roughly speaking, the reason of such instability is due to the interaction becoming too singular as all the three particles are close to each other. We note that this pathology is absent in dimension one, where perturbation theory of quadratic forms can be used, and in dimension two, where the renormalized two-body boundary condition is sufficient to avoid the collapse (see e.g. [3, 4]).

It is worth to underline that the construction of a self-adjoint and bounded from below Hamiltonian for three, or more, interacting bosons with zero-range forces in dimension three is a challenging open problem in Mathematical Physics. Following a suggestion contained in [11], it has been recently studied [1, 6, 10] a regularized version of the Hamiltonian for a system of three bosons (see also [5] for the case of N bosons interacting with an impurity). The main idea is to introduce a three-body repulsion that reduces to zero the strength of the contact interaction between

D. Ferretti
Mathematics Area, GSSI - Gran Sasso Science Institute, L'Aquila, Italy
e-mail: daniele.ferretti@gssi.it

A. Teta (✉)
Department of Mathematics G. Castelnuovo, University of Rome "La Sapienza", Rome, Italy
e-mail: teta@mat.uniroma1.it

M. Correggi, M. Falconi (eds.), *Quantum Mathematics I*,
Springer INdAM Series 57, https://doi.org/10.1007/978-981-99-5894-8_8

two particles if the third particle approaches the common position of the first two. On the other hand, when the third particle is far enough, the usual two-body point interaction is restored. The result is that the regularized Hamiltonian is self-adjoint and bounded from below if the strength γ of the three-body interaction is larger than a threshold value γ_c.

The aim of this paper is to describe the construction of such regularized Hamiltonian following the approach developed in [1] and also to prove two further results. More precisely, in Sect. 2 we introduce the notation and we formulate the main result of [1], essentially based on the analysis of a suitable quadratic form Q.

In Sect. 3 we prove that the threshold value γ_c obtained in [1] is optimal, in the sense that for $\gamma < \gamma_c$ the quadratic form Q is unbounded from below.

In Sect. 4 we give a different proof of the main result in [1] based on a new approach in position space. The proof is surely less general since it is valid only for $\gamma > \gamma_c'$, where $\gamma_c' > \gamma_c$. On the other hand it has the advantage to be much simpler and to show that the choice of the three-body force is not arbitrary but it is dictated by the inherent singularity of the problem.

2 Regularized Hamiltonian

Let us consider a system composed of three identical spinless bosons of mass $\frac{1}{2}$ in three dimensions and let us fix the center of mass reference frame so that $\boldsymbol{x}_1$, $\boldsymbol{x}_2$ and $\boldsymbol{x}_3 = -\boldsymbol{x}_1 - \boldsymbol{x}_2$ represent the Cartesian coordinates of the three particles. We also introduce the Jacobi coordinates

$$\begin{cases} \boldsymbol{r}_k := \frac{1}{2}\sum_{i,j=1}^{3} \epsilon_{ijk}(\boldsymbol{x}_i - \boldsymbol{x}_j), \\ \boldsymbol{\rho}_k := \frac{3}{2}\boldsymbol{x}_k - \frac{1}{2}\sum_{\ell=1}^{3}\boldsymbol{x}_\ell, \end{cases} \qquad k \in \{1, 2, 3\} \tag{1}$$

where ϵ_{ijk} is the Levi-Civita symbol, so that one has the following identities

$$\begin{cases} \boldsymbol{r}_{k\pm1} = -\frac{1}{2}\boldsymbol{r}_k \mp \boldsymbol{\rho}_k, \\ \boldsymbol{\rho}_{k\pm1} = \pm\frac{3}{4}\boldsymbol{r}_k - \frac{1}{2}\boldsymbol{\rho}_k, \end{cases} \qquad k \in \mathbb{Z}/_{\{3\}}. \tag{2}$$

Denoting by $\boldsymbol{x} = \boldsymbol{r}_1 = \boldsymbol{x}_2 - \boldsymbol{x}_3$ and $\boldsymbol{y} = \boldsymbol{\rho}_1 = \boldsymbol{x}_1 - \frac{\boldsymbol{x}_2+\boldsymbol{x}_3}{2}$, the Hilbert space of the system is

$$L^2_{\text{sym}}(\mathbb{R}^6) := \left\{\psi \in L^2(\mathbb{R}^6) \,\middle|\, \psi(\boldsymbol{x}, \boldsymbol{y}) = \psi(-\boldsymbol{x}, \boldsymbol{y}) = \psi\left(\tfrac{1}{2}\boldsymbol{x} + \boldsymbol{y}, \tfrac{3}{4}\boldsymbol{x} - \tfrac{1}{2}\boldsymbol{y}\right)\right\}. \tag{3}$$

Indeed, notice that the symmetry conditions in (3) corresponds to the exchange of particles 2, 3 and 1, 2 that implies also the condition $\psi(\boldsymbol{x}, \boldsymbol{y}) = \psi\left(\frac{1}{2}\boldsymbol{x} - \boldsymbol{y}, -\frac{3}{4}\boldsymbol{x} - \frac{1}{2}\boldsymbol{y}\right)$, associated with the exchange of particles 3, 1. If the bosons interact via zero-range forces, then the system is described, at least formally, by the Hamiltonian

$$\hat{\mathcal{H}} = -\Delta_x - \tfrac{3}{4}\,\Delta_y + \mu\,\delta(x) + \mu\,\delta(\tfrac{1}{2}x + y) + \mu\,\delta(\tfrac{1}{2}x - y) \tag{4}$$

where $\mu \in \mathbb{R}$ is a coupling constant and we denote by $\mathcal{H}_0$ the free Hamiltonian of the system, i.e.

$$\mathcal{H}_0 := -\Delta_x - \tfrac{3}{4}\,\Delta_y, \quad \mathscr{D}(\mathcal{H}_0) = H^2(\mathbb{R}^6) \cap L^2_{\mathrm{sym}}(\mathbb{R}^6). \tag{5}$$

In order to define a rigorous counterpart of $\hat{\mathcal{H}}$, one needs to build a perturbation of the free Hamiltonian supported on the coincidence hyperplanes

$$\pi_k := \left\{ (\boldsymbol{r}_k, \boldsymbol{\rho}_k) \in \mathbb{R}^6 \mid \boldsymbol{r}_k = \boldsymbol{0} \right\}, \qquad \pi := \textstyle\bigcup_{k=1}^{3} \pi_k \tag{6a}$$

or, equivalently,

$$\begin{gathered}\pi_1 := \left\{ (x, y) \in \mathbb{R}^6 \mid x = \boldsymbol{0} \right\}, \quad \pi_2 := \left\{ (x, y) \in \mathbb{R}^6 \mid y = -\tfrac{1}{2}x \right\}, \\ \pi_3 := \left\{ (x, y) \in \mathbb{R}^6 \mid y = \tfrac{1}{2}x \right\}. \end{gathered} \tag{6b}$$

In other words, we look for a self-adjoint and bounded from below extension in $L^2_{\mathrm{sym}}(\mathbb{R}^6)$ of the following symmetric, densely defined, and closed (with respect to the graph norm of $\mathcal{H}_0$) operator

$$\dot{\mathcal{H}}_0 := \mathcal{H}_0\big|_{\mathscr{D}(\dot{\mathcal{H}}_0)}, \qquad \mathscr{D}(\dot{\mathcal{H}}_0) := H^2_0(\mathbb{R}^6 \setminus \pi) \cap L^2_{\mathrm{sym}}(\mathbb{R}^6). \tag{7}$$

In particular, we are interested in the family of self-adjoint extensions studied in [1] (see also [10]) which, at least formally, are characterized by the boundary condition

$$\psi(x, y) = \frac{\xi(y)}{|x|} + \alpha(y)\xi(y) + o(1), \qquad |x| \to 0, \tag{8}$$

where α is a position dependent parameter given by

$$\alpha : y \longmapsto -\frac{1}{\mathfrak{a}} + \frac{\gamma}{|y|}\,\theta(|y|) \tag{9}$$

with $\mathfrak{a}$ the two-body scattering length, γ a positive parameter representing the strength of the regularization and θ a real measurable function with compact support such that

$$1 - \frac{s}{b} \le \theta(s) \le 1 + \frac{s}{b}, \qquad s \ge 0 \tag{10}$$

for some $b > 0$. Notice that assumption (10) forces the function θ to be continuous at zero, with $\theta(0) = 1$. Moreover, the simplest choice for θ is the characteristic function of the ball of radius b centered in the origin. We also stress that, due to the symmetry constraints of $L^2_{\text{sym}}(\mathbb{R}^6)$, the boundary condition (8) implies

$$\psi(\boldsymbol{x}, \boldsymbol{y}) = \frac{\xi(\boldsymbol{x})}{\left|\boldsymbol{y} + \frac{1}{2}\boldsymbol{x}\right|} + \alpha(\boldsymbol{x})\xi(\boldsymbol{x}) + o(1), \qquad \boldsymbol{y} \to -\tfrac{1}{2}\boldsymbol{x},$$

$$\psi(\boldsymbol{x}, \boldsymbol{y}) = \frac{\xi(-\boldsymbol{x})}{\left|\boldsymbol{y} - \frac{1}{2}\boldsymbol{x}\right|} + \alpha(-\boldsymbol{x})\xi(-\boldsymbol{x}) + o(1), \qquad \boldsymbol{y} \to \tfrac{1}{2}\boldsymbol{x}.$$

Observe that for $\gamma = 0$ Eq. (8) reduces to the standard TMS boundary condition, which leads to the Thomas effect. Then, for $\gamma > 0$ we are introducing a three-body repulsion meant to regularize the ultraviolet singularity occurring when the positions of all particles coincide. However, since $\operatorname{supp}\theta$ is a compact, the usual two-body point interaction is restored when the third particle is far enough.

The procedure adopted in [1] for the rigorous construction of the Hamiltonian is the following: one first introduces the quadratic form Q in $L^2_{\text{sym}}(\mathbb{R}^6)$ describing, at least formally, the expectation value of the energy of our three-body system. Then one defines a suitable form domain $\mathscr{D}(Q)$ and proves that $Q, \mathscr{D}(Q)$ is closed and bounded from below. Finally, the Hamiltonian is defined as the unique self-adjoint and bounded from below operator associated to the quadratic form.

In order to define the quadratic form $Q, \mathscr{D}(Q)$, we first introduce an auxiliary hermitian quadratic form Φ^λ in $L^2(\mathbb{R}^3)$ given by [1, equation (3.1)] for $\lambda > 0$, i.e.

$$\Phi^\lambda := \Phi^\lambda_{\text{diag}} + \Phi^\lambda_{\text{off}} + \Phi_{\text{reg}} + \Phi_0, \qquad \mathscr{D}(\Phi^\lambda) = H^{1/2}(\mathbb{R}^3), \tag{11}$$

where

$$\Phi^\lambda_{\text{diag}}[\xi] := 12\pi \int_{\mathbb{R}^3} d\boldsymbol{p}\, \sqrt{\tfrac{3}{4}p^2 + \lambda}\, |\hat{\xi}(\boldsymbol{p})|^2, \tag{12a}$$

$$\Phi^\lambda_{\text{off}}[\xi] := -\frac{12}{\pi} \int_{\mathbb{R}^6} d\boldsymbol{p} d\boldsymbol{q}\, \frac{\overline{\hat{\xi}(\boldsymbol{p})}\, \hat{\xi}(\boldsymbol{q})}{p^2 + q^2 + \boldsymbol{p}\cdot\boldsymbol{q} + \lambda}, \tag{12b}$$

$$\Phi_{\text{reg}}[\xi] := \frac{6\gamma}{\pi} \int_{\mathbb{R}^6} d\boldsymbol{p} d\boldsymbol{q}\, \frac{\overline{\hat{\xi}(\boldsymbol{p})}\, \hat{\xi}(\boldsymbol{q})}{|\boldsymbol{p} - \boldsymbol{q}|^2}, \tag{12c}$$

$$\Phi_0[\xi] := 12\pi \int_{\mathbb{R}^3} d\boldsymbol{y}\, \beta(\boldsymbol{y})|\xi(\boldsymbol{y})|^2, \qquad \beta : \boldsymbol{y} \longmapsto -\frac{1}{\mathfrak{a}} + \gamma\, \frac{\theta(y) - 1}{y}. \tag{12d}$$

By assumption (10), one has $\beta \in L^\infty(\mathbb{R}^3)$ and therefore Φ_0 is bounded. The proof of the fact that Φ^λ is well defined in $H^{1/2}(\mathbb{R}^3)$ is relatively standard and it is given in [1, proposition 3.1]. The more relevant point concerning Φ^λ is that it is coercive for λ large enough as long as $\gamma > \gamma_c$, with

$$\gamma_c = \frac{4}{3} - \frac{\sqrt{3}}{\pi} \approx 0.782004. \tag{13}$$

The proof is given in [1, proposition 3.6] and it is based on a rather long and non trivial analysis performed in the momentum representation. The conclusion is that there exists $\lambda_0 > 0$ such that Φ^λ is closed and bounded from below by a positive constant for each $\lambda > \lambda_0$ and $\gamma > \gamma_c$. Therefore one can uniquely define a self-adjoint, positive and invertible operator Γ^λ in $L^2(\mathbb{R}^3)$ such that

$$\Phi^\lambda[\xi] = \langle \xi, \Gamma^\lambda \xi \rangle_{L^2(\mathbb{R}^3)}, \qquad \forall \xi \in D \tag{14}$$

with $D = \mathscr{D}(\Gamma^\lambda)$ a dense subspace independent of λ. Furthermore, defining the continuous[1] operator

$$\begin{aligned} \tau : \mathscr{D}(\mathcal{H}_0) &\longrightarrow L^2(\mathbb{R}^3), \\ \varphi &\longmapsto 12\pi\, \varphi|_{\pi_1} \end{aligned} \tag{15}$$

satisfying $\operatorname{ran}(\tau) = H^{1/2}(\mathbb{R}^3)$ and $\ker(\tau) = \mathscr{D}(\dot{\mathcal{H}}_0)$, one can check that the injective operator $G(z) := (\tau R_{\mathcal{H}_0}(\bar{z}))^* \in \mathscr{B}(L^2(\mathbb{R}^3), L^2_{\text{sym}}(\mathbb{R}^6))$ with $z \in \rho(\mathcal{H}_0)$ is represented in the Fourier space by

$$\widehat{(G(z)\xi)}(\boldsymbol{k}, \boldsymbol{p}) = \sqrt{\frac{2}{\pi}}\, \frac{\hat{\xi}(\boldsymbol{p}) + \hat{\xi}\left(\boldsymbol{k} - \frac{1}{2}\boldsymbol{p}\right) + \hat{\xi}\left(-\boldsymbol{k} - \frac{1}{2}\boldsymbol{p}\right)}{k^2 + \frac{3}{4}p^2 - z}. \tag{16}$$

We are now in position to introduce the quadratic form in $L^2_{\text{sym}}(\mathbb{R}^6)$ [1, definition 2.1]

$$\begin{aligned} \mathscr{D}(Q) :=& \Big\{ \psi \in L^2_{\text{sym}}(\mathbb{R}^6) \;\Big|\; \psi = \phi_\lambda + G(-\lambda)\xi,\; \phi_\lambda \in H^1(\mathbb{R}^6), \\ &\quad \xi \in H^{1/2}(\mathbb{R}^3),\; \lambda > 0 \Big\}, \\ Q[\psi] :=& \|\mathcal{H}_0^{1/2}\phi_\lambda\|^2 + \lambda\|\phi_\lambda\|^2 - \lambda\|\psi\|^2 + \Phi^\lambda[\xi]. \end{aligned} \tag{17}$$

Using the properties of Φ^λ and $G(-\lambda)$, it is now easy to show that the above quadratic form is closed and bounded from below if $\gamma > \gamma_c$. Hence it uniquely

[1] Here $\mathscr{D}(\mathcal{H}_0)$ must be intended as a Hilbert subspace of $L^2_{\text{sym}}(\mathbb{R}^6)$ endowed with the graph norm of $\mathcal{H}_0$.

defines a self-adjoint and lower semi-bounded operator $\mathcal{H}$ which, by definition, is the Hamiltonian of the three bosons system.

Following an equivalent approach, one can consider the densely defined and closed operator $\Gamma(z) : D \subset L^2(\mathbb{R}^3) \longrightarrow L^2(\mathbb{R}^3)$, given by

$$\Gamma(z) := \Gamma^\lambda - (\lambda + z)\, G(\bar{z})^* G(-\lambda), \qquad \lambda > \lambda_0,\ z \in \rho(\mathcal{H}_0) \tag{18}$$

which represents a sort of analytic continuation of Γ^λ, D. Actually, one can prove that $\Gamma(z)$ fulfils

$$\Gamma(z)^* = \Gamma(\bar{z}), \qquad \forall z \in \rho(\mathcal{H}_0), \tag{19a}$$

$$\Gamma(z) - \Gamma(w) = (z - w)\, G(\bar{z})^* G(w), \qquad \forall w, z \in \rho(\mathcal{H}_0), \tag{19b}$$

$$\forall z \in \mathbb{C} : \ \Re(z) < -\lambda_0 \ \vee \ \Im(z) > 0, \quad 0 \in \rho(\Gamma(z)). \tag{19c}$$

These properties imply, according to e.g. [13] (see also [2, theorem 2.19]), that for any $z \in \mathbb{C}$ such that $\Gamma(z)$ has a bounded inverse, the operator

$$R(z) = (\mathcal{H}_0 - z)^{-1} + G(z)\Gamma(z)^{-1} G(\bar{z})^* \tag{20}$$

defines the resolvent of a self-adjoint and bounded from below operator which coincides with the Hamiltonian $\mathcal{H}$ obtained with the approach based on the quadratic form and $\left\{z \in \mathbb{C} \,\middle|\, \Re(z) < -\lambda_0 \ \vee \ \Im(z) > 0\right\} \subseteq \rho(\mathcal{H})$. Moreover, one can verify that $\mathcal{H}$ coincides with $\mathcal{H}_0$ on $\mathscr{D}(\dot{\mathcal{H}}_0)$, satisfies the boundary condition (8) in the L^2 sense (see [1, remark 4.1]) and it is characterized by

$$\begin{aligned} &\mathscr{D}(\mathcal{H}) = \left\{\psi \in \mathscr{D}(Q) \,\middle|\, \phi_z \in \mathscr{D}(\mathcal{H}_0),\ \xi \in D,\ \Gamma(z)\xi = \tau\phi_z\right\}, \\ &\mathcal{H}\psi = \mathcal{H}_0\phi_z + z\,G(z)\xi. \end{aligned} \tag{21}$$

3 Optimality of γ_c

In this section we prove the optimality of the threshold parameter γ_c defined by (13). More precisely our goal is to prove the following theorem.

Theorem 3.1 *Whenever $\gamma < \gamma_c$, the quadratic form Q given by (17) is unbounded from below.*

In order to achieve the result, we shall adapt the ideas contained in [7, section 5]. Denote for short $G^\lambda := G(-\lambda)$ for any $\lambda > 0$ and let $\{u_n\}_{n\in\mathbb{N}} \subset \mathscr{D}(Q)$ be a sequence of trial functions given by

$$u_n(\boldsymbol{x}, \boldsymbol{y}) = (G^\lambda \eta_n)(\boldsymbol{x}, \boldsymbol{y}), \tag{22}$$

$$\eta_n(\boldsymbol{y}) = n^2 f(n\,\boldsymbol{y}), \qquad f \in H^{1/2}(\mathbb{R}^3). \tag{23}$$

We stress that, by an explicit estimate due to (16), one finds

$$\inf_{n\in\mathbb{N}} \|G^\lambda \eta_n\|_{L^2_{\text{sym}}(\mathbb{R}^6)} \gneq 0. \tag{24}$$

Indeed,

$$\begin{aligned}\|G^\lambda \eta_n\|^2 &= \frac{2}{\pi}\int_{\mathbb{R}^6} d\boldsymbol{k}d\boldsymbol{p}\, \frac{1}{n^2}\frac{\left|\hat{f}(\frac{\boldsymbol{p}}{n}) + \hat{f}\left(\frac{\boldsymbol{k}}{n} - \frac{\boldsymbol{p}}{2n}\right) + \hat{f}\left(\frac{\boldsymbol{k}}{n} + \frac{\boldsymbol{p}}{2n}\right)\right|^2}{\left(k^2 + \frac{3}{4}p^2 + \lambda\right)^2} \\ &= \frac{2}{\pi}\int_{\mathbb{R}^6} d\boldsymbol{\kappa}d\boldsymbol{q}\, \frac{\left|\hat{f}(\boldsymbol{q}) + \hat{f}\left(\boldsymbol{\kappa} - \frac{\boldsymbol{q}}{2}\right) + \hat{f}\left(\boldsymbol{\kappa} + \frac{\boldsymbol{q}}{2}\right)\right|^2}{\left(\kappa^2 + \frac{3}{4}q^2 + \frac{\lambda}{n^2}\right)^2} > \|G^\lambda f\|^2.\end{aligned}$$

Our goal is to show that whenever γ is smaller than the threshold value γ_c given by (13), one has

$$\lim_{n\to+\infty} Q[u_n] = -\infty. \tag{25}$$

According to (17), we have

$$Q[u_n] = -\lambda\, \|G^\lambda \eta_n\| + \Phi^\lambda[\eta_n] \le -\lambda\, \|G^\lambda f\| + \Phi^\lambda[\eta_n] \tag{26}$$

and therefore the theorem is proven if we exhibit some $f \in H^{1/2}(\mathbb{R}^3)$ such that

$$\lim_{n\to+\infty} \Phi^\lambda[\eta_n] = -\infty. \tag{27}$$

Lemma 3.1 *Let Φ^λ and η_n be given by (11) and (23), respectively. Then, one has*

$$\Phi^\lambda[\eta_n] = n^2(\Phi^0_{\text{diag}} + \Phi^0_{\text{off}} + \Phi_{\text{reg}})[f] + O(n).$$

Proof First of all, we can neglect the bounded component Φ_0, since

$$\begin{aligned}\Phi_0[\eta_n] &= 12\pi n^4 \int_{\mathbb{R}^3} d\boldsymbol{y}\, \beta(\boldsymbol{y})\, |f(n\,\boldsymbol{y})|^2 = 12\pi\, n \int_{\mathbb{R}^3} d\boldsymbol{t}\, \beta(\tfrac{\boldsymbol{t}}{n})\, |f(\boldsymbol{t})|^2 \\ &\le 12\pi\, n\, \|\beta\|_{L^\infty(\mathbb{R}^3)}\, \|f\|^2_{L^2(\mathbb{R}^3)} \implies \Phi_0[\eta_n] = O(n), \qquad n \to +\infty.\end{aligned}$$

Next, rescaling properly the variables in computing $\Phi^{\lambda}_{\text{diag}}[\eta_n]$, one gets

$$\begin{aligned}\Phi^{\lambda}_{\text{diag}}[\eta_n] &= 12\pi\, n \int_{\mathbb{R}^3} d\boldsymbol{\kappa}\, \sqrt{\tfrac{3}{4}\, n^2\kappa^2 + \lambda}\; |\hat{f}(\boldsymbol{\kappa})|^2 \\ &= 6\sqrt{3}\,\pi\, n^2 \int_{\mathbb{R}^3} d\boldsymbol{\kappa}\; |\boldsymbol{\kappa}|\, |\hat{f}(\boldsymbol{\kappa})|^2 \\ &\quad + 12\pi\, n \int_{\mathbb{R}^3} d\boldsymbol{\kappa} \left(\sqrt{\tfrac{3}{4}\, n^2\kappa^2 + \lambda} - \sqrt{\tfrac{3}{4}}\, n\kappa \right) |\hat{f}(\boldsymbol{\kappa})|^2 \\ &= n^2 \Phi^{0}_{\text{diag}}[f] + o(n).\end{aligned}$$

Indeed, exploiting the elementary inequality $\sqrt{a^2+b^2} - |a| \le |b|$, we can use the dominated convergence theorem to obtain

$$\lim_{n\to+\infty} \int_{\mathbb{R}^3} d\boldsymbol{\kappa} \left(\sqrt{\tfrac{3}{4}\, n^2\kappa^2 + \lambda} - \sqrt{\tfrac{3}{4}}\, n\kappa \right) |\hat{f}(\boldsymbol{\kappa})|^2 = 0.$$

Concerning the regularizing contribution, one simply has

$$\begin{aligned}\Phi_{\text{reg}}[\eta_n] &= \frac{6\gamma}{\pi} n^2 \int_{\mathbb{R}^3} d\boldsymbol{p} \int_{\mathbb{R}^3} d\boldsymbol{q}\; \frac{\overline{\hat{f}(\boldsymbol{p})}\, \hat{f}(\boldsymbol{q})}{|\boldsymbol{p}-\boldsymbol{q}|^2} \\ &= n^2 \Phi_{\text{reg}}[f].\end{aligned}$$

Finally, we compute $\Phi^{\lambda}_{\text{off}}[\eta_n]$

$$\begin{aligned}\Phi^{\lambda}_{\text{off}}[\eta_n] &= -\frac{12}{\pi} n^2 \int_{\mathbb{R}^6} d\boldsymbol{p} d\boldsymbol{q}\; \frac{\overline{\hat{f}(\boldsymbol{p})}\, \hat{f}(\boldsymbol{q})}{p^2 + q^2 + \boldsymbol{p}\cdot\boldsymbol{q} + \frac{\lambda}{n^2}} \\ &= -\frac{12}{\pi} n^2 \int_{\mathbb{R}^6} d\boldsymbol{p} d\boldsymbol{q}\; \frac{\overline{\hat{f}(\boldsymbol{p})}\, \hat{f}(\boldsymbol{q})}{p^2 + q^2 + \boldsymbol{p}\cdot\boldsymbol{q}} + \\ &\quad + \frac{12}{\pi} \lambda \int_{\mathbb{R}^6} d\boldsymbol{p} d\boldsymbol{q}\; \frac{\overline{\hat{f}(\boldsymbol{p})}\, \hat{f}(\boldsymbol{q})}{\left(p^2 + q^2 + \boldsymbol{p}\cdot\boldsymbol{q} + \frac{\lambda}{n^2}\right)\left(p^2 + q^2 + \boldsymbol{p}\cdot\boldsymbol{q}\right)}.\end{aligned}$$

Defining the integral operator in $L^2(\mathbb{R}^3)$ given by

$$(P_n\hat{\varphi})(\boldsymbol{p}) := \frac{12}{\pi} \lambda \int_{\mathbb{R}^3} d\boldsymbol{q}\; \frac{\hat{\varphi}(\boldsymbol{q})}{\left(p^2 + q^2 + \boldsymbol{p}\cdot\boldsymbol{q} + \frac{\lambda}{n^2}\right)\left(p^2 + q^2 + \boldsymbol{p}\cdot\boldsymbol{q}\right)}, \tag{28}$$

we can write

$$\Phi^{\lambda}_{\text{off}}[\eta_n] = n^2 \Phi^0_{\text{off}}[f] + \int_{\mathbb{R}^3} d\boldsymbol{p}\, \overline{\hat{f}(\boldsymbol{p})}\, (P_n \hat{f})(\boldsymbol{p}).$$

We notice that P_n is a Hilbert-Schmidt operator and

$$\begin{aligned}
\|P_n\|^2_{\mathscr{B}(L^2(\mathbb{R}^3))} &\leq \frac{124\lambda^2}{\pi^2} \int_{\mathbb{R}^6} d\boldsymbol{p} d\boldsymbol{q}\, \frac{1}{\left(p^2 + q^2 + \boldsymbol{p}\cdot\boldsymbol{q} + \frac{\lambda}{n^2}\right)^2 \left(p^2 + q^2 + \boldsymbol{p}\cdot\boldsymbol{q}\right)^2} \\
&\leq \frac{124\lambda^2}{\pi^2} \int_{\mathbb{R}^6} d\boldsymbol{p} d\boldsymbol{q}\, \frac{4}{\left(\frac{p^2+q^2}{2} + \frac{\lambda}{n^2}\right)^2 (p^2 + q^2)^2} \\
&= 496\pi\lambda^2 \int_0^{+\infty} dk\, \frac{k}{\left(\frac{k^2}{2} + \frac{\lambda}{n^2}\right)^2} \\
&= 496\pi\lambda n^2.
\end{aligned}$$

Using the above estimate, we find

$$\Phi^{\lambda}_{\text{off}}[\eta_n] = n^2 \Phi^0_{\text{off}}[f] + O(n)$$

and the lemma is proven. □

In light of Lemma 3.1, it is straightforward to see that (27) is achieved as soon as we exhibit a function $f \in H^{1/2}(\mathbb{R}^3)$ such that, whenever $\gamma < \gamma_c$, there holds

$$\Phi^0_{\text{diag}}[f] + \Phi^0_{\text{off}}[f] + \Phi_{\text{reg}}[f] < 0.$$

A relevant feature of the previous lemma is that the leading order of $\Phi^{\lambda}[\eta_n]$ as n goes to infinity does not depend on λ and, therefore, we have reduced the problem to the study of the hermitian quadratic form evaluated in $\lambda = 0$ which is diagonalizable.

In the next lemma we exhibit a trial function that allows us to prove our result.

Lemma 3.2 *Let γ_c be defined by (13), assume $\gamma < \gamma_c$ and let us consider the family of trial functions $f_\beta \in H^{1/2}(\mathbb{R}^3)$ such that*

$$\hat{f}_\beta(\boldsymbol{p}) = \tfrac{1}{p^2} \exp\!\left(-\tfrac{p^\beta + p^{-\beta}}{2}\right), \qquad \beta > 0.$$

Then there exists $\beta_0 > 0$ such that for any $\beta \in (0, \beta_0)$ we have

$$\left(\Phi^0_{\text{diag}} + \Phi^0_{\text{off}} + \Phi_{\text{reg}}\right)[f_\beta] < 0.$$

Proof We stress that our trial functions are entirely lying in the s-wave subspace, therefore we have

$$\Phi^0_{\text{diag}}[f_\beta] = 48\pi^2\sqrt{\tfrac{3}{4}}\int_0^{+\infty} dp\; p^3\,|\hat{f}_\beta(p)|^2, \tag{29a}$$

$$\Phi^0_{\text{off}}[f_\beta] = -96\pi\int_0^{+\infty} dp\; p\int_0^{+\infty} dq\; q\;\overline{\hat{f}_\beta(p)}\;\hat{f}_\beta(q)\ln\left(\frac{p^2+q^2+pq}{p^2+q^2-pq}\right), \tag{29b}$$

$$\Phi_{\text{reg}}[f_\beta] = 24\pi\gamma\int_0^{+\infty} dp\; p\int_0^{+\infty} dq\; q\;\overline{\hat{f}_\beta(p)}\;\hat{f}_\beta(q)\ln\left(\frac{p^2+q^2+2pq}{p^2+q^2-2pq}\right), \tag{29c}$$

where we have used the identity[2]

$$\int_{\mathbb{R}^6} d\boldsymbol{p}d\boldsymbol{q}\; g\left(p, q, \tfrac{\boldsymbol{p}\cdot\boldsymbol{q}}{pq}\right) = 8\pi^2\int_0^{+\infty} dp\; p^2\int_0^{+\infty} dq\; q^2\int_{-1}^{1} du\; g(p,q,u) \tag{30}$$

holding for any integrable function $g : \mathbb{R}^2_+ \times [-1, 1] \longrightarrow \mathbb{C}$. According to, e.g. [1, lemma 3.4], the quantities in Eqs. (29) can be diagonalized through the unitary transformation

$$\begin{aligned} \mathcal{M} &: L^2(\mathbb{R}_+,\, p^2\sqrt{p^2+1}\,dp) \longrightarrow L^2(\mathbb{R}),\\ \psi &\longmapsto \psi^\sharp(x) = \frac{1}{\sqrt{2\pi}}\int_{\mathbb{R}} dt\; e^{-itx}e^{2t}\psi(e^t) \end{aligned} \tag{31}$$

yielding (see [1, lemmata 3.4, 3.5])

$$\Phi^0_{\text{diag}}[f_\beta] = 48\sqrt{\tfrac{3}{4}}\,\pi^2\int_{\mathbb{R}} dx\; |\hat{f}^\sharp_\beta(x)|^2, \tag{32a}$$

$$\Phi^0_{\text{off}}[f_\beta] = -48\pi^2\int_{\mathbb{R}} dx\; |\hat{f}^\sharp_\beta(x)|^2\,\frac{4\sinh\left(\frac{\pi}{6}x\right)}{x\cosh\left(\frac{\pi}{2}x\right)}, \tag{32b}$$

$$\Phi_{\text{reg}}[f_\beta] = 48\pi^2\int_{\mathbb{R}} dx\; |\hat{f}^\sharp_\beta(x)|^2\,\frac{\gamma\tanh\left(\frac{\pi}{2}x\right)}{x}. \tag{32c}$$

Let us introduce the bounded and continuous (except at the point $x = 0$) function

$$S(x) := \frac{\sqrt{3}}{2} + \frac{\gamma\sinh\left(\frac{\pi}{2}x\right) - 4\sinh\left(\frac{\pi}{6}x\right)}{x\cosh\left(\frac{\pi}{2}x\right)} \tag{33}$$

[2] Equation (30) is an application of the addition formula for the spherical harmonics in the s-wave.

so that we have

$$\left(\Phi^0_{\text{diag}} + \Phi^0_{\text{off}} + \Phi_{\text{reg}}\right)[f_\beta] = 48\pi^2 \int_{\mathbb{R}} dx \, |\hat{f}^\sharp_\beta(x)|^2 \, S(x), \tag{34}$$

with

$$\lim_{x\to 0} S(x) = \tfrac{\sqrt{3}}{2} - \frac{2\pi}{3} + \frac{\pi}{2}\gamma = \frac{\pi}{2}(\gamma - \gamma_c) < 0. \tag{35}$$

Roughly speaking, the integral in (34) is negative if we choose the trial function such that the support of $\hat{f}^\sharp_\beta$ is sufficiently concentrated in a neighborhood of zero. More precisely, considering the explicit expression of $\hat{f}_\beta$, we have[3]

$$\hat{f}^\sharp_\beta(x) = \frac{1}{\beta}\, \hat{h}\left(\tfrac{x}{\beta}\right) \tag{36}$$

where $h(p) = e^{-\cosh p}$ with $h \in \mathcal{S}(\mathbb{R})$. Then

$$\begin{aligned} \int_{\mathbb{R}} dx \, |\hat{f}^\sharp_\beta(x)|^2 S(x) &= \frac{1}{\beta^2} \int_{\mathbb{R}} dx \, |\hat{h}(x/\beta)|^2 \, S(x) \\ &= \frac{1}{\beta} \int_{\mathbb{R}} dx \, |\hat{h}(x)|^2 \, S(\beta x). \end{aligned}$$

By dominated convergence theorem we obtain

$$\lim_{\beta\to 0^+} \int_{\mathbb{R}} dx \, |\hat{h}(x)|^2 \, S(\beta x) = \|h\|^2_{L^2(\mathbb{R})} \lim_{x\to 0} S(x) < 0.$$

Hence, the lemma is proven by noticing that the previous integral is continuous in $\beta > 0$ and therefore, the quadratic form $\left(\Phi^0_{\text{diag}} + \Phi^0_{\text{off}} + \Phi_{\text{reg}}\right)[f_\beta]$ is negative for any β small enough. □

Proof of Theorem 3.1 Let $\hat{f}_\beta$ be the trial function given in Lemma 3.2 with $\beta < \beta_0$ and consider the following sequence of charges

$$\hat{\eta}^\beta_n(\boldsymbol{p}) = \frac{1}{n}\, \hat{f}_\beta\left(\frac{\boldsymbol{p}}{n}\right).$$

By Lemma 3.1, we know that

$$\Phi^\lambda[\eta^\beta_n] = n^2(\Phi^0_{\text{diag}} + \Phi^0_{\text{off}} + \Phi_{\text{reg}})[f_\beta] + O(n), \qquad n \to +\infty$$

and then $\Phi^\lambda[\eta^\beta_n] \longrightarrow -\infty$ as n grows to infinity. □

[3] We stress that $\hat{f}^\sharp_\beta(x) = \sqrt{\frac{2}{\pi}}\, \frac{1}{\beta}\, \mathrm{K}_{\mathrm{i}x/\beta}(1)$ because of the integral representation for the Macdonald function K_ν given by [8, p. 384, 3.547 4].

4 Analysis in Position Space

In this section, we give a different proof of the coercivity of Φ^λ based on the representation of Φ^λ in position space. In particular, this approach allows to identify the negative contribution of the quadratic form $\Phi^\lambda_{\text{off}}$ and therefore to justify the choice of the regularization term Φ_{reg}.

In the next proposition, we write the quadratic form Φ^λ defined in (11) in the position-space representation.

Proposition 4.1 *For any $\xi \in H^{1/2}(\mathbb{R}^3)$ and $\lambda > 0$ one has*

$$\Phi^\lambda_{\text{diag}}[\xi] = 12\pi\sqrt{\lambda}\,\|\xi\|^2 + \frac{2\sqrt{3}\lambda}{\pi}\int_{\mathbb{R}^6} d\boldsymbol{x}d\boldsymbol{y}\,\frac{|\xi(\boldsymbol{x})-\xi(\boldsymbol{y})|^2}{|\boldsymbol{x}-\boldsymbol{y}|^2}\,\mathrm{K}_2\Big(\sqrt{\tfrac{4\lambda}{3}}\,|\boldsymbol{x}-\boldsymbol{y}|\Big), \tag{37a}$$

$$\Phi^\lambda_{\text{off}}[\xi] = -\frac{8\sqrt{3}\lambda}{\pi}\int_{\mathbb{R}^6} d\boldsymbol{x}d\boldsymbol{y}\,\frac{\overline{\xi(\boldsymbol{x})}\,\xi(\boldsymbol{y})}{y^2+x^2+\boldsymbol{x}\cdot\boldsymbol{y}}\,\mathrm{K}_2\Big(\sqrt{\tfrac{4\lambda}{3}}\sqrt{y^2+x^2+\boldsymbol{x}\cdot\boldsymbol{y}}\Big), \tag{37b}$$

$$\Phi_{\text{reg}}[\xi] = 12\pi\,\gamma\int_{\mathbb{R}^3} d\boldsymbol{x}\,\frac{|\xi(\boldsymbol{x})|^2}{|\boldsymbol{x}|} \tag{37c}$$

where $\mathrm{K}_\nu : \mathbb{R}_+ \longrightarrow \mathbb{C}$ is the modified Bessel function of the second kind (also known as Macdonald function) and order $\nu \in \mathbb{C}$.

Proof Identity (37a) is a consequence of (12a) and [9, section 7.12, (5)], while (37c) is obtained by comparing (12c) with the identity

$$\int_{\mathbb{R}^3} d\boldsymbol{r}\,\frac{|f(\boldsymbol{r})|^2}{r} = \frac{1}{2\pi^2}\int_{\mathbb{R}^6} d\boldsymbol{p}d\boldsymbol{q}\,\frac{\overline{\hat{f}(\boldsymbol{p})}\,\hat{f}(\boldsymbol{q})}{|\boldsymbol{p}-\boldsymbol{q}|^2}, \qquad \forall\, f \in H^{1/2}(\mathbb{R}^3). \tag{38}$$

Concerning the proof of (37b), we consider (12b) for $\varphi \in \mathcal{S}(\mathbb{R}^3)$ and we observe that we have uniformly in $\lambda > 0$

$$\boldsymbol{\sigma} \longmapsto \frac{1}{\sigma^2+\tau^2+\boldsymbol{\tau}\cdot\boldsymbol{\sigma}+\lambda} \in L^2(\mathbb{R}^3, d\boldsymbol{\sigma}), \qquad \text{for } \boldsymbol{\tau} \neq 0.$$

Therefore, by Plancherel's theorem we find

$$\Phi^\lambda_{\text{off}}[\varphi] = -\frac{12}{\pi}\int_{\mathbb{R}^3} d\boldsymbol{\tau}\,\overline{\hat{\varphi}(\boldsymbol{\tau})}\int_{\mathbb{R}^3} d\boldsymbol{x}\,\frac{\varphi(\boldsymbol{x})}{(2\pi)^{3/2}}\int_{\mathbb{R}^3} d\boldsymbol{\sigma}\,\frac{e^{-i\boldsymbol{x}\cdot\boldsymbol{\sigma}}}{\tau^2+\sigma^2+\boldsymbol{\tau}\cdot\boldsymbol{\sigma}+\lambda}.$$

Using the change of coordinates $\boldsymbol{\sigma} = \boldsymbol{q} - \frac{\boldsymbol{\tau}}{2}$, we obtain

$$\begin{aligned}\Phi^\lambda_{\text{off}}[\varphi] &= -\frac{12}{\pi}\int_{\mathbb{R}^3} d\boldsymbol{\tau}\,\overline{\hat{\varphi}(\boldsymbol{\tau})}\int_{\mathbb{R}^3} d\boldsymbol{x}\,\frac{e^{\frac{\boldsymbol{\tau}\cdot\boldsymbol{x}}{2}\mathrm{i}}}{(2\pi)^{3/2}}\varphi(\boldsymbol{x})\int_{\mathbb{R}^3} d\boldsymbol{q}\,\frac{e^{-\mathrm{i}\boldsymbol{q}\cdot\boldsymbol{x}}}{\frac{3}{4}\tau^2+q^2+\lambda}\\
&= -\frac{24\pi}{(2\pi)^{3/2}}\int_{\mathbb{R}^3} d\boldsymbol{\tau}\,\overline{\hat{\varphi}(\boldsymbol{\tau})}\int_{\mathbb{R}^3} d\boldsymbol{x}\,\frac{\varphi(\boldsymbol{x})}{|\boldsymbol{x}|}\,e^{\frac{\boldsymbol{\tau}\cdot\boldsymbol{x}}{2}\mathrm{i}-\sqrt{\frac{3}{4}\tau^2+\lambda}\,|\boldsymbol{x}|}\\
&= -\frac{24\pi}{(2\pi)^{3/2}}\int_{\mathbb{R}^3} d\boldsymbol{x}\,\frac{\varphi(\boldsymbol{x})}{|\boldsymbol{x}|}\int_{\mathbb{R}^3} d\boldsymbol{\tau}\,\overline{\hat{\varphi}(\boldsymbol{\tau})}\,e^{\frac{\boldsymbol{\tau}\cdot\boldsymbol{x}}{2}\mathrm{i}-\sqrt{\frac{3}{4}\tau^2+\lambda}\,|\boldsymbol{x}|}.\end{aligned}$$

Since uniformly in $\lambda > 0$

$$\boldsymbol{\tau}\longmapsto e^{\frac{\boldsymbol{\tau}\cdot\boldsymbol{x}}{2}\mathrm{i}-\sqrt{\frac{3}{4}\tau^2+\lambda}\,|\boldsymbol{x}|}\in L^2(\mathbb{R}^3, d\boldsymbol{\tau}),\qquad \text{for } x\neq 0$$

we use again Plancherel's theorem to obtain

$$\begin{aligned}\Phi^\lambda_{\text{off}}[\varphi] &= -\frac{24\pi}{(2\pi)^3}\int_{\mathbb{R}^3} d\boldsymbol{x}\,\frac{\varphi(\boldsymbol{x})}{|\boldsymbol{x}|}\int_{\mathbb{R}^3} d\boldsymbol{y}\,\overline{\varphi(\boldsymbol{y})}\int_{\mathbb{R}^3} d\boldsymbol{\tau}\,e^{\mathrm{i}\boldsymbol{\tau}\cdot(\boldsymbol{y}+\frac{\boldsymbol{x}}{2})-\sqrt{\frac{3}{4}\tau^2+\lambda}\,|\boldsymbol{x}|}\\
&= -\frac{12}{\pi}\int_{\mathbb{R}^3} d\boldsymbol{x}\,\frac{\varphi(\boldsymbol{x})}{|\boldsymbol{x}|}\int_{\mathbb{R}^3} d\boldsymbol{y}\,\frac{\overline{\varphi(\boldsymbol{y})}}{|\boldsymbol{y}+\frac{\boldsymbol{x}}{2}|}\int_0^{+\infty} d\tau\,\tau\sin\!\left(\tau|\boldsymbol{y}+\tfrac{\boldsymbol{x}}{2}|\right)e^{-\sqrt{\frac{3}{4}\tau^2+\lambda}\,|\boldsymbol{x}|}.\end{aligned}$$

The last integral can be explicitly computed using the formula (see, e.g., [8, p. 491, 3.914.6])

$$\begin{aligned}&\int_0^{+\infty} dx\,x\sin(bx)\,e^{-\beta\sqrt{x^2+\gamma^2}}\\
&\quad=\frac{b\beta\gamma^2}{\beta^2+b^2}\,\mathrm{K}_2\!\left(\gamma\sqrt{\beta^2+b^2}\right),\qquad \forall b\in\mathbb{R} \text{ and } \beta,\gamma>0 \end{aligned}\tag{39}$$

and therefore identity (37b) is proven for $\varphi\in\mathcal{S}(\mathbb{R}^3)$. By a density argument[4] the result is extended to any $\varphi\in H^{1/2}(\mathbb{R}^3)$. □

Before proceeding, let us briefly recall some elementary properties of $\mathrm{K}_2(\cdot)$:

$$x^2\mathrm{K}_2(x) \text{ is decreasing in } x>0, \tag{40a}$$

$$\mathrm{K}_2(x)=\sqrt{\frac{\pi}{2}}\,e^{-x}\left[\frac{1}{\sqrt{x}}+O\!\left(\frac{1}{x^{3/2}}\right)\right],\qquad \text{as } x\to+\infty, \tag{40b}$$

$$\mathrm{K}_2(x)=\frac{2}{x^2}-\frac{1}{2}+O\!\left(x^2\ln x\right),\qquad \text{as } x\to 0^+. \tag{40c}$$

[4] Because of Propositions 4.2 and 4.3, one can obtain a control in the $H^{1/2}$ norm.

In particular, notice that (40a) and (40c) imply

$$\mathrm{K}_2(x) \le \frac{2}{x^2}, \qquad \forall x > 0. \tag{40d}$$

In the next proposition we show the relevant fact that the negative contribution of $\Phi^\lambda_{\mathrm{off}}$ can be explicitly characterized.

Proposition 4.2 *For any $\varphi \in H^{1/2}(\mathbb{R}^3)$ and $\lambda > 0$ one has*

$$\begin{aligned}\Phi^\lambda_{\mathrm{off}}[\varphi] = &- 24\pi \int_{\mathbb{R}^3} dx \, \frac{e^{-\sqrt{\lambda}|x|}}{|x|} |\varphi(x)|^2 + \\ &+ \frac{4\sqrt{3}\lambda}{\pi} \int_{\mathbb{R}^6} dx dy \, \frac{|\varphi(x) - \varphi(y)|^2}{y^2 + x^2 + x\cdot y} \mathrm{K}_2\Big(\sqrt{\tfrac{4\lambda}{3}}\sqrt{y^2+x^2+x\cdot y}\Big).\end{aligned} \tag{41}$$

Proof Let us decompose the expression given by (37b) as follows

$$\begin{aligned}\Phi^\lambda_{\mathrm{off}}[\varphi] = -\frac{8\sqrt{3}\lambda}{\pi}\Bigg[&\int_{\mathbb{R}^3} dy\, |\varphi(y)|^2 \int_{\mathbb{R}^3} dx \, \frac{\mathrm{K}_2\Big(\sqrt{\frac{4\lambda}{3}}\sqrt{y^2+x^2+x\cdot y}\Big)}{y^2+x^2+x\cdot y} + \\ &+ \int_{\mathbb{R}^3} dy\, \overline{\varphi(y)} \int_{\mathbb{R}^3} dx \, \frac{\varphi(x) - \varphi(y)}{y^2+x^2+x\cdot y} \mathrm{K}_2\Big(\sqrt{\tfrac{4\lambda}{3}}\sqrt{y^2+x^2+x\cdot y}\Big)\Bigg],\end{aligned}$$

Then, we evaluate the first term in the right hand side. In Proposition 4.1 we have seen that the function

$$\begin{aligned} \hat{f}^\lambda_x : \mathbb{R}^3 &\longrightarrow \mathbb{R}_+, \qquad x, \lambda \in \mathbb{R}_+, \\ \tau &\longmapsto \frac{e^{-x\sqrt{\frac{3}{4}\tau^2 + \lambda}}}{x} \end{aligned} \tag{42}$$

is such that

$$f^\lambda_{|x|}\Big(y + \frac{x}{2}\Big) = \sqrt{\frac{8}{3\pi}}\, \lambda \, \frac{\mathrm{K}_2\Big(\sqrt{\frac{4\lambda}{3}}\sqrt{y^2+x^2+x\cdot y}\Big)}{y^2+x^2+x\cdot y}. \tag{43}$$

Notice the symmetry in the exchange $x \longleftrightarrow y$. Then,

$$\int_{\mathbb{R}^3} dx\, f^\lambda_{|x|}(y + \tfrac{x}{2}) = \int_{\mathbb{R}^3} dx\, f^\lambda_{|y|}(x + \tfrac{y}{2}) = \int_{\mathbb{R}^3} dz\, f^\lambda_{|y|}(z) = (2\pi)^{3/2} \hat{f}^\lambda_{|y|}(0).$$

Therefore we find

$$\frac{\lambda}{\sqrt{3}\pi^2}\int_{\mathbb{R}^3} d\boldsymbol{x}\; \frac{\mathrm{K}_2\Big(\sqrt{\frac{4\lambda}{3}}\sqrt{y^2+x^2+\boldsymbol{x}\cdot\boldsymbol{y}}\Big)}{y^2+x^2+\boldsymbol{x}\cdot\boldsymbol{y}} = \frac{e^{-\sqrt{\lambda}|\boldsymbol{y}|}}{|\boldsymbol{y}|}, \qquad \forall\,\lambda>0,\ \boldsymbol{y}\in\mathbb{R}^3\setminus\{\boldsymbol{0}\}. \tag{44}$$

According to (44), we obtain

$$\begin{aligned}\Phi^{\lambda}_{\text{off}}[\varphi] = &-24\pi\int_{\mathbb{R}^3} d\boldsymbol{y}\,|\varphi(\boldsymbol{y})|^2\,\frac{e^{-\sqrt{\lambda}|\boldsymbol{y}|}}{|\boldsymbol{y}|} + \\ &+\frac{8\sqrt{3}\lambda}{\pi}\int_{\mathbb{R}^6} d\boldsymbol{x}d\boldsymbol{y}\,\frac{\overline{\varphi(\boldsymbol{y})}\,[\varphi(\boldsymbol{y})-\varphi(\boldsymbol{x})]}{y^2+x^2+\boldsymbol{x}\cdot\boldsymbol{y}}\,\mathrm{K}_2\Big(\sqrt{\tfrac{4\lambda}{3}}\sqrt{y^2+x^2+\boldsymbol{x}\cdot\boldsymbol{y}}\Big).\end{aligned} \tag{45}$$

It is now sufficient to notice that the symmetry in exchanging $\boldsymbol{x} \longleftrightarrow \boldsymbol{y}$ allows us to write

$$\begin{aligned}&\int_{\mathbb{R}^3} d\boldsymbol{y}\,\overline{\varphi(\boldsymbol{y})}\int_{\mathbb{R}^3} d\boldsymbol{x}\,\frac{\varphi(\boldsymbol{y})-\varphi(\boldsymbol{x})}{y^2+x^2+\boldsymbol{x}\cdot\boldsymbol{y}}\,\mathrm{K}_2\Big(\sqrt{\tfrac{4\lambda}{3}}\sqrt{y^2+x^2+\boldsymbol{x}\cdot\boldsymbol{y}}\Big) = \\ &\qquad = \frac{1}{2}\int_{\mathbb{R}^6} d\boldsymbol{x}d\boldsymbol{y}\,\frac{|\varphi(\boldsymbol{y})-\varphi(\boldsymbol{x})|^2}{y^2+x^2+\boldsymbol{x}\cdot\boldsymbol{y}}\,\mathrm{K}_2\Big(\sqrt{\tfrac{4\lambda}{3}}\sqrt{y^2+x^2+\boldsymbol{x}\cdot\boldsymbol{y}}\Big).\end{aligned}$$

and the proposition is proved. □

Thanks to Proposition 4.2, it is not hard to find lower and upper bounds for Φ^{λ}. To this end, it is convenient to introduce the Gagliardo semi-norm of the Sobolev space $H^{1/2}(\mathbb{R}^3)$, defined as

$$[u]^2_{\frac{1}{2}} := \int_{\mathbb{R}^6} d\boldsymbol{x}d\boldsymbol{y}\,\frac{|u(\boldsymbol{x})-u(\boldsymbol{y})|^2}{|\boldsymbol{x}-\boldsymbol{y}|^4}, \qquad u\in H^{1/2}(\mathbb{R}^3), \tag{46}$$

so that $\|u\|^2_{H^{1/2}(\mathbb{R}^3)} = \|u\|^2 + [u]^2_{\frac{1}{2}}$. In terms of the Fourier transform of u we also have (see e.g., [9, section 7.12 (4)])

$$[u]^2_{\frac{1}{2}} = 2\pi^2\int_{\mathbb{R}^3} d\boldsymbol{k}\,|\boldsymbol{k}|\,|\hat{u}(\boldsymbol{k})|^2. \tag{47}$$

Proposition 4.3 *For any given $\varphi\in H^{1/2}(\mathbb{R}^3)$, one has*

$$\Phi^{\lambda}[\varphi] \geq \Phi_0[\varphi] + \Phi^{\lambda}_{\text{diag}}[\varphi] + 3\min\{0,\gamma-2\}[\varphi]^2_{\frac{1}{2}}, \tag{48a}$$

$$\Phi^{\lambda}[\varphi] \leq \Phi_0[\varphi] + \Phi^{\lambda}_{\text{diag}}[\varphi] + \left(3\gamma + \frac{96\sqrt{3}}{\pi}\right)[\varphi]^2_{\frac{1}{2}}. \tag{48b}$$

Proof The lower bound is obtained by neglecting the positive part of $\Phi^{\lambda}_{\text{off}}$

$$\Phi^{\lambda}_{\text{off}}[\varphi] + \Phi_{\text{reg}}[\varphi] \geq 12\pi \int_{\mathbb{R}^3} d\boldsymbol{y}\, \frac{|\varphi(\boldsymbol{y})|^2}{|\boldsymbol{y}|} \left(\gamma - 2e^{-\sqrt{\lambda}|\boldsymbol{y}|}\right)$$

and by considering the following inequalities

$$\inf_{\boldsymbol{y}\in\mathbb{R}^3} \left\{\gamma - 2e^{-\sqrt{\lambda}|\boldsymbol{y}|}\right\} \geq \min\{0, \gamma - 2\}, \tag{49}$$

$$\int_{\mathbb{R}^3} d\boldsymbol{x}\, \frac{|\varphi(\boldsymbol{x})|^2}{|\boldsymbol{x}|} \leq \frac{1}{4\pi} [\varphi]^2_{\frac{1}{2}}. \tag{50}$$

Notice that (50) is a consequence of the Hardy-Rellich inequality (see[5] [14])

$$\int_{\mathbb{R}^3} d\boldsymbol{x}\, \frac{|u(\boldsymbol{x})|^2}{|\boldsymbol{x}|} \leq \frac{\pi}{2} \int_{\mathbb{R}^3} d\boldsymbol{k}\, |\boldsymbol{k}|\, |\hat{u}(\boldsymbol{k})|^2, \qquad \forall u \in H^{1/2}(\mathbb{R}^3) \tag{51}$$

compared with (47). In order to obtain the upper bound, we recall (46) to get

$$\begin{aligned}
&\int_{\mathbb{R}^6} d\boldsymbol{x}d\boldsymbol{y}\, \frac{|\varphi(\boldsymbol{y}) - \varphi(\boldsymbol{x})|^2}{y^2 + x^2 + \boldsymbol{x}\cdot\boldsymbol{y}} \mathrm{K}_2\left(\sqrt{\tfrac{4\lambda}{3}}\sqrt{y^2 + x^2 + \boldsymbol{x}\cdot\boldsymbol{y}}\right) \leq \\
&\leq [\varphi]^2_{\frac{1}{2}} \sup_{(\boldsymbol{x},\,\boldsymbol{y})\in\mathbb{R}^6} \frac{|\boldsymbol{x} - \boldsymbol{y}|^4}{y^2 + x^2 + \boldsymbol{x}\cdot\boldsymbol{y}} \mathrm{K}_2\left(\sqrt{\tfrac{4\lambda}{3}}\sqrt{y^2 + x^2 + \boldsymbol{x}\cdot\boldsymbol{y}}\right).
\end{aligned}$$

We make use of (40d) and get rid of the dependence on the angles in evaluating the sup, since

$$\begin{cases} x^2 + y^2 - 2\boldsymbol{x}\cdot\boldsymbol{y} \leq 2(x^2 + y^2), \\ x^2 + y^2 + \boldsymbol{x}\cdot\boldsymbol{y} \geq \frac{x^2+y^2}{2}, \end{cases} \implies \frac{|\boldsymbol{x} - \boldsymbol{y}|^2}{x^2 + y^2 + \boldsymbol{x}\cdot\boldsymbol{y}} \leq 4.$$

Hence,

$$\begin{aligned}
&\int_{\mathbb{R}^6} d\boldsymbol{x}d\boldsymbol{y}\, \frac{|\varphi(\boldsymbol{y}) - \varphi(\boldsymbol{x})|^2}{y^2 + x^2 + \boldsymbol{x}\cdot\boldsymbol{y}} \mathrm{K}_2\left(\sqrt{\tfrac{4\lambda}{3}}\sqrt{y^2 + x^2 + \boldsymbol{x}\cdot\boldsymbol{y}}\right) \leq \\
&\qquad\qquad \leq \frac{3}{2\lambda} [\varphi]^2_{\frac{1}{2}} \sup_{(\boldsymbol{x},\,\boldsymbol{y})\in\mathbb{R}^6} \frac{|\boldsymbol{x} - \boldsymbol{y}|^4}{\left(y^2 + x^2 + \boldsymbol{x}\cdot\boldsymbol{y}\right)^2} = \frac{24}{\lambda} [\varphi]^2_{\frac{1}{2}}.
\end{aligned}$$

[5] There is a typo in [14, equation (1.4)]: a power 2 on the Euler Gamma function in the numerator is missing.

We stress that in the last step we have an equality since the argument of the supremum in $\mathbb{R}^6$ attains its upper bound in the hyperplane $\boldsymbol{x} + \boldsymbol{y} = \boldsymbol{0}$. So far, we have obtained for any $\varphi \in H^{1/2}(\mathbb{R}^3)$

$$\Phi^{\lambda}_{\text{off}}[\varphi] \le -24\pi \int_{\mathbb{R}^3} d\boldsymbol{y}\, \frac{|\varphi(\boldsymbol{y})|^2}{|\boldsymbol{y}|} e^{-\sqrt{\lambda}|\boldsymbol{y}|} + \frac{96\sqrt{3}}{\pi} [\varphi]^2_{\frac{1}{2}}. \tag{52}$$

We complete the proof simply by neglecting the negative contribution. □

The major difficulties in the proof of the coercivity of Φ^λ in momentum space obtained in [1] lie in the search of a lower bound. On the other hand, in position space such estimate, provided in Proposition 4.3, turns out to be much easier. However, some accuracy is lost in this framework. Indeed, adopting (48a) to obtain an estimate from below for Φ^λ, one gets

$$\Phi^\lambda[\xi] \ge \Phi^\lambda_{\text{diag}}[\xi] - 3\max\{0, 2-\gamma\}[\xi]^2_{\frac{1}{2}} + 12\pi \operatorname*{essinf}_{\boldsymbol{y}\in\mathbb{R}^3} \beta(\boldsymbol{y})\, \|\xi\|^2_{L^2(\mathbb{R}^3)} \tag{53}$$

or, equivalently

$$\Phi^\lambda[\xi] \ge \int_{\mathbb{R}^3} d\boldsymbol{p} \left[12\pi\sqrt{\tfrac{3}{4}p^2+\lambda} - 6\pi^2\max\{0,2-\gamma\}p + 12\pi \operatorname*{essinf}_{\boldsymbol{y}\in\mathbb{R}^3}\beta(\boldsymbol{y})\right] |\xi(\boldsymbol{p})|^2. \tag{54}$$

The function in square brackets attains its minimum at

$$p_{\min} = \frac{2\pi\max\{0,2-\gamma\}\sqrt{\lambda}}{\sqrt{9-3\pi^2\max\{0,2-\gamma\}^2}},$$

provided

$$3-\pi^2\max\{0,2-\gamma\}^2 > 0 \iff \gamma > 2 - \frac{\sqrt{3}}{\pi} =: \gamma_c'. \tag{55}$$

Indeed, plugging the value of $p_{\min}$ in the right hand side of (54), one gets

$$\Phi^\lambda[\xi] \ge \left(4\sqrt{3}\pi\sqrt{\lambda}\sqrt{3-\pi^2\max\{0,2-\gamma\}^2} + 12\pi \operatorname*{essinf}_{\boldsymbol{y}\in\mathbb{R}^3}\beta(\boldsymbol{y})\right)\|\xi\|^2_{L^2(\mathbb{R}^3)} \tag{56}$$

that is positive for

$$\lambda > \frac{3\min\{0,\operatorname{essinf}\beta\}^2}{3-\pi^2\max\{0,2-\gamma\}^2}. \tag{57}$$

As mentioned above, $\gamma_c' \approx 1.44867$ is not optimal, since $\gamma_c' \gneq \gamma_c$.

References

1. Basti, G., Cacciapuoti, C., Finco, D., Teta, A.: Three-body Hamiltonian with regularized zero-range interactions in dimension three. Ann. Henri Poincaré (2022). https://doi.org/10.1007/s00023-022-01214-9
2. Cacciapuoti, C., Fermi, D., Posilicano, A.: On inverses of Kreĭn's $\mathcal{Q}$-functions. Rend. Mat. Appl. **39**, 229–240 (2018)
3. Dell'Antonio, G.F., Figari, R., Teta, A.: Hamiltonians for systems of N particles interacting through point interactions. Ann. Inst. H. Poincare Phys. Theor. **60**, 253–290 (1994)
4. Dimock, J., Rajeev, S.G.: Multi-particle Schrödinger operators with point interactions in the plane. J. Phys. A **37**, 9157–9173 (2004)
5. Ferretti, D., Teta, A.: Zero-Range Hamiltonian for a Bose Gas with an Impurity. Complex Anal. Oper. Theory **17**, 55 (2023). https://doi.org/10.1007/s11785-023-01358-4
6. Figari, R., Teta, A.: On the Hamiltonian for three bosons with point interactions. In: Hilbert A., Mastrogiacomo E., Mazzucchi S., Rudiger B., Ugolini S. (eds.) Quantum and Stochastic Mathematical Physics: Sergio Albeverio, Adventures of a Mathematician. Springer Proceedings in Mathematics & Statistics, vol. 377, pp. 141–159, Springer (2023)
7. Finco, D., Teta, A.: Quadratic forms for the fermionic unitary gas model. Rep. Math. Phys. **69**, 131–159 (2012)
8. Gradshteyn, I.S., Ryzhik, I.M.: Table of Integrals, Series, and Products, 7th edn. Academic Press, Amsterdam (2007)
9. Lieb, E.H., Loss, M.: Analysis, 2nd edn. American Mathematical Society, Providence (2001)
10. Michelangeli, A.: Models of zero-range interaction for the bosonic trimer at unitarity. Rev. Math. Phys. **33**, 2150010 (2021)
11. Minlos, R.A., Faddeev, L.: On the point interaction for a three-particle system in Quantum Mechanics. Soviet Phys. Dokl. **6**, 1072–1074 (1962)
12. Minlos, R.A., Faddeev, L.: Comment on the problem of three particles with point interactions. Soviet Phys. Jetp. **14**, 1315–1316 (1962)
13. Posilicano, A.: A Krein-like formula for singular perturbations of self-adjoint operators and applications. J. Funct. Anal. **183**, 109–147 (2001)
14. Yafaev, D.: Sharp constants in the Hardy-Rellich inequalities. J. Funct. Anal. **168**, 121–144 (1999)

On the Magnetic Laplacian with a Piecewise Constant Magnetic Field in $\mathbb{R}^3_+$

Emanuela L. Giacomelli

1 Introduction

We study a Schrödinger operator in $\mathbb{R}^3_+ := \{\mathbf{x} \in \mathbb{R}^3 \,|\, \mathbf{x} = (x_1, x_2, x_3),\ x_2 > 0\}$ with a magnetic field admitting a piecewise constant strength and a uniform direction. We review some recent results obtained in [3] about the bottom of the spectrum. Such an operator is interesting to be considered in the theory of superconductivity. We first introduce it and later we motivate the last assertion.

1.1 The Main Result

We denote by $\mathbf{B}_{\alpha,\gamma,a}$ a piecewise constant magnetic field having uniform direction. More precisely, we suppose $\mathbf{B}_{\alpha,\gamma,\alpha}$ to have an intensity equal to 1 in the region $\mathcal{D}^1_\alpha$ and equal to a in $\mathcal{D}^2_\alpha$, where

$$\mathcal{D}^1_\alpha = \big\{\mathbb{R}^3 \ni \mathbf{x} = \rho(\cos\theta \sin\phi, \sin\theta \sin\phi, \cos\phi) \,|\, \rho \in (0,\infty),\\ \theta \in (0,\alpha),\ \phi \in (0,\pi)\big\},$$

$$\mathcal{D}^2_\alpha = \big\{\mathbb{R}^3 \ni \mathbf{x} = \rho(\cos\theta \sin\phi, \sin\theta \sin\phi, \cos\phi) \,|\, \rho \in (0,\infty),\\ \theta \in (\alpha,\pi),\ \phi \in (0,\pi)\big\}.$$

E. L. Giacomelli (✉)
Department of Mathematics, LMU Münich, München, Germany
e-mail: emanuela.giacomelli@math.lmu.de

M. Correggi, M. Falconi (eds.), *Quantum Mathematics I*,
Springer INdAM Series 57, https://doi.org/10.1007/978-981-99-5894-8_9

For any $a \in [-1, 1) \setminus \{0\}$, $\alpha \in (0, \pi)$ and $\gamma \in [0, \pi/2]$, we define

$$\mathbf{B}_{\alpha,\gamma,a} = (\cos\alpha \sin\gamma, \sin\alpha \sin\gamma, \cos\gamma)\big(\mathbb{1}_{\mathcal{D}^1_\alpha} + a\mathbb{1}_{\mathcal{D}^2_\alpha}\big). \tag{1}$$

Note that by symmetry considerations, we can restrict the angle γ to the interval $[0, \pi/2]$. Moreover, instead of taking into account a generic piecewise constant magnetic field having two different uniform intensities in the regions $\mathcal{D}^1_\alpha$ and $\mathcal{D}^2_\alpha$, it is possible (by scaling arguments) to reduce the study to the case of a field having intensity equal to $\mathbb{1}_{\mathcal{D}^1_\alpha} + a\mathbb{1}_{\mathcal{D}^2_\alpha}$ for some $a \in [-1, 1)$. The condition $a \neq 0$ we imposed above is a technical restriction which we need to simplify our analysis.

Let now $\mathbf{A}_{\alpha,\gamma,a} \in H^1_{\mathrm{loc}}(\mathbb{R}^3_+, \mathbb{R}^3)$ be a vector potential such that $\mathrm{curl}\mathbf{A}_{\alpha,\gamma,a} = \mathbf{B}_{\alpha,\gamma,a}$ (see Sect. 2.2.1 for an explicit expression of $\mathbf{A}_{\alpha,\gamma,a}$). In this paper we take into account the magnetic Neumann realization of the following self-adjoint operator in $\mathbb{R}^3_+$

$$\mathcal{L}_{\alpha,\gamma,a} = -(\nabla - i\mathbf{A}_{\alpha,\gamma,a})^2, \tag{2}$$

with a domain given by

$$\begin{aligned}\mathrm{Dom}(\mathcal{L}_{\alpha,\gamma,a}) = \big\{u \in L^2(\mathbb{R}^3_+) \quad : \quad & (\nabla - i\mathbf{A}_{\alpha,\gamma,a})^n u \in L^2(\mathbb{R}^3_+),\\ & \text{for } n \in \{1, 2\},\ (\nabla - i\mathbf{A}_{\alpha,\gamma,a})u \cdot (0, 1, 0)|_{\partial\mathbb{R}^3_+} = 0\big\}.\end{aligned} \tag{3}$$

We denote by $\lambda_{\alpha,\gamma,a}$ the bottom of the spectrum of $\mathcal{L}_{\alpha,\gamma,a}$, i.e.,

$$\lambda_{\alpha,\gamma,a} := \inf \mathrm{sp}(\mathcal{L}_{\alpha,\gamma,a}). \tag{4}$$

Note that the case $a = 1, \alpha = \pi/2$ corresponds to the Lu-Pan/Helffer-Morame model (see e.g., [23, 29]). If in addition one has $\gamma = 0$, the problem of characterizing $\lambda_{\alpha,\gamma,a}$ can be reduced to the study of the so called de Gennes operator (a harmonic oscillator on the half-axis with Neumann condition at the origin). Moreover, in our analysis we exclude the values $\alpha = 0, \pi$ since in this case there is no discontinuity jump and the magnetic field is constant over $\mathbb{R}^3_+$.

The main result proved in [3] is in the theorem below.

Theorem 1 (Bottom of the Spectrum of $\mathcal{L}_{\alpha,\gamma,a}$) *Let $a \in [-1, 1) \setminus \{0\}$, $\alpha \in (0, \pi)$, $\gamma \in [0, \pi/2]$ and $\nu_0 = \arcsin(\sin\alpha \sin\gamma)$. Let $\lambda_{\alpha,\gamma,a}$ as in (4). It holds*

$$\lambda_{\alpha,\gamma,a} \le \min(\beta_a, |a|\zeta_{\nu_0}), \tag{5}$$

where β_a and ζ_{ν_0} are as in (10) and (13). Moreover, if the inequality in (5) is strict, then there exists $\tau_\star \in \mathbb{R}$ such that

$$\lambda_{\alpha,\gamma,a} = \underline{\sigma}(\alpha, \gamma, a, \tau_\star),$$

where $\underline{\sigma}(\alpha, \gamma, a, \tau_\star)$ is an eigenvalue of the operator $\underline{\mathcal{L}}_{\alpha,\gamma,a}(\tau_\star)$ defined in (16).

Remark 1 (On the Bound for $\lambda_{\alpha,\gamma,a}$) Note that for $a \in (0, 1)$, $\alpha \in (0, \pi)$ and $\gamma \in [0, \pi/2]$ one has $\beta_a = a$ and $\zeta_{\nu_0} \leq 1$ (see (11) and (14)); from Theorem 1 we then get $\lambda_{\alpha,\gamma,a} \leq |a|\zeta_{\nu_0}$. Moreover, when $a = -1$, we know that (see again (11)) $\beta_{-1} = \Theta_0 \sim 0.59$; being in this case $\zeta_{\nu_0} \in (\Theta_0, 1)$ (see (14)), Theorem 1 implies that $\lambda_{\alpha,\gamma,-1} \leq \Theta_0$.

1.2 Motivation

The motivations to study the operator $\mathcal{L}_{\alpha,\gamma,a}$ go back to the phenomenon of superconductivity, which was discovered in 1911 by H. Kamerlingh Onnes in Leiden. Here we put the focus on the breakdown of superconductivity in presence of an external magnetic field [37]. In general, superconductors can be divided into two types, according to how the breakdown occurs. For type-I, superconductivity is abruptly destroyed via a first order phase transition. In 1957 Abrikosov deduced the existence of a class of materials which exhibit a different behavior, i.e., some of their superconducting properties are preserved when submitted to a suitably large magnetic field. Physically, these two classes can be identified by the value of a parameter κ, also known as the Ginzburg-Landau parameter (i.e., a value proportional to the inverse of the penetration depth and typical of the material). The value κ is smaller than $1/\sqrt{2}$ for type-I superconductors and larger than $1/\sqrt{2}$ for the so called type-II superconductors. We consider here extreme type-II superconductors, i.e., we assume that the Ginzburg-Landau parameter satisfies the condition $\kappa \gg 1$.

In general, it is well-known [20] that a superconducting material exposed to a strong magnetic field with intensity h_{ex} loses permanently its superconducting properties (i.e., goes to the normale state) when h_{ex} exceeds some critical value. Determining this critical value is not an easy task and strongly depends on the geometry of the sample. Below we underline the main ideas toward this characterization, and emphasizing where the operator we introduced in (2) is expected to play a role.

1.2.1 The Ginzburg-Landau Theory

To study the transition to the normal state, it is convenient to use the Ginzburg-Landau theory [19] which, in general, allows to describe the behavior of a type-II superconductor exposed to an external magnetic field $\mathbf{B}$ (such that $|\mathbf{B}| = h_{\text{ex}}$) in a temperature close to the critical one. Let $\Omega \subset \mathbb{R}^3$ be a domain, the GL functional is defined by[1]

$$\mathcal{G}_{\Omega,\kappa}[\psi, \mathbf{A}] = \int_{\Omega} |\nabla + i\text{h}_{\text{ex}}\mathbf{A})\psi|^2 - \frac{1}{2}\kappa^2(2|\psi|^2 - |\psi|^4) + h_{\text{ex}} \int_{\mathbb{R}^3} |\text{curl}\mathbf{A} - 1|^2.$$

[1] Note that in some cases the GL model reduces to a two dimensional one: this happens when for example the external magnetic field is supposed to be perpendicular to the cross section of a superconducting wire.

Here ψ is the order parameter ($|\psi|^2$ denotes the density of Cooper pairs) and $h_{ex}\mathbf{A}$ is the induced magnetic vector potential. We recall that we take into account extreme type II-superconductors, which means that $\kappa \to \infty$.

Using the GL theory it is possible to study the phase transitions which occur in a type-II superconductor: these can be described by identifying three increasing critical values of the magnetic field. When the first critical value H_{c_1} is reached, superconductivity is lost in the bulk of the sample at isolated points (see e.g. [18, 38]). Between the second and third critical fields, i.e., in the regime $H_{c_2} \leq h_{ex} \leq H_{c_3}$, superconductivity survives only close to the boundary of the sample (see e.g.,[11–15, 21, 31]) . Above the third critical field H_{c_3}, the sample goes back to its normal state (see e.g., [8, 16, 17, 22, 24, 28]).

In this framework the normal state corresponds to the choice $(\psi, \mathbf{A}) = (0, \mathbf{F})$ where $\mathbf{F}$ is such that curl$\mathbf{F} = 1$, i.e., three are no Cooper pairs and the external field penetrates completely the sample. It is then natural to expect that to characterize the value of H_{c_3}, the first term in GL functional is playing the main role, i.e., one should study the magnetic Laplacian:

$$-(\nabla - ih_{ex}\mathbf{A})^2.$$

As suggested in [18, Chapter 13], we take into account external magnetic fields of intensity proportional to the GL parameter κ, i.e., $h_{ex} = \kappa\sigma$ for some $\sigma > 0$. Note that since $\kappa \to \infty$, the intensity of the magnetic field is high. Under this choices, the study of H_{c_3} is naturally linked to a semiclassical limit $h \to 0$:

$$-(\nabla - i(\kappa\sigma)\mathbf{A})^2 = -(\kappa\sigma)^2(h\nabla - i\mathbf{A})^2, \qquad h := (\kappa\sigma)^{-1}. \tag{6}$$

1.2.2 A Semiclassical Problem

Analysing the semiclassical problem in (6) is important to prove the localization of the GL minimizing order parameter in order to characterize the transition to the normal state. Many works indeed have been dedicated to the study of the operator $(-ih\nabla - \mathbf{A})^2$ in the limit $h \to 0$ deriving an asymptotics of the first eigenvalue and proving the localization of the associated eigenfunctions. In dimensions $d = 2, 3$, the first eigenvalue of $(h\nabla - i\mathbf{A})^2$ behaves at first order as $h\mathcal{E}(\mathbf{B}, \Omega)$, where $\mathcal{E}(\mathbf{B}, \Omega)$ is the smallest eigenvalue of a given model operator which strongly depends on the dimension d, on the geometry of Ω and on the shape of the magnetic field. In other words, to study the semiclassical problem it is useful to first take into account specific effective models.

Below we list some well-known situations and we underline where the 3D Schrödinger operator we study here is expected to appear. In the case of a uniform external magnetic, if the domain $\Omega \subset \mathbb{R}^d$ $(d = 2, 3)$ is smooth, one has to deal with two model operators: one defined over $\mathbb{R}^d$ (when working in the interior of Ω) and the other living in $\mathbb{R}^d_+$ (to work near the boundary $\partial\Omega$). For such situations we refer to [18, 22, 28] for dimension $d = 2$ and to [16, 18, 20, 23, 29] in the 3D setting. Moreover, it is well-know that in presence of singularities (e.g., corners, wedges) along the boundary of Ω, one has to introduce another operator to work close to the singularities. This operator is defined over an infinite angular sector or an infinite wedges (according to the dimension), see [8, 10] for the 2D case and [10, 29, 32–34] for $d = 3$. For non-uniform magnetic field there are, in general, less results available in the literature. We mention here [1, 2, 6] for piecewise constant magnetic fields in dimension $d = 2$ and [35, 36] for smoothly varying magnetic fields both in dimension $d = 2$ and $d = 3$.

The transition to the normal state in the case of a smooth domain $\Omega \subset \mathbb{R}^3$ with an external magnetic field which is piecewise constant is completely open. The study of the 3D Schrödinger operator introduced above can be seen as a first step towards this characterization. To give an idea of that, we take into account an external magnetic such that

$$\mathbf{B}(x) = (\mathbb{1}_{x_2>0} + a\mathbb{1}_{x_2<0})(x)(0, 0, 1) \qquad \text{for } x = (x_1, x_2, x_3),$$

and a domain $\Omega \subset \mathbb{R}^3$ with smooth boundary which intersects transversally the plane (x_1x_3), we call this intersection the *discontinuity plane* and we denote it by S. Moreover, we set $\Gamma := \partial\Omega \cap S$ to be the discontinuity curve. In this setting superconductivity is expected to nucleate close to Γ right before disappearing. To rigorously prove this, it is convenient to use the model operator $\mathcal{L}_{\alpha,\gamma,a}$ introduced above to work in regions localized at the boundary along Γ. Indeed, the fact that localization occurs at the boundary forces us to work with a model operator defined on $\mathbb{R}^3_+$. Moreover, to work close to Γ requires to take into account a model operator with a discontinuous magnetic field.

2 Proof of Theorem 1

In this section we give the main ideas for the study of the bottom of the spectrum of

$$\mathcal{L}_{\alpha,\gamma,a} = -(\nabla - i\mathbf{A}_{\alpha,\gamma,a}), \qquad \alpha \in (0, \pi), \gamma \in \left[0, \frac{\pi}{2}\right], \quad a \in [-1, 1) \setminus \{0\},$$

i.e., $\lambda_{\alpha,\gamma,a} = \inf \mathrm{sp}(\mathcal{L}_{\alpha,\gamma,a})$. In the following we make a specific choice of the vector potential $\mathbf{A}_{\alpha,\gamma,a}$ which allows us to use a partial Fourier transform to compare our operator with 2D operators.

2.1 The Reference 2D Operators

As mentioned above, it turns out that it is possible to compare $\mathcal{L}_{\alpha,\gamma,a}$ with 2D operators. We will study the spectrum of such 2D operators making use of two additional models. The first one is a 2D operator with discontinuous magnetic field and the second one is Schrödinger operator in $\mathbb{R}^3_+$ having a uniform magnetic field. We introduce them in what follows and we refer to [3] (and references therein) for more details.

2.1.1 Magnetic Laplacian with a Piecewise Constant Magnetic Field in 2D

Let $\mathbf{A}_a \in H^1_{\text{loc}}(\mathbb{R}^2, \mathbb{R}^2)$ be such that

$$\text{curl}\mathbf{A}_a(x) = \big(\mathbb{1}_{x_2>0} + a\mathbb{1}_{x_2<0}\big)(x), \qquad x \in \mathbb{R}^2, \quad a \in [-1, 1) \setminus \{0\}.$$

In particular, one can fix the gauge and set

$$\mathbf{A}_a(x) = \begin{cases} (-x_2, 0) & \text{if } x_2 > 0, \\ (-ax_2, 0) & \text{if } x_2 < 0. \end{cases}$$

Consider the magnetic Neumann realization of

$$\mathcal{L}_a := \big(-\nabla - i\mathbf{A}_a\big)^2, \tag{7}$$

with domain

$$\text{dom}(\mathcal{L}_a) := \big\{u \in L^2(\mathbb{R}^2) \quad : \quad (\nabla - i\mathbf{A}_a)^n u \in L^2(\mathbb{R}^2), \ n = 1, 2\big\}.$$

We denote by β_a the bottom of the spectrum, i.e.,

$$\beta_a = \inf \text{sp}(\mathcal{L}_a). \tag{8}$$

The operator $\mathcal{L}_a$ as well as the value β_a were widely studied (see e.g., [4, 5, 25, 26]). Here we just recall that $\mathcal{L}_a$ can be decomposed by one dimensional fiber operators via a partial Fourier transform, i.e.,

$$\mathcal{L}_a = \int_{\xi \in \mathbb{R}}^{\oplus} \mathfrak{h}_a(\xi)\, d\xi, \qquad \mathfrak{h}_a(\xi) = \begin{cases} -\frac{d^2}{dt^2} + (t - \xi)^2 & \text{for } t > 0, \\ -\frac{d^2}{dt^2} + a(t - \xi)^2 & \text{for } t < 0. \end{cases} \tag{9}$$

As a consequence, denoting by $\mu_a(\xi)$ the bottom of the spectrum of $\mathfrak{h}_a(\xi)$, one has

$$\beta_a = \inf_{\xi \in \mathbb{R}} \mu_a(\xi). \tag{10}$$

We now recall some properties of β_a which we need later:

$$\beta_a = a \quad \text{for} \quad 0 < a < 1, \quad \beta_{-1} = \Theta_0, \quad |a|\Theta_0 < \beta_a < |a| \quad \text{for} \quad -1 < a < 0. \tag{11}$$

For more details and for some properties of $\mu_a(\cdot)$, see [3, Section 2.1].

2.1.2 Magnetic Laplacian with Constant Magnetic Field in $\mathbb{R}^3_+$

We now take into account a uniform magnetic field $\mathbf{B}_\nu$ with unit strength on $\mathbb{R}^3_+$, where ν denotes the angle between $\mathbf{B}_\nu$ and the plane (x_1x_3). We can explicitly write

$$\mathbf{B}_\nu = (0, \sin\gamma, \cos\gamma).$$

We can then take into account the magnetic Neumann realization of

$$H_\nu = -(\nabla - i\mathbf{A}_\nu)^2 \qquad \text{in } L^2(\mathbb{R}^3_+), \tag{12}$$

where $\mathbf{A}_\nu \in H^1_{\text{loc}}(\mathbb{R}^3_+, \mathbb{R}^3)$ is such that $\text{curl}\mathbf{A}_\nu = \mathbf{B}_\nu$. Fixing the gauge, we can set $\mathbf{A}_\nu(x) = (x_3 \sin\nu - x_2 \cos\nu, 0, 0)$ for $x = (x_1, x_2, x_3) \in \mathbb{R}^3_+$. We denote the bottom of the spectrum of H_ν by

$$\zeta_\nu := \inf \text{sp}(H_\nu). \tag{13}$$

This model operator is studied in [28–30]. We refer to [23, 29] for a collection of some useful properties of ζ_ν. Here, we only recall that

$$\zeta_0 = \Theta_0, \qquad \zeta_{\frac{\pi}{2}} = 1, \qquad \zeta_\nu \in (\Theta_0, 1) \quad \forall \nu \in (0, \pi/2). \tag{14}$$

2.2 *Ideas for the Proof of Theorem 1*

We now summarize the strategy of the proof of Theorem 1 done in [3]. More precisely, first we reduce the study of $\lambda_{\alpha,\gamma,a}$ to the one of the bottom of the spectrum of 2D operators (Sect. 2.2.1), then we collect the main properties we need on the spectrum of the aforementioned 2D operators (Sect. 2.2.2) and in Sect. 2.2.3 we give an idea of the final proof.

2.2.1 Reduction to 2D Operators

We do a partial Fourier transform to decompose (see Sect. 2.2.1) $\mathcal{L}_{\alpha,\gamma,a}$. To do that, it is convenient to fix the gauge. Thus, from now on we suppose that the vector potential $\mathbf{A}_{\alpha,\gamma,a}$ is such that $\mathbf{A}_{\alpha,\gamma,a} = (A_1, A_2, A_3)$, with

$$A_1 = 0,$$

$$A_2 = \begin{cases} \cos\gamma \left(x_1 - x_2 (1-a) \cot\alpha\right) & \text{for } x \in \mathcal{D}^1_\alpha, \\ a\cos\gamma\, x_1 & \text{for } x \in \mathcal{D}^2_\alpha, \end{cases}$$

$$A_3 = \begin{cases} \sin\gamma \left(x_2 \cos\alpha - x_1 \sin\alpha\right) & \text{for } x \in \mathcal{D}^1_\alpha, \\ a\sin\gamma \left(x_2 \cos\alpha - x_1 \sin\alpha\right) & \text{for } x \in \mathcal{D}^2_\alpha. \end{cases}$$

Note that this choices for A_1, A_2, A_3 ensure that $\mathbf{A}_{\alpha,\gamma,a} \in H^1_{\mathrm{loc}}(\mathbb{R}^3_+, \mathbb{R}^3)$ and imply that the operator $\mathcal{L}_{\alpha,\gamma,a}$ is translation invariant with respect to the x_3 coordinate. We can then use a partial Fourier transform in the x_3 variable to decompose $\mathcal{L}_{\alpha,\gamma,a}$ via fiber operators living in $\mathbb{R}^2_+$. More precisely, we can write

$$\mathcal{L}_{\alpha,\gamma,a} = \int_{\tau\in\mathbb{R}}^{\oplus} \underline{\mathcal{L}}_{\alpha,\gamma,a}(\tau)\, d\tau, \tag{15}$$

where

$$\underline{\mathcal{L}}_{\alpha,\gamma,a}(\tau) = -(\nabla - i\underline{\mathbf{A}}_{\alpha,\gamma,a})^2 + V_{\alpha,\gamma,a}(\tau). \tag{16}$$

Below we explain our notations. First, we set D^1_α, D^2_α to be the orthogonal projections of the regions $\mathcal{D}^1_\alpha$, $\mathcal{D}^2_\alpha$ over the plane (x_1x_2). The magnetic potential $\underline{\mathbf{A}}_{\alpha,\gamma,a}$ is the projection of $\mathbf{A}_{\alpha,\gamma,a}$ on $\mathbb{R}^2_+$, i.e., $\underline{\mathbf{A}}_{\alpha,\gamma,a} = (\underline{A}_1, \underline{A}_2)$ with $\underline{A}_1 = 0$ and

$$\underline{A}_2 = \begin{cases} \cos\gamma(x_1 - (1-a)\cot\alpha x_2) & \text{for } (x_1, x_2) \in D^1_\alpha, \\ a\cos\gamma & \text{for } (x_1, x_2) \in D^2_\alpha. \end{cases} \tag{17}$$

Moreover, $\underline{\mathbf{A}}_{\alpha,\gamma,a}$ is such that

$$\mathrm{curl}\underline{\mathbf{A}}_{\alpha,\gamma,a} = \underline{s}_{\alpha,a}\cos\gamma, \qquad \underline{s}_{\alpha,a} = \mathbb{1}_{D^1_\alpha} + a\mathbb{1}_{D^2_\alpha}. \tag{18}$$

Finally the potential $V_{\alpha,\gamma,a}(\tau)$ appearing in (16) is an electric potential which is defined through the projection of $\mathbf{B}_{\alpha,\gamma,a}$ on $\mathbb{R}^2_+$. More precisely, we denote the aforementioned projection by $\underline{\mathbf{B}}_{\alpha,\gamma,a}$ and, explicitly, we have

$$\underline{\mathbf{B}}_{\alpha,\gamma,a} = (\cos\alpha, \sin\alpha\sin\gamma)\underline{s}_{\alpha,a} \equiv (\underline{b}_1, \underline{b}_2). \tag{19}$$

The electric potential is then given by

$$V_{\alpha,\gamma,a}(\tau) = (x_1\underline{b}_2 - x_2\underline{b}_1 - \tau)^2. \tag{20}$$

From (15) it turns out that

$$\lambda_{\alpha,\gamma,a} = \inf_\tau \underline{\sigma}_{\alpha,\gamma,a}(\tau),$$

where we denoted by $\underline{\sigma}_{\alpha,\gamma,a}(\tau)$ the bottom of the spectrum of the operator $\underline{\mathcal{L}}_{\alpha,\gamma,a}(\tau)$. As a consequence, we reduced the study of $\lambda_{\alpha,\gamma,a}$ to the one of the map $\tau \mapsto \underline{\sigma}(\alpha, \gamma, a)$ (which can be proven to be C^∞, see e.g., [18, 27]).

2.2.2 Spectrum of the 2D Operators

Here we recall two results we need about the spectrum of the 2D reduced operator $\underline{\mathcal{L}}_{\alpha,\gamma,a}(\tau)$ for fixed[2] $\alpha \in (0,\pi)$, $\gamma \in (0,\pi/2]$, $a \in [-1,1) \setminus \{0\}$, $\tau \in \mathbb{R}$.

Proposition 1 (Bottom of the Essential Spectrum) *Let $a \in [-1,1) \setminus \{0\}$, $\alpha \in (0,\pi)$, $\gamma \in (0,\pi/2]$ and $\tau \in \mathbb{R}$. Let*

$$\underline{\sigma}_{\text{ess}}(\alpha,\gamma,a,\tau) = \inf \text{sp}_{ess}(\underline{\mathcal{L}}_{\alpha,\gamma,a}). \tag{21}$$

It holds

$$\underline{\sigma}_{ess}(\alpha,\gamma,a,\tau) = \inf_{\xi\in\mathbb{R}} \left(\mu_a(\tau\sin\gamma + \xi\cos\gamma) + (\xi\sin\gamma - \tau\cos\gamma)^2\right), \tag{22}$$

where $\mu_a(\cdot)$ is as in (10).

Proposition 2 (Behavior of $\underline{\sigma}(\alpha,\gamma,a,\tau)$ for Large τ) *Let $\alpha \in (0,\pi)$ and $\gamma \in (0,\pi/2]$. It holds:*

1. *For $a \in [-1,0)$:*

$$\lim_{\tau\to-\infty} \underline{\sigma}(\alpha,\gamma,a,\tau) = +\infty, \qquad \lim_{\tau\to+\infty} \underline{\sigma}(\alpha,\gamma,a,\tau) = |a|\zeta_{\nu_0}.$$

2. *For $a \in (0,1)$,*

$$\lim_{\tau\to-\infty} \underline{\sigma}(\alpha,\gamma,a,\tau) = a\zeta_{\nu_0}, \qquad \lim_{\tau\to+\infty} \underline{\sigma}(\alpha,\gamma,a,\tau) = \zeta_{\nu_0}.$$

The proofs of Propositions 1 and 2 are based on the study of two auxiliary operators: one is useful to work near the boundary of $\mathbb{R}^2_+$ away from the discontinuity line (i.e., the intersection between the plane of equation $x_1\sin\alpha - x_2\cos\alpha = 0$ and $\mathbb{R}^2_+$), meanwhile the other is an effective operator useful when working close to the discontinuity. We refer to [3, Section 3] for more details. Here, we only mention that Proposition 1, is an application of Persson's Lemma.

2.2.3 Conclusion of the Proof of Theorem 1

Once Propositions 1 and 2 are established, the proof of Theorem 1 is quite simple. As mentioned before, one can distinguish between $\gamma = 0$ and $\gamma \neq 0$. In the first case, it is immediate to get that

$$\lambda_{\alpha,0,a} \leq |a|\Theta_0, \tag{23}$$

[2] The case $\gamma = 0$ can be treated directly, this is why suppose $\gamma \neq 0$ in this section.

by following what was proven in [1, Section 3]. Combining (23) with the fact that $\zeta_0 = \Theta_0$ (see (14)) and that $\beta_a \geq |a|\Theta_0$ (see (11), one has

$$\lambda_{\alpha,0,a} \leq \min(\beta_a, |a|\zeta_0). \tag{24}$$

We refer to [3, Section 4] for more details.

In the case $\gamma \neq 0$ the proof is more involved. In particular, from Proposition 2, we get that

$$\underline{\sigma}(\alpha, \gamma, a, \tau) \leq |a|\zeta_{\nu_0}. \tag{25}$$

We can now distinguish between $a \in [-1, 0)$ and $a \in (0, 1)$. In the second case, i.e., $a \in (0, 1)$, there is nothing to prove. Indeed, one has that $\beta_a = a$ for $a \in (0, 1)$ (see (11)) and $\zeta_{\nu_0} \leq 1$ (see (14)). This allows to conclude the proof of (5) for a positive. In the case $\gamma \neq 0$, $a \in [-1, 0)$, we have to work a bit more. From Proposition 1 and choosing a particular value[3] of $\tau = \tau_*$, one has

$$\underline{\sigma}(\alpha, \gamma, a, \tau_*) \leq \beta_a. \tag{26}$$

Combining (26) with (25), the estimate in (5) holds. We now discuss the case of a strict inequality. From Proposition 2, we have

$$\begin{aligned} \inf_{\tau} \underline{\sigma}(\alpha, \gamma, a, \tau) &= \lambda_{\alpha,\gamma,a} < |a|\zeta_{\nu_0} \\ &= \min\big(\lim_{\tau\to-\infty} \underline{\sigma}(\alpha, \gamma, a, \tau), \lim_{\tau\to+\infty} \underline{\sigma}(\alpha, \gamma, a, \tau)\big), \end{aligned} \tag{27}$$

which implies that $\inf_\tau \underline{\sigma}(\alpha, \gamma, a, \tau)$ is attained at some $\tau_\star \in \mathbb{R}$. Moreover, from Proposition 1 (see [3, Corollary 3.6]) we know that

$$\inf_{\tau\in\mathbb{R}} \underline{\sigma}_{ess}(\alpha, \gamma, a, \tau) \geq \beta_a.$$

Thus, we get

$$\lambda_{\alpha,\gamma,a} = \underline{\sigma}(\alpha, \gamma, a, \tau_\star) < \beta_a \leq \underline{\sigma}_{ess}(\alpha, \gamma, a, \tau_\star), \tag{28}$$

which implies that $\lambda_{\alpha,\gamma,a}$ is an eigenvalue of $\underline{\mathcal{L}}_{\alpha,\gamma,a}(\tau_\star)$.

Remark 2 In [3, Proposition 1.4] we also provide a condition on (α, γ, a) such that the strict inequality in (5) is realized.

Remark 3 We consider cases of $(\alpha, \gamma, a, \tau)$ where the infimum of the spectrum of $\underline{\mathcal{L}}_{\alpha,\gamma,a}(\tau)$ is an eigenvalue below the essential spectrum (see Remark 2). One

[3] One has to take $\tau_* = \xi_a \sin\gamma$, where ξ_a is the minimum of $\mu_a(\cdot)$ introduced in (10).

can prove an Agmon-estimate result showing the decay of the corresponding eigenfunction, for large values of $|x|$. More precisely, let $a \in [-1, 1) \setminus \{0\}$, $\alpha \in (0, \pi)$, $\gamma \in (0, \pi/2]$ and $\tau \in \mathbb{R}$. Consider the case where $\underline{\sigma}(\alpha, \gamma, a, \tau) < \underline{\sigma}_{ess}(\alpha, \gamma, a, \tau)$. Let $v_{\alpha,\gamma,a,\tau}$ be the normalized eigenfunction corresponding to $\underline{\sigma}(\alpha, \gamma, a, \tau)$. For all $\eta \in \sqrt{\underline{\sigma}_{ess}(\alpha, \gamma, a, \tau) - \underline{\sigma}(\alpha, \gamma, a, \tau)}$, there exists a constant C (depending on η and α) such that

$$\underline{Q}^{\tau}_{\alpha,\gamma,a}(e^{\eta\phi} v_{\alpha,\gamma,a,\tau}) \leq C,$$

where $\phi(x) = |x|$, for $x \in \mathbb{R}^2_+$ and $\underline{Q}^{\tau}_{\alpha,\gamma,a}$ is the quadratic form associated to $\underline{\mathcal{L}}_{\alpha,\gamma,a}(\tau)$ in $\mathbb{R}^2_+$. For the proof, we refer the reader to similar results in [7, Theorem 9.1] and [9].

Acknowledgments The author acknowledges the support of the Istituto Nazionale di Alta Matematica "F. Severi", through the Intensive Period "INdAM Quantum Meetings (IQM22)".

References

1. Assaad, W.: The breakdown of superconductivity in the presence of magnetic steps. Commun. Contemp. Math (2020). https://doi.org/10.1142/S0219199720500054
2. Assaad, W.: Magnetic steps on the threshold of the normal state. J. Math. Phys. **61**(10), 101508 (2020)
3. Assaad, W., Giacomelli, E.L.: A 3D-Schrödinger operator under magnetic steps with semiclassical applications. Discrete Contin. Dyn. Syst. **43**(2), (2023)
4. Assaad, W., Kachmar, A.: Lowest energy band function for magnetic steps. J. Spectr. Theory **12**(2), 813–833 (2022)
5. Assaad, W., Kachmar, A., Persson-Sundqvist, M.: The distribution of superconductivity near a magnetic barrier. Commun. Math. Phys. **366**(1), 269–332 (2019)
6. Assaad, W., Helffer, B., Kachmar, A.: Semi-classical eigenvalue estimates under magnetic steps (2021). arXiv preprint arXiv:2108.03964
7. Bonnaillie, V.: Analyse mathématique de la supraconductivité dans un domaine à coins: méthodes semi-classiques et numériques. PhD thesis, Université Paris Sud-Paris XI (2003)
8. Bonnaillie-Noël, V., Fournais, S.: Superconductivity in Domains with Corners. Rev. Math. Phys. **19**, 607–637 (2007)
9. Bonnaillie-Noël, V., Dauge, M., Popoff, N., Raymond, N.: Discrete spectrum of a model schrödinger operator on the half-plane with Neumann conditions. ZAMP **63**(2), 203–231 (2012)
10. Bonnaillie-Noël, V., Dauge, M., Popoff, N.: Ground state energy of the magnetic Laplacian on general three-dimensional corner domains. Mem. SMF **145** (2015)
11. Correggi, M., Giacomelli, E.L.: Surface superconductivity in presence of corners. Rev. Math. Phys. **29**, 1750005 (2017)
12. Correggi, M., Giacomelli, E.L.: Effects of corners in surface superconductivity. Calc. Var. **60**, 236 (2021)
13. Correggi, M., Giacomelli, E.L.: Almost flat angles in surface superconductivity. Nonlinearity **34**, 7633–7661, (2021)

14. Correggi, M., Rougerie, N.: On the Ginzburg-Landau functional in the surface superconductivity regime. Commun. Math. Phys. **332**, 1297–1343 (2014); Erratum: Commun. Math. Phys. **338**, 1451–1452 (2015)
15. Correggi, M., Rougerie, N.: Boundary behavior of the Ginzburg-Landau order parameter in the surface superconductivity regime. Arch. Ration. Mech. Anal. **219**, 553–606 (2015)
16. Fournais, S., Helffer, B.: On the third critical field in Ginzburg–Landau theory. Commun. Math. Phys. **266**(1), 153–196 (2006)
17. Fournais, S., Helffer, B.: Strong diamagnetism for general domains and application. Ann. Inst. Fourier **57**(7), 2389–2400 (2007)
18. Fournais, S., Helffer, B.: Spectral Methods in Surface Superconductivity, vol. 77. Springer Science & Business Media, Berlin (2010)
19. Ginzburg, V.L., Landau, L.D.: On the theory of superconductivity. Zh. Eksp. Teor. Fiz. **20**, 1064–1082 (1950)
20. Giorgi, T., Phillips, D.: The breakdown of superconductivity due to strong fields for the Ginzburg–Landau model. SIAM J. Math. Anal. **30**, 341–359 (1999)
21. Helffer, B., Kachmar, A.: The density of superconductivity in domains with corners. Lett. Math. Phys. **108**, 2169–2187 (2018)
22. Helffer, B., Morame, A.: Magnetic bottles in connection with superconductivity. J. Funct. Anal. **185**(2), 604–680 (2001)
23. Helffer, B., Morame, A.: Magnetic bottles for the Neumann problem: curvature effects in the case of dimension 3 (general case). Ann. Sci. Éc. Norm. Supér. **37**(1), 105–170 (2004)
24. Helffer, B., Pan, X.B.: Upper critical field and location of surface nucleation of superconductivity. Ann. Inst. H.Poincaré Anal. Non Linaíre **20**(1), 145–181 (2003)
25. Hislop, P.D., Soccorsi, E.: Edge states induced by Iwatsuka Hamiltonians with positive magnetic fields. J. Math. Anal. Appl. **422**(1), 594–624 (2015)
26. Hislop, P.D., Popoff, N., Raymond, N., Sundqvist, M.: Band functions in the presence of magnetic steps. Math. Models Methods Appl. Sci. **26**(1), 161–184 (2016)
27. Kato, T.: Perturbation Theory for Linear Operators. Springer, New York (1966)
28. Lu, K., Pan, X.B.: Eigenvalue problems of Ginzburg–Landau operator in bounded domains. J. Math. Phys. **40**(6), 2647–2670 (1999)
29. Lu, K., Pan, X.B.: Surface nucleation of superconductivity in 3-dimension. J. Differ. Equ. **168**(2), 386–452 (2000)
30. Morame, A., Truc, F.: Remarks on the spectrum of the Neumann problem with magnetic field in the half-space. J. Math. Phys. **46**(1), 012105 (2005)
31. Pan, X.B.: Surface superconductivity in applied magnetic fields above H_{c_2}. Commun. Math. Phys. **228**, 327–370 (2002)
32. Pan, X.-B.: Upper critical field for superconductors with edges and corners. Calc. Var. Partial Differ. Equ. **14**, 447–482 (2002)
33. Popoff, N.: The schrödinger operator on an infinite wedge with a tangent magnetic field. J. Math. Phys. **54**(4), 041507 (2013)
34. Popoff, N.: The model magnetic Laplacian on wedges. J. Spectr. Theory **5**(3), 617–661 (2015)
35. Raymond, N.: Sharp asymptotics for the Neumann Laplacian with variable magnetic field: case of dimension 2. Ann. Henri Poincaré **10**(1), 95–122 (2009)
36. Raymond, N.: From the Laplacian with variable magnetic field to the electric Laplacian in the semiclassical limit. J. Anal. PDE **6**(6), 1289–1326 (2013)
37. Saint James, D., de Gennes, P.G.: Onset of superconductivity in decreasing fields. Phys. Lett. **7**(5), 306–308 (1963)
38. Sandier, E., Serfaty, S.: Vortices in the Magnetic Ginzburg–Landau Model. Progress in Nonlinear Partial Differential Equations and their Applications. Birkhäuser-Boston, Basel (2007)

Quantum Systems at the Brink

Dirk Hundertmark, Michal Jex, and Markus Lange

1 Introduction

Except for the famous Wigner-von Neumann potentials [48], bound states of quantum systems are usually found below the energies of scattering states. The bound state energies and the scattering energies are separated by the ionization threshold corresponding to the essential spectrum threshold. Above this threshold, the particles cease to be bound and move to infinity. Below the threshold, the binding energy, i.e., the difference between the ionization threshold and the energy of the bound state, is positive. Since the minimal energy cost to move a particle to infinity is given by the binding energy and since regular perturbation theory predicts that the energy changes only little under small perturbations the quantum system is stable under small perturbations. As long as the binding energy stays positive the corresponding eigenfunctions are still bound, i.e., they do not suddenly disappear.

D. Hundertmark
Department of Mathematics, Institute for Analysis, Karlsruhe Institute of Technology, Karlsruhe, Germany

Department of Mathematics, Altgeld Hall, University of Illinois at Urbana-Champaign, Urbana, IL, USA
e-mail: dirk.hundertmark@kit.edu

M. Jex (✉)
Department of Physics, Faculty of Nuclear Sciences and Physical Engineering, Czech Technical University in Prague, Prague, Czech Republic

CEREMADE, Université Paris-Dauphine, PSL Research University, Paris, France
e-mail: michal.jex@fjfi.cvut.cz

M. Lange
Institute for AI-Safety and Security, German Aerospace Center (DLR), Sankt Augustin & Ulm, Germany
e-mail: markus.lange@dlr.de

M. Correggi, M. Falconi (eds.), *Quantum Mathematics I*,
Springer INdAM Series 57, https://doi.org/10.1007/978-981-99-5894-8_10

Imagine a parameter of the quantum system being tuned such that the energy of a bound state, e.g., the ground state energy, approaches the ionization threshold. At this critical value, the perturbation theory in the parameter breaks down. Moreover, at this threshold there is no energy penalty for moving the quantum particle to infinity anymore. So it is unclear what happens *exactly at* this binding–unbinding transition: Does the bound state disappear, i.e., the quantum particle can move to infinity and the eigenstate of the quantum system spreads out more and more and dissolves, or does the bound state still exist at the critical parameter and then suddenly disappears (see, for example, the discussion in [29]). Consider a Schrödinger operators of the form

$$H_\lambda = -\frac{1}{2m}\Delta - V_\lambda(x) + U(x) \tag{1}$$

where $-\frac{1}{2m}\Delta$ is the kinetic energy, U a non-zero repulsive part of the potential and $-V_\lambda$ a compactly supported attractive part of the potential depending on a parameter λ. This operator describes one-particle models, however with slight modifications it can also describe interacting many-particle systems. The well-known WKB asymptotics, see also the work of Agmon [2], shows that the eigenfunction ψ_λ corresponding to a discrete eigenvalue E_λ of the operator (1) falls off exponentially with the distance to the origin, i.e.,

$$\psi_\lambda \sim \exp\left(-\sqrt{2m\Delta E_\lambda}|x|\right)$$

for $|x| \to \infty$ where $\Delta E_\lambda \geq 0$ is the binding energy, i.e., the distance of the eigenvalue E_λ to the bottom of the essential spectrum of H_λ. Such a decay estimate *does not provide any useful information at critical coupling* when $\Delta E_\lambda = 0$. Even worse, all rigorous approaches for decay estimates of eigenfunctions usually provide upper bounds of the form

$$|\psi(x)| \leq C_\delta \exp\left(-(\sqrt{2m\Delta E_\lambda} - \delta)|x|\right) \tag{2}$$

for all small enough $\delta > 0$ with a constant C_δ which diverges in the limit $\delta \to 0$, see e.g. [2]. Therefore, in order to be able to prove bounds on the asymptotic behavior of bound states, which still yields useful information when the binding energy vanishes, a new approach is needed. It can not require a gap between the eigenvalue and the threshold of the essential spectrum to work.

The new method developed in [23], which is presented in the next section, can be viewed as a higher order correction to the WKB method. The main ingredient is still a suitable energy estimate. However, our approach to energy estimates is based on the idea that a *positive long–range repulsive part* of the potential can stabilize a quantum system. Such a long range positive part allows us to gain extra flexibility, in particular, it remove the necessity of positive binding energy, i.e., a safety distance with respect to the bottom of the essential spectrum. The underlying intuition is

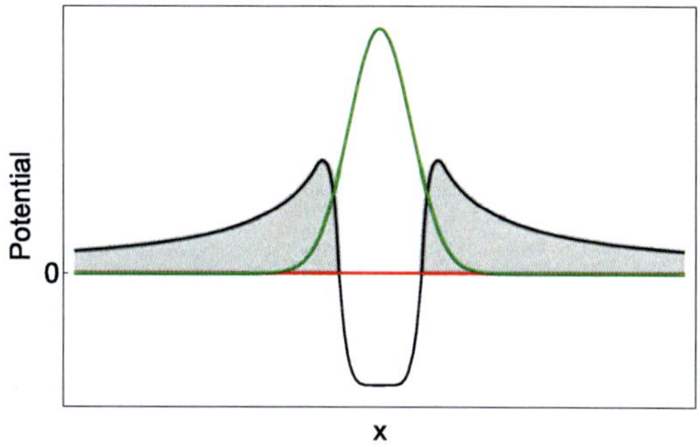

Fig. 1 Sketch of tunneling problem for the ground state at zero energy: black line corresponds to the potential, red line to the energy level, green to the eigenstate and grey area to classically forbidden region

that if the binding energy ΔE_λ vanishes as the parameter λ approaches a critical value, the bound state can only disappear when it tunnels through the positive tail of potential, see Fig. 1. If this tunneling probability is zero, the ground state cannot disappear, hence the quantum system stays bounded at the critical coupling. This behavior is also predicted by numerical calculations [12, 13, 19, 42]. Our method makes this intuition precise, including upper bounds on the asymptotic behavior of the corresponding eigenfunctions at the ionization threshold.

Before we present our approach let us shortly mention some known results for the existence and non-existence of threshold eigenvalues. Early results on existence or non-existence of threshold eigenvalues go back to [1, 26, 28, 30–32, 37, 39, 40, 44, 49]. In [9] it was noted that a long range Coulomb part can create zero energy eigenstates, see also [36, 50]. An analysis of eigenstates and resonances at the threshold for the case of certain nonlocal operators recently appeared in [27]. The references presented above are by no means exhaustive.

The main result of the paper, presented in Theorem 1, yields decay estimates for bound states of quantum systems which *do not require* that the binding energy ΔE_λ is positive if a suitable long-range repulsive part of the potential is present.

2 The Method

For simplicity of the exposition we will only consider one–particle Schrödinger operators H_λ of the form (1) in the following. We also assume that the potentials V_λ and the long range repulsive part U are in the Kato–class, see [4, 11] or [45] for the definition. This ensures that the potentials are infinitesimally form bounded with respect to the kinetic energy $-\Delta$, so the Schrödinger operator (1) is well–defined with the help of quadratic form methods, [41, 46]. The ionisation threshold $\Sigma_\lambda = \inf \sigma_{\mathrm{ess}}(H_\lambda)$ is given by the bottom of the essential spectrum. We also assume that the potential vanishes at infinity, in which case $\sigma_{\mathrm{ess}}(H_\lambda) = [0, \infty)$, i.e., $\Sigma_\lambda = 0$.

Theorem 1 *Each normalized eigenfunction ψ_λ corresponding to an eigenenergy $E_\lambda \leq 0$ of H_λ satisfies*

$$|\psi_\lambda| \lesssim \exp\left(-F - \frac{1}{2}\ln\left(\Delta E_\lambda + U - \frac{|\nabla F|^2}{2m}\right)\right) \tag{3}$$

with $\Delta E_\lambda = -E_\lambda$ being the binding energy, and F being any function which is bounded from below and satisfies

$$\frac{|\nabla F|^2}{2m} < \Delta E_\lambda + U \tag{4}$$

for all $|x| \geq R > 0$. Here $U \geq 0$ is the repulsive part of the potential.

Remark 1 Choosing $F(x) = \mu|x|$ yields $|\nabla F(x)|^2 = \mu^2$. Note that $\Delta E_\lambda + U - \frac{|\nabla F|^2}{2m} \geq \Delta E_\lambda - \frac{\mu^2}{2m}$ since $U \geq 0$. Thus in the subcritical case, when the binding energy is positive, upper bounds of the form (2), which, for one–particle operators, coincide with the result of Agmon [2], follow immediately from Theorem 1.

In the critical case, when the binding energy vanishes, a non-zero repulsive part U is indispensable since otherwise (4) can never be satisfied. However, in contrast to the usual WKB asymptotics our bound provides detailed information on how well the quantum system is localized at critical coupling, when a repulsive part U is present. The logarithmic expression in the exponent of (3) corresponds to a polynomial correction of the asymptotic behavior and in all relevant cases it is of smaller order than F.

The existence of the eigenstate is a necessary assumption in Theorem 1. On the other hand, as shown in [23, 24], the existence of an eigenstate for the critical case follows from bounds of the form (3) together with tightness arguments in the form of, e.g., [21].

Proof In the following we will, for notational simplicity, drop the dependence of the Schrödinger operator, the wave function, and the eigenenergy on the parameter λ.

Starting Point Consider a self-adjoint operator H given in (1) with and a normalized eigenvector ψ satisfying

$$H\psi = E\psi$$

where E is the corresponding eigenvalue below or at the threshold of the essential spectrum.

1st Step Let $0 \leq \chi \leq 1$ be a smooth real-valued function satisfying

$$\chi(x) = \begin{cases} 0, & \text{for } |x| \leq 1 \\ 1, & \text{for } |x| \geq 2 \end{cases} . \tag{5}$$

The scaled functions given by $\chi_R(x) = \chi(x/R)$ for $R > 0$ smoothly localize in the region $\{|x| \geq R\}$. Note that $\mathrm{supp}\nabla\chi_R$ is localized in the annulus $\{R \leq |x| \leq 2R\}$.

Let F be another smooth and bounded real–valued function for which also $|\nabla F|$ is bounded. With $\xi = \chi_R e^F$ one calculates from the eigenvalue equation

$$\mathrm{Re}\langle(\xi)^2\psi, H\psi\rangle = E\langle(\xi)^2\psi, \psi\rangle = E\|\xi\psi\|^2 .$$

2nd Step Using a variant [11, 16] of the IMS localization formula [25, 35, 43], we obtain

$$E\|\chi_R e^F\psi\|^2 = \mathrm{Re}\langle(\xi)^2\psi, H\psi\rangle = \langle\xi\psi, H\xi\psi\rangle - \frac{1}{2m}\langle\psi, |\nabla\xi|^2\psi\rangle .$$

Clearly, $\nabla\xi = \nabla(\chi_R e^F) = \nabla\chi_r e^F + \chi_R\nabla F e^F$, so

$$|\nabla\xi|^2 \leq \Big(|\nabla\chi_R|^2 + 2\chi_R|\nabla\chi_R||\nabla F|\Big)\,e^{2F} + |\nabla F|^2\xi^2 .$$

Note that the good part $G = \big(|\nabla\chi_R|^2 + 2\chi_R\nabla\chi_R\cdot\nabla F\big)\,e^{2F}$ has compact support, because the support of $\nabla\chi_R$ is compact for any $R > 0$. Rearranging the terms, we obtain

$$\Big\langle\chi_R e^F\psi, \Big(H - E - \frac{1}{2m}|\nabla F|^2\Big)\chi_R e^F\psi\Big\rangle \leq \frac{1}{2m}\langle\psi, G\psi\rangle . \tag{6}$$

The usual argument now uses Persson's theorem [38] for the bottom of the essential spectrum

$$\Sigma = \inf\sigma_{\mathrm{ess}}(H) = \lim_{R\to\infty}\{\langle\varphi, H\varphi\rangle : \|\varphi\| = 1, \mathrm{supp}\,(\varphi) \subset B_R^c\} \tag{7}$$

where $B_R^c = \{|x| \geq R\}$. Thus, since we assume that $\Sigma = 0$, for any $\delta > 0$ there exist $R_\delta < \infty$ such that

$$\langle\varphi, H\varphi\rangle > (\Sigma - \delta)\langle\varphi, \varphi\rangle = -\delta\langle\varphi, \varphi\rangle$$

for all φ with support outside a centered ball of radius R_δ. So with $R = R_\delta$, we get from (6)

$$\Big\langle\chi_R e^F\psi, \Big(-\delta - E - \frac{1}{2m}|\nabla F|^2\Big)\chi_R e^F\psi\Big\rangle \leq \frac{1}{2m}\langle\psi, G\psi\rangle$$

but one needs positivity of $-\delta - E - \frac{1}{2m}|\nabla F|^2$ and this requires $E < -\delta$, i.e., a safety distance of the negative eigenvalue to the essential spectrum.

Instead, we use the assumption that the potential is given by $-V + U$, where V has compact support and U is positive. Chosing R so large that the support of V is contained in $\{|x| \leq R\}$, we have

$$\langle \chi_R e^F \psi, H \chi_R e^F \psi \rangle = \langle \chi_R e^F \psi, (-\Delta + U) \chi_R e^F \psi \rangle \geq \langle \chi_R e^F \psi, U \chi_R e^F \psi \rangle$$

and using this in (6) one arrives at

$$\Big\langle \chi_R e^F \psi, \Big(\Delta E + U - \frac{1}{2m} |\nabla F|^2 \Big) \chi_R e^F \psi \Big\rangle \leq \frac{1}{2m} \langle \psi, G \psi \rangle . \tag{8}$$

where $\Delta E = -E$ is the binding energy. We want to use this energy inequality to prove exponential bounds on ψ, but for this we need that F is growing.

3rd Step In order to overcome the requirement that F is bounded, we regularize it. Let F be smooth, bounded from below and let ∇F be bounded. Adding a constant to F, we can assume that $F \geq 0$. This also does not change the gradient of F. Then for any $\varepsilon > 0$ the function

$$F_\varepsilon = \frac{F}{1 + \varepsilon F}$$

is smooth and bounded. Since $\nabla F_\varepsilon = (1 + \varepsilon F)^{-2} \nabla F$ also ∇F_ε is bounded. Let ξ_ε and G_ε be defined as above with F replaced by F_ε. Clearly $F_\varepsilon \leq F$ and $|\nabla F_\varepsilon| \leq |\nabla F|$ for all $\varepsilon \geq 0$. Hence $G_\varepsilon \leq G$ and

$$|\nabla \xi_\varepsilon|^2 \leq G_\varepsilon + |\nabla F_\varepsilon|^2 \xi_\varepsilon^2 \leq G + |\nabla F|^2 \xi_\varepsilon^2$$

for all $\varepsilon \geq 0$. The argument leading to (8) then shows

$$\Big\langle \chi_R e^{F_\varepsilon} \psi, \Big(\Delta E + U - \frac{1}{2m} |\nabla F|^2 \Big) \chi_R e^{F_\varepsilon} \psi \Big\rangle \leq \frac{1}{2m} \langle \psi, G \psi \rangle \leq K \|\psi\|^2 . \tag{9}$$

with $K = \frac{1}{2m} \sup_{R \leq |x| \leq 2R} G(x) < \infty$, since G is supported inside $\{R \leq |x| \leq 2R\}$.

Note that

$$\Big\langle \chi_R e^{F_\varepsilon} \psi, \Big(\Delta E + U - \frac{1}{2m} |\nabla F|^2 \Big) \chi_R e^{F_\varepsilon} \psi \Big\rangle = \Big\| \chi_R e^{F_\varepsilon + \frac{1}{2} \ln \left(\Delta E + U - \frac{1}{2m} |\nabla F|^2 \right)} \psi \Big\|^2 .$$

The monotone convergence theorem and (9) yield

$$\begin{aligned} &\Big\| \chi_R e^{F + \frac{1}{2} \ln \left(\Delta E + U - \frac{1}{2m} |\nabla F|^2 \right)} \psi \Big\|^2 \\ &\quad = \lim_{\varepsilon \to 0} \Big\langle \chi_R e^{F_\varepsilon} \psi, \Big(\Delta E + U - \frac{1}{2m} |\nabla F|^2 \Big) \chi_R e^{F_\varepsilon} \psi \Big\rangle \leq K \|\psi\|^2 < \infty . \end{aligned}$$

for any normalized eigenfunction ψ with energy $E \leq 0$. This proves an L^2 exponential bound on ψ, i.e., the function

$$x \mapsto \exp\left(F(x) + \frac{1}{2}\ln\left(\Delta E + U - \frac{1}{2m}|\nabla F|^2\right)\right)\psi(x)$$

is in L^2 under the condition that all exponential weights F satisfy (4). The claimed pointwise bound on ψ then follows from such an L^2 bound using subsolution estimates of [45], see, e.g., the discussion in [23, Corollary 5.4].

3 Examples

In this section we consider illustrative examples of a quantum particle in a potential well with a long range Coulomb repulsion term. In the first example the tunable parameter is the depth of the potential well. We will see that the bound from Theorem 1 fits very well with the explicitly calculated asymptotic behavior of the ground state of such a system. In a second example we tune the strength of the repulsion term. In the last example we illustrate that a long range repulsion term is crucial at critical coupling.

First Example In dimension 3 let us consider

$$H_\lambda = -\Delta - \lambda\, 1_{\{|x|\leq 1\}} + \frac{1_{\{|x|>1\}}}{|x|}. \tag{10}$$

Here we chose $m = \frac{1}{2}$ for convenience. In this case $U(x) = 1/|x|$ for $|x| \geq 1$. It can be easily shown that there exists a critical value λ_{cr} s.t. for $\lambda > \lambda_{\mathrm{cr}}$, the Hamiltonian H_λ has at least one bound state and for $\lambda < \lambda_{\mathrm{cr}}$ there are none. Furthermore, for this system we have $\Sigma = 0$ and $\lambda_{\mathrm{cr}} \approx 0.634366$.

Take $F(x) = 2b|x|^{1/2}$. Then $\nabla F(x) = b|x|^{-1/2}$ and

$$U(x) - |\nabla F(x)|^2 = \frac{1-b^2}{|x|} > 0$$

whenever $b^2 < 1$ and $|x| \geq 1$. Thus Theorem 1 shows the upper bound

$$|\psi_\lambda| \lesssim e^{-2b|x|^{1/2} + \frac{1}{2}\ln|x|} \tag{11}$$

for large $|x|$ and all eigenstates with energy $E_\lambda \leq 0$ and all $0 < b < 1$. This is a stretched exponential decay.

One can make the bound tighter by choosing a more general radial weight function. With a slight abuse of notation, we set $F(x) = F(|x|)$. Then $\nabla F(x) =$

$F'(|x|)x/|x|$ and the borderline case allowed, or better, just not allowed by condition (4) is

$$\Delta E + U(r) - |F'(r)|^2 = 0$$

with $r = |x|$. Hence we want to solve the equation $F'(r) = \sqrt{\Delta E + U(r)}$. For $a, b \geq 0$ let $F_{a,b}$ be given by

$$\begin{aligned} F_{a,b}(r) &= \int_0^r \left(a + \frac{b}{s}\right)^{1/2} ds \\ &= \left(a + \frac{b}{r}\right)^{1/2} r + \frac{b}{\sqrt{a}} \operatorname{arcsinh}\left[\sqrt{\frac{ar}{b}}\right]. \end{aligned} \tag{12}$$

It is easy to check that the derivative in r of the right hand side is given by integrand $(a + b/r)^{1/2}$. Splitting $U(r) = 1/|r| = \delta/r + (1-\delta)/r$, for $0 < \delta < 1$, suggests to take $a = \Delta E_\lambda$ and $b = 1 - \delta$. Theorem 1 then gives the upper bound

$$|\psi_\lambda(x)| \lesssim e^{-F_{\Delta E_\lambda, 1-\delta}(|x|) + \frac{1}{2} \ln |x|} . \tag{13}$$

for the ground state of H_λ with $\Delta E_\lambda \geq 0$ and any $0 < \delta < 1$. In the subcritical case, where the binding energy $\Delta E_\lambda > 0$, the first part on the right hand side of (12) corresponds to exponential fall-off with exponential weight $\sqrt{\Delta E_\lambda}|x|$ (recall that we put $m = 1/2$), which is exactly the prediction of the WKB method, and the second one is the polynomial correction since $\operatorname{arcsinh}[y] = \ln(y + \sqrt{y^2 + 1})$.

Note that

$$\lim_{a \to 0} F_{a,b}(r) = 2\sqrt{br}$$

so in the limit where the binding energy vanishes we recover the bound (11) from (13). See Fig. 2 for an illustration.

One can further improve upon the upper bound, by trying an ansatz of the form

$$F(r) = F_{a,b}(r) - K|x|^\kappa \tag{14}$$

for any $K > 0$ and $0 < \kappa < 1/2$. It is straigtforward to check that with $a = \Delta E_\lambda$ and $b = 1$, this ansatz satisfies (4) for all large $|x|$ and all $\Delta E_\lambda \geq 0$.

For vanishing binding energy, i.e., at $\lambda = \lambda_{cr}$, a matching lower bound for the ground state, which can be chosen to be strictly positive, of the form

$$e^{-2\sqrt{|x|} - K|x|^\kappa} \lesssim \psi_{\lambda_{cr}}(x)$$

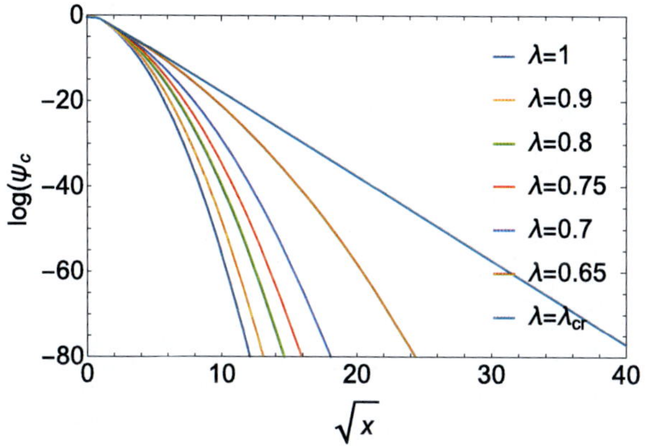

Fig. 2 Scaled plot of normalized ground states for the Hamiltonian (10) with varying parameter λ for $x \in [0, 1600]$. The convergence of the ground states for $\lambda \searrow \lambda_{\text{cr}} \approx 0.63$ is visible. Note that in this choice of scale the parabolic curves correspond to the ground state decaying asymptotically as $\exp(-c|x|)$, as is predicted by the WKB method, when the parameter $\lambda > \lambda_{\text{cr}}$. For $\lambda = \lambda_{\text{cr}}$ the nearly straight line indicates that the ground state decays like $\exp(-2\sqrt{|x|})$

for any $K > 0$ and $0 < \kappa < 1/2$, was obtained in [23] using a subharmonic comparison lemma [3, 17]. Explicit calculations show that the eigenfunction has asymptotic behavior in the form

$$\psi_{\lambda_{\text{cr}}}(x) \sim C \frac{e^{-2\sqrt{|x|}}}{|x|^{3/4}}$$

for large $|x|$ which is in perfect agreement with our result.

Remark 2 In general the existence or non–existence of ground states at critical coupling depends crucially on the dimension of the considered problem, see [24] for more details.

Second Example We consider again an operator describing a quantum particle in a potential well with a repulsion term everywhere outside that well. However we do not decrease the depth of the well but increase the repulsion term. We start with an operator having a long range Coulomb repulsion term in three dimensions

$$H_c = -\Delta - 1_{\{|x| \leq 1\}}(|x|) + 1_{\{1<|x|\}}(|x|)\frac{c}{|x|}. \tag{15}$$

Increasing the repulsive term, i.e., increasing the parameter c, the eigenfunctions become more localized up to the numerically calculated critical value $c_{\text{cr}} \approx 3.11693$, see Fig. 3.

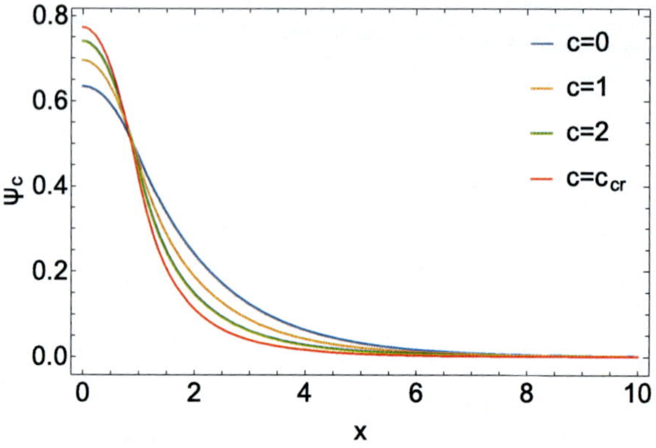

Fig. 3 Plot of the normalized ground state eigenfunction for the model (15) for several values of c

This operator has essential spectrum $\sigma_{\text{ess}}(H) = [0, \infty)$ for any c and it has negative energy ground state for sufficiently small positive $c < c_{\text{cr}}$. The argument from Example 1 shows that eigenfunctions ψ_c of H_c with energy $E_c \leq 0$ decay as

$$|\psi_c(x)| \lesssim \exp\left(-F_{|E_c|,c}(x) + \kappa|x|^{\delta}\right)$$

for any $\kappa > 0$ and $0 < \delta < 1/2$, where $F_{|E_c|,c}$ is given by (12) with the choice $a = |E_c|$ and $b = c$.

At critical coupling $c = c_{\text{cr}}$ the operator (15) has a normalizable ground state with eigenvalue 0. For this it is crucial to have a long range repulsive term. Without long range repulsion the eigenfunctions will delocalize more and more for $c \nearrow c_{\text{cr}}$, as is illustrated in the next example.

Third Example To see the importance of long range behavior of the repulsive potential we consider next a Hamiltonian with only a finite size repulsive barrier

$$\widetilde{H}_c = -\Delta - 1_{\{|x|\leq 1\}}(|x|) + c1_{\{1<|x|<2\}}(|x|), \tag{16}$$

again in dimension three. Note that the value 2 is artificial and has no particular importance. If we start to increase the parameter c up to the critical value $\widetilde{c}_{\text{cr}} \approx 2.7938776$, we see that far away from the critical value the increase of c leads to the localization of the wavefunction even by a short range potential. However for $c \geq 2.5$ the wavefunction starts to spread further and further and for $c = 2.78$ the fall-off of the function is hardly visible, see Fig. 4. Using results of [24] it is easy to see that 0 is not an eigenvalue of the operator given in (16) for $c = \widetilde{c}_{\text{cr}}$.

The presented plots highlight the physical intuition that the wavefunction has to tunnel through the repulsive barrier in order to leave the potential well and delocalize. However the long range Coulomb repulsion is too *sticky* for the wavefunction to delocalize even at the critical value and hence we are able to prove fall-off behavior at the threshold of the essential spectrum.

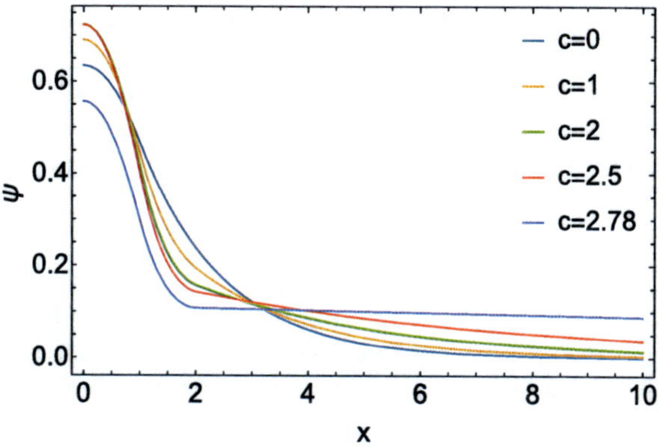

Fig. 4 Plot of the normalized ground state eigenfunction for the model (16) for several values of c. It illustrates, that as c approaches $\widetilde{c}_{cr}$ the wavefunction delocalizes

4 Outlook

Our method is not restricted to a Coulomb type long range part of the potential nor to the case of one-particle models. A variety of physical systems can be handled. For example, it is easy to check that for a long range repulsive potential U, which is radial, say, any exponential weight F of the form

$$F(r) = \delta \int_{r_0}^{r} \sqrt{U(s)}\, ds \tag{17}$$

for some $r_0 \geq 0$ and $0 < \delta < 1$ will satisfy (4). To yield a useful upper bound one need that $\lim_{r\to\infty} F(r) = \infty$, i.e., the integral $\int_{r_0}^{r} \sqrt{U(s)}\, ds$ should diverge in the limit $r \to \infty$. For power law repulsive potentials of the form $U(r) = c_1 r^{-\alpha}$ this shows that one needs $\alpha \leq 2$. Since for vanishing binding energy $\Delta E = 0$, the correction term satisfy

$$-\frac{1}{2} \ln \left(U(r) - \frac{|\nabla F(r)|^2}{2m} \right) \sim c_2 \ln r$$

for some (computable) constant c_2 and all large r, we get a useful upper bound for any $c_1 > 0$ when $\alpha < 2$. If $\alpha = 2$, i.e., the repulsive part $U(r)$ decays like a Hardy type potential, we also need that c_1 is large enough.

Of particular importance are multi–particle systems, such as N electron atoms with a nucleus of charge Z. For such atomic systems ground states exist once $N < Z + 1$, due to a classical result by Zhislin [51]. For $N > 2Z + 1$, no such states exist [33]. Hence, for any fixed number N there is a critical charge $Z_c(N)$ such that for $Z > Z_c(N)$ bound states exist and for charges $Z < Z_c(N)$ the quantum system has no bound state. Note that $Z_c(N)$ does not have to be a whole number.

For helium-like systems, a variational calculation of Bethe[8] shows that $Z_c(2) < 1$. Numerically, it is known [5] that $Z_c(2) \sim 0.91$. The existence and absence of an eigenstate for the simplest nontrivial example of helium-like systems for $Z = Z_c(2)$, was studied extensively by M. and T. Hoffmann-Ostenhof and Simon [18]. They derived the existence of an eigenstate at critical coupling $Z_c(2)$ for a singlet state and conjectured its fall-off behavior to be subexponential [18]. This conjectured fall-off behavior of threshold eigenstates was used for example in [10, 14, 20, 34]. Using our method we recently proved in [23] that the conjecture made in [18] is correct.

For general atoms, the existence of a ground state at critical coupling was studied in the Born-Oppenheimer approximation in [7] and without it under the additional condition $Z_c(N) \in (N-2, N-1)$ in [15]. These results establish the existence of an eigenstate, but the derived decay bounds are far from what is physically expected [20].

Our approach relies mostly on energy estimates which, when combined with a geometrically inspired lower bounds for the multiparticle potentials of atomic systems, see e.g. [47], are also applicable to many-particle systems. In particular, our method is applicable to atomic systems under the additional assumption that $N - K > Z_c(N)$, where K is the number of electrons leaving the atom as Z decreases below $Z_c(N)$. A preprint with a proof of concept is available on the arXiv [22].

For very large atoms, it is undoubtedly necessary to use, at least for the inner electrons, the corresponding relativistic equations to obtain the correct results. Our method relies mainly on the IMS localization formula. Thus using known results for pseudo-relativistic quantum systems [6], it should be possible to adapt our method to systems with pseudo-relativistic electrons. Moreover, calculations suggest that our method is also valid within Hartree-Fock and Density Functional Theory (DFT). This is especially interesting due to the fact that these theories are inherently nonlinear.

Acknowledgments Funded by the Deutsche Forschungsgemeinschaft (DFG, German Research Foundation)—Project-ID 258734477—SFB 1173 (Dirk Hundertmark). This project has received funding from the European Research Council (ERC) under the European Union's Horizon 2020 research and innovation programme (grant agreement MDFT No. 725528) (Michal Jex). Michal Jex also received financial support from the Ministry of Education, Youth and Sport of the Czech Republic under the Grant No. RVO 14000. This project has received funding from the European Research Council (ERC) under the European Union's Horizon 2020 research and innovation programme (ERC StG MaMBoQ, grant agreement No. 802901) (Markus Lange).

References

1. Agmon, S.: Lower bounds for solutions of Schrödinger equations. J. Anal. Math. **23**, 1–25 (1970). https://doi.org/10.1007/BF02795485
2. Agmon, S.: Lectures on exponential decay of solutions of second-order elliptic equations: bounds on eigenfunctions of N-body Schrödinger operators. In: Mathematical Notes, vol. 29. Princeton University Press, Princeton (1982)

3. Agmon, S.: Bounds on exponential decay of eigenfunctions of Schrödinger operators. In: Schrödinger Operators (Como, 1984). Lecture Notes in Math., vol. 1159, pp. 1–38. Springer, Berlin (1985). https://doi.org/10.1007/BFb0080331
4. Aizenman, M., Simon, B.: Brownian motion and Harnack inequality for Schrödinger operators. Commun. Pure Appl. Math. **35**(2), 209–273 (1982). https://doi.org/10.1002/cpa.3160350206
5. Baker, J.D., Freund, D.E., Hill, R.N., Morgan III, J.D.: Radius of convergence and analytic behavior of the 1 z expansion. Phys. Rev. A **41**(3), 1247 (1990)
6. Barbaroux J.M., Hartig M.C., Hundertmark D., Vugalter, S.: Van der Waals-London interaction of atoms with pseudorelativistic kinetic energy. Analysis & PDE **15**(6), 1375–1428 (2022). https://doi.org/10.2140/apde.2022.15.1375
7. Bellazzini, J., Frank, R.L., Lieb, E.H., Seiringer, R.: Existence of ground states for negative ions at the binding threshold. Rev. Math. Phys. **26**(1), 1350021, 18 (2014). https://doi.org/10.1142/S0129055X13500219
8. Bethe, H.: Berechnung der Elektronenaffinität des Wasserstoffs. Z. Phys. **57**(11–12), 815–821 (1929)
9. Bollé, D., Gesztesy, F., Schweiger, W.: Scattering theory for long-range systems at threshold. J. Math. Phys. **26**(7), 1661–1674 (1985). https://doi.org/10.1063/1.526963
10. Burton, H.G.A.: Hartree–Fock critical nuclear charge in two-electron atoms. J. Chem. Phys. **154**(11), 111103 (2021). https://doi.org/10.1063/5.0043105
11. Cycon, H.L., Froese, R.G., Kirsch, W., Simon, B.: Schrödinger Operators with Application to Quantum Mechanics and Global Geometry, study edn. Texts and Monographs in Physics. Springer, Berlin (1987)
12. Dubau, J., Ivanov, I.A.: Numerical calculation of the complex energy of the resonance of a two-electron atom with nuclear charge below the threshold value. J. Phys. B: At. Mol. Opt. Phys. **31**(15), 3335–3344 (1998). https://doi.org/10.1088/0953-4075/31/15/007
13. Estienne, C.S., Busuttil, M., Moini, A., Drake, G.W.F.: Critical nuclear charge for two-electron atoms. Phys. Rev. Lett. **112**, 173001 (2014). https://doi.org/10.1103/PhysRevLett.112.173001
14. Grabowski, P.E., Burke, K.: Quantum critical benchmark for electronic structure theory. Phys. Rev. A **91**(3), 032501 (2015)
15. Gridnev, D.K., Garcia, M.E.: Rigorous conditions for the existence of bound states at the threshold in the two-particle case. J. Phys. A **40**(30), 9003–9016 (2007). https://doi.org/10.1088/1751-8113/40/30/022
16. Griesemer, M.: Exponential decay and ionization thresholds in non-relativistic quantum electrodynamics. J. Funct. Anal. **210**(2), 321–340 (2004). https://doi.org/10.1016/j.jfa.2003.06.001
17. Hoffmann-Ostenhof, T.: A comparison theorem for differential inequalities with applications in quantum mechanics. J. Phys. A **13**(2), 417–424 (1980)
18. Hoffmann-Ostenhof, M., Hoffmann-Ostenhof, T., Simon, B.: A multiparticle Coulomb system with bound state at threshold. J. Phys. A **16**(6), 1125–1131 (1983)
19. Hogreve, H.: On the (de) stabilization of quantum mechanical binding by potential barriers. Phys. Lett. A **201**(2–3), 111–118 (1995)
20. Hogreve, H.: On the maximal electronic charge bound by atomic nuclei. J. Phys. B: At. Mol. Opt. Phys. **31**(10), L439–L446 (1998). https://doi.org/10.1088/0953-4075/31/10/001
21. Hundertmark, D., Lee, Y.R.: On non-local variational problems with lack of compactness related to non-linear optics. J. Nonlinear Sci. **22**(1), 1–38 (2012). https://doi.org/10.1007/s00332-011-9106-1
22. Hundertmark, D., Jex, M., Lange, M.: Quantum systems at The Brink. existence and decay rates of bound states at thresholds; atoms, p. 14 (2019). Preprint arXiv:1908.05016
23. Hundertmark, D., Jex, M., Lange, M.: Quantum systems at the brink: Helium–type systems, p. 62 (2021). Preprint arXiv:1908.04883v2
24. Hundertmark, D., Jex, M., Lange, M.: Quantum systems at The Brink: existence of bound states, critical potentials and dimensionality, Forum of Mathematics, Sigma 11, e61 (2023). https://doi.org/10.1017/fms.2023.39

25. Ismagilov, R.S.: Conditions for the semiboundedness and discreteness of the spectrum in the case of one-dimensional differential operators. Dokl. Akad. Nauk SSSR **140**, 33–36 (1961)
26. Jensen, A., Kato, T.: Spectral properties of Schrödinger operators and time-decay of the wave functions. Duke Math. J. **46**(3), 583–611 (1979)
27. Kaleta, K., Lőrinczi, J.: Zero-energy bound state decay for non-local Schrödinger operators. Commun. Math. Phys. **374**(3), 2151–2191 (2020). https://doi.org/10.1007/s00220-019-03515-3
28. Kenig, C.E.: Restriction theorems, Carleman estimates, uniform Sobolev inequalities and unique continuation. In: Harmonic Analysis and Partial Differential Equations (El Escorial, 1987). Lecture Notes in Mathematics, vol. 1384, pp. 69–90. Springer, Berlin (1989). https://doi.org/10.1007/BFb0086794
29. Klaus, M., Simon, B.: Coupling constant thresholds in nonrelativistic quantum mechanics. I. Short-range two-body case. Ann. Phys. **130**(2), 251–281 (1980). https://doi.org/10.1016/0003-4916(80)90338-3
30. Knowles, I.: On the location of eigenvalues of second-order linear differential operators. Proc. R. Soc. Edinb. A **80**(1–2), 15–22 (1978). https://doi.org/10.1017/S030821050001009X
31. Knowles, I.: On the number of L^2-solutions of second order linear differential equations. Proc. R. Soc. Edinb. A **80**(1–2), 1–13 (1978). https://doi.org/10.1017/S0308210500010088
32. Lieb, E.H.: Thomas-Fermi and related theories of atoms and molecules. Rev. Mod. Phys. **53**(4), 603–641 (1981). https://doi.org/10.1103/RevModPhys.53.603
33. Lieb, E.H.: Bound on the maximum negative ionization of atoms and molecules. Phys. Rev. A **29**, 3018–3028 (1984). https://doi.org/10.1103/PhysRevA.29.3018
34. Mirtschink, A., Umrigar, C.J., Morgan, J.D., Gori-Giorgi, P.: Energy density functionals from the strong-coupling limit applied to the anions of the he isoelectronic series. J. Chem. Phys. **140**(18), 18A532 (2014). https://doi.org/10.1063/1.4871018
35. Morgan III, J.D.: Schrödinger operators whose potentials have separated singularities. J. Oper. Theory **1**(1), 109–115 (1979)
36. Nakamura, S.: Low energy asymptotics for Schrödinger operators with slowly decreasing potentials. Commun. Math. Phys. **161**(1), 63–76 (1994)
37. Newton, R.G.: Nonlocal interactions; the generalized Levinson theorem and the structure of the spectrum. J. Math. Phys. **18**(8), 1582–1588 (1977). https://doi.org/10.1063/1.523466
38. Persson, A.: Bounds for the discrete part of the spectrum of a semi-bounded Schrödinger operator. Math. Scand. **8**, 143–153 (1960). https://doi.org/10.7146/math.scand.a-10602
39. Ramm, A.G.: Sufficient conditions for zero not to be an eigenvalue of the Schrödinger operator. J. Math. Phys. **28**(6), 1341–1343 (1987). https://doi.org/10.1063/1.527817
40. Ramm, A.G.: Conditions for zero not to be an eigenvalue of the Schrödinger operator. II. J. Math. Phys. **29**(6), 1431–1432 (1988). https://doi.org/10.1063/1.527935
41. Reed, M., Simon, B.: Methods of Modern Mathematical Physics. IV. Analysis of Operators. Academic, New York-London (1978)
42. Sergeev, A.V., Kais, S.: Resonance states of atomic anions. Int. J. Quantum Chem. **82**(5), 255–261 (2001). https://doi.org/10.1002/qua.1047
43. Sigal, I.M.: Geometric methods in the quantum many-body problem. Nonexistence of very negative ions. Commun. Math. Phys. **85**(2), 309–324 (1982)
44. Simon, B.: Large time behavior of the L^p norm of Schrödinger semigroups. J. Funct. Anal. **40**(1), 66–83 (1981). https://doi.org/10.1016/0022-1236(81)90073-2
45. Simon, B.: Schrödinger semigroups. Bull. Amer. Math. Soc. (New Series) **7**(3), 447–526 (1982). https://doi.org/bams/1183549767
46. Teschl, G.: Mathematical Methods in Quantum Mechanics. Graduate Studies in Mathematics, vol. 157, 2nd edn. American Mathematical Society, Providence (2014). https://doi.org/10.1090/gsm/157
47. Uchiyama, J.: Finiteness of the number of discrete eigenvalues of the schrödinger operator for a three particle system. Publ. Res. Inst. Math. Sci. **5**(1), 51–63 (1969). https://doi.org/10.2977/prims/1195194752

48. von Neumann, J., Wigner, E.P.: Über merkwürdige diskrete Eigenwerte. In: The Collected Works of Eugene Paul Wigner, pp. 291–293. Springer, Berlin (1993)
49. Yafaev, D.R.: The virtual level of the Schrödinger equation. Zap. Naučn. Sem. Leningrad. Otdel. Mat. Inst. Steklov. (LOMI) **51**, 203–216, 220 (1975)
50. Yafaev, D.R.: The low energy scattering for slowly decreasing potentials. Commun. Math. Phys. **85**(2), 177–196 (1982)
51. Zhislin, G.M.: Discussion of the spectrum of schrödinger operators for systems of many particles. Trudy Moskovskogo matematiceskogo obscestva **9**, 81–120 (1960)

Lowest Eigenvalue Asymptotics in Strong Magnetic Fields with Interior Singularities

Ayman Kachmar and Xing-Bin Pan

1 Introduction

Throughout this chapter, Ω is a bounded domain in $\mathbb{R}^2$ with a C^1 boundary $\partial\Omega$ that has a finite number of connected components. Given a vector field $\mathbf{A} : \Omega \to \mathbb{R}^2$ and a positive parameter σ, we consider the magnetic Laplacian,

$$-\nabla_{\sigma\mathbf{A}}^2 := -\Delta + i\sigma\,\mathrm{div}\mathbf{A} + 2i\sigma\mathbf{A}\cdot\nabla + \sigma^2|\mathbf{A}|^2\,. \tag{1}$$

By imposing a boundary condition, like Dirichlet or Neumann, we can associate to the magnetic Laplacian a self-adjoint realization in $L^2(\Omega, \mathbb{C})$, the space of square integrable complex-valued functions on Ω. In fact, we introduce the quadratic form

$$q_{\sigma\mathbf{A}}(u) = \int_\Omega |(\nabla - i\sigma\mathbf{A})u|^2\,dx \tag{2}$$

which is closed on the magnetic Sobolev space

$$H^1_{\sigma\mathbf{A}}(\Omega, \mathbb{C}) = \{u \in L^2(\Omega, \mathbb{C}) \ : \ (\nabla - i\sigma\mathbf{A})u \in L^2(\Omega, \mathbb{C}^2)\}\,, \tag{3}$$

A. Kachmar (✉)
School of Science and Engineering, The Chinese University of Hong Kong, Shenzhen, Shenzhen, China
e-mail: akachmar@cuhk.edu.cn

X.-B. Pan
The Chinese University of Hong Kong (Shenzhen), Shenzhen, China

School of Science and Engineering, East China Normal University, Shanghai, China
e-mail: panxingbin@cuhk.edu.cn

M. Correggi, M. Falconi (eds.), *Quantum Mathematics I*,
Springer INdAM Series 57, https://doi.org/10.1007/978-981-99-5894-8_11

thereby it yields by the Friedrichs theorem the Neumann self-adjoint realization denoted by $P^N_{\sigma\mathbf{A}}$. Moreover, $q_{\sigma\mathbf{A}}$ is also closed on the space

$$H^1_{0,\sigma\mathbf{A}}(\Omega,\mathbb{C}) = \{u \in H^1_{\sigma\mathbf{A}}(\Omega,\mathbb{C}) \;:\; u = 0 \text{ on } \partial\Omega\} \tag{4}$$

and so we get the Dirichlet realization denoted by $P^D_{\sigma\mathbf{A}}$. The lowest eigenvalues of $P^N_{\sigma\mathbf{A}}$ and $P^D_{\sigma\mathbf{A}}$ can be expressed in the variational form (min-max principle)

$$\lambda^N(\sigma\mathbf{A}) = \inf_{\substack{u\in H^1_{\sigma\mathbf{A}}(\Omega,\mathbb{C})\\ u\neq 0}} \frac{q_{\sigma\mathbf{A}}(u)}{\|u\|^2_{L^2(\Omega,\mathbb{C})}} \quad\text{and}\quad \lambda^D(\sigma\mathbf{A}) = \inf_{\substack{u\in H^1_{0,\sigma\mathbf{A}}(\Omega,\mathbb{C})\\ u\neq 0}} \frac{q_{\sigma\mathbf{A}}(u)}{\|u\|^2_{L^2(\Omega,\mathbb{C})}}. \tag{5}$$

A part from the mathematical interest in studying the strong field limit, $\sigma \to +\infty$, of the lowest eigenvalue of $-\nabla^2_{\sigma\mathbf{A}}$, with Dirichlet or Neumann boundary conditions, the lowest eigenvalue stores information related to phase transitions occurring in models of superconductivity and liquid crystals [8].

The foregoing limit is understood to a large extent when the potential $\mathbf{A}$ is smooth [21] and to a lesser extent when the magnetic field curl $\mathbf{A}$ is a step function [3, 10], but we would like to discuss here what happens when the regularity of $\mathbf{A}$, or even the regularity of the associated magnetic field curl $\mathbf{A}$, is altered in other ways. More precisely, how does the lowest eigenvalue of the magnetic Laplacian feels the lack of regularity in the magnetic potential or the magnetic field ? Can one compare the effect of singularities of the magnetic field to the effect of domain topology, as is observed for the Aharonov-Bohm effect ?

One motivation comes from a 3D model of liquid crystals which involves $\mathbb{S}^2$-valued vector fields with *constant direction*, thereby leading to examples where the magnetic potential $\mathbf{A}$ and the magnetic field curl $\mathbf{A}$ have singularities. The typical situation of a cylindrical container leads to a magnetic potential[1] with constant direction and unit length; if the potential is non-uniform, then it has to be singular at certain points. In other occasions, one encounters a possibly discontinuous magnetic field but living in the Sobolev space $H^1(\Omega,\mathbb{R}^2)$.

The present chapter addresses the effect of various examples of non-smooth magnetic fields on the lowest eigenvalue of the magnetic Laplacian, starting with the analysis of the magnetic Sobolev space, followed by presenting strong field asymptotics of the lowest eigenvalue, and finishing with some open questions.

The chapter is organized as follows. In Sect. 2, we recall the investigation of the magnetic Sobolev space introduced in (3); in particular, we signal out situations where it reduces to the Sobolev space $H^1(\Omega,\mathbb{C})$.

In Sect. 3, we revisit a classical lower bound for the quadratic form in (2), and generalize it to other situations.

[1] In that context this is called a *director field* [13].

In Sect. 4, we inspect magnetic potentials with singularities, but with a unit length. We derive asymptotics for the lowest eigenvalue both for the Dirichlet and Neumann realizations. In the presence of various singularities, we study the problem of minimizing their cost.

In Sect. 5, we revisit the asymptotics of the lowest eigenvalue for the Dirichlet realization, when the magnetic field is in the Sobolev space $H^1(\Omega, \mathbb{R})$. We conclude by discussing the case of the Neumann realization, the case of square integrable magnetic fields, and other questions related to the Aharonov-Bohm magnetic potential.

2 Magnetic Sobolev Space

The definition of $H^1_{\sigma\mathbf{A}}(\Omega, \mathbb{C})$ in (3) makes sense when, for example, $\mathbf{A} \in L^2(\Omega, \mathbb{R}^2)$. In fact, for $u \in L^2(\Omega, \mathbb{C})$, Hölder's inequality yields that $\mathbf{A}u \in L^1(\Omega, \mathbb{C}^2)$ and so we can define the distribution $(\nabla - i\sigma\mathbf{A})u \in \mathcal{D}'(\Omega, \mathbb{C}^2)$. One has to be more careful when introducing $H^1_{\sigma\mathbf{A}}(\Omega, \mathbb{C})$ in the case where $\mathbf{A} \in L^p(\Omega, \mathbb{R}^2)$ and $1 \leq p < 2$ (see [15, Sec. 2]).

2.1 The Diamagnetic Inequality

If $\mathbf{A} \in L^2(\Omega, \mathbb{R}^2)$ and $u \in H^1_{\sigma\mathbf{A}}(\Omega, \mathbb{C})$, we have the following point-wise inequality [8, Thm. 2.1.1]

$$|\nabla|u|| \leq |(\nabla - i\sigma\mathbf{A})u| \text{ a.e. in } \Omega\,. \tag{6}$$

As a consequence of it, we observe that $H^1_{\sigma\mathbf{A}}(\Omega, \mathbb{C}) \subset L^r(\Omega, \mathbb{C})$ for all $r \in [1, +\infty)$. In fact, if $u \in H^1_{\sigma\mathbf{A}}(\Omega, \mathbb{C})$, then (6) yields that $|u| \in H^1(\Omega)$ where $H^1(\Omega)$ embeds in $L^r(\Omega)$ for all $r \in [1, +\infty)$, by the Sobolev embedding theorem in $\mathbb{R}^2$.

If moreover we have that $\mathbf{A} \in L^q(\Omega, \mathbb{R}^2)$, for some $q > 2$, then we get that $\mathbf{A}u \in L^2(\Omega, \mathbb{C}^2)$ by Hölder's inequality and then

$$\nabla u = (\nabla - i\mathbf{A})u + \mathbf{A}u \in L^2(\Omega, \mathbb{C}^2)\,.$$

In this case the magnetic and non-magnetic Sobolev spaces coincide, i.e.

$$H^1_{\sigma\mathbf{A}}(\Omega, \mathbb{C}) = H^1(\Omega, \mathbb{C})\,.$$

Note that, by Sobolev embedding, we get $\mathbf{A} \in L^q(\Omega, \mathbb{R}^2)$ for some $q > 2$, if we know that[2] $\mathbf{A} \in W^{1,p}(\Omega, \mathbb{R}^2)$ for some $p \in (1, 2]$.

2.2 *Examples*

The magnetic potential defined for $x = (x_1, x_2) \in \mathbb{R}^2$ by

$$\mathbf{A}(x) = \Big(-\frac{x_2}{|x|}, \frac{x_1}{|x|}\Big)$$

is singular at 0. However, it belongs to $L^\infty(\mathbb{R}^2, \mathbb{R}^2)$, since $|\mathbf{A}(x)| = 1$ for $x \neq 0$. Consequently, $H^1_{\sigma\mathbf{A}}(\Omega, \mathbb{C}) = H^1(\Omega, \mathbb{C})$.

Let us discuss the example of an Aharonov-Bohm magnetic potential which has a more complex singularity at 0. Assume that $0 \in \Omega$ and consider the magnetic potential defined for $x = (x_1, x_2) \in \mathbb{R}^2$ by

$$\mathbf{A}(x) = \Big(\frac{-x_2}{2\pi|x|^2}, \frac{x_1}{2\pi|x|^2}\Big).$$

Note that $\mathbf{A} \in L^p(\Omega, \mathbb{R}^2)$ for all $1 \le p < 2$. However, $\mathbf{A} \in L^2(U, \mathbb{R}^2)$ when U is relatively compact in $\Omega^* := \Omega \setminus \{0\}$. Thus, given $u \in L^2(\Omega, \mathbb{C})$, $\mathbf{A}u$ defines a distribution on Ω^* (not on Ω) and so does $f(u, \sigma\mathbf{A}) := (\nabla - i\sigma\mathbf{A})u \in \mathcal{D}'(\Omega^*; \mathbb{C}^2)$. The space $H^1_{\sigma\mathbf{A}}(\Omega, \mathbb{C})$ is initially defined as follows

$$H^1_{\sigma\mathbf{A}}(\Omega, \mathbb{C}) = \{u \in L^2(\Omega, \mathbb{C}) \;:\; f(u, \sigma\mathbf{A}) \in L^2(\Omega^*, \mathbb{C}^2)\}.$$

Comparing with the definition in (3), we observe that $H^1_{\sigma\mathbf{A}}(\Omega, \mathbb{C}) = H^1_{\sigma\mathbf{A}}(\Omega^*, \mathbb{C})$. Now, if $u \in H^1_{\sigma\mathbf{A}}(\Omega^*, \mathbb{C})$, the diamagnetic inequality yields that $u \in L^p(\Omega^*, \mathbb{C}) = L^p(\Omega, \mathbb{C})$ and hence $\mathbf{A}u \in L^1(\Omega, \mathbb{C}^2) \subset \mathcal{D}'(\Omega, \mathbb{C}^2)$. Consequently, $(\nabla - i\sigma\mathbf{A})u$ becomes a distribution on Ω and $H^1_{\sigma\mathbf{A}}(\Omega, \mathbb{C})$ can be expressed in the usual form as in (3).

3 Lower Bound on the Quadratic Form

It is quite pleasant to have a lower bound on the quadratic form $q_{\sigma\mathbf{A}}(u)$, introduced in (2), holding for a wide class of functions u and vector fields $\mathbf{A}$. We will recall such a useful bound which follows by a tricky computation, but also requires some conditions on $\mathbf{A}$ and u.

[2] The Soboelv space $W^{1,p}(\Omega, \mathbb{R}^2)$ consists of functions in $L^p(\Omega, \mathbb{R}^2)$ with gradient in $L^p(\Omega, \mathbb{R}^2)$.

Proposition 1 *Assume that* $u \in C_c^\infty(\mathbb{R}^2, \mathbb{C})$ *and* $\mathbf{A} \in C^\infty(\mathbb{R}^2, \mathbb{R}^2)$. *Then, for all* $\sigma > 0$,

$$\int_{\mathbb{R}^2} |(\nabla - i\sigma\mathbf{A})u|^2 dx \geq \sigma \int_{\mathbb{R}^2} B(x)|u(x)|^2\, dx\,,$$

where $B := \mathit{curl}\,\mathbf{A}$.

The proof of Proposition 1 follows from the following tricky formula [8, Lem. 1.4.1]:

$$(\partial_{x_2} - i\sigma A_2)(\partial_{x_1} - i\sigma A_1)u - (\partial_{x_1} - i\sigma A_1)(\partial_{x_2} - i\sigma A_2)u = i\sigma Bu$$

which yields, after taking the inner product with u and integrating by parts[3]

$$-2i\mathrm{Re}\,\langle(\partial_{x_1} - i\sigma A_1)u, (\partial_{x_2} - i\sigma A_2)u\rangle_{L^2(\Omega,\mathbb{C})} = \sigma\langle Bu, u\rangle_{L^2(\Omega,\mathbb{C})}\,.$$

The inequality in Proposition 1 now becomes a consequence of Hölder's inequality.

By a standard density argument, Proposition 1 can be generalized as follows.

Proposition 2 *Assume that* $u \in H_0^1(\Omega, \mathbb{C})$ *and* $\mathbf{A} \in W^{1,p}(\Omega, \mathbb{C}^2)$ *with* $1 < p < +\infty$. *Then for all* $\sigma > 0$,

$$q_{\sigma\mathbf{A}}(u) \geq \sigma \int_\Omega B(x)|u(x)|^2\, dx\,,$$

where $B := \mathit{curl}\,\mathbf{A}$ *and* $q_{\sigma\mathbf{A}}(u)$ *is introduced in* (2).

Proof Consider sequences $(\mathbf{A}_n)_{n\geq1} \subset C^\infty(\Omega, \mathbb{R}^2)$ and $(u_n)_{n\geq1} \subset C_c^\infty(\Omega, \mathbb{C}^2)$ such that

$$\|\mathbf{A}_n - \mathbf{A}\|_{W^{1,p}(\Omega,\mathbb{R}^2)} \to 0 \quad \text{and} \quad \|u_n - u\|_{H^1(\Omega,\mathbb{C})} \to 0\,.$$

Using the Sobolev embeddings of $W^{1,p}(\Omega)$ and $H^1(\Omega)$ in $L^q(\Omega)$ and $L^r(\Omega)$ respectively, where $q = \frac{2p}{2-p} > 2$ and $r \in [1, +\infty)$, we get by Hölder's inequality,

$$\lim_{n\to+\infty} \|(\mathbf{A}_n - \mathbf{A})u\|_{L^2(\Omega,\mathbb{C}^2)} = 0 \quad \text{and} \quad \lim_{n\to+\infty} \int_\Omega (\mathrm{curl}\,\mathbf{A}_n - \mathrm{curl}\,\mathbf{A})|u(x)|^2 dx = 0.$$

To finish the proof, we apply Proposition 1 on $\mathbf{A}_n$ and u_n then we take the limit $n \to +\infty$.

[3] The support of the function u is compact, hence contained in some disk D. The boundary term resulting from the integration by parts vanishes since u vanishes on the boundary of D.

Remark 1 If u does not vanish on the boundary then the lower bound in Proposition 2 fails in general. In fact, if $B(x) = 1$ then one has [12]

$$q_{\sigma\mathbf{A}}(u) \geq \Theta_0\sigma\big(1+\varepsilon(\sigma)\big)\int_\Omega |u(x)|^2\,dx$$

where $\lim\limits_{\sigma\to+\infty}\varepsilon(\sigma) = 0$ and $\Theta_0 \in (\frac{1}{2}, 1)$ is a constant that will be introduced in (14). The foregoing lower bound is optimal in the limit of large σ.

Remark 2 A particular example of a vector field satisfying the hypothesis in Proposition 2 is

$$\mathbf{F}(x) = \frac{x^\perp}{|x|}, \tag{7}$$

where $x = (x_1, x_2) \in \mathbb{R}^2$ and $x^\perp = (-x_2, x_1)$. In fact, $\mathbf{F} \in W^{1,p}(\Omega, \mathbb{R}^2)$ for all $p \in [1, 2)$ and the magnetic field generated by $\mathbf{F}$ is

$$\operatorname{curl}\mathbf{F}(x) = \frac{1}{|x|}. \tag{8}$$

4 Magnetic Potentials in Liquid Crystals

Generally speaking, a vector field may have various types of singularity, for instance, point, line or higher dimensional singularity. The typical point singularity at a point x^0 is of the form

$$\mathbf{A} = Q\frac{x - x^0}{|x - x^0|}, \tag{9}$$

where Q is an orthogonal matrix-valued function. Such a magnetic potential is of particular interest in the theory of liquid crystals because it obeys the constraint $|\mathbf{A}| = 1$ and it arises naturally as a local approximation of the isolated dislocation of the director field (see [17]). A point singularity in 2 dimensions can be viewed as a straight line singularity in cylindrical domains. In general, line singularities could be much more complicated.

Molecules in liquid crystals obey two types of order, an axial order in the nematic phase and a layering structure in the smectic phase. Nematic/smectic phase transition occurs at a threshold temperature. Smectics are generally observed at lower temperatures and their nucleation is related to the lowest eigenvalue of the operator $-\nabla^2_{q\mathbf{n}}$, where $\mathbf{n}$ is the minimizer of the Oseen-Frank nematic energy [2, 13, 18]. In the one constant approximation of the Oseen-Frank energy in 3 dimensions, under the smooth boundary datum, the minimizer $\mathbf{n}$ is smooth inside the domain except for a finite number of singular points, and near each singular point x^0, $\mathbf{n}$ approaches $\mathbf{A}$, where $\mathbf{A}$ is the field in (9) for some orthogonal matrix Q (see [5, 17, 22]).

4.1 Singularity at One Point

We will focus on the magnetic potential

$$\mathbf{F}^Q(x) = Q\frac{x}{|x|} \tag{10}$$

where Q is a 2×2 orthogonal matrix. The matrix Q will be either a reflection (in which case $Q = \begin{pmatrix} p & -q \\ q & p \end{pmatrix}$ with $p^2 + q^2 = 1$) or a rotation (in which case $Q = \begin{pmatrix} p & q \\ q & -p \end{pmatrix}$ with $p^2 + q^2 = 1$).

In the reflection case, we see that

$$\operatorname{curl} \mathbf{F}^Q = \frac{q}{|x|} \tag{11}$$

is non-vanishing (for $(p, q) \neq (\pm 1, 0)$). While, in the rotation case, we see that

$$\operatorname{curl} \mathbf{F}^Q = \frac{2px_1x_2 - q(x_1^2 - x_2^2)}{|x|^3} \tag{12}$$

vanishes on the set

$$\mathfrak{z} = \{2px_1x_2 - q(x_1^2 - x_2^2) = 0\}.$$

Writing $(p, q) = (\cos\alpha, \sin\alpha)$ and $(x_1, x_2) = (r\cos\theta, r\sin\theta)$, we can express the zero set $\mathfrak{z}$ in polar coordinates as follows, $\mathfrak{z} = \{\sin(2\theta - \alpha) = 0\}$. Hence, $\mathfrak{z}$ consists of two perpendicular straight lines intersecting at the singular point of $\mathbf{F}^Q$, $x = 0$ (these straight lines are $\{\theta = \frac{\alpha}{2}\}$ and $\{\theta = \frac{\pi}{2} + \frac{\alpha}{2}\}$).

The behavior associated with non-vanishing magnetic fields is completely different from the one with vanishing magnetic fields, see for instance [1, 4, 7, 11, 12, 20]. So we have to deal separately with the cases of Q being a reflection (non-vanishing magnetic field) and a rotation (vanishing magnetic field). However, the present text focuses on the case where Q is a reflection and excludes the rotation case.

Given (11) and the property of gauge invariance,[4] we can assume without loss of generality, that $q = 1$. In this way we are led to consider the following magnetic potential $\mathbf{F}$ introduced in (7). For $a \in \Omega$, we consider

$$\mathbf{A}(x) = \mathbf{F}(x-a) \quad \text{and} \quad B(x) = \operatorname{curl}\mathbf{A}(x) = \frac{1}{|x-a|}. \tag{13}$$

Theorem 1 *Let the magnetic potential* $\mathbf{A}$ *and the magnetic field* B *be as in* (13). *The lowest eigenvalues introduced in* (5) *satisfy, as* $\sigma \to +\infty$,

$$\lambda^N(\sigma\mathbf{A}) = \sigma \min\Big(m(B,\Omega), \Theta_0 m(B,\partial\Omega)\Big) + o(\sigma),$$

$$\lambda^D(\sigma\mathbf{A}) = \sigma m(B,\Omega) + o(\sigma),$$

where

$$m(B,\Omega) = \inf_{x\in\Omega} |B(x)| \quad \text{and} \quad m(B,\partial\Omega) = \inf_{x\in\partial\Omega} |B(x)|.$$

The constant Θ_0 appearing in Theorem 1 is universal and satisfies $\frac{1}{2} < \Theta_0 < 1$. It can be introduced as follows (see [12] and [8, Sec. 3.2 & 4.3])

$$\Theta_0 = \inf_{\substack{u\in H^1_{\mathbf{A}_0}(\mathbb{R}\times\mathbb{R}_+,\mathbb{C})\\ u\neq 0}} \frac{\|(\nabla - i\mathbf{A}_0)u\|^2_{L^2(\mathbb{R}\times\mathbb{R}_+,\mathbb{C}^2)}}{\|u\|^2_{L^2(\mathbb{R}\times\mathbb{R}_+,\mathbb{C})}}, \tag{14}$$

where

$$\mathbf{A}_0(x) = \frac{1}{2}(-x_2, x_1) \text{ for } x = (x_1, x_2) \in \mathbb{R}^2. \tag{15}$$

Note that Theorem 1 is consistent with the known results where the magnetic field $B = \operatorname{curl}\mathbf{A}$ is continuous [8, Thm. 8.1.1]. The proof of Theorem 1 does not differ from the case where B is continuous, except that we have to be a bit careful near the singularity point a. As we shall see, the presence of the singularity will increase the energy much above $m(B,\Omega)$, by Proposition 2.

Proof *(of Theorem 1)*
Lower Bound
Let us start with the Neumann realization. Let $\varepsilon \in (0,1)$ be a constant such that

$$\frac{1}{\varepsilon} > m(B,\Omega) \quad \text{and} \quad \overline{D(a,2\varepsilon)} \subset \Omega, \tag{16}$$

where $D(a,r) := \{x \in \mathbb{R}^2 \,:\, |x-a| < r\}$ is the open disc of center a and radius r.

[4] The lowest eigenvalues with magnetic potentials $\mathbf{A}$ and $\mathbf{A}' := \mathbf{A} - \nabla\chi$ are equal, by the unitary transformation $(u,\mathbf{A}) \mapsto (u' := e^{i\chi}u, \mathbf{A}' = \mathbf{A} - \nabla\chi)$, for any function $\chi \in H^1(\Omega)$.

Let $u \in H^1(\Omega, \mathbb{C})$ and $u \not\equiv 0$, so that $\|u\|_{L^2(\Omega,\mathbb{C})} > 0$. We will write estimates of the quadratic form $q_{\sigma\mathbf{A}}(u)$ (see (2) and (13)) that hold uniformly with respect to $u \in H^1(\Omega, \mathbb{C})$, ε and $\sigma \in \mathcal{N}_\infty$, where $\mathcal{N}_\infty$ is some neighborhood of $+\infty$.

We define the two open sets

$$\Omega_{1,\varepsilon} = \{x \in \Omega \, : \, |x-a| > \varepsilon\} \quad \text{and} \quad \Omega_{2,\varepsilon} = \{x \in \Omega \, : \, |x-a| < 2\varepsilon\}.$$

Then we consider the partition of unity

$$\varphi_{1,\varepsilon}^2 + \varphi_{2,\varepsilon}^2 = 1,$$

where, for $i \in \{1, 2\}$,

$$\mathrm{supp}\varphi_{i,\varepsilon} \subset \Omega_{i,\varepsilon},$$

and

$$|\nabla\varphi_{i,\varepsilon}| \leq C\varepsilon^{-1}.$$

Let

$$u_{i,\varepsilon} = \varphi_{i,\varepsilon} u \quad \text{for } i \in \{1, 2\}.$$

Then we have the following decomposition of the quadratic form (the first identity below is the celebrated IMS formula):

$$\begin{aligned} q_{\sigma\mathbf{A}}(u) &= q_{\sigma\mathbf{A}}(u_{1,\varepsilon}) + q_{\sigma\mathbf{A}}(u_{2,\varepsilon}) - \sum_{i=1}^{2} \int_\Omega |\nabla\varphi_{i,\varepsilon}|^2 |u|^2 \, dx \\ &\geq q_{\sigma\mathbf{A}}(u_{1,\varepsilon}) + q_{\sigma\mathbf{A}}(u_{2,\varepsilon}) - \frac{C}{\varepsilon^2} \int_\Omega |u|^2 \, dx . \end{aligned}$$

Since $u_{2,\varepsilon}$ is supported in $\Omega_{2,\varepsilon} \subset \{|x-a| < 2\varepsilon\}$, we get by (13) and Proposition 2,

$$q_{\sigma\mathbf{A}}(u_{2,\varepsilon}) \geq \sigma \int_{\Omega_{2,\varepsilon}} \frac{1}{|x-a|} |u_{2,\varepsilon}|^2 \, dx \geq \frac{\sigma}{\varepsilon} \|u_{2,\varepsilon}\|_{L^2(\Omega,\mathbb{C})}^2 . \tag{17}$$

Since $\mathbf{A}$ is smooth on $\Omega_{1,\varepsilon}$ and $u_{1,\varepsilon} = 0$ on $\partial D(a, \varepsilon)$, then (see the argument in [8, Sec. 8.2.2])

$$q_{\sigma\mathbf{A}}(u_{1,\varepsilon}) = \int_{\Omega_{1,\varepsilon}} |(\nabla - i\sigma\mathbf{A})u_{1,\varepsilon}|^2 \, dx \geq \left(\sigma\mathfrak{d}_\varepsilon - C_\varepsilon\sigma^{3/4}\right) \int_{\Omega_{1,\varepsilon}} |u_{2,\varepsilon}|^2 \, dx , \tag{18}$$

where $C_\varepsilon > 0$ is a constant independent of σ and

$$\mathfrak{d}_\varepsilon = \min\Big(\inf_{x\in\Omega_{1,\varepsilon}} |B(x)|\,,\ \Theta_0 \inf_{x\in\partial\Omega} |B(x)|\Big) \geq \min\big(m(B,\Omega), \Theta_0 m(B,\partial\Omega)\big)\,.$$

Hence, collecting (18), (17) and (16), we obtain the lower bound

$$q_{\sigma\mathbf{A}}(u) \geq \Big(\sigma \min\big(m(B,\Omega), \Theta_0 m(B,\partial\Omega)\big) - C_\varepsilon \sigma^{3/4} - \frac{C}{\varepsilon^2}\Big)\int_\Omega |u|^2\,dx\,.$$

The min-max principle then yields (recall that ε is fixed by (16)),

$$\lambda^N(\sigma\mathbf{A}) \geq \sigma \min\big(m(B,\Omega), \Theta_0 m(B,\partial\Omega)\big) + O(\sigma^{3/4})\,.$$

For the Dirichlet case, we should consider the boundary condition $u = 0$ on $\partial\Omega$, and use the lower bound in Proposition 2 which yields

$$q_{\sigma\mathbf{A}}(u) \geq \int_\Omega B(x)|u(x)|^2 dx \geq m(B,\Omega)\|u\|^2_{L^2(\Omega,\mathbb{C})}\,.$$

Now we conclude by the min-max principle the non-asymptotic lower bound,

$$\lambda^D(\sigma\mathbf{A}) \geq \sigma m(B,\Omega)\,.$$

Upper Bound

Step 1. ***The Dirichlet realization***

Fix an arbitrary point $x_0 \in \Omega \setminus \{a\}$. There exists $\varepsilon_0 > 0$ such that $D(x_0,\varepsilon_0) \subset \Omega \setminus \{a\}$. Define the function

$$u_{x_0,\sigma}(x) = \pi^{-1/2}\sqrt{\sigma B(x_0)}\,\chi(x)\exp\Big(-\frac{1}{2}\sigma B(x_0)\,|x-x_0|^2\Big)\,, \tag{19}$$

where $\chi \in C_c^\infty(\Omega)$ satisfies that $\chi = 1$ in $D(x_0,\varepsilon_0/2)$, $\mathrm{supp}\chi \subset D(x_0,\varepsilon_0)$, and $0 \leq \chi \leq 1$ in Ω. Since $\mathbf{A}$ is smooth on $D(x_0,\varepsilon_0)$, we can expand it by Taylor's formula to order 2 near x_0 and get a function $\varphi_0 \in C^\infty\big(\overline{D(x_0,\varepsilon_0)}\big)$ such that (see [8, p. 11])

$$\mathbf{A}(x) - \nabla\varphi_0 = \mathbf{A}^{\mathrm{lin}}(x-x_0) + O(|x-x_0|^2)\,, \tag{20}$$

where

$$\mathbf{A}^{\mathrm{lin}}(x) = \frac{1}{2}B(x_0)\big(-x_2, x_1\big)\,.$$

Now we set

$$w_{x_0,b}(x) = e^{i\sigma\varphi_0(x)}u_{x_0,\sigma}(x).$$

Then we can check easily that (see [8, Sec. 1.4.2 & Eq. (1.36)])

$$q_{\sigma\mathbf{A}}(u_{x_0,\sigma}) \le \Big(\sigma B(x_0) + O(\sigma^{1/2})\Big)\|u_{x_0,\sigma}\|^2_{L^2(\Omega,\mathbb{C})}.$$

The min-max principle yields

$$\lambda^D(\sigma\mathbf{A}) \le \sigma B(x_0) + O(\sigma^{1/2}).$$

Dividing by σ, taking $\limsup_{\sigma\to+\infty}$ then minimizing over $x_0 \in \Omega \setminus \{a\}$, we get

$$\lambda^D(\sigma\mathbf{A}) \le \sigma\, m(B,\Omega) + o(\sigma).$$

Step 2. ***The Neuamnn realization***
Since the Dirichlet form domain is contained in that of Neumann, we infer from (5),

$$\lambda^N(\sigma\mathbf{A}) \le \lambda^D(\sigma\mathbf{A}).$$

So, in light of Step 1, it is enough to prove the upper bound

$$\lambda^N(\sigma\mathbf{A}) \le \sigma\Theta_0 m(B,\partial\Omega) + o(\sigma).$$

Choose $x_0 \in \partial\Omega$ such that

$$B(x_0) = \inf_{x\in\partial\Omega} B(x).$$

Again, we can select a smooth function φ_0 such that, in a (boundary) neighborhood[5] V_{x_0} of x_0,

$$\mathbf{A}(x) = \mathbf{A}^{\mathrm{lin}}(x-x_0) + \nabla\varphi_0 + \mathbf{r}(x) \quad \text{where } \mathbf{r}(x) = O(|x-x_0|^2).$$

Then, for a function u supported in V_{x_0},

$$\begin{aligned} q_{\sigma\mathbf{A}}(e^{i\varphi_0}u) = \int_{V_{x_0}} |(\nabla - i\sigma\mathbf{A}^{\mathrm{lin}}(x-x_0))u|^2\,dx + \int_{V_{x_0}} |\mathbf{r}u|^2\,dx \\ + \langle(\nabla - i\sigma\mathbf{A}^{\mathrm{lin}}(x-x_0))u, \mathbf{r}u\rangle_{L^2(V_{x_0})}. \end{aligned}$$

[5] We can take the boundary neighborhood of the form $V_{x_0} = D(x_0, r) \cap \Omega$.

But, since $\partial\Omega$ is smooth, we can select the function u such that (see [8, Sec. 8.2.1])

$$\int_{V_{x_0}} |(\nabla - i\sigma \mathbf{A}^{\mathrm{lin}}(x - x_0))u|^2\,dx \le \Big(\Theta_0 \sigma B(x_0) + o(\sigma)\Big) \int_\Omega |u|^2\,dx$$

$$\int_{V_{x_0}} |x - x_0|^4 |u|^2\,dx = o(\sigma) \int_{V_{x_0}} |u|^2\,dx\,.$$

The min-max principle then yields the desired upper bound on the eigenvalue $\lambda^N(\sigma\mathbf{A})$.

Remark 3 The proof of Theorem 1 yields a lower bound on the quadratic form which allows us to extract localization properties of ground states via Agmon estimates (see [8, Sec. 7.2.6]). The ground states of $\lambda^D(\sigma\mathbf{A})$ are localized near the set $C_0 := \{x \in \Omega \,:\, |B(x)| = m(B,\Omega)\}$. As for the ground states of $\lambda^N(\sigma\mathbf{A})$, they are localized near C_0 if $m(B,\Omega) < \Theta_0 m(B,\partial\Omega)$; if $\Theta_0 m(B,\partial\Omega) < m(B,\Omega)$, then the localization occurs near the set $C_1 := \{x \in \partial\Omega \,:\, |B(x)| = m(B,\partial\Omega)\}$ (see [8, Sec. 8.2.3]).

Remark 4 The upper bounds

$$\lambda^D(\sigma\mathbf{A}) \le \sigma\, m(B,\Omega) + o(\sigma)$$

$$\lambda^N(\sigma\mathbf{A}) \le \sigma \min\Big(m(B,\Omega), \Theta_0 m(B,\partial\Omega)\Big) + o(\sigma)$$

continue to hold for any magnetic potential $\mathbf{A}$ generating a magnetic field $B = \operatorname{curl}\mathbf{A}$ that satisfies[6] $B \in C(\overline{\Omega} \setminus \{a_1, \cdots, a_N\})$ and $B > 0$ on $\overline{\Omega}$, where $a_1, \cdots, a_N \in \Omega$.

4.2 *Finitely Many Singular Points*

We will introduce a magnetic potential with finitely many singular points in the domain Ω as superposition of the potentials with one singular point. Consider an integer $N \ge 1$, real numbers $m_1, \cdots, m_N \in (0, 1]$, and distinct points $a_1, \cdots, a_N$ in Ω. We put $\mathsf{m} = (m_1, \cdots, m_N)$ and $\mathsf{a} = (a_1, \cdots, a_N)$. Define the magnetic potential with N interior singular points

$$\mathbf{A}_{N,\mathsf{m},\mathsf{a}}(x) = \frac{1}{N}\sum_{i=1}^{N} m_i \mathbf{F}(x - a_i)\,, \tag{21}$$

[6] The only difference is that the remainder $O(|x - x_0|^2)$ in (20) will be replaced by $o(|x - x_0|)$.

which generates the following magnetic field

$$B_{N,\mathsf{m},\mathsf{a}}(x) = \operatorname{curl} \mathbf{A}_{N,\mathsf{m},\mathsf{a}}(x) = \frac{1}{N}\sum_{i=1}^{N}\frac{m_i}{|x-a_i|}\,. \tag{22}$$

The proof of Theorem 1 can be easily adapted when we take $\mathbf{A} = \mathbf{A}_{N,\mathsf{m},\mathsf{a}}$.

Theorem 2 *Let $B = curl\,\mathbf{A}$ and $\mathbf{A} = \mathbf{A}_{N,\mathsf{m},\mathsf{a}}$ be as in (21). The lowest eigenvalues introduced in (5) satisfy, as $\sigma \to +\infty$,*

$$\lambda^N(\sigma\mathbf{A}) = \sigma \min\Big(m(B,\Omega), \Theta_0 m(B,\partial\Omega)\Big) + o(\sigma)\,,$$

$$\lambda^D(\sigma\mathbf{A}) = \sigma m(B,\Omega) + o(\sigma)\,,$$

where

$$m(B,\Omega) = \inf_{x\in\Omega}|B(x)| \quad \text{and} \quad m(B,\partial\Omega) = \inf_{x\in\partial\Omega}|B(x)|\,.$$

It is natural to address the following

Question For a given N, what are the locations of the singularities $a_1, \cdots, a_N$ so that the ground state energies $\lambda^N(\sigma\mathbf{A}_{N,\mathsf{m},\mathsf{a}})$ and $\lambda^D(\sigma\mathbf{A}_{N,\mathsf{m},\mathsf{a}})$ are minimal ?

In light of Theorem 2, we opt to minimize the limit quantities

$$\begin{aligned} \mathfrak{e}(N,\mathrm{m},\mathrm{a},\Omega) &:= m(B_{N,\mathsf{m},\mathsf{a}},\Omega) \text{ and} \\ \mathfrak{e}'(N,\mathrm{m},\mathrm{a},\Omega) &:= \min\Big(m(B_{N,\mathsf{m},\mathsf{a}},\Omega), \Theta_0 m(B_{N,\mathsf{m},\mathsf{a}},\partial\Omega)\Big)\,. \end{aligned} \tag{23}$$

To that end we introduce

$$\mathfrak{e}_*(N,\mathrm{m},\Omega) = \inf_{\mathrm{a}\in\mathsf{S}_\Omega}\mathfrak{e}(N,\mathrm{m},\mathrm{a},\Omega) \quad \text{and} \quad \mathfrak{e}'_*(N,\mathrm{m},\Omega) = \inf_{\mathrm{a}\in\mathsf{S}_\Omega}\mathfrak{e}'(N,\mathrm{m},\mathrm{a},\Omega) \tag{24}$$

where

$$\mathsf{S}_\Omega = \{(a_1,\cdots,a_N)\in\overline{\Omega}^N \ :\ a_i\neq a_j \text{ for } i\neq j\}\,, \tag{25}$$

and introduce the sets of minimal points

$$\begin{aligned} \mathcal{M}_*(N,\mathsf{m},\Omega) &= \{\mathrm{a}\in\overline{\Omega}^N \ :\ \mathfrak{e}(N,\mathrm{m},\mathrm{a},\Omega) = \mathfrak{e}_*(N,\mathrm{m},\Omega)\} \quad \text{and} \\ \mathcal{M}'_*(N,\mathsf{m},\Omega) &= \{\mathrm{a}\in\overline{\Omega}^N \ :\ \mathfrak{e}'(N,\mathrm{m},\mathrm{a},\Omega) = \mathfrak{e}'_*(N,\mathrm{m},\Omega)\}\,, \end{aligned} \tag{26}$$

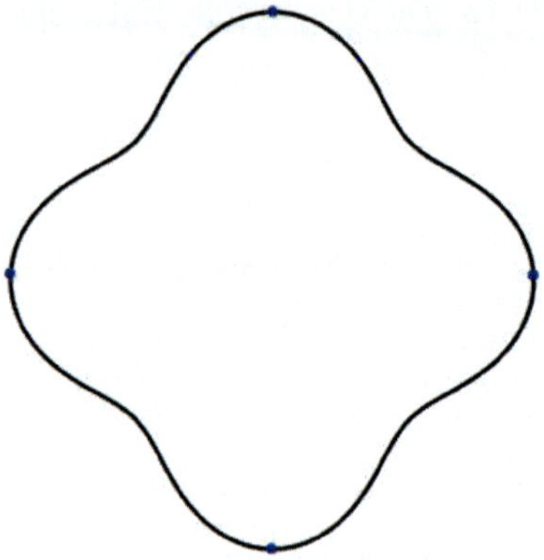

Fig. 1 A domain with several 'poles'

which describe the location of the minimum points. We will find, in Proposition 3, the values of $\mathfrak{e}_*(N, \mathrm{m}, \Omega)$ and $\mathfrak{e}'_*(N, \mathrm{m}, \Omega)$, and the location of the singular points, where the infimum in (24) is achieved, is on the boundary, at the 'pole(s)' of the domain (see Fig. 1). The precise result is

Proposition 3 *For all* $N \geq 1$ *and* $\mathsf{m} = (m_1, \cdots, m_N) \in (0, \infty)^N$, *we have*

$$\mathfrak{e}_*(N, \mathrm{m}, \Omega) = \frac{1}{N\,\mathrm{diam}(\Omega)} \sum_{i=1}^{N} m_i \quad \text{and} \quad \mathfrak{e}'_*(N, \mathrm{m}, \Omega) = \frac{\Theta_0}{N\,\mathrm{diam}(\Omega)} \sum_{i=1}^{N} m_i\,, \tag{27}$$

and

$$\begin{aligned}&\mathcal{M}_*(N, \mathsf{m}, \Omega) = \mathcal{M}'_*(N, \mathsf{m}, \Omega)\\ =&\{(a_1, \cdots, a_N) \in (\partial\Omega)^N \ : \ \exists\, y \in \partial\Omega,\ \forall\, i \in \{1, \cdots, N\},\ |a_i - y| = \mathrm{diam}(\Omega)\}\,,\end{aligned}$$

where $\mathrm{diam}(\Omega)$ *denotes the diameter of* Ω.

We may call the elements of the set $\mathcal{M}_*(N, \mathsf{m}, \Omega)$ the 'poles' of the domain Ω. We emphasize that

$$\mathcal{M}_*(N, \mathsf{m}, \Omega) \subset (\partial\Omega)^N.$$

Let us look at some particular examples:

- When $\Omega = D(0, 1)$ is the unit disk,

$$\mathcal{M}_*(N, \mathsf{m}, \partial\Omega) = \partial\Omega \times \cdots \times \partial\Omega \quad (N\text{-folds}).$$

- For the ellipse case, $\Omega = \{(x, y) : \frac{x^2}{c^2} + \frac{y^2}{d^2} < 1\}$,

$$\mathcal{M}_*(N, \mathsf{m}, \Omega) = \{-P, P\}^N,$$

supposing that $c > d > 0$, where $P = (c, c)$ and $-P = (-c, -c)$.
- For other domains, the set $\mathcal{M}_*(N, \mathsf{m}, \Omega)$ may have a more complex form, in particular, the symmetric domain displayed in Fig. 1, which has four 'poles'.

Proof *(of Proposition 3)*

Choose $a, b \in \partial\Omega$ such that $\mathrm{dist}(a, b) = \mathrm{diam}(\Omega)$. For every $\varepsilon \in (0, \frac{1}{2N}\mathrm{diam}(\Omega)]$ and $i \in \{1, \cdots, N\}$, let

$$a_i = a + i\varepsilon\nu$$

where ν is the *interior* unit normal vector of $\partial\Omega$. If ε is small enough, then all the points a_i are in Ω, $\mathsf{a} := (a_1, \cdots, a_N) \in \mathsf{S}_\Omega$ (see (25)) and

$$\mathfrak{e}_*(N, \mathsf{m}, \Omega) \le \mathfrak{e}_*(N, \mathsf{m}, \mathsf{a}, \Omega) \le B_{N,\mathsf{m},\mathsf{a}}(b)\,,$$

where $\mathfrak{e}_*(N, \mathsf{m}, \mathsf{a}, \Omega)$ is introduced in (24) and $B_{N,\mathsf{m},\mathsf{a}}$ is introduced in (22). By the triangle inequality, $|b - a_i| \ge |b - a| - N\varepsilon > 0$, hence

$$B_{N,\mathsf{m},\mathsf{a}}(b) \le \frac{1}{|b - a| - N\varepsilon}\frac{1}{N}\sum_{i=1}^{N} m_i\,.$$

As $\varepsilon \to 0_+$, we get

$$\mathfrak{e}_*(N, \mathsf{m}, \Omega) \le \frac{1}{|b - a|}\frac{1}{N}\sum_{i=1}^{N} m_i = \frac{1}{\mathrm{diam}(\Omega)}\frac{1}{N}\sum_{i=1}^{N} m_i\,,$$

since $|b - a| = \mathrm{diam}(\Omega)$.

On the other hand, for all $\mathsf{a} = (a_1, \cdots, a_N) \in \mathsf{S}_\Omega$,

$$|x - a_i| \le \mathrm{diam}(\Omega) \quad (x \in \overline{\Omega})\,.$$

In light of (22), we infer

$$\forall\, x \in \overline{\Omega}, \quad |B_{N,\mathsf{m},\mathsf{a}}(x)| \ge \frac{1}{\mathrm{diam}(\Omega)}\frac{1}{N}\sum_{i=1}^{N} m_i\,.$$

Minimizing over $x \in \overline{\Omega}$ then over $\mathsf{a} \in \mathsf{S}_\Omega$, we get

$$\frac{1}{\mathrm{diam}(\Omega)}\frac{1}{N}\sum_{i=1}^{N} m_i \leq \mathfrak{e}_*(N, \mathsf{m}, \Omega)\,.$$

This finishes the proof of the part in Proposition 3 concerning $\mathfrak{e}_*(N, \mathsf{m}, \Omega)$. The statement concerning $\mathfrak{e}'_*(N, \mathsf{m}, \Omega)$ can be handled similarly.

5 Magnetic Field in Sobolev Space

So far, we considered in Sect. 4, a magnetic potential $\mathbf{A}$ generating the magnetic field $B(x) = \frac{1}{|x-a|}$ which is singular at $a \in \Omega$. Other singularities may occur when the magnetic field is in the Sobolev space $H^1(\Omega, \mathbb{R})$. For instance, the magnetic field

$$B(x) = \left(\ln \frac{R}{|x-a|}\right)^p \quad \text{where } 0 < p < 1 \text{ and } R > \mathrm{diam}(\Omega)\,, \tag{28}$$

is singular at a, despite that $B \in H^1(\Omega, \mathbb{R})$, which suggests that the singularity here is slightly less than the one encountered in Sect. 4.

In general, given $B \in H^1(\Omega, \mathbb{R})$, we can extend B to a compactly supported function in $H^1(\mathbb{R}^2, \mathbb{R})$ and introduce the magnetic potential

$$\mathbf{A}(x) = \left(\int_0^1 sB(sx)ds\right)\begin{bmatrix} -x_2 \\ x_1 \end{bmatrix} \tag{29}$$

generating the magnetic field $\mathrm{curl}\,\mathbf{A} = B$, which is not singular in Ω since $\mathbf{A} \in H^1(\Omega, \mathbb{R}^2)$ and after a gauge transformation it yields a potential in $H^2(\Omega, \mathbb{R}^2)$, see [8, Prop. D.2.5].

5.1 The Dirichlet Realization

The presence of a magnetic field in the Sobolev space, $H^1(\Omega, \mathbb{R})$, is related to [8, Open problem 9], [19, Open problem 2.2.9] and [9]. The case when the magnetic field does not not vanish in Ω was studied in [16] for the Dirichlet realization.

Theorem 3 (cf. Theorem 1.2 in [16]) *Assume that* $\mathbf{A}$ *is given by* (29), *where* $B \in H^1(\Omega)$ *satisfies*

$$m_0(B, \Omega) := \inf_{x\in\Omega}\mathrm{ess}\, B(x) > 0\,.$$

Then, the lowest eigenvalue of the Dirichlet Laplacian, introduced in (5), satisfies as $\sigma \to +\infty$,

$$\lambda_1^D(\sigma \mathbf{A}) = \sigma\, m_0(B, \Omega) + o(\sigma)\,.$$

Let us give a sketch of the proof of Theorem 3. First, by Proposition 2, we have the non asymptotic lower bound

$$\lambda_1^D(\sigma \mathbf{A}) \geq \sigma\, m_0(B, \Omega)\,. \tag{30}$$

So the point is to prove the asymptotic upper bound

$$\lambda_1^D(\sigma \mathbf{A}) \leq \sigma\, m_0(B, \Omega) + o(\sigma)\,. \tag{31}$$

This follows if we construct a family of trial states, $(w_\varepsilon)_{\varepsilon\in(0,1]}$, such that, for all $\varepsilon \in (0, 1]$, we have

$$q_{\sigma\mathbf{A}}(w_\varepsilon) \leq \sigma\, (m_0(B, \Omega) + \varepsilon + R_\varepsilon(\sigma))\, \|w_\varepsilon\|^2_{L^2(\Omega,\mathbb{C})},$$

where $q_{\sigma\mathbf{A}}$ is introduced in (2) and $\lim\limits_{\sigma\to+\infty} R_\varepsilon(\sigma) = 0$.

The idea behind the construction of w_ε is to replace $B(x)$ by $f(x, \sigma^{-3/8})$, its average on the disc $D(x, \sigma^{-3/8})$, then to find an appropriate gauge function φ^ε so that, locally, $\mathbf{A} - \nabla\varphi$ is a good approximation of a magnetic potential generating a uniform magnetic field,[7] of strength $f(x_\varepsilon, \sigma^{-3/8})$, for an appropriately chosen point x_ε in Ω.

Consider the open set in $\mathbb{R}^2 \times \mathbb{R}_+$, $\tilde{\Omega} = \{(x, r) \in \Omega \times \mathbb{R}_+ \;:\; D(x, r) \subset \Omega\}$. We introduce the two functions on $\tilde{\Omega}$, defined as follows,

$$f(x, r) = \frac{1}{|D(x, r)|}\int_{D(x,r)} B(z)\, dz \quad \text{and} \quad g(x, r) = \frac{1}{|D(x, r)|}\int_{D(x,r)} |\nabla B(z)|^2 dz\,.$$

By the Lebesgue differentiation theorem, the limits

$$f(x) := \lim_{r\to 0} f(x, r) \quad \text{and} \quad g(x) := \lim_{r\to 0} g(x, r)$$

exist almost everywhere. Moreover

$$B(x) = f(x) \quad \text{and} \quad |\nabla B(x)|^2 = g(x) \quad \text{a.e.}$$

---

7 In general, when localizing in a disc of radius $\sigma^{-\rho}$, one encounters two types of errors, $O(\sigma^{2\rho})$ resulting from the localization cut-off, and $O(\sigma^{3-6\rho})$ resulting from the approximation of the magnetic field; optimizing we get $2\rho = 3 - 6\rho$ and therefore $\rho = 3/8$; see [16, Sec. 5.1 & Eq. (5.4)].

For instance, there exists a set N of (area) measure 0 such that $m_0(B,\Omega) = \inf_{x\in\Omega\setminus N} f(x)$ and, for any $\varepsilon > 0$, we can pick $x_\varepsilon \in \Omega \setminus N$ such that

$$m_0(B,\Omega) \leq \lim_{r\to 0}\left(\frac{1}{|D(x_\varepsilon,r)|}\int_{D(x_\varepsilon,r)} B(z)\,dz\right) = f(x_\varepsilon) \leq m_0(B,\Omega)+\varepsilon \quad (32)$$

and

$$\lim_{r\to 0}\left(\frac{1}{|D(x_\varepsilon,r)|}\int_{D(x_\varepsilon,r)} |\nabla B(z)|^2 dz\right) = g(x_\varepsilon) < +\infty\,.$$

We pick $\sigma_\varepsilon \geq 1$ such that $D(x_\varepsilon, \sigma_\varepsilon^{-3/8}) \subset \Omega$. In the sequel $\sigma \geq \sigma_\varepsilon$. Consider the two vector fields

$$\mathbf{A}^\varepsilon(x) = 2\int_0^1 sB\big(s(x-x_\varepsilon)+x_\varepsilon\big)\mathbf{A}_0(x-x_\varepsilon)ds\,,$$
$$\mathbf{A}_0^\varepsilon(x) = f(x_\varepsilon,\sigma^{-3/8})\mathbf{A}_0(x-x_\varepsilon)\,,$$

where $\mathbf{A}_0$ is the vector field in (15). Notice that, $\operatorname{curl}\mathbf{A}^\varepsilon = B$ and $\operatorname{curl}\mathbf{A}_0^\varepsilon = f(x_\varepsilon,\sigma^{-3/8})$ in $D(x_\varepsilon,\sigma^{-3/8})$. We can find a function φ^ε in $H^1\big(D(x_\varepsilon,\sigma^{-3/8})\big)$ such that

$$\mathbf{A}^\varepsilon(x) - \nabla\varphi^\varepsilon = \mathbf{A} \text{ on } D(x_\varepsilon,\sigma^{-3/8})\,.$$

Moreover, we have the following inequality (cf. [16, Thm. 1.1])

$$\int_{D(x_\varepsilon,\sigma^{-3/8})} |\mathbf{A}^\varepsilon(x) - \mathbf{A}_0^\varepsilon(x)|^2 dx \leq 8\pi\sigma^{-9/4} g(x_\varepsilon,\sigma^{-3/8})\,. \quad (33)$$

We introduce the trial state

$$w_\varepsilon(x) = e^{-i\sigma\varphi^\varepsilon(x)} v_\varepsilon(x)$$

where (compare with (19))

$$v_\varepsilon(x) = \pi^{-1/2}\sqrt{\sigma f(x_\varepsilon,\sigma^{-3/8})}\,\chi_\varepsilon(\sigma^{3/8}x)\exp\left(-\frac{1}{2}\sigma f(x_\varepsilon,\sigma^{-3/8})|x-x_0|^2\right),$$

and $\chi_\varepsilon \in C_c^\infty(\mathbb{R}^2)$ satisfies,

$$\chi_\varepsilon = 1 \text{ in } D(x_\varepsilon,1/2),\ \operatorname{supp}\chi_\varepsilon \subset D(x_\varepsilon,1) \text{ and } 0 \leq \chi_\varepsilon \leq 1 \text{ in } \mathbb{R}^2\,.$$

Then,

$$\|w_\varepsilon\|^2_{L^2(\Omega,\mathbb{C})} = 1 + o(\sigma^{-1})$$

and, by (33) and Cauchy's inequality,

$$q_{\sigma\mathbf{A}}(w_\varepsilon) \leq \sigma f(x_\varepsilon, \sigma^{-3/8}) + C_\varepsilon \sigma^{7/8}. \tag{34}$$

The min-max principle and (32) yields

$$\limsup_{\sigma\to+\infty}\left(\frac{\lambda^D(\sigma\mathbf{A})}{\sigma}\right) \leq f(x_\varepsilon) \leq m_0(B,\Omega) + \varepsilon$$

for all $\varepsilon \in (0, 1]$. Sending ε to 0 we get the desired upper bound in (31).

5.2 *The Neumann Realization*

The analogue of Theorem 3 for the Neumann realization is not tackled yet. By comparing the Dirichlet and Neumann eigenvalues in (5), we get

$$\lambda^N(\sigma\mathbf{A}) \leq \sigma\, m_0(B,\Omega) + o(\sigma).$$

Let us introduce the essential infimum of B with respect to the arc-length measure along $\partial\Omega$, $m_0(B,\partial\Omega) = \inf_{x\in\partial\Omega}\operatorname{ess} B(x)$, and assume that $m_0(B,\partial\Omega) > 0$. We can repeat the constructions in the derivation of the upper bound in (31), by working in a neighborhood of the boundary $\partial\Omega$, where one can use the adapted Frenet coordinates (defined by the transversal distance to and the arc-length distance along $\partial\Omega$). In this way, we can establish the upper bound (compare with Theorem 1)

$$\lambda^N(\sigma\mathbf{A}) \leq \Theta_0\sigma\, m_0(B,\partial\Omega) + o(\sigma).$$

Consequently, we have

$$\limsup_{\sigma\to+\infty}\frac{\lambda^N(\sigma\mathbf{A})}{\sigma} \leq \min\big(m_0(B,\Omega), \Theta_0 m_0(B,\partial\Omega)\big)$$

and one can ask whether the following matching lower bound holds

$$\liminf_{\sigma\to+\infty}\frac{\lambda^N(\sigma\mathbf{A})}{\sigma} \geq \min\big(m_0(B,\Omega), \Theta_0 m_0(B,\partial\Omega)\big). \tag{35}$$

The lower bound in (35) seems related to a lower bound on

$$q_{\sigma\mathbf{A}}(u) - \Theta_0 \int_\Omega B(x)|u(x)|^2 dx\,,$$

which is an analogue of the one in Proposition 2 but for functions that do not vanish on the boundary $\partial\Omega$. Such a lower bound is out of reach at the moment.

Another possibility is to study the limit profile of an actual ground state, u_σ, of $\lambda^N(\sigma\mathbf{A})$, which is quite clear in the setting of a smooth magnetic field [8, Sec. 11.4], but the study is missing in the setting where the magnetic field only satisfies $B \in H^1(\Omega, \mathbb{R})$.

6 Other Magnetic Fields

6.1 Square Integrable Magnetic Fields

The lower bound in (30) continues to hold, by Proposition 2, when $B \in L^2(\Omega)$, since we can always find $\mathbf{A} \in H^1(\Omega, \mathbb{R}^2)$ satisfying $\operatorname{curl}\mathbf{A} = B$ (see [8, Prop. D.2.1]). An open question is whether the matching upper bound in (31) holds too under this relaxed assumption.

We discuss here two examples of a square integrable magnetic field not in the Sobolev space $H^1(\Omega)$. For instance, for a fixed $a \in \Omega$, the magnetic field (compare with (7))

$$B(x) = \frac{1}{|x-a|^p} \quad \text{for } 0 < p < 1\,, \tag{36}$$

is singular at $a \in \Omega$ and satisfies $B \in L^2(\Omega)$. It is generated by the magnetic potential

$$\mathbf{A}(x) = \frac{1}{(4-2p)|x-a|^p}\mathbf{A}_0(x-a)\,,$$

where $\mathbf{A}_0$ is introduced in (15). Since $B \in C^1(\overline{\Omega} \setminus \{a\})$, we get as in the proof of Theorem 1,

$$\limsup_{\sigma\to+\infty} \frac{\lambda^D(\sigma\mathbf{A})}{\sigma} \le m_0(B,\Omega)\,,$$

thereby Theorem 3 holds for the square integrable magnetic field in (36).

Another example is that of a magnetic step [3], where

$$B(x) = b\mathbf{1}_{\Omega_1} + \mathbf{1}_{\Omega_2} \tag{37}$$

where $b \in (0, 1)$ is a constant, Ω_1, Ω_2 are non-empty pairwise disjoint open subsets of Ω such that their closures cover $\overline{\Omega}$. Note that $m_0(B, \Omega) = b$. Picking $x_0 \in \Omega_1$ and considering the quasi-mode (as in (19), $D(x_0, \varepsilon_0) \subset \Omega_1$ for some $\varepsilon_0 > 0$),

$$u_{x_0,\sigma}(x) = \pi^{-1/2}\sqrt{\sigma b}\,\chi(x)\exp\left(-\frac{1}{2}\sigma b\,|x - x_0|^2\right)$$

we get that

$$\lambda^D(\sigma \mathbf{A}) \leq \sigma b + o(\sigma) = \sigma\, m_0(B, \Omega) + o(\sigma)\,,$$

hence Theorem 3 continues to hold for this example too.

Note that, in [3], for the magnetic field in (37) and the Neumann realization, an interesting example where $b \approx -1$ and Ω_1, Ω_2 are separated by a smooth curve that intersects $\partial\Omega$ transversely, the lowest eigenvalue $\lambda^N(\sigma \mathbf{A})$ satisfies

$$\lim_{\sigma \to +\infty} \frac{\lambda^N(\sigma \mathbf{A})}{\sigma} < \min\left(m_0(B, \Omega), \Theta_0 m_0(B, \partial\Omega)\right)$$

thereby violating (35), but when B changes sign. It would be interesting to get the same result for a particular $b \in (0, 1)$.

6.2 *Aharonov-Bohm Fields*

Loosely speaking, a magnetic field defined on a domain Ω with singularity in a subset $S \subset \Omega$ can be seen as a magnetic field without singularity defined in the punched domain $\Omega \setminus S$, and the singular set S can be viewed as holes. The celebrated Aharonov-Bohm effect is a nice example, where the non-trivial topology of the domain interacts with the magnetic flux.

Let us introduce the Aharonov-Bohm magnetic potential

$$\mathbf{F}^{\mathrm{AB}}(x) = \left(-\frac{x_2}{2\pi|x|^2}, \frac{x_1}{2\pi|x|^2}\right) \tag{38}$$

which satisfies $\operatorname{curl}\mathbf{A} = \delta_0$ in the distributional sense (i.e. in $\mathcal{D}'(\Omega)$). Here δ_0 is the Dirac measure at 0.

It can be approximated formally via the regularized potential

$$\mathbf{F}^{\varepsilon}(x) = \begin{cases} \mathbf{F}^{\mathrm{AB}}(x) & \text{if } |x| > \varepsilon \\ \frac{1}{\pi\varepsilon^2}\mathbf{A}_0(x) & \text{if } |x| < \varepsilon \end{cases}$$

where $\mathbf{A}_0$ is introduced in (15). The regularized magnetic potential generates the following magnetic field

$$B^{\varepsilon}(x) = \operatorname{curl} \mathbf{F}^{\varepsilon}(x) = \frac{1}{\pi \varepsilon^2} \mathbf{1}_{D(0,\varepsilon)} ,$$

which approximates δ_0 in $\mathcal{D}'(\Omega)$.

Let $a \in \Omega$ and consider

$$\mathbf{A}(x) = \mathbf{F}^{\mathrm{AB}}(x - a) \quad \text{and} \quad \mathbf{A}^{\varepsilon}(x) = \mathbf{F}^{\varepsilon}(x - a) .$$

Then, by [15, Thm. Eq. (5.17)], for every $\sigma > 0$, the lowest eigenvalue for the Neumann realization satisfies, as $\varepsilon \to 0_+$,

$$\lambda^N(\sigma \mathbf{A}^{\varepsilon}) = \lambda^N(\sigma \mathbf{A}) + o(1) .$$

The same formula continues to hold for the Dirichlet realization. Here, the function

$$\sigma \mapsto \lambda^N(\sigma \mathbf{A})$$

is 2π-periodic and vanishes if and only if σ is an integer multiple of π (see [14] and [15, Thm. 1.4]). It would be interesting to establish these formulas uniformly with respect to $\sigma \in \mathbb{R}_+$, or when σ depends on ε and is large (i.e. $\sigma \to +\infty$ as $\varepsilon \to 0$).

Moreover, as in [6], another interesting case could be when $\mathbf{A}(x) = \mathbf{A}_0(x) + \mathbf{F}^{\varepsilon}(x - a)$ (resp. $\mathbf{A}(x) = \mathbf{A}_0(x) + \mathbf{F}^{\mathrm{AB}}(x - a)$), which induces the magnetic field

$$\operatorname{curl} \mathbf{A}(x) = 1 + \frac{1}{\pi \varepsilon^2} \mathbf{1}_{D(0,\varepsilon)} \quad (\text{resp. } \operatorname{curl} \mathbf{A}(x) = 1 + \delta_0) .$$

This might be helpful in violating Theorem 3, when working under the hypothesis that B only belongs to $L^2(\Omega)$.

Acknowledgments AK acknowledges the support of the Istituto Nazionale di Alta Matematica “F. Severi”, through the Intensive Period “INdAM Quantum Meetings (IQM22)” and of the Center for Advanced Mathematical Sciences at the American University of Beirut. Pan was partially supported by the National Natural Science Foundation of China grant no. 12071142, the Natural Science Foundation of Guangdong Province grant no. 2023A1515012868, the Natural Science Foundation of Shenzhen grant no. JCYJ20220530143801004, and the Shenzhen Municipal Grant No. GXWD20201231105722002. The authors would like to thank the anonymous referee for their valuable suggestions.

References

1. Abatangelo, L., Felli, V., Noris, B., Nys, M.: Estimates for eigenvalues of Aharonov-Bohm operators with varying poles and non-half-integer circulation. Anal. PDE **11**(7), 1741–1785 (2018)
2. Almog, Y.: Thin boundary layers of chiral smectics. Calc. Var. PDE. **33**(3), 299–328 (2008)
3. Assaad, W.: The breakdown of superconductivity in the presence of magnetic steps. Commun. Contemp. Math. **23**(2), 53 pp. (2021), art. no. 2050005
4. Attar, K.: Pinning with a variable magnetic field of the two-dimensional Ginzburg-Landau model. Non-Linear Anal. TMA **139**, 1–54 (2016)
5. Brezis, H., Coron, J.-M., Lieb, E.H.: Harmonic maps with defects. Commun. Math. Phys. **107**(4), 649–705 (1986)
6. Correggi, M., Fermi, D.: Magnetic perturbations of anyonic and Aharonov-Bohm Schrödinger operators. J. Math. Phys. **62**(3), 032101, 24 pp. (2021)
7. Dauge, M., Miqueu, J.P., Raymond, N.: On the semiclassical Laplacian with magnetic field having self-intersecting zero set. J. Spectr. Theory **10**(4), 1211–1252 (2020)
8. Fournais, S., Helffer, B.: Spectral Methods in Surface Superconductivity. Progress in Nonlinear Differential Equations and Their Applications, vol. 77. Birkhäuser Boston Inc., Boston (2010)
9. Fournais, S., Kachmar, A., Pan, X.B.: Existence of surface smectic states of liquid crystals. J. Funct. Anal. **274**(3), 900–958 (2018)
10. Giacomelli, E.L.: On the magnetic Laplacian with a piecewise constant magnetic field in $\mathbb{R}^3_+$. In: Quantum Mathematics I. Springer Nature Singapore, Singapore (this volume)
11. Helffer, B., Kachmar, A.: The Ginzburg-Landau functional with vanishing magnetic field. Arch. Ration. Mech. Anal. **218**(1), 55–122 (2015)
12. Helffer, B., Morame, A.: Magnetic bottles in connection with superconductivity. J. Funct. Anal. **185**(2), 604–680 (2001)
13. Helffer, B., Pan, X.B.: Reduced Landau-de Gennes functional and surface smectic state of liquid crystals. J. Funct. Anal. **255**(11), 3008–3069 (2008)
14. Helffer, B., Hoffmann-Ostenhof, M., Hoffmann-Ostenhof, T., Owen, M.P.: Nodal sets for ground states of Schrödinger operators with zero magnetic field in non-simply connected domains. Commun. Math. Phys. **202**(3), 629–649 (1999)
15. Kachmar, A., Pan, X.B.: Oscillatory patterns in the Ginzburg-Landau model driven by the Aharonov-Bohm potential. J. Funct. Anal. **279**(10) (2020), art. no. 108718
16. Kachmar, A., Wehbe, M.: Averaging of magnetic fields and applications. Commun. Contemp. Math. **25**(2), 2150108, 25 pp. (2023). https://doi.org/10.1142/S021919972150108X
17. Lin, F.H.: A remark on the map $x/|x|$. C. R. Acad. Sci. Paris Sér. I Math. **305**(12), 529–531 (1987)
18. Pan, X.B.: Surface superconductivity in applied magnetic fields above H_{C_2}. Commun. Math. Phys. **228**(2), 228–370 (2002)
19. Pan, X.B.: Analogies between superconductors and liquid crystals: nucleation and critical fields. In: Advanced Studies in Pure Mathematics (ASPM) 47-2 (2007). Asymptotic Analysis and Singularities, pp. 479–518
20. Pan, X.B., Kwek, K.H.: Schrödinger operators with non-degenerately vanishing magnetic fields in bounded domains. Trans. Amer. Math. Soc. **354**(10), 4201–4227 (2002)
21. Raymond, N.: Bound States of the Magnetic Schrödinger Operator. EMS Tracts in Mathematics, vol. 27. European Mathematical Society, Zürich (2017)
22. Schoen, R., Uhlenbeck, K.: Boundary regularity and the Dirichlet problem for harmonic maps. J. Differ. Geom. **17**(2), 307–335 (1982)

Part III
Effective Nonlinear Models

Invariant Measures as Probabilistic Tools in the Analysis of Nonlinear ODEs and PDEs

Zied Ammari, Shahnaz Farhat, and Vedran Sohinger

1 Introduction

The concept of a Gibbs measure originates from statistical physics and is related to the thermal equilibrium of a dynamical system. Nowadays, it is quite widespread in the literature, appearing in different forms and in various subjects. For instance in Probability, gaussian random fields are (free) Gibbs measures; while in Ergodic theory the Sinai–Ruelle–Bowen measures are an appropriate form of local Gibbs measures. On the other hand, in Mathematical Physics, Gibbs measures correspond to the Euclidean (φ_d^4) field theory. There is of course a large body of literature on each of these topics. For a general overview, we refer the reader to the following three monographs [7, 18, 19].

In this short note, we consider Gibbs measures for finite and infinite dimensional Hamiltonian systems and emphasise their statistical mechanical properties. On the other hand, we explain how Gibbs measures (likewise any other invariant measure) can be used as a tool for constructing global solutions for some nonlinear PDEs almost surely. For more than two decades, probabilistic tools have been used with remarkable success in order to study nonlinear dispersive PDEs at low regularity (see for instance [4–6, 11, 15, 17] and the references therein). Most of these results consider specific equations and combine local well-posedness analysis with probabilistic arguments. Our purpose here is to observe that almost sure existence (possibly without uniqueness) of global solutions for an abstract initial value problem is a general feature that is the sole consequence of existence of an

Z. Ammari (✉) · S. Farhat
Univ Rennes, [UR1], CNRS, IRMAR - UMR 6625, Rennes, France
e-mail: zied.ammari@univ-rennes1.fr; shahnaz.farhat@univ-rennes1.fr

V. Sohinger
Mathematics Institute, University of Warwick, Coventry, UK
e-mail: V.Sohinger@warwick.ac.uk

M. Correggi, M. Falconi (eds.), *Quantum Mathematics I*,
Springer INdAM Series 57, https://doi.org/10.1007/978-981-99-5894-8_12

"invariant" measure. Invariance here is not meant in the sense that the flow preserves the measure, since we may not a priori have a well-defined one, but instead invariant measures are defined to be stationary solutions for a related Liouville equation. The results reported here are based on our two articles [2] and [1].

2 Gibbs Measures

2.1 Finite Dimensions

Hamiltonian systems: Consider a finite dimensional phase-space $E \equiv \mathbb{R}^{2n}$ and a Hamiltonian function $h : E \to \mathbb{R}$ which is of class $\mathscr{C}^1$ or $\mathscr{C}^2$. A natural symplectic structure is introduced on E by choosing a skew-symmetric matrix J such that $J^2 = -I_{2n}$ and defining a symplectic form:

$$\sigma(u, v) = \langle u, Jv \rangle . \tag{1}$$

A Hamiltonian dynamical system is then defined through the vector field $X : E \to E$,

$$X(u) = J\nabla h(u) , \qquad \forall u \in E, \tag{2}$$

and the field equation:

$$\dot{u}(t) = X(u(t)) . \tag{3}$$

The later initial value problem is complemented with an initial condition $u(t_0) = u_0 \in E$ at a given time t_0. In the case where $h \in \mathscr{C}^1$, existence of local solutions for the equation (3) is provided by the Peano Theorem. On the other hand, when $h \in \mathscr{C}^2$, the Cauchy–Lipschitz theorem guarantees the uniqueness and existence of local (maximal) solutions. Remark that completeness of the vector field X is in general not guaranteed without further assumptions, which means that some initial conditions may not give rise to global in time solutions for (3).

In order to define a Gibbs measure for the above Hamiltonian system, we assume that

$$z_\beta := \int_E e^{-\beta h(u)}\, dL < +\infty , \tag{4}$$

for some $\beta > 0$. Here, dL denotes the Lebesgue measure on E.

Definition 1 (Gibbs Measure) The Gibbs measure of the Hamiltonian system (2)-(3), at inverse temperature $\beta > 0$, is the Borel probability measure given by

$$\mu_\beta = \frac{e^{-\beta h(\cdot)}\, dL}{\int_E e^{-\beta h(u)}\, dL} \equiv \frac{1}{z_\beta} e^{-\beta h(\cdot)}\, dL\,. \tag{5}$$

Theorem 1 (Invariance) *If the Hamiltonian vector field X is complete then the Gibbs measure μ_β is invariant with respect to the Hamiltonian flow. More precisely, let ϕ_t denote the global Hamiltonian flow of (3) then for all $t \in \mathbb{R}$,*

$$(\phi_t)_\sharp \mu_\beta = \mu_\beta\,,$$

or equivalently, for all Borel subsets B of E,

$$\mu_\beta\left((\phi_t)^{-1}(B)\right) = \mu_\beta\,(B)\,.$$

Proof It follows from the Liouville theorem and the conservation of energy:

$$\frac{d}{dt} h(\phi_t(u)) = \langle \nabla h(\phi_t(u)),\, J\nabla h(\phi_t(u))\rangle = 0\,.$$

Gibbs measures are more than invariant measures. They satisfy more properties reflecting their statistical stability. In fact, Gibbs measures are the unique probability measures verifying the Gibbs variational principle and the Classical Kubo-Martin-Schwinger (KMS) condition stated below.

Theorem 2 (Gibbs Variational Principle) *The Gibbs measure μ_β is the unique minimizer of the entropy (free-energy) functional:*

$$\mathcal{E}(\nu) = \int_E \varrho \log\left(\varrho\right) dL + \beta \int_E h\, d\nu\,, \tag{6}$$

where ν is any probability measure such that $\nu = \varrho dL$.

Proof Note that by convention $\mathcal{E}(\nu) = +\infty$ if the r.h.s of (6) is non integrable. A simple computation then yields

$$\mathcal{E}(\nu) = \int_E \frac{d\nu}{d\mu_\beta} \log\left(\frac{d\nu}{d\mu_\beta}\right) d\mu_\beta - \log(z_\beta).$$

Hence, Jensen's inequality (for $x\log(x)$) proves that μ_β is the unique minimizer.

In the sequel, we denote by $\mathcal{P}(E)$ the set of all Borel probability measures on E.

Definition 2 We say that a probability measure $\mu \in \mathcal{P}(E)$ satisfies the classical Kubo–Martin–Schwinger (KMS) condition if and only if

$$\int_E \{F, G\}(u)\, d\mu = \beta \int_E \{F, h\}(u)\, G(u)\, d\mu\,, \tag{7}$$

for all compactly supported smooth functions $F, G \in \mathscr{C}_c^\infty(E)$.

Theorem 3 (KMS Principle) *The Gibbs measure μ_β is the unique probability measure satisfying the KMS condition* (7).

Proof See [1, Theorem 4.2]

It is worth noticing that in the above result, we do not require μ in (7) to be absolutely continuous with respect to the Lebesgue measure.

2.2 *Infinite Dimensions*

The latter results extend to the following framework of infinite dimensional Hamiltonian systems. Consider a positive operator $A : D(A) \subseteq H \to H$ such that,

$$\exists c > 0, \quad A \geq c\mathbb{1}\,. \tag{8}$$

A Hamiltonian dynamical system is defined using the quadratic function,

$$h_0 : D(A^{1/2}) \to \mathbb{R}, \qquad h_0(u) = \frac{1}{2}\langle u, Au\rangle\,. \tag{9}$$

In this case, the vector field is a linear operator $X_0 : D(A) \to H$,

$$X_0(u) = -iAu,$$

and the linear field equation governing the dynamics of the system is:

$$\dot{u}(t) = X_0(u(t)) = -iAu(t). \tag{10}$$

We suppose furthermore that the operator A admits a compact resolvent. Hence, there exists an orthonormal basis in H of eigenvectors $\{e_j\}_{j\in\mathbb{N}}$ of A associated with eigenvalues $\{\lambda_j\}_{j\in\mathbb{N}}$ such that for all $j \in \mathbb{N}$,

$$Ae_j = \lambda_j\, e_j. \tag{11}$$

We make the following assumption:

$$\exists s \geq 0: \quad \sum_{j=1}^{\infty} \frac{1}{\lambda_j^{1+s}} < +\infty, \tag{12}$$

and henceforth work with fixed s satisfying (12).

Here the Hilbert space H is complex. But sometimes, we will use it with its real structure and in this case we denote it by $H_{\mathbb{R}}$. In particular, the family $\{e_j, ie_j\}_{j\in\mathbb{N}}$ is an O.N.B of $H_{\mathbb{R}}$ endowed with its inner product $\langle\cdot,\cdot\rangle_{H_{\mathbb{R}}} := \Re\mathrm{e}\langle\cdot,\cdot\rangle$.

Weighted Sobolev Spaces The operator A introduces a weighted Sobolev space constructed as follows. For any $r \in \mathbb{R}$, define the inner product:

$$\forall x, y \in \mathcal{D}(A^{\frac{r}{2}}), \qquad \langle x, y\rangle_{H^r} := \langle A^{r/2}x, A^{r/2}y\rangle .$$

Then for $r \geq 0$:

- H^r is the Hilbert space $(\mathcal{D}(A^{r/2}), \langle\cdot,\cdot\rangle_{H^r})$.
- H^{-r} denotes the completion of the pre-Hilbert space $(\mathcal{D}(A^{-r/2}), \langle\cdot,\cdot\rangle_{H^{-r}})$.
- One has the canonical continuous and dense embedding (Hilbert rigging):

$$H^r \subseteq H \subseteq H^{-r} .$$

We remark that H^{-r} identifies also with the dual space of H^r relatively to the inner product of H.

Gaussian Measures The free Gibbs measure written formally as

$$\mu_{\beta,0} \equiv \frac{e^{-\beta h_0(\cdot)}\,du}{\int e^{-\beta h_0(u)}\,du},$$

is rigorously defined as a Gaussian measure on the Hilbert space H^{-s}.

The set of all Borel probability measures on H^{-s} is denoted by $\mathcal{P}(H^{-s})$.

Definition 3

1. The mean-vector of $\mu \in \mathcal{P}(H^{-s})$ is the vector $m \in H^{-s}$ such that:

$$\langle f, m\rangle_{H_{\mathbb{R}}^{-s}} = \int_{H^{-s}} \langle f, u\rangle_{H_{\mathbb{R}}^{-s}}\,d\mu, \qquad \forall f \in H^{-s}.$$

If $m = 0$, one says that μ is a centred measure.

2. The covariance operator of $\mu \in \mathcal{P}(H^{-s})$ is a linear operator $Q : H_{\mathbb{R}}^{-s} \to H_{\mathbb{R}}^{-s}$ such that:

$$\langle f, Q\, g\rangle_{H_{\mathbb{R}}^{-s}} = \int_{H^{-s}} \langle f, u - m\rangle_{H_{\mathbb{R}}^{-s}} \langle u - m, g\rangle_{H_{\mathbb{R}}^{-s}}\, d\mu\,, \quad \forall f, g \in H^{-s}.$$

3. $\mu \in \mathcal{P}(H^{-s})$ is Gaussian if $B \mapsto \mu(\{y \in H^{-s} : \langle x, y\rangle_{H_{\mathbb{R}}^{-s}} \in B\})$ are Gaussian measures on $\mathbb{R}$ for all $x \in H^{-s}$.

The following result is well-known (see e.g. [1] and references therein).

Theorem 4 *There exists a unique centred Gaussian measure on H^{-s}, denoted $\mu_{\beta,0}$, such that its covariance operator is $\beta^{-1}A^{-(1+s)}$, i.e.: for all $f, g \in H^{-s}$*

$$\frac{1}{\beta}\langle f, A^{-(1+s)} g\rangle_{H_{\mathbb{R}}^{-s}} = \int_{H^{-s}} \langle f, u\rangle_{H_{\mathbb{R}}^{-s}} \langle u, g\rangle_{H_{\mathbb{R}}^{-s}}\, d\mu_{\beta,0}\,. \tag{13}$$

Moreover, the characteristic function of $\mu_{\beta,0}$ is given for any $v \in H^{-s}$ by

$$\int_{H^{-s}} e^{i\langle v,u\rangle_{H_{\mathbb{R}}^{-s}}}\, d\mu_{\beta,0}(u) = e^{-\frac{1}{2\beta}\langle v, A^{-(1+s)} v\rangle_{H_{\mathbb{R}}^{-s}}}\,. \tag{14}$$

Note that centred Gaussian measures are Gibbs measures over infinite dimensional spaces related to the linear Hamiltonian system (9)–(10).

Nonlinear Hamiltonian Systems Consider the linear operator A satisfying the assumptions in (8), (11) and (12). Take a nonlinear functional $h^I : H^{-s} \to \mathbb{R}$ satisfying for some $\beta > 0$ and some $p \in [1, \infty)$:

$$e^{-\beta h^I(\cdot)} \in L^p(\mu_{\beta,0})\,, \tag{15a}$$

$$h^I \in \mathbb{D}^{1,p}(\mu_{\beta,0})\,, \tag{15b}$$

where $\mathbb{D}^{1,p}(\mu_{\beta,0})$ denotes the Gross-Sobolev spaces recalled below in Definition 5. The nonlinear Hamiltonian system that we would like to study is described by the Hamiltonian function:

$$h(u) = \frac{1}{2}\langle u, Au\rangle + h^I(u) = h_0(u) + h^I(u), \tag{16}$$

and the vector field:

$$X(u) = -iAu - i\nabla h^I(u) = X_0(u) + X^I(u)\,, \tag{17}$$

where ∇h is the Malliavin derivative of the functional h (recalled below in Lemma 1). Thus, a dynamical system is defined through the autonomous field equation:

$$\dot{u}(t) = X(u(t)),$$

or equivalently, in the interaction representation, through the non-autonomous field equation:

$$\dot{u}(t) = e^{itA} X^I(e^{-itA} u(t)) \equiv X(t, u(t)).$$

Gross-Sobolev Spaces In order to introduce these spaces, one needs a good differential calculus on a convenient space of test functions which in our case is the space of smooth cylindrical functions.

Definition 4 Let $\{f_j\}_{j\in\mathbb{N}}$ be an O.B.N. of $H_{\mathbb{R}}$. Consider for $n \in \mathbb{N}$, the mapping $\pi_n : H^{-s} \to \mathbb{R}^{2n}$,

$$\pi_n(x) = (\langle x, f_1\rangle_{H_{\mathbb{R}}}, \dots, \langle x, f_{2n}\rangle_{H_{\mathbb{R}}}). \tag{18}$$

Define the space of smooth cylindrical functions $\mathscr{C}^{\infty}_{c,cyl}(H^{-s})$ as the set of all functions $F : H^{-s} \to \mathbb{R}$ such that

$$F = \varphi \circ \pi_n \tag{19}$$

for some $n \in \mathbb{N}$ and $\varphi \in \mathscr{C}^{\infty}_c(\mathbb{R}^{2n})$.

On such functions, the gradient of F at the point $u \in H^{-s}$ is given by

$$\nabla F(u) = \sum_{j=1}^{2n} \partial_j \varphi(\pi_n(u))\, f_j\,, \tag{20}$$

where $\partial_j \varphi$ are the partial derivatives with respect to the $2n$ coordinates of φ.

Lemma 1 (Malliavin Derivative) *The following linear operator is closable:*

$$\nabla : \mathscr{C}^{\infty}_{c,cyl}(H^{-s}) \subset L^p(\mu_{\beta,0}) \longrightarrow L^p(\mu_{\beta,0}; H^{-s}),$$

$$F = \varphi \circ \pi_n \longmapsto \nabla F = \sum_{j=1}^{2n} \partial_j \varphi(\pi_n(\cdot))\, f_j\,.$$

The Malliavin derivative is closure of such linear operators (still denoted by ∇).

Definition 5 (Gross–Sobolev Spaces) The Gross-Sobolev space $\mathbb{D}^{1,p}(\mu_{\beta,0})$ is the closure domain of the Malliavin derivative ∇ with respect to the norm:

$$\|F\|^p_{\mathbb{D}^{1,p}(\mu_{\beta,0})} := \|F\|^p_{L^p(\mu_{\beta,0})} + \|\nabla F\|^p_{L^p(\mu_{\beta,0};H^{-s})} \,. \tag{21}$$

In the case $p = 2$, we remark that $\mathbb{D}^{1,2}(\mu_{\beta,0})$ is a Hilbert space.

Definition 6 (Gibbs Measure) Assume (8), (11), (12) and suppose that (15) holds true for some $p \in [1, \infty)$. The Gibbs measure of the Hamiltonian dynamical system (16)-(17), at inverse temperature $\beta > 0$, is the probability measure on H^{-s} given by:

$$\mu_\beta = \frac{e^{-\beta h^I(\cdot)}\, d\mu_{\beta,0}}{\int_{H^{-s}} e^{-\beta h^I(u)}\, d\mu_{\beta,0}} \equiv \frac{1}{z_\beta}\, e^{-\beta h^I(\cdot)}\, d\mu_{\beta,0} \,. \tag{22}$$

We remark that μ_β is well-defined whenever (15a) is satisfied at least for $p = 1$. The other assumption (15b) is useful for defining the dynamical system (16)–(17).

In this infinite dimensional framework, the completeness (even almost surely) of the vector fields X or $X(t, \cdot)$ as well as the invariance of the Gibbs measure by the Hamiltonian flow are far from obvious questions. In particular, the conservation of energy cannot be exploited since the Hamiltonian function h does not make sense on the spaces H^{-s}. We will come back to these issues in Sect. 3.

2.3 Statistical Properties

As in the finite dimensional case (Sect. 2.1), the Gibbs measure μ_β in (22) is characterized by an entropy variational principle. However, in the present case it is necessary to use the notion of relative entropy.

Theorem 5 (Gibbs Variational Principle) *Define the relative entropy for all $\mu, \nu \in \mathcal{P}(H^{-s})$ such that $\nu \ll \mu$ as:*

$$E(\nu|\mu) := \int_{H^{-s}} \frac{d\nu}{d\mu} \log\left(\frac{d\nu}{d\mu}\right) d\mu \,.$$

Assume that (15) holds true for $p = 2$. Then the Gibbs measure μ_β is the unique minimizer of the entropy (free-energy) functional

$$\mathcal{E}_{\mu_{\beta,0}}(\nu) = E(\nu|\mu_{\beta,0}) + \beta \int_{H^{-s}} h^I d\nu \,, \tag{23}$$

among all ν such that $\frac{d\nu}{d\mu_{\beta,0}} \in L^2(\mu_{\beta,0})$.

Proof Similarly to the finite dimensional case, we have

$$\mathcal{E}_{\mu_{\beta,0}}(\nu) = E(\nu|\mu_\beta) - \log(z_\beta)\,,$$

where z_β is given by (22). Thanks to Jensen's inequality, one knows that $E(\nu|\mu_\beta)$ is non-negative with $E(\nu|\mu_\beta) = 0$ if and only if $\nu = \mu_\beta$.

In order to prove a classical KMS principle similar to the one in Theorem 3, we define a Poisson structure over the spaces H^{-s} as follows. Consider:

- The algebra of smooth bounded cylindrical functions $\mathscr{C}^\infty_{b,cyl}(H^{-s})$.
- $F, G \in \mathscr{C}^\infty_{b,cyl}(H^{-s})$ such that: $\forall u \in H^{-s}$,

$$F(u) = \varphi \circ \pi_n(u)\,, \qquad G(u) = \psi \circ \pi_m(u)\,, \tag{24}$$

  with $\varphi \in \mathscr{C}^\infty_b(\mathbb{R}^{2n})$ and $\psi \in \mathscr{C}^\infty_b(\mathbb{R}^{2m})$ for some $n, m \in \mathbb{N}$. Here, we recall that $\mathscr{C}^\infty_b$ consists of smooth functions all of whose derivatives are bounded.

Definition 7 (Poisson Bracket) For all such $F, G \in \mathscr{C}^\infty_{b,cyl}(H^{-s})$, the Poisson bracket of F and G is defined by

$$\{F, G\}(u) := \sum_{j=1}^{\min(n,m)} \partial_j^{(1)}\varphi(\pi_n(u))\, \partial_j^{(2)}\psi(\pi_m(u)) - \partial_j^{(1)}\psi(\pi_m(u))\, \partial_j^{(2)}\varphi(\pi_n(u))\,. \tag{25}$$

The classical KMS condition was introduced by Gallavotti and Verboven [13] in order to characterize the Gibbs measures of infinite systems of statistical mechanics. It was inspired by the Kubo-Martin-Schwinger work for quantum systems [8].

Definition 8 (Classical KMS Condition) A measure $\mu \in \mathcal{P}(H^{-s})$ satisfies the classical KMS condition, at inverse temperature β, for the Hamiltonian system (16)-(17) if and only if for all $F, G \in \mathscr{C}^\infty_{c,cyl}(H^{-s})$,

$$\int_{H^{-s}} \{F, G\}(u)\, d\mu = \beta \int_{H^{-s}} \langle \nabla F(u), X(u)\rangle\, G(u)\, d\mu\,, \tag{26}$$

with the Poisson bracket $\{\cdot, \cdot\}$ defined as in (25).

The following result is proved in [1].

Theorem 6 (KMS Principle) *Assume that the assumption* (15) *is true for* $p = 2$. *Let* $\mu \in \mathcal{P}(H^{-s})$ *be such that* $\mu \ll \mu_{\beta,0}$ *and suppose that*

$$\frac{d\mu}{d\mu_{\beta,0}} \in \mathbb{D}^{1,2}(\mu_{\beta,0})\,.$$

Then μ satisfies the classical KMS condition for the Hamiltonian system (16)–(17) *if and only if μ is the Gibbs measure μ_β. i.e.,*

$$\mu = \frac{e^{-\beta h^I}\,\mu_{\beta,0}}{\int_{H^{-s}} e^{-\beta h^I(u)} d\mu_{\beta,0}} = \mu_\beta\,.$$

Proof Using Malliavin calculus, see [1, Theorem 4.11], one can show that the Gibbs measure μ_β given in Definition 6 satisfies the KMS condition (26) with the vector field X as in (17).

For the opposite implication, take $\mu \in \mathcal{P}(H^{-s})$ such that the KMS condition (26) is satisfied and $\mu = \varrho\,\mu_{\beta,0}$ for some density $\varrho \in \mathbb{D}^{1,2}(\mu_{\beta,0})$. Thanks to (26), one can show

$$\nabla\varrho + \beta\varrho\,\nabla h^I = 0\,, \tag{27}$$

as an identity in $L^1(\mu_{\beta,0}; H^{-s})$. We notice that there is no loss of generality in assuming $\varrho > 0$ almost surely and then we show

$$\log(\varrho) \in \mathbb{D}^{1,2}(\mu_{\beta,0}) \quad \text{and} \quad \nabla\log(\varrho) = \frac{\nabla\varrho}{\varrho}\,.$$

Hence, (27) gives

$$\nabla\big(\log(\varrho) + \beta h^I\big) = \frac{\nabla\varrho}{\varrho} + \beta\nabla h^I = 0\,.$$

By Malliavin calculus, see [1, Proposition A.4], one knows that if $F \in \mathbb{D}^{1,2}(\mu_{\beta,0})$ such that $\nabla F = 0$ for $\mu_{\beta,0}$-almost surely then F is a constant. So, we get

$$\log(\varrho) + \beta h^I = c\,,$$

$\mu_{\beta,0}$-almost surely. Using the normalization of the density ϱ, we obtain

$$\varrho = \frac{e^{-\beta h^I}}{\int_{H^{-s}} e^{-\beta h^I(u)} d\mu_{\beta,0}}\,.$$

3 Nonlinear PDEs

In this section, we discuss one of the main applications of Gibbs measures to the analysis of nonlinear dispersive PDEs, namely the construction of low regularity global solutions almost surely. This was first proved rigorously by Bourgain [4–6], building on the work of Lebowitz–Rose–Speer [14] and Zhidkov [20]. In this framework, Gibbs measures can be used as substitutes for conservation laws at low regularity. There is a vast literature on this subject. For an overview, we refer the reader to [10, 15, 17] and the references therein.

3.1 Bourgain's Method

We now briefly describe Bourgain's method and compare it to the applications of our results from [1, Section 5]. In this discussion, we set for simplicity of notation $\beta = 1$, unless it is otherwise specified. Let us first consider the *defocusing NLS* equation on the spatial domain $\mathbb{T}^d$ for $d = 1, 2$

$$\begin{cases} i\partial_t u_t(x) = \big(-\Delta + \mathbb{1}\big)u_t(x) + |u_t(x)|^{2q}\, u_t(x) \\ u_0(x) = \varphi(x) \in H^\sigma(\mathbb{T}^d)\,. \end{cases} \tag{28}$$

Here $q \in [1, 2]$ when $d = 1$. When $d = 2$, we only consider the *cubic problem*, which corresponds to $q = 1$. We take the operator $A = -\Delta + \mathbb{1}$ and $H^\sigma = H^\sigma(\mathbb{T}^d)$ with $\sigma = -s$ for s satisfying the condition (12). Then the Gaussian measure $\mu_{1,0}$ is well-defined by Theorem 4 and one can recast (28) as the Hamiltonian system in (16)-(17). When $d = 2$, the nonlinearity needs to be renormalized by Wick ordering, see (31)–(32) below. The Hamiltonian function takes the form

$$h(u) = \frac{1}{2}\int_{\mathbb{T}^d} \bar{u}(x)(\mathbb{1} - \Delta)u(x)\, dx + h^I(u)\,. \tag{29}$$

- For $d = 1$, let $s = 0$ and

$$h^I(u) = \frac{1}{4}\int_{\mathbb{T}} |u(x)|^4\, dx\,. \tag{30}$$

- For $d = 2$, let $s > 0$ and

$$\begin{aligned} h^I(u) &= \lim_n \frac{1}{4}\int \big(|P_n u(x)|^4 - 4\mathbb{E}_{\mu_{1,0}}[|P_n u(x)|^2]\, |P_n u(x)|^2 \\ &\quad + 2\mathbb{E}^2_{\mu_{1,0}}[|P_n u(x)|^2]\big)\, dx, \end{aligned}$$

$$\begin{aligned} h^I(u) &\equiv \frac{1}{4}\int \Big(|u(x)|^4 - \underbrace{4\mathbb{E}_{\mu_{1,0}}[|u(\cdot)|^2]\, |u(\cdot)|^2}_{\text{mass renormalization}} + \underbrace{2\mathbb{E}^2_{\mu_{1,0}}[|u(\cdot)|^2]}_{\text{energy renormalization}}\Big)\, dx \\ &=: \frac{1}{4}\int_{\mathbb{T}^2} :|u|^4:\, dx\,, \end{aligned} \tag{31}$$

where P_n is the orthogonal projection over the first n (ordered) eigenfunctions of A. The convergence (31) holds in $L^p(\mu_{1,0})$ for $p \in [1, \infty)$ as is shown by an application of Wick's theorem. On the torus, we note that $\mathbb{E}^2_{\mu_{1,0}}[|u(x)|^2]$ does not depend on x. The nonlinearity $\mathcal{N}(u)(t) = |u_t|^2 u_t$ in (28) is then replaced by the *Wick-ordered nonlinearity*

$$\mathcal{N}^{\mathrm{Wick}}(u)(t) = 2\frac{\partial h^I}{\partial \bar{u}}(u_t) = |u_t|^2 u_t - 2\left(\int_{\mathbb{T}^2} |u_t|^2 \, dx\right) u_t . \tag{32}$$

When $d = 2, 3$, one also considers the *Hartree equation*

$$\begin{cases} i\partial_t u_t(x) = (-\Delta + \mathbb{1})u_t(x) + (V * |u_t|^2)\, u_t(x) \\ u_0(x) = \varphi(x) \in H^{-s}(\mathbb{T}^d), \end{cases} \tag{33}$$

where $V : \mathbb{T}^d \to \mathbb{R}$ is an even integrable function whose Fourier coefficients satisfy

$$\begin{cases} 0 \le \hat{V}(k) \le \frac{c}{\langle k \rangle^{\epsilon}} & \text{for } d = 2 \\ 0 \le \hat{V}(k) \le \frac{c}{\langle k \rangle^{2+\epsilon}} & \text{for } d = 3, \end{cases} \tag{34}$$

for some fixed $\epsilon > 0$ and $c > 0$. In particular, it is assumed that V is of positive type (i.e. that $\hat{V}$ is pointwise nonnegative). For $d = 2$, let $s > 0$ and for $d = 3$, let $s > \frac{1}{2}$. As before, it is necessary to apply a Wick-ordering renormalization and consider

$$\begin{aligned} h^I(u) &= \lim_n \frac{1}{4} \int \left(|P_n u(x)|^2 - 4\mathbb{E}_{\mu_{1,0}}[|P_n u(x)|^2]\right) V(x-y) \\ &\quad \times \left(|P_n u(y)|^2 - 4\mathbb{E}_{\mu_{1,0}}[|P_n u(y)|^2]\right) dx\, dy \\ &= \frac{1}{4} \int_{\mathbb{T}^d} \int_{\mathbb{T}^d} :|u(x)|^2: \, V(x-y) \, :|u(y)|^2: \, dx\, dy \\ &= \frac{1}{4} \int_{\mathbb{T}^d} \int_{\mathbb{T}^d} \Big(|u(x)|^2 V(x-y) |u(y)|^2 - \underbrace{2\hat{V}(0)\, \mathbb{E}_{\mu_{1,0}}[|u(\cdot)|^2]\, |u(x)|^2}_{\text{mass renormalization}} \\ &\quad + \underbrace{\widehat{V}(0)\, \mathbb{E}^2_{\mu_{1,0}}[|u(\cdot)|^2]}_{\text{energy renormalization}} \Big). \end{aligned} \tag{35}$$

The Wick ordering of the nonlinearity in (33) is now defined analogously as in (32), starting from (35).

Note that the Wick-ordered nonlinearity in (35) is nonnegative, whereas (31) is not. In [5], this difficulty is overcome by use of the *Nelson trick* [16]. In all the cases discussed above, one obtains that μ_1 given by (22) satisfies $\mu_1 \ll \mu_{1,0}$. With notation as above, the following series of results was proved by Bourgain.

Theorem 7 (Bourgain [4–6]) *The above NLS equations admit global solutions on H^σ almost surely with respect to the Gibbs measure μ_1. Here, we take $\sigma < 1 - \frac{d}{2}$. Furthermore, μ_1 is invariant under the flow of the NLS.*

We recall that $\sigma = -s$ for s satisfying (12). In particular, when $d = 1$, in Theorem 7 we can take $\sigma = s = 0$. In the above examples, the vector field is given by

$$X(u) = -iAu - i\nabla h^I(u)\,,$$

where the gradient ∇h^I is understood as the Malliavin derivative from Lemma 1 above. In [1, Section 5], we show the following result, which allows us to make a link between Bourgain's method and the analysis in [1].

Proposition 1 *With h^I given as in (30), (31), (35) above, we have for all $p \in [1, \infty)$ and $\beta > 0$*

$$e^{-\beta h^I(\cdot)} \in L^p(\mu_{\beta,0}) \qquad \textit{and} \qquad h^I \in \mathbb{D}^{1,p}(\mu_{\beta,0})\,. \tag{36}$$

We note that it makes sense to study the above problems for *focusing* problems, i.e. when the nonlinear term does not give a positive-definite contribution to the Hamiltonian. The above analysis carries over [4, 6, 14] provided that one truncates in the mass when $d = 1$ or in the Wick-ordered when $d = 2, 3$. When $d = 1$, the focusing problem was studied in [1, Section 5.3] by means of a *local KMS condition.* We note that the decay assumption on the Fourier coefficients (34) was used in [6] for the focusing problem. In the defocusing problem, the condition when $d = 3$ was recently relaxed in [12]. It is known that, when $d = 2$, the focusing cubic NLS does not admit a well-defined Gibbs measure [9].

In the remainder of the section, we give a brief summary of the strategy used to prove Theorem 7 and compare it with the methods used in [1]. For simplicity, we consider (28) with $d = 1$. The first step is to compare (28) with a suitable finite-dimensional approximation

$$\begin{cases} i\partial_t u_t^n(x) = \big(-\Delta + \mathbb{1}\big)u_t^n(x) + P_n\big(|u_t^n(x)|^{2q}\, u_t^n\big)(x) \\ u_0^n(x) := P_n u_0(x) \in \mathscr{C}^\infty(\mathbb{T})\,. \end{cases} \tag{37}$$

The equation (37) is globally well-posed. Moreover, it is shown in the *approximation lemma* [4, Lemma 2.27], using the analysis of the flow of the periodic NLS developed in [3], that the flow of (37) approximates that of (28) as $n \to \infty$. One notes that (37) is a Hamiltonian system, with Hamiltonian given by $h(u^n)$, with notation as in (29)–(30). The (truncated) Gibbs measure μ_1^n associated with the flow of (37) (with $\beta = 1$) is defined analogously as in (22), where we now project everything to finitely many modes. By Liouville's theorem and conservation of the Hamiltonian, μ_1^n is invariant under the flow of (37). Furthermore, one has $\mu_1^n = (P_n)_\sharp \mu_1 \rightharpoonup \mu_1$.

When analysing (37), one of the main steps is to note that one has a uniform in n local well-posedness theory in $H^\sigma(\mathbb{T})$, for $\sigma \in (0, 1/2)$, which we fix throughout the discussion. This follows by using the arguments from [3]. More precisely, one notes that, for $K > 0$, and u_0 satisfying

$$\|u_0^n\|_{H^\sigma(\mathbb{T})} \leq K\,, \tag{38}$$

(37) admits local solutions on a time interval $[-T, T]$ with $T \sim K^{-\delta}$ for some $\delta > 0$ depending on σ. Moreover, one has

$$\sup_{t\in[-T,T]} \|u_t^n\|_{H^\sigma(\mathbb{T})} \lesssim K\,.$$

One wants to analyse what is the probability, with respect to μ_1^n, that (38) does not occur. To this end, using concentration inequalities for Gaussian random variables, one can prove that

$$\mu_1^n\big(\|u_0^n\|_{H^\sigma(\mathbb{T})} > K\big) \lesssim e^{-cK^2}\,. \tag{39}$$

Denote by S_T^n the flow map of (37) for time T defined on $\Omega_{\sigma,K}^n$, which denotes the set of initial data u_0^n satisfying (38). For $\tau > 0$, we let

$$\Omega_{\sigma,K}^n(\tau) := \Omega_{\sigma,K}^n \cap (S_T^n)^{-1}(\Omega_{\sigma,K}^n) \cap \cdots \cap (S_T^n)^{-\lfloor \tau/T\rfloor}(\Omega_{\sigma,K}^n)\,. \tag{40}$$

We note that for $u_0^n \in \Omega_{\sigma,K}^n(\tau)$ and $0 \leq t \leq \tau$, we have

$$\|u^n(t)\|_{H^\sigma} \lesssim K\,. \tag{41}$$

Furthermore, by using the invariance of μ_1^n under the flow of (37) as well as a union bound and (39) in (40), we deduce that

$$\mu_1^n\big(P_N H^\sigma \setminus \Omega_{\sigma,K}^n(\tau)\big) \lesssim \tau K^\delta e^{-cK^2}\,. \tag{42}$$

Using (41) and (42), the approximation lemma [4, Lemma 2.27], and considering a dyadic sequence of times, it is shown in [4, Lemma 4.4] that, given $\epsilon > 0$, there exists $\mathcal{G}_\epsilon \subset H^\sigma(\mathbb{T})$ such that the following properties hold.

(i) $\mu_1\big(H^\sigma \setminus \mathcal{G}_\epsilon\big) < \epsilon$.
(ii) For all $u_0 \in \mathcal{G}_\epsilon$ and $t \in \mathbb{R}$, we have

$$\|u_t\|_{H^\sigma(\mathbb{T})} \lesssim \left(\frac{1+|t|}{\epsilon}\right)^{\sigma+}.$$

From this result, one deduces the existence of global solutions of (28) evolving from initial data belonging to a set of full μ_1 measure. One then has all the tools to show the invariance of μ_1 under the flow of (28) [4, Section 4]. For $t \in \mathbb{R}$, we let S_t denote the time evolution of (28). By the earlier discussion, this is defined almost everywhere. For $\mathcal{K} \subset H^\sigma$ a compact set, one shows by using the approximation lemma [4, Lemma 2.27] and the invariance of μ_1^n under the flow of (37) that

$$\mu_1(S_t K) \geq \mu_1(K). \tag{43}$$

The reverse inequality in (43) follows by time-reversibility of the flow. Finally, the general claim follows by inner regularity of the measure μ_1.

Our methods in [1] differ in the sense that we do not need to approximate the flow of (28) with (37). In particular, we do not use variants of the approximation lemma [4, Lemma 2.27]. Instead, we use global measure-theoretic techniques.

3.2 *General Principle*

In this paragraph, we underline a general principle that enlightens in some sense the above discussion about the construction of global solutions by means of Gibbs or invariant measures. Although the result in Theorem 8 below holds true in a more general context, it is convenient here to consider the finite and infinite dimensional frameworks given respectively in Sects. 2.1 and 2.2. Recall that the Hamiltonian functions and the vector fields are respectively:

$$h : E \to \mathbb{R} \in \mathscr{C}^1(E) \quad \text{and} \quad X(u) = J\nabla h(u), \tag{44}$$

and

$$h(u) = \frac{1}{2}\langle u, Au\rangle + h^I(u) \quad \text{and} \quad X(u) = -iAu - i\nabla h^I(u), \tag{45}$$

with A satisfying (11)–(12) and the nonlinearity h^I verifying (15a) for $p = 2$.

Definition 9 (Invariance) A probability measure $\mu \in \mathcal{P}(H^{-s})$ (resp. $\mu \in \mathcal{P}(E)$) is an invariant measure for the Hamiltonian system (45) (resp. (44)) if and only if for all $F \in \mathscr{C}^\infty_{c,cyl}(H^{-s})$ (resp. $F \in \mathscr{C}^\infty_c(E)$),

$$\int_{H^{-s}} \langle \nabla F(u), X(u)\rangle \, d\mu = 0 \quad (\text{resp.} \int_E \langle \nabla F(u), X(u)\rangle \, d\mu = 0).$$

The following result illustrates the general principle:

$$(\mu \text{ invariant measure}) \Longrightarrow (\text{existence of global solutions } \mu\text{-almost surely}).$$

Theorem 8 (Ammari-Farhat-Sohinger) *Let μ be an invariant measure for the Hamiltonian system* (45) *(resp.* (44)*). Assume in the case* (44) *that $\nabla h \in L^1(\mu)$ and in the case* (45) *suppose that $\nabla h^I \in L^1(\mu)$. Then the initial value problem*

$$\dot{u}(t) = -ie^{itA}\nabla h^I(e^{-itA}u(t)) \quad (resp.\ \ \dot{u}(t) = J\nabla h(u(t)),$$

admits global solutions for μ-almost all initial conditions in H^{-s} (resp. E).

Such a result applies of course to Gibbs measures on H^{-s} or E and in particular to the NLS equations as in Sect. 3.1. It shows also that the statistical properties of Gibbs measures do not play an important role in the construction of global solutions. But instead the invariance of the measure (according to Definition 9) is crucial. A detailed proof can be found in [2].

References

1. Ammari, Z., Sohinger, V.: Gibbs measures as unique KMS equilibrium states of nonlinear Hamiltonian PDEs. Rev. Mat. Iberoam. **39**(1), 29–90 (2023)
2. Ammari, Z., Farhat, S., Sohinger, V.: Almost sure existence of global solutions for general initial value problems. https://arxiv.org/abs/2305.17789
3. Bourgain, J.: Fourier transform restriction phenomena for certain lattice subsets and applications to nonlinear evolution equations. I. Schrödinger equations. Geom. Funct. Anal. **3**(2), 107–156 (1993)
4. Bourgain, J.: Periodic nonlinear Schrödinger equation and invariant measures. Commun. Math. Phys. **166**(1), 1–26 (1994)
5. Bourgain, J.: Invariant measures for the 2D-defocusing nonlinear Schrödinger equation. Commun. Math. Phys. **176**(2), 421–445 (1996)
6. Bourgain, J.: Invariant measures for the Gross-Pitaevskii equation. J. Math. Pures Appl. **76**, 649–702 (1997)
7. Bowen, R.: Equilibrium States and the Ergodic Theory of Anosov Diffeomorphisms. Lecture Notes in Mathematics, vol. 470. Springer, Berlin-New York (1975)
8. Bratteli, O., Robinson, D.W.: Operator Algebras and Quantum Statistical Mechanics, vol. 2. Texts and Monographs in Physics, 2nd edn. Springer, Berlin (1997). Equilibrium states. Models in quantum statistical mechanics
9. Brydges, D.C., Slade, G.: Statistical mechanics of the 2-dimensional focusing nonlinear Schrödinger equation. Commun. Math. Phys. **182**(2), 485–504 (1996)
10. Burq, N., Tzvetkov, N.: Random data Cauchy theory for supercritical wave equations. II. A global existence result. Invent. Math. **173**(3), 477–496 (2008)
11. Burq, N., Thomann, N., Tzvetkov, N.: Remarks on the Gibbs measures for nonlinear dispersive equations. Ann. Fac. Sci. Toulouse Math. (6) **27**(3), 527–597 (2018)
12. Deng, Y., Nahmod, A.R., Yue, H.: Invariant Gibbs measure and global strong solutions for the Hartree NLS equation in dimension three. J. Math. Phys. **62**(3), Paper No. 031514, 39 pp. (2021)
13. Gallavotti, G., Verboven, E.: On the classical KMS boundary condition. Nuovo Cimento B (11) **28**(1), 274–286 (1975)
14. Lebowitz, J.L., Rose, H.A., Speer, E.R.: Statistical mechanics of the nonlinear Schrödinger equation. J. Stat. Phys. **50**(3–4), 657–687 (1988)
15. Nahmod, A.R., Staffilani, G.: Randomness and nonlinear evolution equations. Acta Math. Sin. (Engl. Ser.) **35**(6), 903–932 (2019)

16. Nelson, E.: The free Markoff field J. Funct. Anal. **12**, 211–227 (1973)
17. Oh, T., Thomann, L.: A pedestrian approach to the invariant Gibbs measures for the 2-d defocusing nonlinear Schrödinger equations. Stoch. Partial Differ. Equ. Anal. Comput. **6**(3), 397–445 (2018)
18. Preston, C.: Random Fields. Lecture Notes in Mathematics, vol. 534. Springer, Berlin-New York (1976)
19. Simon, B.: The $P(\phi)_2$ Euclidean (quantum) Field Theory. Princeton Series in Physics. Princeton University Press, Princeton (1974)
20. Zhidkov, P.E.: An invariant measure for the nonlinear Schrödinger equation. Dokl. Akad. Nauk SSSR **317**(3), 543–546 (1991)

Two Comments on the Derivation of the Time-Dependent Hartree–Fock Equation

Niels Benedikter and Davide Desio

1 Interacting Fermi Gases at High Density

In condensed matter, one-, two-, and three-dimensional quantum systems are realized. In a basic approximation, an ordinary piece of metal can be modelled as a gas of interacting fermions in three dimensions; transistor-like semiconductor structures can in first approximation be considered as a two-dimensional electron gas; and the one-dimensional electron gas may be used as a simplified model of a carbon nanotube. Mathematically even these simple models are difficult to study because a quantum system of N particles is described by a vector in the antisymmetrized tensor product of N copies of $L^2(\mathbb{R}^d)$. As N is easily of the order of 10^4 and more likely up to 10^{23}, numerical methods quickly find their limits in the analysis of the many-body Schrödinger equation. One way of overcoming this difficulty is the use of effective equations: in idealized physical regimes the Schrödinger equation may be approximated by equations involving fewer degrees of freedom. For fermions, Hartree–Fock theory is such an approximation: one considers initial data given as an antisymmetrized elementary tensor (a Slater determinant) and then projects [9, 23] the many-body Schrödinger evolution on the submanifold of antisymmetrized elementary tensors. In the present note we show that the quantitative error estimates proved in [7] for the Hartree–Fock equation in dimension $d = 3$ generalize to all space dimensions, and we reformulate the proof using an explicit formula for the unitary implementation of a particle-hole

N. Benedikter (✉)
Dipartimento di Matematica, Università degli studi di Milano, Milano, Italy
e-mail: niels.benedikter@unimi.it

D. Desio
Dipartimento di Fisica, Università degli studi di Milano, Milano, Italy
e-mail: davide.desio@studenti.unimi.it

M. Correggi, M. Falconi (eds.), *Quantum Mathematics I*,
Springer INdAM Series 57, https://doi.org/10.1007/978-981-99-5894-8_13

transformation, thus casting it in a form completely analogous to the coherent state method of [27] for bosons.

In the following paragraphs we will introduce the many-body Schrödinger equation, the scaling regime, reduced density matrices, and the Hartree–Fock equation.

1.1 Fundamental Description: The Schrödinger Equation

The fundamental theory is given by the Hamiltonian (with a coupling constant $\lambda \in \mathbb{R}$)

$$H_N := -\sum_{i=1}^{N} \Delta_i + \lambda \sum_{1 \le i < j \le N} V(x_i - x_j)\,, \tag{1}$$

a self-adjoint operator on the antisymmetric subspace $L^2_{\mathrm{a}}(\mathbb{R}^{dN})$ of $L^2(\mathbb{R}^d)^{\otimes N} \simeq L^2(\mathbb{R}^{dN})$, i.e., functions $\psi \in L^2(\mathbb{R}^{dN})$ satisfying

$$\psi(x_1, x_2, \ldots, x_N) = \operatorname{sgn}(\sigma)\psi(x_{\sigma(1)}, x_{\sigma(2)}, \ldots, x_{\sigma(N)}) \quad \text{for } \sigma \in \mathcal{S}_N \tag{2}$$

where $\mathcal{S}_N$ is the group of all permutations of N objects. The evolution of initial data $\psi_0 \in L^2_{\mathrm{a}}(\mathbb{R}^{dN})$ is given by the Schrödinger equation

$$i\partial_t \psi_t = H_N \psi_t\,. \tag{3}$$

Our goal is to approximate solutions of Eq. (3) by the time-dependent Hartree–Fock equation. Considering an appropriate scaling of the system parameters with the particle number N, one can prove estimates on the difference asymptotically as $N \to \infty$. In the next paragraph we discuss our choice of such a scaling regime.

1.2 Coupled Mean-Field and Semiclassical Scaling Regime

No approximation applies to all physical situations. The situation we consider is the scaling limit introduced by [24, 29] for deriving the Vlasov equation from quantum mechanics. In this setting the density of the system is large but the interaction between any pair of particles weak, so that mean-field like behaviour may be expected. To derive the precise choice of parameters we consider for the moment the torus $\mathbb{T}^d := \mathbb{R}^d/2\pi\mathbb{Z}^d$ instead of $\mathbb{R}^d$. The simplest fermionic wave functions are antisymmetrized elementary tensors (i.e., Slater determinants)

$$\begin{aligned} \psi(x_1, x_2, \ldots x_N) &= f_1 \wedge \cdots \wedge f_N(x_1, \ldots, x_N) \\ &= (N!)^{-1/2} \det\big(f_j(x_i)\big)_{i,j=1,\ldots,N}\,. \end{aligned} \tag{4}$$

Ignoring for the moment the interaction V, the ground state is the Slater determinant of N plane waves $f_j(x) := (2\pi)^{-d/2} e^{ik_j \cdot x}$ where

$$k_j \in B_\mathrm{F} := \{k \in \mathbb{Z}^d : |k| \leq k_\mathrm{F}\} .$$

If instead of using N as independent parameter we use the Fermi momentum $k_\mathrm{F} > 0$, i.e., define $N := |B_\mathrm{F}|$ as a function of k_F, then the Slater determinant of the plane waves with $k_j \in B_\mathrm{F}$ is the unique minimizer of the non-interacting Hamiltonian. Since $k_\mathrm{F} \sim N^{1/d}$, the total kinetic energy becomes

$$\langle \psi, \Big(-\sum_{i=1}^{N} \Delta_i \Big) \psi \rangle = \sum_{k \in B_\mathrm{F}} |k|^2 \sim N^{1+\frac{2}{d}} \qquad \text{as } k_\mathrm{F} \to \infty . \tag{5}$$

Now let us bring back the interaction into the game, and consider its expectation value in the same Slater determinant of plane waves. To have a large-N limit in which neither kinetic nor interaction energy (as a sum over pairs being of order λN^2) dominates, we set

$$\lambda := N^{\frac{2}{d}-1} .$$

The particles most affected by the interaction are those close to the surface of the Fermi ball B_F, i.e., with momenta $|k| \sim k_\mathrm{F} \sim N^{1/d}$. Like their momentum, also their velocity is of order $N^{1/d}$. Therefore we study times of order $N^{-1/d}$; the accordingly rescaled equation is

$$i N^{1/d} \partial_t \psi_t = \Big(\sum_{i=1}^{N} -\Delta_i + N^{\frac{2}{d}-1} \sum_{1 \leq i < j \leq N} V(x_i - x_j) \Big) \psi_t .$$

Introducing an effective Planck constant

$$\hbar := N^{-1/d}$$

and multiplying the entire equation by $\hbar^2$, we obtain the rescaled Schrödinger equation we study in this note:

$$i\hbar \partial_t \psi_t = \Big(\sum_{i=1}^{N} -\hbar^2 \Delta_i + \frac{1}{N} \sum_{1 \leq i < j \leq N} V(x_i - x_j) \Big) \psi_t . \tag{6}$$

Other scaling limits, with weaker interaction or shorter time scale, have been considered in [1–3, 21, 25].

1.3 Reduced Density Matrices

Given an N-particle observable $\mathcal{A}$, i.e., a self-adjoint operator on $L^2_{\mathrm{a}}(\mathbb{R}^{dN})$, its expectation value in a state $\psi \in L^2_{\mathrm{a}}(\mathbb{R}^{dN})$ can be written with a trace over $L^2_{\mathrm{a}}(\mathbb{R}^{dN})$ in Dirac's bra-ket notation as

$$\langle \psi, \mathcal{A}\psi \rangle = \mathrm{tr}_N \left(|\psi\rangle\langle\psi|\mathcal{A} \right) .$$

Simpler observables are the averages of one-particle observables: if A is an operator on $L^2(\mathbb{R}^d)$ and A_j means A acting on the j-th of N tensor factors, $A_j := 1 \otimes \cdots \otimes 1 \otimes A \otimes 1 \otimes \cdots \otimes 1$, the expectation value can be written with a partial trace over $N-1$ tensor factors as

$$\frac{1}{N}\sum_{j=1}^{N} \langle \psi, A_j \psi \rangle = \langle \psi, A_1 \psi \rangle = \mathrm{tr}_1 \left(A \, \mathrm{tr}_{N-1} |\psi\rangle\langle\psi| \right) .$$

The one-particle reduced density matrix, an operator on the one-particle space $L^2(\mathbb{R}^d)$, is

$$\gamma^{(1)}_\psi := N \, \mathrm{tr}_{N-1} |\psi\rangle\langle\psi| . \tag{7}$$

As a trace class operator, the spectral theorem permits to decompose it as

$$\gamma^{(1)}_\psi = \sum_{j \in \mathbb{N}} \lambda_j |\varphi_j\rangle\langle\varphi_j| , \qquad \varphi_j \in L^2(\mathbb{R}^d) , \qquad \lambda_j \in \mathbb{R} .$$

In particular we may speak of its integral kernel and its "diagonal" (representing the density of particles in position space), defined by

$$\gamma^{(1)}_\psi(x;y) := \sum_{j \in \mathbb{N}} \lambda_j \varphi_j(x)\overline{\varphi_j(y)} , \qquad \gamma^{(1)}_\psi(x;x) := \sum_{j \in \mathbb{N}} \lambda_j |\varphi_j(x)|^2 .$$

A Slater determinant $\psi(x_1, x_2, \ldots x_N) = (N!)^{-1/2} \det(\varphi_j(x_i))$ is an example of a quasi-free state, and as such uniquely (up to a phase factor) determined by its one-particle reduced density matrix. The one-particle reduced density matrix of a Slater determinant is a rank-N projection, i.e., of the λ_j in the spectral decomposition N have value 1 and the rest are 0.

1.4 Effective Description: Hartree–Fock Theory

In Hartree–Fock theory, attention is restricted to Slater determinants, with the choice of the orbitals φ_j to be optimized. Projecting the time-dependent Schrödinger equation locally onto the tangent space of this submanifold (i.e., applying the Dirac–Frenkel principle, see [9, 23]) one obtains the time-dependent Hartree–Fock equations (a system of N non-linear coupled equations)

$$i\hbar\partial_t\varphi_{j,t} = -\hbar^2\Delta\varphi_{j,t} + \frac{1}{N}\sum_{i=1}^{N}\Big(V * |\varphi_{i,t}|^2\Big)\varphi_{j,t} - \frac{1}{N}\sum_{i=1}^{N}\Big(V * (\varphi_{j,t}\overline{\varphi_{i,t}})\Big)\varphi_{i,t}\,. \tag{8}$$

Using the one-particle density matrix $\omega_{N,t} := \sum_{j=1}^{N}|\varphi_{j,t}\rangle\langle\varphi_{j,t}|$ they take the form

$$i\hbar\partial_t\omega_{N,t} = [-\hbar^2\Delta + (V * \rho_t) - X_t, \omega_{N,t}]\,. \tag{9}$$

The term $V * \rho_t$ with $\rho_t(x) := \omega_{N,t}(x;x)$ is a multiplication operator called the direct term. The exchange term X_t is defined by its integral kernel $X_t(x;x') = V(x-x')\omega_{N,t}(x;x')$.

Given a rank-N projection operator ω_N as initial data, the solution of Eq. (9) is for all times a rank-N projection operator. From its spectral decomposition, fixing the phase ambiguity appropriately, one obtains the N orbitals solving Eq. (8).

2 Main Result

Let X be the one-particle position operator on $L^2(\mathbb{R}^d)$, i.e., the multiplication operator $X\psi(x) = x\psi(x)$ for $x \in \mathbb{R}^d$. Let $P := -i\hbar\nabla$ be the one-particle momentum operator.

Let ω_N be a sequence of rank-N projection operators on $L^2(\mathbb{R}^d)$. Let $\psi_{N,0}$ be the Slater determinant uniquely (up to a phase factor) determined by ω_N. Let $\gamma_{N,t}^{(1)}$ be the one-particle reduced density matrix of the solution $\psi_{N,t} := e^{-iH_N t/\hbar}\psi_{N,0}$ of the Schrödinger equation. Let $\omega_{N,t}$ be the solution of the Hartree–Fock equation Eq. (9) with initial data ω_N.

We have now introduced everything necessary to state our main result, according to which $\omega_{N,t}$ is a good approximation to $\gamma_{N,t}^{(1)}$.

Theorem 1 (Validity of the Hartree–Fock Equation) *Let $d \in \mathbb{N}$. Consider $V \in L^1(\mathbb{R}^d)$ with Fourier transform satisfying*

$$q_0 := \int \mathrm{d}p(1+|p|)^2|\hat{V}(p)| < \infty.$$

Assume there exist $C_X > 0$ and $C_P > 0$ such that for all $i \in \mathbb{N} \cap [1,d]$ and for all $N \in \mathbb{N}$ we have

$$\sup_{\alpha\in\mathbb{R}^d}\frac{\|[e^{i\alpha\cdot X}, \omega_N]\|_{\mathrm{tr}}}{1+|\alpha|} \le N\hbar\, C_X\,, \qquad \|[P, \omega_N]\|_{\mathrm{tr}} \le N\hbar\, C_P\,. \tag{10}$$

(The latter estimate is to be read in ℓ^2-sense with respect to the components of the momentum operator, i.e., $\|[P,\omega_N]\|_{\mathrm{tr}} = (\sum_{i=1}^{d} \|[P_i,\omega_N]\|_{\mathrm{tr}}^2)^{1/2}$.)

Then for all $t \in \mathbb{R}$ and for $N \in \mathbb{N}$ sufficiently large we have

$$\|\gamma^{(1)}_{N,t} - \omega_{N,t}\|_{\mathrm{tr}} \leq \sqrt{N} 6 \exp\left(2^3 \frac{C_X + C_P}{\max\{2,q_0\}} e^{2\max\{2,q_0\}|t|}\right). \tag{11}$$

The trace norm estimate of order $N^{1/2}$ is to be compared to the triangle inequality which would yield $2N$. As in [7], the result may be generalized to k-particle reduced density matrices; and as in [6] it can be generalized to relativistic massive particles.

Remark 1 The assumption Eq. (10) is realized by the Fermi ball (see Eqs. (4) and (5)), which however is stationary under the Hartree–Fock evolution (for $\hat{V} \geq 0$ it is even the global minimizer [11, Theorem A.1]). The assumption is also realized by some examples with non-trivial Hartree–Fock evolution such as the ground state of non-interacting fermions in a harmonic trap [5] or even a general trapping potential [19]. Actually, in [5] a bound was shown for $\|[X_i,\omega_N]\|_{\mathrm{tr}}$ instead of $\sup_{\alpha\in\mathbb{R}^d} \|[e^{i\alpha\cdot X},\omega_N]\|_{\mathrm{tr}}(1+|\alpha|)^{-1}$. These are related by

$$\begin{aligned}
[\omega_N, e^{i\alpha\cdot X}] &= e^{i\alpha\cdot X} \int_0^1 \mathrm{d}\lambda \, \frac{\mathrm{d}}{\mathrm{d}\lambda} \left(e^{-i\alpha\cdot X\lambda} \omega_N e^{i\alpha\cdot X\lambda}\right) \\
&= e^{i\alpha\cdot X} \int_0^1 \mathrm{d}\lambda \, e^{-i\alpha\cdot X\lambda} [\omega_N, i\alpha\cdot X] e^{i\alpha\cdot X\lambda},
\end{aligned}$$

so (as shown similarly also in [19, Corollary 1.3])

$$\begin{aligned}
\sup_{\alpha\in\mathbb{R}^d} \frac{\mathrm{tr}|[\omega_N, e^{i\alpha\cdot X}]|}{1+|\alpha|} &\leq \sup_{\alpha\in\mathbb{R}^d} \frac{1}{1+|\alpha|} \mathrm{tr}|[\omega_N, \alpha\cdot X]| \\
&\leq \sup_{\alpha\in\mathbb{R}^d} \frac{1}{1+|\alpha|} \sum_{j=1}^{d} |\alpha_j| \, \mathrm{tr}|[\omega_N, X_j]| \\
&\leq \sup_{\alpha\in\mathbb{R}^d} \frac{|\alpha|}{1+|\alpha|} \left[\sum_{j=1}^{d} \left(\mathrm{tr}|[\omega_N, X_j]|\right)^2\right]^{1/2} = \|[\omega_N, X]\|_{\mathrm{tr}}.
\end{aligned}$$

In [26, 28], a theorem similar to Theorem 1 has been proved for more singular interaction potentials but for initial data which is stationary under the time-dependent Hartree–Fock equation. The Hartree–Fock equation has also been derived for initial data given by a mixed state [8]. This has been generalized to singular interaction potentials, including the Coulomb potential and the gravitational attraction in [14, 15]. The validity of the Hartree–Fock equation has been derived for extended Fermi gases in three dimensions by [20]. Next-order corrections (the random phase approximation) and a Fock space norm approximation, however only for approximately bosonic collective excitations of the stationary Fermi ball, have been obtained in [13], based on the collective bosonization method developed in [4, 10–12]. A non-collective bosonization method has recently been developed in [16–18]. For a discussion of different levels of dynamical approximation, see the review [5].

3 Proof of Theorem 1

Let us quickly fix some notation. Fermionic Fock space is defined as

$$\mathcal{F} := \mathbb{C} \oplus \bigoplus_{n=1}^{\infty} L^2_{\mathrm{a}}(\mathbb{R}^{dn}) \, .$$

For $f, g \in L^2(\mathbb{R}^d)$, the well-known creation and annihilation operators $a^*(f)$ and $a(g)$ satisfy the canonical anticommutator relations

$$\{a(f), a^*(g)\} = \langle f; g \rangle \, , \qquad \{a(f), a(g)\} = 0 = \{a^*(f), a^*(g)\} \, .$$

In the fermionic case these operators satisfy for all $\psi \in \mathcal{F}$ the bounds

$$\|a(f)\psi\|_{\mathcal{F}} \leq \|f\|_{L^2(\mathbb{R}^d)} \|\psi\|_{\mathcal{F}} , \qquad \|a^*(f)\psi\|_{\mathcal{F}} \leq \|f\|_{L^2(\mathbb{R}^d)} \|\psi\|_{\mathcal{F}} .$$

The particle number operator is denoted by $\mathcal{N}$. The vacuum is $\Omega = (1, 0, 0, 0, \ldots)$, the (up to a phase) unique vector in the null space of all annihilation operators. This implies $\mathcal{N}\Omega = 0$. Moreover, given any operator A on $L^2(\mathbb{R}^d)$ with integral kernel $A(x; y)$, its second quantization written using the operator valued distributions associated to the creation and annihilation operators is

$$\mathrm{d}\Gamma(A) := \int \mathrm{d}x \mathrm{d}y \, A(x; y) a^*_x a_y \, .$$

The following lemma collects standard bounds; see [7, Section 3] for proofs.

Lemma 1 (Bounds for Second Quantization) *Let $\psi \in \mathcal{F}$ and let A be an operator on $L^2(\mathbb{R}^d)$. Then we have*

$$\|\mathrm{d}\Gamma(A)\psi\|_{\mathcal{F}} \leq \|A\|_{\mathrm{op}} \|\mathcal{N}\psi\|_{\mathcal{F}} , \tag{12}$$

$$\|\mathrm{d}\Gamma(A)\psi\|_{\mathcal{F}} \leq \|A\|_{\mathrm{HS}} \|\mathcal{N}^{1/2}\psi\|_{\mathcal{F}} , \tag{13}$$

$$\|\mathrm{d}\Gamma(A)\psi\|_{\mathcal{F}} \leq \|A\|_{\mathrm{tr}} \|\psi\|_{\mathcal{F}} . \tag{14}$$

Moreover, if A has an integral kernel $A(x; y)$, then

$$\|\int \mathrm{d}x\mathrm{d}y A(x; y) a_x a_y \psi\|_{\mathcal{F}} \leq \|A\|_{\mathrm{HS}} \|\mathcal{N}^{1/2}\psi\|_{\mathcal{F}}, \tag{15}$$

$$\|\int \mathrm{d}x\mathrm{d}y A(x; y) a_x^* a_y^* \psi\|_{\mathcal{F}} \leq 2\|A\|_{\mathrm{HS}} \|(\mathcal{N}+1)^{1/2}\psi\|_{\mathcal{F}}, \tag{16}$$

and

$$\|\int \mathrm{d}x\mathrm{d}y A(x; y) a_x a_y \psi\|_{\mathcal{F}} \leq 2\|A\|_{\mathrm{tr}} \|\psi\|_{\mathcal{F}}, \tag{17}$$

$$\|\int \mathrm{d}x\mathrm{d}y A(x; y) a_x^* a_y^* \psi\|_{\mathcal{F}} \leq 2\|A\|_{\mathrm{tr}} \|\psi\|_{\mathcal{F}}. \tag{18}$$

Finally, note that the definition of the one-particle reduced density matrix may be generalized to $\psi \in \mathcal{F}$ by setting

$$\gamma_{\psi}^{(1)}(x; y) := \langle \psi, a_y^* a_x \psi \rangle . \tag{19}$$

In fact, if $\psi \in L^2_{\mathrm{a}}(\mathbb{R}^{dN})$ is considered as a subspace of Fock space, then this $\gamma_{\psi}^{(1)}$ is exactly the integral kernel of the operator defined in Eq. (7).

3.1 Implementation of Particle-Hole Transformations

Let $(\varphi_j)_{j=1}^N$ be an orthonormal system in $L^2(\mathbb{R}^d)$. In [7] an explicit formula for R_N is absent. Therefore, to do any computations, a formula for the conjugation of the number operator with R_N was used to compute the time derivative in Lemma 5. In the present proof, we introduce the following explicit definition which allows us to instead compute the generator of fluctuations as done for the bosonic case in [27]:

$$R_N := \prod_{j=1}^{N} \left(a^*(\varphi_j) + a(\varphi_j)\right) . \tag{20}$$

This is a unitary map on Fock space which maps the vacuum on a Slater determinant,

$$R_N \Omega = \prod_{j=1}^{N} a^*(\varphi_j)\Omega = (N!)^{-1/2} \det(\varphi_j(x_i)) ,$$

and satisfies

$$R_N a^*(\varphi_j) R_N^* = \begin{cases} (-1)^{N+1} a(\varphi_j) \text{ for } j \leq N \\ (-1)^N a^*(\varphi_j) \text{ for } j > N \,. \end{cases} \tag{21}$$

The formula Eq. (20) is an implementation of a particle-hole transformation as constructed by abstract Bogoliubov theory in [7]. We got aware of this formula from [22, Eq. (57)].

Moreover it is convenient to introduce the operators

$$Q_N := \sum_{j=1}^{N} |\varphi_j\rangle\langle\overline{\varphi_j}| \,, \qquad P_N := 1 - \sum_{j=1}^{N} |\varphi_j\rangle\langle\varphi_j| \,, \tag{22}$$

where $\overline{\varphi_j}$ is the complex conjugation of $\varphi_j \in L^2(\mathbb{R}^d)$. The action of the particle-hole transformation on creation and annihilation operators can then be computed to be

$$R_N^* a_x R_N = (-1)^N \left(a(P_{N,x}) - a^*(Q_{N,x}) \right) \tag{23}$$

$$R_N^* a_x^* R_N = (-1)^N \left(a^*(P_{N,x}) - a(Q_{N,x}) \right) \,, \tag{24}$$

where $Q_N(x; y)$ and $P_N(x; y)$ are (formal) integral kernels of the operators Q_N and P_N, and $Q_{N,x}(y) := Q_N(y; x)$, $P_{N,x}(y) := P_N(y; x)$ for all $y \in \mathbb{R}^d$.

We are going to use Eq. (20) to construct a unitary fluctuation dynamics as in [27]. The proof of the main theorem will then be obtained by an application of the Grönwall lemma, following the strategy of [7].

3.2 *Many-Body Analysis*

The Hamiltonian H_N may be represented on Fock space as

$$\mathcal{H}_N := \hbar^2 \int \mathrm{d}x \, \nabla_x a_x^* \nabla_x a_x + \frac{1}{2N} \int \mathrm{d}x\mathrm{d}y \, V(x-y) a_x^* a_y^* a_y a_x \,.$$

In fact, considering $L^2_{\mathrm{a}}(\mathbb{R}^{dN})$ as a subspace of $\mathcal{F}$, we have $\mathcal{H}_N \restriction_{L^2_{\mathrm{a}}(\mathbb{R}^{dN})} = H_N$. Since we consider only initial data in the N-particle subspace and the evolution preserves particle numbers (i.e., $[\mathcal{N}, \mathcal{H}_N] = 0$) we can use $\mathcal{H}_N$ in the place of H_N.

Let $\omega_{N,t}$ be the solution of the time-dependent Hartree–Fock equation (Eq. (9); for a discussion of the well-posedness see, e. g., [9]) with initial data ω_N. Let $\varphi_{j,t}$, with $j = 1, 2 \ldots N$ be the corresponding orthonormal systems of orbitals, and

$R_{N,t}$ the correspondingly constructed particle-hole transformation as in Eq. (20). We define the unitary fluctuation dynamics

$$\mathcal{U}_N(t,s) := R^*_{N,t} e^{-i(t-s)\mathcal{H}_N/\hbar} R_{N,s} \,. \tag{25}$$

The advantage of introducing the fluctuation dynamics $\mathcal{U}_N$ is the following representation of the difference that we want to estimate:

Lemma 2 (Trace Norm Difference) *Let ω_N be a rank-N projection operator and $\omega_{N,t}$ its Hartree–Fock evolution. Let $R_{N,0}$ and $R_{N,t}$ be the corresponding particle-hole transformations. Let moreover $\psi_{N,0} := R_{N,0}\Omega$ and $\psi_{N,t} := e^{-i\mathcal{H}_n t/\hbar}\psi_{N,0}$ its many-body Schrödinger evolution. Then for all $t \in \mathbb{R}$ we have*

$$\|\gamma^{(1)}_{N,t} - \omega_{N,t}\|_{\mathrm{tr}} \leq \left(2 + 4\sqrt{N}\right) \langle \mathcal{U}_N(t,0)\Omega, (\mathcal{N}+1)\mathcal{U}_N(t,0)\Omega\rangle \,.$$

The proof of Lemma 2 is unchanged from [7, Section 4].

As in Eq. (22), also for the Hartree–Fock evolved orbitals $\varphi_{j,t}$ we define

$$Q_{N,t} := \sum_{j=1}^N |\varphi_{j,t}\rangle\langle\overline{\varphi_{j,t}}| \,, \qquad P_{N,t} := 1 - \sum_{j=1}^N |\varphi_{j,t}\rangle\langle\varphi_{j,t}| \,.$$

The novelty of the present note lies in the use of the explicit formula Eq. (20) for computing the time derivative of $\langle \mathcal{U}_N(t,0)\Omega, (\mathcal{N}+1)\mathcal{U}_N(t,0)\Omega\rangle$. The computation is then essentially identical to that given for bosons in the derivation of the Hartree equation by the coherent states method of [27], simply with the Weyl operators $W(\sqrt{N}\varphi_t)$ replaced by $R_{N,t}$. The computation results in the following new lemma.

Lemma 3 (Generator of Fluctuations) *Given $\mathcal{U}_N(t;s)$ by Eq. (25), we define the generator of fluctuations $\mathcal{L}_N(t)$ by*

$$i\hbar\partial_t \mathcal{U}_N(t;s) = \mathcal{L}_N(t)\mathcal{U}_N(t;s) \,.$$

Then we have

$$\mathcal{L}_N(t) = \left(\mathcal{A}_N(t) + \mathcal{B}_N(t) + C_N(t) + \text{h.c.}\right) + \mathcal{M}_N(t) \,, \tag{26}$$

where

$$\mathcal{A}_N(t) := \frac{1}{2N}\int \mathrm{d}x\mathrm{d}y V(x-y) a^*(P_{N,t,x})a^*(P_{N,t,y})a^*(Q_{N,t,y})a^*(Q_{N,t,x})$$

$$\mathcal{B}_N(t) := \frac{1}{N}\int \mathrm{d}x\mathrm{d}y V(x-y) a^*(P_{N,t,x})a^*(P_{N,t,y})a^*(Q_{N,t,x})a(P_{N,t,y})$$

$$C_N(t) := \frac{1}{N}\int \mathrm{d}x\mathrm{d}y V(x-y) a^*(P_{N,t,x})a^*(Q_{N,t,x})a^*(Q_{N,t,y})a(Q_{N,t,y})$$

and where $\mathcal{M}_N(t)$ is an operator (a sum of quadratic and quartic products of a- and a^*-operators) that commutes with the particle number operator:

$$[\mathcal{M}_N(t), \mathcal{N}] = 0 \qquad \text{for all } N \in \mathbb{N} \text{ and all } t \in \mathbb{R} .$$

Proof In this proof $\mathcal{M}_N(t)$ may change from line to line without further comment. Obviously

$$\mathcal{L}_N(t) = (i\hbar\partial_t R^*_{N,t})R_{N,t} + R^*_{N,t}\mathcal{H}_N(t)R_{N,t} .$$

The contribution of $R^*_{N,t}\mathcal{H}_N R_{N,t}$ is easily computed using Eq. (23), expanding all the products and using the canonical anticommutator relations to obtain an expression completely in normal order (i.e., with creation operators to the left of annihilation operators). One finds

$$\begin{aligned} &R^*_{N,t}\hbar^2 \int \mathrm{d}x \nabla_x a^*_x \nabla_x a_x R_{N,t} \\ &= \sum_{j=1}^{N} a^*(\hbar^2\Delta\varphi_{j,t})a^*(\varphi_{j,t}) - \sum_{k,j=1}^{N} \langle\varphi_{j,t}, \hbar^2\Delta\varphi_{k,t}\rangle a^*(\varphi_{j,t})a^*(\varphi_{k,t}) \\ &+ \text{h.c.} + \mathcal{M}_N(t) \end{aligned} \tag{27}$$

and

$$\begin{aligned} &R^*_{N,t}\frac{1}{2N}\int \mathrm{d}x\mathrm{d}y V(x-y)a^*_x a^*_y a_y a_x R_{N,t} \\ &= \frac{1}{2N}\int dxdy\, V(x-y) \\ &\times \Big[a^*(P_{t,x})a^*(P_{t,y})a^*(Q_{t,y})a^*(Q_{t,x}) + 2a^*(P_{t,x})a^*(P_{t,y})a^*(Q_{t,x})a(P_{t,y}) \\ &+ 2a^*(P_{t,x})a^*(Q_{t,x})a^*(Q_{t,y})a(Q_{t,y}) - 2\langle Q_{t,y}, Q_{t,y}\rangle a^*(P_{t,x})a^*(Q_{t,x}) \\ &+ 2\langle Q_{t,y}, Q_{t,x}\rangle a^*(P_{t,x})a^*(Q_{t,y})\Big] + \text{h.c.} + \mathcal{M}_N(t) . \end{aligned} \tag{28}$$

The summand involving the time derivative is slightly more complicated to compute. We define

$$R(h) := a^*(h) + a(h)$$

for $h \in L^2(\mathbb{R}^d)$ and observe that $\{R(\varphi_{l,t}), R(\varphi_{k,t})\} = 2\delta_{l,k}$. Since $R_{N,t} = \prod_{k=1}^{N} R(\varphi_{k,t})$, by a slightly lengthy but straightforward computation

$$
\begin{aligned}
&(i\hbar\partial_t R^*_{N,t})R_{N,t} \\
&= i\hbar R(\partial_t\varphi_{N,t})R(\varphi_{N,t}) + i\hbar\sum_{j=1}^{N-1}\prod_{k=0}^{j-1} R(\varphi_{N-k,t})R(\partial_t\varphi_{N-j,t})\prod_{m=N-j}^{N} R(\varphi_{m,t}) \\
&= \sum_{k=1}^{N} i\hbar R(\partial_t\varphi_{k,t})R(\varphi_{k,t}) - 2i\hbar\sum_{k=1}^{N}\sum_{j=1}^{k-1}\mathrm{Re}\langle\varphi_{k,t}, \partial_t\varphi_{k-j,t}\rangle R(\varphi_{k,t})R(\varphi_{k-j,t}) \\
&= \sum_{k=1}^{N} i\hbar R(\partial_t\varphi_{k,t})R(\varphi_{k,t}) - i\hbar\sum_{k=1}^{N}\sum_{\substack{j=1\\ j\neq k}}^{N}\langle\varphi_{k,t}, \partial_t\varphi_{j,t}\rangle R(\varphi_{k,t})R(\varphi_{j,t}) \\
&= \sum_{k=1}^{N} a^*(i\hbar\partial_t\varphi_{k,t})a^*(\varphi_{k,t}) - \sum_{k=1}^{N}\sum_{j=1}^{N}\langle\varphi_{k,t}, i\hbar\partial_t\varphi_{j,t}\rangle a^*(\varphi_{k,t})a^*(\varphi_{j,t}) + \text{h.c.} \\
&\quad + \mathcal{M}_N(t)\,.
\end{aligned}
$$

In the last step we made use of $a^*(\varphi_{k,t})a^*(\varphi_{k,t}) = 0$. Thus

$$
\begin{aligned}
&(i\hbar\partial_t R^*_{N,t})R_{N,t} \\
&= \sum_{k=1}^{N} a^*(i\hbar\partial_t\varphi_{k,t})a^*(\varphi_{k,t}) - \sum_{k=1}^{N}\sum_{j=1}^{N}\langle\varphi_{k,t}, i\hbar\partial_t\varphi_{j,t}\rangle a^*(\varphi_{k,t})a^*(\varphi_{j,t}) \\
&\quad + \text{h.c.} + \mathcal{M}_N(t).
\end{aligned} \tag{29}
$$

Summing Eqs. (27), (28), and (29), the Hartree–Fock equation Eq. (9) implies the cancellation of all the quadratic (containing products of two creation or annihilation operators) terms that do not commute with $\mathcal{N}$. The remaining terms are as claimed in Eq. (26). □

3.3 *Propagation of Commutator Bounds*

The following lemma propagates the bounds on the commutators from the initial data to all times. Though stated in [7, Proposition 3.4] only for $d = 3$, the proof is without modifications valid for any $d \in \mathbb{N}$. This lemma refers only to the Hartree–Fock evolution.

Lemma 4 (Propagation of Commutator Bounds) *Let V and $\omega_{N,t}$ satisfy the same assumptions as in Eq. (1). Then for all $t \in \mathbb{R}$ and all $N \in \mathbb{N}$ we have*

$$\sup_{\alpha\in\mathbb{R}^d} \frac{\|[e^{i\alpha\cdot X}, \omega_{N,t}]\|_{\mathrm{tr}}}{1+|\alpha|} \le N\hbar(C_X + C_P)e^{2\max\{2,q_0\}|t|}, \tag{30}$$

$$\|[P, \omega_{N,t}]\|_{\mathrm{tr}} \le N\hbar(C_X + C_P)e^{2\max\{2,q_0\}|t|}. \tag{31}$$

The exponential time dependence may not be optimal; however, for our proof the important aspect of these bounds is that we gain at all times a factor $\hbar$ with respect to the naive bound $\|[e^{i\alpha\cdot X}, \omega_N]\|_{\mathrm{tr}} \le \|e^{i\alpha\cdot X}\omega_N\|_{\mathrm{tr}} + \|\omega_N e^{i\alpha\cdot X}\|_{\mathrm{tr}} = 2\|\omega_N\|_{\mathrm{tr}} = 2N$.

3.4 *Conclusion of the Proof*

With the following lemma we are back at [7, Proposition 3.3].

Lemma 5 *With $\mathcal{A}_N(t)$, $\mathcal{B}_N(t)$, and $C_N(t)$ as defined in Lemma 3, we have*

$$\begin{aligned} &i\hbar\frac{\mathrm{d}}{\mathrm{d}t}\langle \mathcal{U}_N(t,0)\Omega, (\mathcal{N}+1)\mathcal{U}_N(t,0)\Omega\rangle \\ &= -2i\,\mathrm{Im}\langle \mathcal{U}_N(t,0)\Omega, (4\mathcal{A}_N(t) + 2\mathcal{B}_N(t) + 2C_N(t))\,\mathcal{U}_N(t,0)\Omega\rangle\,. \end{aligned} \tag{32}$$

Proof Obviously

$$i\hbar\frac{\mathrm{d}}{\mathrm{d}t}\langle \mathcal{U}_N(t,0)\Omega, (\mathcal{N}+1)\mathcal{U}_N(t,0)\Omega\rangle = \langle \mathcal{U}_N(t,0)\Omega, [\mathcal{N}, \mathcal{L}_N(t)]\mathcal{U}_N(t,0)\Omega\rangle\,.$$

With the explicit formula for $\mathcal{L}_N(t)$ from Lemma 3 the result is obtained. □

One now writes $V(x-y)$ in Eq. (32) in terms of its Fourier transform and then, using Lemmas 4 and 1 one shows as in [7, Lemma 3.5] that

$$\begin{aligned} &\left|\hbar\frac{\mathrm{d}}{\mathrm{d}t}\langle \mathcal{U}_N(t,0)\Omega, (\mathcal{N}+1)\mathcal{U}_N(t,0)\Omega\rangle\right| \\ &\le \hbar\, 2^4(C_X + C_P)e^{2\max\{2,q_0\}|t|}\langle \mathcal{U}_N(t,0)\Omega, (\mathcal{N}+1)\mathcal{U}_N(t,0)\Omega\rangle \end{aligned}$$

for all $t \in \mathbb{R}$, whence the main result follows by Grönwall's lemma.

Acknowledgments NB has been supported by Gruppo Nazionale per la Fisica Matematica (GNFM) in Italy and the European Research Council (ERC) through the Starting Grant FERMIMATH, grant agreement nr. 101040991. The authors acknowledge the support of Istituto Nazionale di Alta Matematica "F. Severi" through the Intensive Period "INdAM Quantum Meetings (IQM22)". The authors do not have any conflicts of interest to disclose.

References

1. Bach, V., Breteaux, S., Petrat, S., Pickl, P., Tzaneteas, T.: Kinetic energy estimates for the accuracy of the time-dependent Hartree-Fock approximation with Coulomb interaction. J. Math. Pures Appl. **105**(1), 1–30 (2016)
2. Bardos, C., Golse, F., Gottlieb, A.D., Mauser, N.J.: Mean field dynamics of fermions and the time-dependent Hartree–Fock equation. J. Math. Pures Appl. **82**(6), 665–683 (2003)
3. Bardos, C., Golse, F., Gottlieb, A.D., Mauser, N.J.: Accuracy of the time-dependent Hartree–Fock approximation for uncorrelated initial states. J. Stat. Phys. **115**(3/4), 1037–1055 (2004)
4. Benedikter, N.: Bosonic collective excitations in Fermi gases. Rev. Math. Phys. **33**(1), 2060009 (2021)
5. Benedikter, N.: Effective dynamics of interacting fermions from semiclassical theory to the random phase approximation. J. Math. Phys. **63**(8), 081101 (2022)
6. Benedikter, N., Porta, M., Schlein, B.: Mean-field dynamics of fermions with relativistic dispersion. J. Math. Phys. **55**(2), 021901 (2014)
7. Benedikter, N., Porta, M., Schlein, B.: Mean-field evolution of Fermionic systems. Commun. Math. Phys. **331**(3), 1087–1131 (2014)
8. Benedikter, N., Jakšić, V., Porta, M., Saffirio, C., Schlein, B.: Mean-field evolution of Fermionic mixed states. Commun. Pure Appl. Math. **69**(12), 2250–2303 (2016)
9. Benedikter, N., Sok, J., Solovej, J.P.: The Dirac–Frenkel principle for reduced density matrices, and the Bogoliubov–de Gennes equations. Ann. Henri Poincaré **19**(4), 1167–1214 (2018)
10. Benedikter, N., Nam, P.T., Porta, M., Schlein, B., Seiringer, R.: Optimal upper bound for the correlation energy of a Fermi gas in the mean-field regime. Commun. Math. Phys. **374**(3), 2097–2150 (2020)
11. Benedikter, N., Nam, P.T., Porta, M., Schlein, B., Seiringer, R.: Correlation energy of a weakly interacting Fermi gas. Invent. Math. **225**(3), 885–979 (2021)
12. Benedikter, N., Porta, M., Schlein, B., Seiringer, R.: Correlation energy of a weakly interacting Fermi gas with large interaction potential (June 2021). Arch. Rational Mech. Anal. **247**(65) (2023). https://doi.org/10.1007/s00205-023-01893-6
13. Benedikter, N., Nam, P.T., Porta, M., Schlein, B., Seiringer, R.: Bosonization of Fermionic many-body dynamics. Ann. Henri Poincaré **23**(5), 1725–1764 (2022)
14. Chong, J.J., Lafleche, L., Saffirio, C.: From many-body quantum dynamics to the Hartree–Fock and Vlasov equations with singular potentials (May 2021). arXiv:2103.10946 [math-ph]
15. Chong, J.J., Lafleche, L., Saffirio, C.: On the l^2 rate of convergence in the limit from Hartree to Vlasov–Poisson equation (Mar 2022).arXiv:2203.11485 [math-ph, physics:quant-ph]
16. Christiansen, M.R., Hainzl, C., Nam, P.T.: The random phase approximation for interacting Fermi gases in the mean-field regime (June 2021). arXiv:2106.11161 [cond-mat, physics:math-ph]
17. Christiansen, M.R., Hainzl, C., Nam, P.T.: The Gell-Mann–Brueckner formula for the correlation energy of electron gas: a rigorous upper bound in the mean-field regime (Aug 2022). Commun. Math. Phys. **401**, 1469–1529 (2023). https://doi.org/10.1007/s00220-023-04672-2
18. Christiansen, M.R., Hainzl, C., Nam, P.T.: On the effective Quasi-Bosonic Hamiltonian of the electron gas: collective excitations and plasmon modes (June 2022). Lett. Math. Phys. **112**, 114 (2022). https://doi.org/10.1007/s11005-022-01607-1
19. Fournais, S., Mikkelsen, S.: An optimal semiclassical bound on commutators of spectral projections with position and momentum operators. Lett. Math. Phys. **110**(12), 3343–3373 (2020)
20. Fresta, L., Porta, M., Schlein, B.: Effective dynamics of extended Fermi gases in the high-density regime (Apr 2022). Commun. Math. Phys. **401**, 1701–1751 (2023). https://doi.org/10.1007/s00220-023-04677-x
21. Fröhlich, J., Knowles, A.: A microscopic derivation of the time-dependent Hartree-Fock equation with Coulomb two-body interaction. J. Stat. Phys. **145**(1), 23 (2011)

22. Lill, S.: Implementing Bogoliubov transformations beyond the Shale-Stinespring condition (Apr 2022). arXiv:2204.13407 [math-ph]
23. Lubich, C.: From Quantum to Classical Molecular Dynamics: Reduced Models and Numerical Analysis. Zurich Lectures in Advanced Mathematics. European Mathematical Society, Zürich (2008)
24. Narnhofer, H., Sewell, G.L.: Vlasov hydrodynamics of a quantum mechanical model. Commun. Math. Phys. **79**(1), 9–24 (1981
25. Petrat, S., Pickl, P.: A new method and a new scaling for deriving Fermionic mean-field dynamics. Math. Phys. Anal. Geom. **19**(1), 3 (2016)
26. Porta, M., Rademacher, S., Saffirio, C., Schlein, B.: Mean field evolution of Fermions with Coulomb interaction. J. Stat. Phys. **166**(6), 1345–1364 (2017)
27. Rodnianski, I., Schlein, B.: Quantum fluctuations and rate of convergence towards mean field dynamics. Commun. Math. Phys. **291**(1), 31–61 (2009)
28. Saffirio, C.: Mean-field evolution of Fermions with singular interaction. In: Cadamuro, D., Duell, M., Dybalski, W., Simonella, S. (eds.) Macroscopic Limits of Quantum Systems, vol. 270, pp. 81–99. Springer International Publishing, Cham (2018)
29. Spohn, H.: On the Vlasov hierarchy. Math. Methods Appl. Sci. **3**(1), 445–455 (1981)

Derivation of the Gross-Pitaevskii Theory for Interacting Fermions in a Trap

Andrea Calignano and Michele Correggi

1 Introduction

The low-temperature behavior of interacting fermions has been widely studied in the physics literature (see, e.g., the monographs [22, 24]), in order to understand phenomena as the occurrence of superconductivity in materials, i.e., a sudden drop of resistivity below a certain critical temperature. A microscopic model for such a phenomenon was proposed in the '50s in [4] by J. Bardeen, L. Cooper and R. Schrieffer, and it is nowadays very well known as the BCS theory: the presence of an attraction between the fermions may be responsible for the formation of (weakly) bound pairs (Cooper pairs) of fermions with opposite spin; such pairs behave in all respect as charged bosons and as such they undergo Bose-Einstein condensation below a certain critical temperature. The emergence of this collective behavior of Cooper pairs is the signature of the occurrence of superconductivity in the material, and it can be understood starting from the minimization of the free energy of the system given by the *BCS energy functional* depending on the two-particle reduced density matrix.

Few years before the appearance of the BCS description of superconductivity, a much more phenomenological macroscopic explanation was provided in [16] by V.L. Ginzburg and L.D. Landau. In the GL theory the superconducting features of the sample are encoded in an order parameter ψ, i.e., a complex wave function minimizing a suitable energy functional, which is supposed to approximate the free energy of the system (see [2, 6–8, 21] and references therein for some recent mathematical results). The connection between the two models was heuristically investigated in [17], but only much more recently a rigorous derivation of GL theory from the BCS model was obtained in [12] (see also the related papers

A. Calignano · M. Correggi (✉)
Dipartimento di Matematica, Politecnico di Milano, Milano, Italy
e-mail: andrea.calignano@polimi.it

M. Correggi, M. Falconi (eds.), *Quantum Mathematics I*,
Springer INdAM Series 57, https://doi.org/10.1007/978-981-99-5894-8_14

[11, 13, 15, 18–20]): it is shown that, in a translational invariant system in presence of slowly varying external potentials and close to the critical temperature for the superconductivity transition, the leading order of the BCS ground state energy is given by the minimum of the GL functional, provided the attraction admits at least a bound state and in the limit of zero ratio between the microscopic scale of the interaction and the macroscopic size of the sample. The zero-temperature analogue of the same result for a fermionic system in a bounded domain was successively obtained in [14], while a similar question for the Bogolubov-Hartree-Fock functional, i.e., the BCS energy functional with the addition of direct and exchange terms, was studied in [5].

The setting we consider here is quite close to the one addressed in [14, 18], i.e., we study the zero-temperature behavior of a gas of interacting fermions, but, unlike the previous references, here we assume the presence of a *confining external potential*. The particles interact via a two-body attraction, which is strong enough to bind two particles together. Naively, one may think that the fermions at low temperature would arrange in bounded pairs, so forming a bosonic gas, which then undergoes BE condensation. However, as in [5, 14, 18], one observes that the possibility to form a two-body bound state is in fact enough to generate the superconductivity transition, even though the gas does not exactly arrange in two-particle bound pairs.

Let us describe the setting more precisely: we set the length scale of the trap to be 1, while the microscopic interaction varies on a scale $h \ll 1$. The parameter h thus describes the ratio between the micro- and macroscopic scales and we study the limit $h \to 0$ of the ground state energy of the BCS energy functional and of any corresponding minimizer. We do not fix the number of particles a priori, but we study the grand-canonical problem in presence of a chemical potential μ.

We stress that the physical setting we are considering is not the typical one of BCS theory in which the formation of Cooper pairs occurs on a scale much larger than the mean interparticle distance. On the contrary, here, the size of bounded pairs is of order h and it is therefore much smaller than the mean distance travelled by fermions, which, as we are going to see, is of order $h^{1/3}$ (the density of particles if of order h^{-1}). There is however a physical regime in which this setting becomes meaningful, namely the BEC/BCS crossover region (see [18]), where for certain values of the two-particle scatting length, the picture is very close to the one considered here. Note also that, as a gas made of almost bosonic pairs, the system is dilute (see also next Remark 1 and the analogous discussions in [10, 23]), because the density times the microscopic volume where the interaction acts non-trivially is of order $h^{-1} \cdot h^3 = h^2 \ll 1$.

1.1 BCS Theory of Superconductivity

In the BCS model all the information about the state of the system is encoded in two variables: the reduced one-particle density matrix γ and the pairing density matrix

α. Hence, the system is fully described by an operator

$$\Gamma = \begin{pmatrix} \gamma & \alpha \\ \bar{\alpha} & 1 - \overline{\gamma} \end{pmatrix}, \qquad 0 \leqslant \Gamma \leqslant 1, \tag{1}$$

acting on $L^2(\mathbb{R}^3) \oplus L^2(\mathbb{R}^3)$. The bar denotes complex conjugation, i.e., the integral kernels of the operators $\overline{\gamma}, \overline{\alpha}$ are $\overline{\gamma(x, y)}$ and $\overline{\alpha(x, y)}$, respectively. For a given BCS state Γ, the BCS functional at $T = 0$ in macroscopic units is given by

$$\mathcal{E}_\mu^{\text{BCS}}[\Gamma] := \operatorname{Tr} \mathfrak{h}\, \gamma + \int_{\mathbb{R}^6} dxdy\, V\left(\tfrac{x-y}{h}\right) |\alpha(x, y)|^2, \tag{2}$$

where the one-body operator $\mathfrak{h} = -h^2 \Delta + h^2 W - \mu$ describes the energy of non-interacting electrons at chemical potential $\mu < 0$. The BCS ground state energy is

$$E_\mu^{\text{BCS}} := \inf_{0 \leqslant \Gamma \leqslant 1} \mathcal{E}_\mu^{\text{BCS}}[\Gamma]. \tag{3}$$

Assumption 1 (Existence of a Ground State) We assume that V is real, radially symmetric, locally integrable and bounded from below. Moreover, the two-particle operator $-\Delta + V$ is assumed to admit a normalized ground state $\alpha_0 \in L^2(\mathbb{R}^6)$ with corresponding energy $-E_0$, $E_0 > 0$, which in particular implies that the negative part of V is non-zero.

Assumption 2 (Spectral Gap) Let α_0 be the ground state as in Assumption 1 above. We assume that $\exists g > 0$ and $0 < \varepsilon < 1$, such that

$$P_{\alpha_0}^{\perp} [-(1 - \varepsilon)\Delta + V + E_0]\, P_{\alpha_0}^{\perp} \geqslant g P_{\alpha_0}^{\perp} \tag{4}$$

where $P_{\alpha_0}^{\perp}$ stands for the projector onto the orthogonal complement of α_0.

Assumption 3 (Trapping Potential) We also assume that $W \in C^1(\mathbb{R}^3)$ is positive and there exist $0 < \beta, c_1, c_2 < +\infty$ such that

$$\begin{cases} c_1 |x|^\beta \leqslant W(x) \leqslant c_2 |x|^\beta, \\ |\nabla W(x)| \leqslant c_2 \beta |x|^{\beta - 1}, \end{cases} \qquad \text{for } |x| \geqslant 1. \tag{5}$$

We stress that for the class of attractive potentials in Assumption 1, one can deduce by standard Agmon estimates (see, e.g.,[1]) the exponential decay of the bound state wave function α_0: there exists $b > 0$ such that

$$\int_{\mathbb{R}^3} dx\, |\alpha_0(x)|^2 e^{2bx} < +\infty. \tag{6}$$

Note also that Assumption 3 allows to Taylor expand

$$W(\eta + \xi/2) = W(\eta) + \tfrac{\xi}{2} \cdot \nabla W(\zeta), \tag{7}$$

with the variable ζ belonging to $(\eta, \eta + \xi/2)$. A special case of a potential satisfying Assumption 3 is obviously given by the harmonic potential. In this case, the two-body Hamiltonian perfectly decouples in relative and centre-of-mass coordinates, which allows to get rid of several error terms in the discussion below.

The condition $0 \leqslant \Gamma \leqslant 1$, which is often call admissibility of Γ, implies that the operator γ is hermitian, i.e. $\gamma(x, y) = \overline{\gamma(y, x)}$ and that α is such that $\overline{\alpha} = \alpha^{\dagger}$. Furthermore, the operators $\gamma, \alpha : L^2(\mathbb{R}^3) \to L^2(\mathbb{R}^3)$ have a specific physical meaning (see, e.g., [3] for a formal derivation): given a many-body fermionic state Ψ, we have

$$\gamma(x, y) = \left\langle a_x^{\dagger} a_y \right\rangle_{\Psi}, \qquad \alpha(x, y) = \left\langle a_x a_y \right\rangle_{\Psi} \tag{8}$$

i.e., they represent the one-particle density matrix of the system and the wave function of a Cooper pair, respectively. Here $a_x^{\dagger}, a_x$ are the fermionic creation and annihilation operators. In fact, in absence of any pairing between the fermions, the system is in the so-called *normal state*, which is characterized by a trivial off-diagonal component, i.e., $\alpha \equiv 0$. The emergence of superconductivity is then associated to a non-trivial α.

1.2 Gross-Pitaevskii Theory

For any $D \in \mathbb{R}$, the Gross-Pitaevskii (GP) energy functional is defined as

$$\mathcal{E}_D^{\mathrm{GP}}(\psi) := \int_{\mathbb{R}^3} d\eta \, \left\{ \tfrac{1}{4} |\nabla \psi|^2 + (W(\eta) - D)|\psi|^2 + g_{\mathrm{BCS}} |\psi|^4 \right\}, \tag{9}$$

where the coefficient $g_{\mathrm{BCS}} > 0$ represents the interaction strength among different pairs, and whose expression in terms of the microscopic quantities is provided in Theorem 1. The GP energy can be proven to be bounded from below for any positive g_{BCS} (see Corollary 1). We denote then the GP ground state energy by

$$E_D^{\mathrm{GP}} := \inf_{\psi \in \mathscr{D}^{\mathrm{GP}}} \mathcal{E}_D^{\mathrm{GP}}(\psi), \tag{10}$$

where $\mathscr{D}^{\mathrm{GP}} = \{\psi \in H^1(\mathbb{R}^3) | W|\psi|^2 \in L^1(\mathbb{R}^3)\}$ is the natural minimization domain for (9). We denote by ψ_* the corresponding minimizer, which can be shown to be unique up to choice of the phase by strict convexity of the functional in $|\psi|^2$.

We point out here that, mathematically speaking, the GP functional introduced above may as well be named *Ginzburg-Landau* functional, although the energy does not look exactly as the usual GL energy, which in a homogeneous sample would read

$$\mathcal{E}^{\mathrm{GL}}[\phi] = \int_{\mathbb{R}^3} d\eta \left\{ \tfrac{1}{4} |\nabla\phi|^2 + \tilde{g}_{\mathrm{BCS}} \left(1 - |\phi|^2\right)^2 \right\}. \tag{11}$$

However, it is possible (see below and the discussion in [9, Sect. 1]) to reduce the minimization of (9) to the one of a functional very close to (11) (in fact, its inhomogeneous counterpart).

Notice that the GP wave function ψ is not normalized in L^2 since we are performing the energy minimization in the grand canonical setting, and therefore we may think that $\|\psi\|_2$ is determined by the value of the chemical potential μ. Let us denote by N such a quantity, i.e., $N := \|\psi_*\|_2^2$, and let f_0 be the positive minimizer of the GP energy

$$\widetilde{\mathcal{E}}^{\mathrm{GP}}[f] = \int_{\mathbb{R}^3} d\eta \left\{ \tfrac{1}{4}|\nabla f|^2 + W(\eta)|f|^2 + g_{\mathrm{BCS}} N |f|^4 \right\},$$

with L^2 norm set equal to 1. Such a minimizer satisfies the variational equation

$$-\tfrac{1}{4}\Delta f_0 + W f_0 + 2 g_{\mathrm{BCS}} N f_0^3 = \mu_0 f_0,$$

for a chemical potential $\mu_0 = \widetilde{E}^{\mathrm{GP}} + gN \|f_0\|_4^4$, where we have set $\widetilde{E}^{\mathrm{GP}} := \inf_{\|\psi\|_2=1} \widetilde{\mathcal{E}}^{\mathrm{GP}}[\psi]$. With the splitting $\psi_* =: \sqrt{N} f_0 \phi_*$ and exploiting the variational equation for f_0, one gets

$$E_D^{\mathrm{GP}} = N \left\{ \widetilde{E}^{\mathrm{GP}} - D + \widetilde{\mathcal{E}}^{\mathrm{GL}}[\phi_*] \right\},$$

where the last term is a weighted Ginzburg-Laudau functional explicitly given by

$$\widetilde{\mathcal{E}}^{\mathrm{GL}}[\phi] = \int_{\mathbb{R}^3} d\eta \; f_0^2 \left\{ \tfrac{1}{4} |\nabla\phi|^2 + \tilde{g}_{\mathrm{BCS}} N f_0^2 \left(1 - |\phi|^2\right)^2 \right\}, \tag{12}$$

and ϕ_* its minimizer.

This makes apparent the connection between the GP and GL functionals, so that, from this point of view, both names are mathematically equivalent to identify (9). There is however a strong physical motivation (see also [18]) for the choice we made, namely the fact that the physical regime we are investigating is a BEC one: as described in Sect. 1, the mechanism behind the emergence of a collective behavior in the low-temperature Fermi gas considered here is not the usual BCS pairing phenomenon, but rather a condensation of fermionic pairs playing the role of bosonic molecules. The pairs have indeed a size of order $h \ll 1$ which is much smaller that the typical distance between the fermionic particles of order of the trap length scale $O(1)$.

2 Main Results

This section contains our main results about the semiclassical expansion of the BCS energy.

Theorem 1 (BCS Energy) *Let $\mu = -E_0 + Dh^2$, for some $D \in \mathbb{R}$ and let Assumption 1 to 2 to 3 be satisfied. Then,*

$$E_\mu^{\mathrm{BCS}} = hE_D^{\mathrm{GP}} + O(h^2), \tag{13}$$

as $h \to 0$, where

$$g_{\mathrm{BCS}} := (2\pi)^3 \int_{\mathbb{R}^3} dp\, (p^2 + E_0)|\hat{\alpha}_0(p)|^4. \tag{14}$$

Moreover, for any approximate ground state Γ of the BCS functional, i.e., such that

$$\mathcal{E}_\mu^{\mathrm{BCS}}[\Gamma] \leqslant E_\mu^{\mathrm{BCS}} + \varepsilon h, \qquad 0 < \varepsilon < +\infty, \tag{15}$$

its off-diagonal element α can be decomposed as

$$\alpha(x, y) = h^{-2}\psi\left(\tfrac{x+y}{2}\right)\alpha_0\left(\tfrac{x-y}{h}\right) + r(x, y), \tag{16}$$

where $\psi \in \mathscr{D}^{\mathrm{GP}}$ satisfies $\mathcal{E}_D^{\mathrm{GP}}(\psi) \leqslant E_D^{\mathrm{GP}} + \varepsilon + o(1)$, α_0 is the ground state of the two-particle operator and the correction r is small in the following sense:

$$\|r\|_{L^2}^2 = O(h), \qquad \|\nabla r\|_{L^2}^2 + \left\| W|r|^2 \right\|_{L^1} = O(h^{-1}). \tag{17}$$

Remark 1 (Diluteness) The expansion (16) together with the heuristics $\gamma \simeq \alpha\overline{\alpha}$ (see Sect. 3.4) suggests that the density of the gas in our setting is proportional[1] to $h^{-1}|\psi|^2$, i.e., the total number of particle is of order h^{-1}. This vindicates the statement about the diluteness of the system since the range of the two-body interaction is $\propto h$ and therefore the diluteness parameter $h^{-1}h^3 = h^2 \ll 1$ is small.

Remark 2 (Properties of α_0) Note that by the estimate (6), $\alpha_0 \in L^1 \cup H^1(\mathbb{R}^3)$, which guarantees that $\hat{\alpha}_0 \in L^\infty(\mathbb{R}^3)$, so that $\hat{\alpha}_0 \in L^p(\mathbb{R}^3)$ for any $p \geqslant 2$ and g_{BCS} is a finite quantity.

Whether the systems is superconducting in the asymptotic regime $h \to 0$ thus depends on the fact that the GP wave function ψ is non-trivial. For the GP minimizer this depends on the value of the coefficient D, which in turn is determined by

[1] In fact, it may be possible to prove a weak version of such a statement as in [14, Proposition 1.11] using the Griffith's argument, i.e., variation w.r.t. to the external potential. However, we omit this discussion here for the sake of brevity.

the chemical potential μ. In fact, one can infer [14, 18] from the properties of the function

$$\mu \mapsto E_{\mu}^{\mathrm{BCS}}, \tag{18}$$

which is continuous, concave, and monotone decreasing, that there exists a unique critical value $\mu_c(h)$ such that below μ_c superconductivity is present and above it the system is in the normal state. The exact definition of $\mu_c(h)$ is the following:

$$\mu_c(h) := \inf\left\{\mu < 0 \mid E_{\mu}^{\mathrm{BCS}} < 0\right\}, \tag{19}$$

i.e., it marks the threshold of the transition from a zero ground state energy (normal state) to a strictly negative one.

Theorem 2 (Critical Chemical Potential) *Under the assumptions of Theorem 1, the critical chemical potential at which the superconductivity phase transition takes place is*

$$\mu_c(h) = -E_0 + E_W h^2 + o(h^2), \tag{20}$$

as $h \to 0$, where E_W is the ground state energy of the one-particle operator $-\frac{1}{4}\Delta + W$.

3 Proofs

The key ingredient to prove Theorems 1 and 2 is given by the following Theorem 1, which provides the link between the BCS and GL functionals.

Proposition 1 (BCS and GP Functionals) *Let $\mu = -E_0 + Dh^2$, $D \in \mathbb{R}$. Then,*

1. *Upper bound: for any $\psi \in \mathscr{D}^{\mathrm{GP}}$ there exists an admissible state Γ_ψ such that*

$$\mathcal{E}_{\mu}^{\mathrm{BCS}}[\Gamma_\psi] \leqslant h\mathcal{E}_D^{\mathrm{GP}}(\psi) + Ch^2\left[1 + \left(\max\left\{\mathcal{E}_D^{\mathrm{GP}}(\psi), 0\right\}\right)^2\right]. \tag{21}$$

2. *Lower bound: let Γ be an admissible* BCS *state such that $\mathcal{E}_{\mu}^{\mathrm{BCS}}[\Gamma] \leqslant C_\Gamma h$. Then, there exists $\psi \in \mathscr{D}^{\mathrm{GP}}(\mathbb{R}^3)$ such that*

$$\mathcal{E}_{\mu}^{\mathrm{BCS}}[\Gamma] \geqslant h\mathcal{E}_D^{\mathrm{GP}}(\psi) - Ch^2. \tag{22}$$

Furthermore, there exists a function r such that the following decomposition holds:

$$\alpha(x, y) = h^{-2}\psi\left(\tfrac{x+y}{2}\right)\alpha_0\left(\tfrac{x-y}{h}\right) + r(x, y). \tag{23}$$

where the remainder r satisfies the bounds

$$\|r\|_{L^2}^2 \leqslant Ch, \qquad \langle r\,|-\Delta + W|\,r\rangle_{L^2} \leqslant Ch^{-1}. \tag{24}$$

Let us then assume that Theorem 1 holds and prove Theorems 1 and 2. The proof of Theorem 1 will be given in next Sects. 3.3 and 3.4 by separately addressing points (a) and (b) of the statement.

Proof *(Theorem 1)* To prove the upper bound, we use the admissible trial state Γ_{ψ_*}, where we recall that ψ_* stands for the minimizer of the GP functional. We then obtain by (21)

$$E_\mu^{\mathrm{BCS}} \leqslant \mathcal{E}_\mu^{\mathrm{BCS}}[\Gamma_{\psi_*}] = h\mathcal{E}_D^{\mathrm{GP}}(\psi_*) + O(h^2) = hE_D^{\mathrm{GP}} + O(h^2), \tag{25}$$

since $\mathcal{E}_D^{\mathrm{GP}}(\psi_*) = E_D^{\mathrm{GP}} \leqslant 0$. In addition to proving a sharp upper bound for the ground state energy, the estimate above also yields the a priori bound $\mathcal{E}_\mu^{\mathrm{BCS}}[\Gamma] \leqslant Ch$ for any approximate minimizer Γ of the BCS energy. Hence, the minimizer satisfies (22), so that we can deduce the estimate from below matching the upper bound, along with the decomposition of α as in (23). □

Proof *(Theorem 2)* We start from the trivial observation that

$$E_D^{\mathrm{GP}} < 0, \quad \Longleftrightarrow \quad D > E_W, \tag{26}$$

where we recall that E_W is the ground state energy of $-\frac{1}{4}\Delta + W$: indeed, if $D > E_W$, it suffices to use $\lambda\psi_W$, $\lambda > 0$, as a trial state for the GP energy, where ψ_W si the normalized ground state of $-\frac{1}{4}\Delta + W$, to get

$$E_D^{\mathrm{GP}} = \lambda(E_W - D) + g_{\mathrm{BCS}}\lambda^2 \|\psi_W\|_{L^4}^4 < 0, \tag{27}$$

for λ small enough. On the other hand, if $D \leqslant E_W$, the functional is trivially positive, since

$$\mathcal{E}_D^{\mathrm{GP}}(\psi) \geqslant (E_W - D)\,\|\psi\|_{L^2}^2. \tag{28}$$

Note also that ψ_* is non-trivial if and only if $E_D^{\mathrm{GP}} < 0$.

Next, we prove the upper bound $\mu_c(h) \leqslant -E_0 + E_W h^2 + o(h^2)$ by showing that, if $\mu = -E_0 + Dh^2$, $D > E_W$, then there exists an admissible BCS state such that

$$\mathcal{E}_\mu^{\mathrm{BCS}}[\Gamma] < 0. \tag{29}$$

By Proposition 1, for any $\psi \in \mathscr{D}^{\mathrm{GP}}$, there exists Γ_ψ admissibile such that

$$h^{-1}\mathcal{E}_\mu^{\mathrm{BCS}}[\Gamma_\psi] = \mathcal{E}_D^{\mathrm{GP}}(\psi) + O(h). \tag{30}$$

This in particular holds true for $\psi = \psi_*$, so that

$$E_\mu^{\mathrm{BCS}} \leqslant E_D^{\mathrm{GP}} h + O(h^2) < 0, \tag{31}$$

if $D > E_W$.

Conversely, we now show that, if $E_\mu^{\mathrm{BCS}} = 0$ for a certain $\mu = -E_0 + Dh^2$, then $D \leqslant E_W$, so completing the proof. By Theorem 1, indeed, if $E_\mu^{\mathrm{BCS}} = 0$, then $E_D^{\mathrm{GP}} = O(h)$ but the GP functional is independent of h and therefore $E_D^{\mathrm{GP}} = 0$, which in turn implies that $D \leqslant E_W$ by (26). □

3.1 GP Functional

We discuss some useful properties of the GP functional (9) and its minimization. We recall that we denote by E^{GP} the infimum of (9) and by ψ_* any associated minimizer.

Proposition 2 (A Priori Bounds on ψ) *There exists $C < +\infty$, depending on $g_{\mathrm{BCS}} > 0$, such that*

$$\|\nabla\psi\|_{L^2}^2 + \langle\psi\,|W|\,\psi\rangle + \|\psi\|_{L^4}^4 + \|\psi\|_{L^2}^2 \leqslant C\left[1 + \max\left\{\mathcal{E}_D^{\mathrm{GP}}(\psi), 0\right\}\right] \tag{32}$$

for all $\psi \in \mathscr{D}^{\mathrm{GP}}$.

Proof We may assume that $D \geqslant 0$ otherwise the result is trivially obtained with $C = \max\left\{|D|^{-1}, g_{\mathrm{BCS}}^{-1}, 4\right\}$. The starting point is the inequality

$$\begin{aligned}\left\langle\psi\left|-\tfrac{1}{4}\Delta + W\right|\psi\right\rangle + g_{\mathrm{BCS}}\,\|\psi\|_{L^4}^4 &\leqslant D\,\|\psi\|_{L^2}^2 + \mathcal{E}_D^{\mathrm{GP}}(\psi)\\ &\leqslant D\,\|\psi\|_2^2 + \max\left\{\mathcal{E}_D^{\mathrm{GP}}(\psi), 0\right\},\end{aligned} \tag{33}$$

which allows to bound from above both the quantities on the l.h.s. in terms of the L^2 norm and the GP energy of ψ. Next, we estimate for R large enough

$$\begin{aligned}\|\psi\|_{L^2}^2 &\leqslant \int_{|x|\leqslant R} dx\,|\psi|^2 + R^{-\beta}\int_{|x|>R} dx\,|x|^\beta|\psi|^2\\ &\leqslant \sqrt{\tfrac{4\pi}{3}}R^{3/2}\,\|\psi\|_{L^4}^2 + CR^{-\beta}\,\langle\psi\,|W|\,\psi\rangle\\ &\leqslant C\left[R^{3/2}g_{\mathrm{BCS}}^{-1}\left(D\,\|\psi\|_{L^2} + \sqrt{E}\right) + R^{-\beta}\left(\sqrt{D}\,\|\psi\|_{L^2}^2 + E\right)\right]\end{aligned}$$

where we have set $E := \max\left\{\mathcal{E}_D^{\mathrm{GP}}(\psi), 0\right\}$ for short. Hence, for $R > (CD)^{1/\beta}$, we get

$$\left(1 - \tfrac{CD}{R^\beta}\right) \|\psi\|_{L^2}^2 - CR^{3/2}D\, \|\psi\|_{L^2} \leqslant C\left(R^{3/2} g_{\mathrm{BCS}}^{-1}\sqrt{E} + R^{-\beta}E\right),$$

which implies

$$\|\psi\|_{L^2}^2 \leqslant \frac{C}{\left(1 - \frac{CD}{R^\beta}\right)^2}\left[\left(1 - \tfrac{CD}{R^\beta}\right)\left(R^{3/2} g_{\mathrm{BCS}}^{-1}\sqrt{E} + R^{-\beta}E\right) + C^2R^3D^2\right] \tag{34}$$

and thus the result. □

Corollary 1 (Boundedness from Below of $\mathcal{E}_D^{\mathrm{GP}}(\psi)$) *For any $g_{\mathrm{BCS}} > 0$, there exists a finite constant $C < +\infty$ such that*

$$E_D^{\mathrm{GP}} \geqslant -C. \tag{35}$$

Proof Again, $E_D^{\mathrm{GP}} = 0$, if $D \leqslant 0$, and there is nothing to prove, so let us assume that $D > 0$. In this case it suffices to observe that $E_D^{\mathrm{GP}} \leqslant 0$, which can be obtained by simply testing the GP energy on the trivial wave function $\psi \equiv 0$. Hence, Proposition 2 implies that $\exists C < +\infty$ such that $\|\psi\|_{L^2}^2 \leqslant C$ for any ψ with non-positive energy, which in turn yields the lower bound $E_D^{\mathrm{GP}} \geqslant -C|D|$ and thus the result. □

The existence of a minimizer ψ_* which is also unique up to gauge transformation can be deduced by standard methods in variational calculus, and any such a minimizer solves the variational equation

$$-\tfrac{1}{4}\Delta\psi_* + (W - D)\psi_* + 2g_{\mathrm{BCS}}|\psi_*|^2\psi_* = 0. \tag{36}$$

Under Assumption 3, one can also show that $\psi_* \in C^3 \cap L^\infty(\mathbb{R}^3)$ and it can be chosen strictly positive.

3.2 *Semiclassical Estimates*

Before attacking the proof of Proposition 1, it is useful to state some technical but standard semiclassical bounds to be used in the rest of the paper.

Proposition 3 (Semiclassical Estimates) *Let $\mu = -E_0 + h^2 D$, $D \in \mathbb{R}$ and let*

$$\alpha_\psi(x, y) := h^{-2}\, \psi\left(\tfrac{x+y}{2}\right) \alpha_0\left(\tfrac{x-y}{h}\right), \tag{37}$$

for any $\psi \in \mathscr{D}^{\mathrm{GP}}$. Then, the following estimates hold as $h \to 0$:

$$\left| \operatorname{Tr} \mathfrak{h}\, \alpha_\psi\, \overline{\alpha_\psi} + \int_{\mathbb{R}^6} dx dy\, V\left(\tfrac{x-y}{h}\right) |\alpha_\psi(x, y)|^2 \right.$$
$$\left. -h \left\langle \psi \left| -\tfrac{1}{4}\Delta + W - D \right| \psi \right\rangle_{L^2(\mathbb{R}^3)} \right| \leqslant \mathrm{A}_0 h^2, \tag{38}$$

where

$$\mathrm{A}_0 = C\left(\left\| W|\psi|^2 \right\|_{L^1} + \|\psi\|_{L^2}^2 \right); \tag{39}$$

$$\left| \operatorname{Tr} \mathfrak{h}\, \alpha_\psi\, \overline{\alpha_\psi}\, \alpha_\psi\, \overline{\alpha_\psi} - h g_{\mathrm{BCS}} \|\psi\|_{L^4}^4 \right|$$
$$\leqslant C h^2 \left[\|\nabla \psi\|_{L^2}^4 + \left\| W|\psi|^2 \right\|_{L^1}^2 + \|\psi\|_{L^2}^4 + A_0 h \right]. \tag{40}$$

Before discussing the proof of the above Proposition, it is convenient to state a technical result about the reduced density α_ψ which is going to be used several times. In the following we will often use the center-of-mass coordinates

$$\eta := \tfrac{1}{2}(x + y), \qquad \xi := x - y, \tag{41}$$

and use the notation

$$\tilde{\alpha}_\psi(\eta, \xi) := \alpha_\psi(x, y). \tag{42}$$

Lemma 1 *Let α_ψ be as (37). Then, for any $n \in \mathbb{N}$ even,*

$$\|\alpha_\psi\|_{\mathfrak{S}^n}^n \leqslant C h^{n-3} \|\psi\|_{L^n}^n \left\| |\hat{\alpha}_0| \right\|_{L^n}^n, \tag{43}$$

$$\|\nabla_\xi \tilde{\alpha}_\psi\|_{\mathfrak{S}^n}^n \leqslant h^{-3} \|\psi\|_{L^n}^n \left\| |\cdot|\, \hat{\alpha}_0 \right\|_{L^n}^n, \tag{44}$$

where $\|\cdot\|_{\mathfrak{S}^n}$ stands for the Schatten norm of order $n \in \mathbb{N}$.

Proof See [5, Lemma 1]. The extension to any $n \in \mathbb{N}$ is obtained by simply observing that, thanks to the monotonicity of Schatten norms, $\left\|\alpha_\psi\right\|_{\mathfrak{S}^\infty} \leqslant \left\|\alpha_\psi\right\|_{\mathfrak{S}^n}$ for any $n \in \mathbb{N}$, which allows to use (43) and (44) repeatedly to extend the result to all natural numbers. □

We are now in position to present the proof of Proposition 3.

Proof *(Proposition 3)* Using the change to center-of-mass and relative coordinates, one gets

$$\begin{aligned}\operatorname{Tr}\mathfrak{h}\,\alpha_\psi\,\overline{\alpha_\psi} + \int_{\mathbb{R}^6} dxdy\, V\left(\tfrac{x-y}{h}\right)|\alpha_\psi(x,y)|^2 \\ = \left\langle \tilde{\alpha}_\psi \left| -\tfrac{1}{4}h^2\Delta_\eta + h^2 W(\eta+\xi/2) - h^2 D \right| \tilde{\alpha}_\psi \right\rangle_{L^2(\mathbb{R}^6)} \\ + \left\langle \tilde{\alpha}_\psi \left| -h^2\Delta_\xi + V(\xi/h) + E_0 \right| \tilde{\alpha}_\psi \right\rangle_{L^2(\mathbb{R}^6)} \\ = \left\langle \tilde{\alpha}_\psi \left| -\tfrac{1}{4}h^2\Delta_\eta + h^2 W(\eta+\xi/2) - h^2 D \right| \tilde{\alpha}_\psi \right\rangle_{L^2(\mathbb{R}^6)}. \qquad (45)\end{aligned}$$

where we used that α_0 is the normalized zero energy eigenvector of the operator $-\Delta + V + E_0$. The result then follows from next Lemma 2.

In order to prove the second estimate, we use the cyclicity of the trace and the symmetry of the Laplacian, to get

$$\begin{aligned}\operatorname{Tr}\Delta\,\alpha_\psi\,\overline{\alpha_\psi}\,\alpha_\psi\,\overline{\alpha_\psi} = \left\langle \alpha_\psi\,\overline{\alpha_\psi}\,\alpha_\psi \left| \tfrac{1}{2}(\Delta_x+\Delta_y)\alpha_\psi \right.\right\rangle_{L^2(\mathbb{R}^6)} \\ = \left\langle \tilde{\omega}_\psi \left| \tfrac{1}{4}\Delta_\eta\tilde{\alpha}_\psi \right.\right\rangle_{L^2(\mathbb{R}^6)} + \left\langle \tilde{\omega}_\psi \left| \Delta_\xi\tilde{\alpha}_\psi \right.\right\rangle_{L^2(\mathbb{R}^6)}, \qquad (46)\end{aligned}$$

where we have set for short $\tilde{\omega}_\psi(\eta,\xi) := (\alpha_\psi\,\overline{\alpha_\psi}\,\alpha_\psi)(x,y)$. Introducing the coordinates

$$X = \tfrac{1}{4}(x_1+x_2+x_3+x_4), \qquad \xi_k = x_{k+1} - x_k, \qquad k = 1,2,3, \qquad (47)$$

and rescaling the relative ones, we obtain

$$\begin{aligned}\operatorname{Tr}(-h^2\Delta + E_0)\,\alpha_\psi\,\overline{\alpha_\psi}\,\alpha_\psi\,\overline{\alpha_\psi} = h\int_{\mathbb{R}^{12}} dXd\xi_1 d\xi_2 d\xi_3\; \psi(X-hs)\overline{\psi(X-ht)}\times \\ \psi(X+hs)\overline{\psi(X+ht)}\,[(-\Delta+E_0)\,\alpha_0](\xi_1)\overline{\alpha_0(\xi_2)}\,\alpha_0(\xi_3)\overline{\alpha_0(\xi_*)} \\ - \tfrac{1}{4}h^2\left\langle \tilde{\omega}_\psi \left| \Delta_\eta\tilde{\alpha}_\psi \right.\right\rangle_{L^2(\mathbb{R}^6)}, \qquad (48)\end{aligned}$$

where $\xi_* := -\xi_1 - \xi_2 - \xi_3$ and s, t are functions of ξ_1, ξ_2, ξ_3, i.e.,

$$s := \tfrac{1}{4}\left(\xi_1 + 2\xi_2 + \xi_3\right), \qquad t := \tfrac{1}{4}\left(\xi_3 - \xi_1\right). \qquad (49)$$

From this expression we are going to extract the quartic term needed to reconstruct the GP functional times h plus higher order contributions. The fundamental theorem of calculus allows to rewrite the first term on the r.h.s. of (48) as

$$\begin{aligned} & h\|\psi\|_{L^4}^4 \int_{\mathbb{R}^9} d\xi_1 d\xi_2 d\xi_3 \; [(-\Delta + E_0)\,\alpha_0(\xi_1)]\,\overline{\alpha_0(\xi_2)}\,\alpha_0(\xi_3)\overline{\alpha_0(\xi_*)} \\ & \quad + h \int_{\mathbb{R}^{12}} dX d\xi_1 d\xi_2 d\xi_3 \int_0^1 d\tau \, \frac{d}{d\tau} \Big(\psi(X - \tau hs)\overline{\psi(X - \tau ht)} \times \\ & \quad \times \psi(X + \tau hs)\overline{\psi(X + \tau ht)} \Big) \, [(-\Delta + E_0)\,\alpha_0(\xi_1)]\,\overline{\alpha_0(\xi_2)}\,\alpha_0(\xi_3)\overline{\alpha_0(\xi_*)} \\ & \qquad =: h g_{\mathrm{BCS}} \|\psi\|_{L^4}^4 + h^2 I_1, \end{aligned} \tag{50}$$

thanks to the explicit computation

$$\begin{aligned} & \int_{\mathbb{R}^9} d\xi_1 d\xi_2 d\xi_3 \; [(-\Delta + E_0)\,\alpha_0(\xi_1)]\,\overline{\alpha_0(\xi_2)}\,\alpha_0(\xi_3)\overline{\alpha_0((\xi_*))} \\ & \qquad = (2\pi)^3 \int_{\mathbb{R}^3} dp \; (p^2 + E_0)|\hat{\alpha}_0(p)|^4. \end{aligned}$$

Hence, (48) yields

$$\operatorname{Tr}(-h^2\Delta + E_0)\,\alpha_\psi\,\overline{\alpha_\psi}\,\alpha_\psi\,\overline{\alpha_\psi} = h g_{\mathrm{BCS}} \|\psi\|_{L^4}^4 + h^2\,(I_1 + I_2)\,, \tag{51}$$

where

$$I_2 := -\tfrac{1}{4} \left\langle \alpha_\psi\,\overline{\alpha_\psi}\,\alpha_\psi \,\middle|\, \Delta_\eta \alpha_\psi \right\rangle_{L^2(\mathbb{R}^6)} . \tag{52}$$

The estimate on the term containing the external potential immediately follows from Lemma 2 using Hölder inequality with exponents $\frac{1}{2}$, $\frac{1}{3}$ and $\frac{1}{6}$:

$$\begin{aligned} & \left|\operatorname{Tr} W\,\alpha_\psi\,\overline{\alpha_\psi}\,\alpha_\psi\,\overline{\alpha_\psi}\right| \leqslant \operatorname{Tr}\left|W^{1/2}\,\alpha_\psi\,\overline{\alpha_\psi}\,\alpha_\psi\,\overline{\alpha_\psi} W^{1/2}\right| \\ & \quad \leqslant \left\|W^{1/2}\alpha_\psi\right\|_{\mathfrak{S}^2} \left\|W^{1/2}\alpha_\psi\right\|_{\mathfrak{S}^6} \left\|\overline{\alpha_\psi}\alpha_\psi\right\|_{\mathfrak{S}^3} \leqslant \left\|W^{1/2}\alpha_\psi\right\|_{\mathfrak{S}^2}^2 \left\|\alpha_\psi\right\|_{\mathfrak{S}^6}^2 \\ & \quad \leqslant Ch\|\psi\|_{L^6}^2 \left\|\hat{\alpha}_0\right\|_{L^6}^2 \left\|W^{1/2}\alpha_\psi\right\|_{L^2}^2 \leqslant C\|\psi\|_{L^6}^2 \left[\left\|W|\psi|^2\right\|_{L^1} + hA_0\right], \end{aligned} \tag{53}$$

by the monotonicity of Schatten norms and Lemma 1. The replacement of $\|\psi\|_{L^6}^2$ with $\|\nabla\psi\|_{L^2}^2 + \|\psi\|_{L^2}^2$ can be done via Sobolev inequality. □

Lemma 2 *Let* α_ψ *be as* (37) *and* A_0 *as in* (39), *then*

$$\left| \left\langle \alpha_\psi \left| W \right| \alpha_\psi \right\rangle_{L^2(\mathbb{R}^6)} - h^{-1} \int_{\mathbb{R}^3} d\eta \; W(\eta) |\psi|^2 \right| \leqslant \mathrm{A}_0. \tag{54}$$

Proof Using center-of-mass and relative coordinates as before, we get by the Taylor expansion (7)

$$\begin{aligned} \tfrac{1}{2} \left\langle \alpha_\psi \left| W(x) + W(y) \right| \alpha_\psi \right\rangle_{L^2(\mathbb{R}^6)} &= \tfrac{1}{2} \left\langle \tilde{\alpha}_\psi \left| W(\eta + \xi/2) \right| \tilde{\alpha}_\psi \right\rangle_{L^2(\mathbb{R}^6)} \\ &\quad + \tfrac{1}{2} \left\langle \tilde{\alpha}_\psi \left| W(\eta - \xi/2) \right| \tilde{\alpha}_\psi \right\rangle_{L^2(\mathbb{R}^6)} \\ = h^{-1} \int_{\mathbb{R}^3} d\eta \; W(\eta) |\psi(\eta)|^2 &+ \tfrac{1}{2} h^{-4} \int_{\mathbb{R}^6} d\eta d\xi \; \xi \cdot \nabla W(\zeta) \, |\psi(\eta)|^2 \, |\alpha_0(\xi/h)|^2 , \end{aligned}$$

where we recall that $\tilde{\alpha}_\psi(\eta, \xi) = \alpha_\psi(x, y)$. Hence, we have only to estimate the last term on the r.h.s. of the expression above: by Assumption 3 on W, we deduce that, since $\zeta \in (\eta, \eta + \xi/2)$,

$$\begin{aligned} &\tfrac{1}{2} h^{-4} \int_{\mathbb{R}^6} d\eta d\xi \; \xi \cdot \nabla W(\zeta) \, |\psi(\eta)|^2 \, |\alpha_0(\xi/h)|^2 \\ &\qquad \leqslant C \int_{\mathbb{R}^6} d\eta d\xi \; |\xi| \left(h^{\beta-1} \, |\xi|^{\beta-1} + |\eta|^{\beta-1} + 1 \right) |\psi(\eta)|^2 \, |\alpha_0(\xi)|^2 \end{aligned} \tag{55}$$

which immediately implies the result, via the trivial bounds $|x|^{\beta-1} \leqslant W(x) + 1$ (again by Assumption 3) and

$$\left\| |\cdot|^{\beta/2} \alpha_0 \right\|_{L^2}^2 \leqslant C, \qquad \left\| |\cdot|^{1/2} \alpha_0 \right\|_{L^2}^2 \leqslant C, \tag{56}$$

which follows from (6). □

Lemma 3 *Let* $\mathrm{I}_1, \mathrm{I}_2$ *as in* (50) *and* (52). *Then, as* $h \to 0, \exists C < +\infty$ *such that*

$$|\mathrm{I}_1| + |\mathrm{I}_2| \leqslant C \|\nabla \psi\|_{L^2}^4. \tag{57}$$

Proof By [5, Proof of Lemma 1], there exist two finite constants C_1, C_2 such that

$$\begin{aligned} |\mathrm{I}_1| &\leqslant C_1 \|\nabla \psi\|_{L^2}^4 \, \||\cdot| \, \alpha_0\|_{L^2} \, \|\alpha_0\|_{L^2} \, \|\alpha_0\|_{L^1} \, \|V \alpha_0\|_{L^1} , \\ |\mathrm{I}_2| &\leqslant C_2 \|\nabla \psi\|_{L^2}^4. \end{aligned}$$

The result then follows from the properties of α_0 (see Remark 2). □

3.3 Energy Upper Bound

The result is obtained by testing the BCS energy functional on a suitable trial state. We define an admissible state Γ_ψ, with off-diagonal element given by α_ψ as in (37) and $\psi \in \mathscr{D}^{\mathrm{GP}}$, and upper left entry

$$\gamma_\psi := \alpha_\psi \overline{\alpha_\psi} + (1 + \lambda h)\, \alpha_\psi \overline{\alpha_\psi} \alpha_\psi \overline{\alpha_\psi}, \tag{58}$$

for some $\lambda \in \mathbb{R}^+$.

Remark 3 (Admissibility) The admissibility requirement makes the correction of order λh necessary. In fact, any correction of order h^β, $0 < \beta \leqslant 1$, would work, if λ is chosen appropriately, but $\beta = 1$ gives the best error bound in our estimates. Indeed, the state is admissible if and only if $\gamma - \gamma^2 - \alpha\overline{\alpha} \geqslant 0$ (see, e.g., [5, Eq. (4.8)]), which, assuming that the quartic correction is proportional to λh^β, yields the condition

$$\lambda h^\beta - (1 + \lambda h^\beta)^2 (\alpha_\psi \overline{\alpha_\psi})^2 - 2(1 + \lambda h^\beta)\, \alpha_\psi \overline{\alpha_\psi} \geqslant 0. \tag{59}$$

Since $\|\alpha_\psi\|_\infty \leqslant \|\alpha_\psi\|_6 \leqslant C h^{1/2}$, this bound implies that we may choose $0 < \beta < 1$, and the latter condition would be satisfied for any value of λ. For $\beta = 1$, on the other hand, one is forced to take the parameter λ large enough, but the inequality may still hold.

We now apply Proposition 3 to get

$$\begin{aligned}
\mathcal{E}_\mu^{\mathrm{BCS}}[\Gamma_\psi] &= \operatorname{Tr} \mathfrak{h}\, \gamma_\psi + \int_{\mathbb{R}^6} dx dy V\left(\tfrac{x-y}{h}\right) |\alpha_\psi(x,y)|^2 \\
&= \operatorname{Tr} \mathfrak{h}\, \alpha_\psi \overline{\alpha_\psi} + \int_{\mathbb{R}^6} dx dy V\left(\tfrac{x-y}{h}\right) |\alpha_\psi(x,y)|^2 + (1+\lambda h) \operatorname{Tr} \mathfrak{h}\, \alpha_\psi \overline{\alpha_\psi} \alpha_\psi \overline{\alpha_\psi} \\
&\leqslant h \int_{\mathbb{R}^3} d\eta \left\{ \tfrac{1}{4} |\nabla \psi|^2 + (W - D)|\psi|^2 + g_{\mathrm{BCS}} |\psi|^4 \right\} \\
&\quad + C h^2 \left[\|\nabla \psi\|_{L^2}^4 + \left\| W |\psi|^2 \right\|_{L^1}^2 + \|\psi\|_{L^2}^4 + 1 \right]
\end{aligned} \tag{60}$$

as $h \to 0$. The upper bound (21) is thus a straightforward consequence of (32).

3.4 Energy Lower Bound

We consider any admissible BCS state Γ satisfying $\mathcal{E}_\mu^{\mathrm{BCS}}[\Gamma] \leqslant C_\Gamma h$, whose existence is ensured by the analysis in the previous Sect. 3.3. The integral kernel of α, the upper-right entry of Γ, can be decomposed as

$$\alpha(x, y) = \alpha_\psi(x, y) + r(x, y) = h^{-2}\,\psi\left(\tfrac{x+y}{2}\right)\alpha_0\left(\tfrac{x-y}{h}\right) + r(x, y). \tag{61}$$

where r is chosen to be orthogonal to α_0:

$$\langle \alpha_0\left(\cdot\,/h\right) | \tilde{r} \rangle_{L^2_\xi(\mathbb{R}^3)} = 0, \tag{62}$$

where $\tilde{r}(\eta, \xi) := r(x, y)$ and the coordinates η, ξ are defined in (41). With such a choice, the order parameter ψ is naturally defined in terms of α as (recall the notation $\tilde{\alpha}(\eta, \xi) := \alpha(x, y)$)

$$\psi(\eta) := h^{-1}\langle \alpha_0\left(\cdot\,/h\right) | \tilde{\alpha} \rangle_{L^2_\xi(\mathbb{R}^3)} = h^{-1}\int_{\mathbb{R}^3} d\xi\ \alpha_0(\xi/h)\tilde{\alpha}(\eta, \xi), \tag{63}$$

Note also that, because of the orthogonality of r to α_0, one immediately gets

$$\|\alpha\|^2_{L^2(\mathbb{R}^6)} = \|\alpha_\psi\|^2_{L^2(\mathbb{R}^6)} + \|r\|^2_{L^2(\mathbb{R}^6)} = h^{-1}\|\psi\|^2_{L^2} + \|r\|^2_{L^2}. \tag{64}$$

The physical meaning of such a decomposition is apparent: α represents the wave function of a pair of particles and it almost factorizes in the coordinates of the center-of-mass reference frame. More precisely, α_0 describes the wave function in the relative coordinate living on the microscopic scale h, while ψ is the wave function in the in center-of-mass coordinate and varies on the macroscopic scale.

We start with a preliminary lower bound on the BCS energy functional in terms of the off diagonal entry α of Γ. Indeed, for any admissible Γ, it can be seen that one can bound $\mathcal{E}^{\mathrm{BCS}}_\mu[\Gamma]$ from below in terms of a functional of α alone.

Lemma 4 *Let $\mu = -E_0 + h^2 D$, $D \in \mathbb{R}$. For any admissible Γ and for h small enough,*

$$\mathcal{E}^{\mathrm{BCS}}_\mu[\Gamma] \geqslant \mathrm{Tr}\,\mathfrak{h}\,\alpha\overline{\alpha} + \mathrm{Tr}\,\mathfrak{h}\,\alpha\overline{\alpha}\alpha\overline{\alpha} + \int_{\mathbb{R}^6} dxdy\ V\left(\tfrac{x-y}{h}\right)|\alpha(x, y)|^2. \tag{65}$$

Proof The proof is given, e.g., in [14, Proposition 6.2]. We spell it in details here for the sake of completeness. The admissibility of Γ, i.e., the condition $0 \leqslant \Gamma \leqslant 1$, is equivalent to

$$\gamma - \gamma^2 - \alpha\overline{\alpha} \geqslant 0. \tag{66}$$

Since for h small enough $\mathfrak{h}$ is positive, as it follows from the trivial bound

$$\mathfrak{h} \geqslant E_0 - Dh^2 > 0, \tag{67}$$

we can use the monotonicity of the trace and apply the above inequality to get the result, since (66) implies that $\gamma \geqslant \alpha\overline{\alpha} + \alpha\overline{\alpha}\alpha\overline{\alpha}$ (see [14, Eq. (6.2)]). □

The next lower bound give more information on the decomposition (61).

Lemma 5 *Let* $\mu = -E_0 + h^2 D$, $D \in \mathbb{R}$, *and let* Γ *an admissible* BCS *state with upper-right entry* α *as in* (61). *Then, there exists a finite constant* C *such that (recall* (39)*), as* $h \to 0$,

$$\operatorname{Tr} \mathfrak{h}\, \alpha\overline{\alpha} + \int_{\mathbb{R}^6} dxdy\, V\left(\tfrac{x-y}{h}\right) |\alpha(x,y)|^2 \geqslant h \left\langle \psi \left| -\tfrac{1}{4}\Delta + W - D \right| \psi \right\rangle_{L^2(\mathbb{R}^3)}$$
$$+ \tfrac{1}{2} g \|r\|^2_{L^2(\mathbb{R}^6)} + h^2 \left[\left\langle \tilde{r} \left| -\tfrac{1}{4}\Delta_\eta - \varepsilon \Delta_\xi + \tfrac{1}{2} W - D \right| \tilde{r} \right\rangle_{L^2(\mathbb{R}^6} - A_0 \right]. \tag{68}$$

Proof Plugging in the operator bound (67), we can immediately get rid of the second term in (65) to obtain

$$\mathcal{E}^{\mathrm{BCS}}_\mu[\Gamma] \geqslant \operatorname{Tr} \mathfrak{h}\, \alpha\overline{\alpha} + \int_{\mathbb{R}^6} dxdy\, V\left(\tfrac{x-y}{h}\right) |\alpha(x,y)|^2 + (E_0 - h^2 D) \|\alpha\|^4_{\mathfrak{S}^4},$$

and the last term can be dropped since it is positive. Next, we estimate the first term, which reads

$$\operatorname{Tr} \mathfrak{h}\, \alpha\overline{\alpha} + \int_{\mathbb{R}^6} dxdy\, V\left(\tfrac{x-y}{h}\right) |\alpha(x,y)|^2$$
$$= \int_{\mathbb{R}^6} d\eta d\xi\, \overline{\tilde{\alpha}}(\eta,\xi) \left(-\tfrac{1}{4} h^2 \Delta_\eta - h^2 \Delta_\xi + h^2 W(\eta + \xi/2) + \right.$$
$$\left. V(\xi/h) - \mu \right) \tilde{\alpha}(\eta,\xi).$$

By plugging in the decomposition (61), we get

$$\operatorname{Tr} \mathfrak{h}\, \alpha\overline{\alpha} = \langle \alpha_\psi \,|\mathfrak{h}|\, \alpha_\psi \rangle + \int_{\mathbb{R}^6} dxdy\, V\left(\tfrac{x-y}{h}\right) |\alpha_\psi(x,y)|^2$$
$$+ \langle r \,|\mathfrak{h}|\, r \rangle + \int_{\mathbb{R}^6} dxdy\, V\left(\tfrac{x-y}{h}\right) |r(x,y)|^2 + 2h^2 \Re \langle \alpha_\psi \,|\, W r \rangle, \tag{69}$$

since the potential W is the only operator which does not factorize in the decomposition $L^2(\mathbb{R}^6) = L^2_\eta(\mathbb{R}^3) \otimes L^2_\xi(\mathbb{R}^3)$. The sum of the first two terms has already been estimated in Proposition 3, so that it just remains to consider the quadratic expression on r and the mixed term.

The mixed term can be controlled by exploiting the Taylor expansion (7) and the orthogonality (62), obtaining

$$2h^2 \left| \Re \langle \alpha_\psi \,|\, W r \rangle \right| = 2 \left| \int_{\mathbb{R}^6} d\eta d\xi\, \xi \cdot \nabla W(\zeta) \overline{\psi}(\eta) \alpha_0(\xi/h) \tilde{r}(\eta,\xi) \right|$$
$$\leqslant C \int_{\mathbb{R}^6} d\eta d\xi\, |\xi| \left(|\xi|^{\beta-1} + |\eta|^\beta + 1 \right) |\psi(\eta)| |\alpha_0(\xi/h)|\, |\tilde{r}(\eta,\xi)|$$

by the trivial bound $|\eta|^{\beta-1} \leqslant |\eta|^{\beta} + 1$. Hence, by Cauchy-Schwarz inequality we get

$$2h^2 \left|\Re \langle \alpha_\psi \,|W r\rangle\right| \leqslant C \, \|\psi\|_{L^2(\mathbb{R}^3)} \|r\|_{L^2(\mathbb{R}^6)} \times \\ \times \left(\int_{\mathbb{R}^3} d\xi \, \left(|\xi|^{2\beta} + |\xi|^2 \right) |\alpha_0(\xi/h)|^2 \right)^{1/2} \\ + C \left(\left\| W|\psi|^2 \right\|^{1/2}_{L^1(\mathbb{R}^3)} + \|\psi\|_{L^2(\mathbb{R}^3)} \right) \left[\left(\int_{\mathbb{R}^6} d\eta d\xi \, |\eta|^{\beta} \, |\tilde{r}|^2 \right)^{1/2} + \|r\|_{L^2(\mathbb{R}^6)} \right] \times \\ \times \left(\int_{\mathbb{R}^3} d\xi \, |\xi|^2 \, |\alpha_0(\xi/h)|^2 \right)^{1/2} \\ \leqslant C h^{5/2} \left(\|\psi\|^2_{L^2} + \|r\|^2_{L^2} + \left\| W|\psi|^2 \right\|_{L^1} + \left\| W|r|^2 \right\|_{L^1} \right)$$

where we have estimated

$$\int_{\mathbb{R}^6} d\eta d\xi \, |\eta|^{\beta} \, |\tilde{r}|^2 \leqslant C \left\| W \, |r|^2 \right\|_{L^1(\mathbb{R}^6)} .$$

The two terms depending on r can then be absorbed in the corresponding positive ones coming from the estimate of $\langle r \,|\mathfrak{h}|\, r\rangle$ by adding a $\frac{1}{2}$ prefactor for h small enough, while the other two can be included in the $A_0 h^2$ remainder up to the change of the constant C in A_0. The quadratic expression in ξ is bounded from below by means of Assumption 2:

$$\int_{\mathbb{R}^3} d\eta \, \left\langle r(\eta,\cdot) \left| -h^2 \Delta_\xi + V(\cdot/h) + E_0 \right| r(\eta,\cdot) \right\rangle_{L^2_\xi(\mathbb{R}^3)} \\ \geqslant \int_{\mathbb{R}^3} d\eta \, \left\langle r(\eta,\cdot) \left| -h^2 \varepsilon \Delta_\xi + g \right| r(\eta,\cdot) \right\rangle_{L^2_\xi(\mathbb{R}^3)} \\ = g \|r\|^2_{L^2(\mathbb{R}^6)} + h^2 \varepsilon \left\| \nabla_\xi r \right\|^2_{L^2(\mathbb{R}^6)} . \qquad (70)$$

□

Lemma 6 *Let $\mu = -E_0 + h^2 D$, $D \in \mathbb{R}$, and let Γ an admissible* BCS *state with upper-right entry α as in (61), such that $\mathcal{E}^{\mathrm{BCS}}_\mu[\Gamma] \leqslant C_\Gamma h$. Then, there exists a finite constant C such that*

$$\left| \mathrm{Tr}\, \mathfrak{h} \, \alpha\overline{\alpha}\alpha\overline{\alpha} - \mathrm{Tr}\, \mathfrak{h} \, \alpha_\psi \overline{\alpha}_\psi \alpha_\psi \overline{\alpha}_\psi \right| \leqslant C h^2 \left(\|\nabla\psi\|^4_{L^2} + A_0^2 \right) . \qquad (71)$$

Proof We first rewrite the quartic term via

$$\alpha\overline{\alpha}\alpha\overline{\alpha} - \alpha_\psi \overline{\alpha_\psi} \alpha_\psi \overline{\alpha_\psi} = r\overline{\alpha}\alpha\overline{\alpha_\psi} + \alpha_\psi \overline{\alpha}\alpha\overline{r} + r\overline{\alpha}\alpha\overline{r} + \alpha_\psi (\overline{\alpha}\alpha - \overline{\alpha_\psi}\alpha_\psi)\overline{\alpha_\psi},$$

so that the cyclicity of trace and triangle inequality yields

$$\left|\operatorname{Tr}\mathfrak{h}\alpha\overline{\alpha}\alpha\overline{\alpha}-\operatorname{Tr}\mathfrak{h}\alpha_\psi\overline{\alpha_\psi}\alpha_\psi\overline{\alpha_\psi}\right| \leqslant \left\|\mathfrak{h}^{1/2}r\overline{\alpha}\alpha\overline{\alpha_\psi}\mathfrak{h}^{1/2}\right\|_{\mathfrak{S}^1}+\left\|\mathfrak{h}^{1/2}\alpha_\psi\overline{\alpha}\alpha\overline{r}\mathfrak{h}^{1/2}\right\|_{\mathfrak{S}^1}$$
$$+\left\|\mathfrak{h}^{1/2}r\overline{\alpha}\alpha\overline{r}\mathfrak{h}^{1/2}\right\|_{\mathfrak{S}^1}+\left\|\mathfrak{h}^{1/2}\alpha_\psi(\overline{\alpha}\alpha-\overline{\alpha_\psi}\alpha_\psi)\overline{\alpha_\psi}\mathfrak{h}^{1/2}\right\|_{\mathfrak{S}^1}. \tag{72}$$

To estimate this four terms, we apply Hölder inequality:

$$\left\|\mathfrak{h}^{1/2}\alpha_\psi\overline{\alpha}\alpha\overline{r}\mathfrak{h}^{1/2}\right\|_{\mathfrak{S}^1}\leqslant\left\|\mathfrak{h}^{1/2}\alpha_\psi\right\|_{\mathfrak{S}^6}\|\alpha\|_{\mathfrak{S}^6}^2\left\|\mathfrak{h}^{1/2}r\right\|_{\mathfrak{S}^2};$$
$$\left\|\mathfrak{h}^{1/2}r\overline{\alpha}\alpha\overline{\alpha_\psi}\mathfrak{h}^{1/2}\right\|_{\mathfrak{S}^1}\leqslant\left\|\mathfrak{h}^{1/2}r\right\|_{\mathfrak{S}^2}\|\alpha\|_{\mathfrak{S}^6}^2\left\|\mathfrak{h}^{1/2}\alpha_\psi\right\|_{\mathfrak{S}^6};$$
$$\left\|\mathfrak{h}^{1/2}r\overline{\alpha}\alpha\overline{r}\mathfrak{h}^{1/2}\right\|_{\mathfrak{S}^1}=\left\|\alpha\overline{r}\mathfrak{h}^{1/2}\right\|_{\mathfrak{S}^2}^2\leqslant\|\alpha\|_{\mathfrak{S}^\infty}^2\left\|\mathfrak{h}^{1/2}r\right\|_{\mathfrak{S}^2}^2;$$
$$\left\|\mathfrak{h}^{1/2}\alpha_\psi(\alpha\overline{\alpha}-\overline{\alpha_\psi}\alpha_\psi)\overline{\alpha_\psi}\mathfrak{h}^{1/2}\right\|_{\mathfrak{S}^1}\leqslant\left\|\alpha\overline{\alpha}-\overline{\alpha_\psi}\alpha_\psi\right\|_{\mathfrak{S}^{3/2}}\left\|\mathfrak{h}^{1/2}\alpha_\psi\right\|_{\mathfrak{S}^6}^2.$$

Plugging the above bounds in (72), we obtain

$$\left|\operatorname{Tr}\mathfrak{h}\alpha\overline{\alpha}\alpha\overline{\alpha}-\operatorname{Tr}\mathfrak{h}\alpha_\psi\overline{\alpha_\psi}\alpha_\psi\overline{\alpha_\psi}\right|\leqslant 2\left\|\mathfrak{h}^{1/2}\alpha_\psi\right\|_{\mathfrak{S}^6}\|\alpha\|_{\mathfrak{S}^6}^2\left\|\mathfrak{h}^{1/2}r\right\|_{\mathfrak{S}^2}$$
$$+\|\alpha\|_{\mathfrak{S}^6}^2\left\|\mathfrak{h}^{1/2}r\right\|_{\mathfrak{S}^2}^2+\left\|\overline{\alpha}\alpha-\overline{\alpha_\psi}\alpha_\psi\right\|_{\mathfrak{S}^{3/2}}\left\|\mathfrak{h}^{1/2}\alpha_\psi\right\|_{\mathfrak{S}^6}^2. \tag{73}$$

By (68) and the condition on the BCS energy of Γ, we deduce the inequality

$$\left(\tfrac{1}{2}g-Dh^2\right)\|r\|_{L^2}^2+h^2\left[\left\langle\tilde{r}\left|-\tfrac{1}{4}\Delta_\eta-\varepsilon\Delta_\xi+\tfrac{1}{2}W\right|\tilde{r}\right\rangle_{L^2(\mathbb{R}^6)}\right]$$
$$\leqslant Ch\left[1+\|\psi\|_{L^2}^2+hA_0\right], \tag{74}$$

with A_0 defined in Proposition 3. This, for h small enough (e.g., smaller than $\sqrt{g/(4D)}$), gives a bound on $\|r\|_{L^2}^2$ as well as its Sobolev norms in terms of the norm of ψ. Hence, we have

$$\left\|\overline{\alpha}\alpha-\overline{\alpha_\psi}\alpha_\psi\right\|_{\mathfrak{S}^{3/2}}=\left\|\overline{\alpha_\psi}r+\overline{r}\alpha_\psi+\overline{r}r\right\|_{\mathfrak{S}^{3/2}}$$
$$\leqslant 2\left\|\alpha_\psi\right\|_{\mathfrak{S}^6}\|r\|_{\mathfrak{S}^2}+\|r\|_{\mathfrak{S}^2}^2\leqslant Ch\left[\|\psi\|_{L^6}^2+\|\psi\|_{L^2}^2+1+hA_0\right],$$

by the monotonicity of Schatten norms, Lemma 1 and (74). Similarly, by Sobolev inequality

$$\|\alpha\|_{\mathfrak{S}^6}^2\leqslant C\left(\left\|\alpha_\psi\right\|_{\mathfrak{S}^6}^2+\|r\|_{\mathfrak{S}^2}^2\right)\leqslant Ch\left[\|\psi\|_{L^6}^2+\|\psi\|_{L^2}^2+1+hA_0\right]$$
$$\leqslant Ch\left[\|\nabla\psi\|_{L^2}^2+\|\psi\|_{L^2}^2+1+hA_0\right]. \tag{75}$$

To conclude, we have to estimate the norms of $\mathfrak{h}^{1/2}\alpha_\psi$ but, for any operator T, one has

$$\left\|\mathfrak{h}^{1/2}T\right\|_{\mathfrak{S}^{2n}} = \left\|T^\dagger\,\mathfrak{h}\,T\right\|_{\mathfrak{S}^n}^{1/2}$$
$$\leqslant \left(h^2\left\|T^\dagger(-\Delta)T\right\|_{\mathfrak{S}^n} + h^2\left\|T^\dagger W T\right\|_{\mathfrak{S}^n} + \mu\left\|T^\dagger T\right\|_{\mathfrak{S}^n}\right)^{1/2}$$
$$\leqslant h\left(\tfrac{1}{2}\left\|\nabla_\eta T\right\|_{\mathfrak{S}^{2n}} + \left\|\nabla_\xi T\right\|_{\mathfrak{S}^{2n}} + \left\|W^{1/2}T\right\|_{\mathfrak{S}^{2n}}\right) + (E_0 - h^2 D)\left\|T\right\|_{\mathfrak{S}^{2n}}.$$

Applying this inequality to estimate the norms above and using once more the monotonicity of Schatten norms, Proposition 3, Lemma 1 and 2 and Sobolev inequality, we obtain

$$\left\|\mathfrak{h}^{1/2}\alpha_\psi\right\|_{\mathfrak{S}^6} \leqslant h\left[\tfrac{1}{2}\left\|\nabla_\eta\tilde{\alpha}_\psi\right\|_{\mathfrak{S}^2} + \left\|\nabla_\xi\tilde{\alpha}_\psi\right\|_{\mathfrak{S}^2} + \left\|W^{1/2}\alpha_\psi\right\|_{\mathfrak{S}^2}\right] + E_0\left\|\alpha_\psi\right\|_{\mathfrak{S}^6}$$
$$\leqslant Ch^{1/2}\left[\left\|W|\psi|^2\right\|_{L^1}^{1/2} + \left\|\psi\right\|_{L^6} + A_0 + h\left\|\nabla\psi\right\|_{L^2}\right]$$
$$\leqslant Ch^{1/2}\left[\left\|\nabla\psi\right\|_{L^2} + \left\|W|\psi|^2\right\|_{L^1}^{1/2} + A_0\right],$$

$$\left\|\mathfrak{h}^{1/2}r\right\|_{\mathfrak{S}^2} \leqslant h\left[\tfrac{1}{2}\left\|\nabla_\eta r\right\|_{\mathfrak{S}^2} + \left\|\nabla_\xi r\right\|_{\mathfrak{S}^2} + \left\|W^{1/2}r\right\|_{\mathfrak{S}^2}\right] + E_0\left\|r\right\|_{\mathfrak{S}^2}$$
$$\leqslant Ch^{1/2}\left[1 + \left\|\psi\right\|_{L^2} + h^{1/2}\sqrt{A_0}\right],$$

as follows from the a priori estimate (74). Putting together all the bounds found so far, we get the result. □

In order to complete the proof of the lower bound, we need a last ingredient.

Lemma 7 *Let $\mu = -E_0 + h^2 D$, $D \in \mathbb{R}$, and let Γ an admissible* BCS *state with upper-right entry α as in (61), such that $\mathcal{E}_\mu^{\mathrm{BCS}}[\Gamma] \leqslant C_\Gamma h$. Then, there exists a finite constant C such that*

$$\int_{\mathbb{R}^3} d\eta\,\left\{|\nabla\psi|^2 + W|\psi|^2 + |\psi|^2 + |\psi|^4\right\} \leqslant C. \tag{76}$$

Proof Let us denote for short

$$\mathcal{E} := \int_{\mathbb{R}^3} d\eta\,\left\{|\nabla\psi|^2 + W|\psi|^2 + |\psi|^2 + |\psi|^4\right\}.$$

Combining Lemma 6 with (40), we get

$$\operatorname{Tr}\mathfrak{h}\,\alpha\overline{\alpha}\alpha\overline{\alpha} \geqslant g_{\mathrm{BCS}} h\left\|\psi\right\|_{L^4}^4 - Ch^2\mathcal{E}^2, \tag{77}$$